Biocatalytic Process Optimization

Biocatalytic Process Optimization

Editors

Chia-Hung Kuo
Chwen-Jen Shieh

MDPI • Basel • Beijing • Wuhan • Barcelona • Belgrade • Manchester • Tokyo • Cluj • Tianjin

Editors
Chia-Hung Kuo
National Kaohsiung University
of Science and Technology
Taiwan

Chwen-Jen Shieh
National Chung-Hsing
University
Taiwan

Editorial Office
MDPI
St. Alban-Anlage 66
4052 Basel, Switzerland

This is a reprint of articles from the Special Issue published online in the open access journal *Catalysts* (ISSN 2073-4344) (available at: https://www.mdpi.com/journal/catalysts/special_issues/biocata_process_optimiz).

For citation purposes, cite each article independently as indicated on the article page online and as indicated below:

LastName, A.A.; LastName, B.B.; LastName, C.C. Article Title. *Journal Name* **Year**, *Volume Number*, Page Range.

ISBN 978-3-03943-915-7 (Hbk)
ISBN 978-3-03943-916-4 (PDF)

© 2020 by the authors. Articles in this book are Open Access and distributed under the Creative Commons Attribution (CC BY) license, which allows users to download, copy and build upon published articles, as long as the author and publisher are properly credited, which ensures maximum dissemination and a wider impact of our publications.

The book as a whole is distributed by MDPI under the terms and conditions of the Creative Commons license CC BY-NC-ND.

Contents

About the Editors . . . . **vii**

Chia-Hung Kuo and Chwen-Jen Shieh
Biocatalytic Process Optimization
Reprinted from: *Catalysts* 2020, 10, 1303, doi:10.3390/catal10111303 . . . . **1**

Chan-Su Rha, Shin-Woo Kim, Kyoung Hee Byoun, Yong Deog Hong and Dae-Ok Kim
Simultaneous Optimal Production of Flavonol Aglycones and Degalloylated Catechins from Green Tea Using a Multi-Function Food-Grade Enzyme
Reprinted from: *Catalysts* 2019, 9, 861, doi:10.3390/catal9100861 . . . . **7**

Bianca Grabner, Yekaterina Pokhilchuk and Heidrun Gruber-Woelfler
DERA in Flow: Synthesis of a Statin Side Chain Precursor in Continuous Flow Employing Deoxyribose-5-Phosphate Aldolase Immobilized in Alginate-Luffa Matrix
Reprinted from: *Catalysts* 2020, 10, 137, doi:10.3390/catal10010137 . . . . **27**

I-Chun Cheng, Jin-Xian Liao, Jhih-Ying Ciou, Li-Tung Huang, Yu-Wei Chen and Chih-Yao Hou
Characterization of Protein Hydrolysates from Eel (*Anguilla marmorata*) and Their Application in Herbal Eel Extracts
Reprinted from: *Catalysts* 2020, 10, 205, doi:10.3390/catal10020205 . . . . **43**

Chengcheng Jiang, Zhen Liu, Jianan Sun, Changhu Xue and Xiangzhao Mao
A Novel Route for Agarooligosaccharide Production with the Neoagarooligosaccharide-Producing β-Agarase as Catalyst
Reprinted from: *Catalysts* 2020, 10, 214, doi:10.3390/catal10020214 . . . . **55**

Yumei Hu, Jian Min, Yingying Qu, Xiao Zhang, Juankun Zhang, Xuejing Yu and Longhai Dai
Biocatalytic Synthesis of Calycosin-7-O-β-D-Glucoside with Uridine Diphosphate–Glucose Regeneration System
Reprinted from: *Catalysts* 2020, 10, 258, doi:10.3390/catal10020258 . . . . **65**

Chen-Fu Chung, Shih-Che Lin, Tzong-Yuan Juang and Yung-Chuan Liu
Shaking Rate during Production Affects the Activity of *Escherichia coli* Surface-Displayed *Candida antarctica* Lipase A
Reprinted from: *Catalysts* 2020, 10, 382, doi:10.3390/catal10040382 . . . . **77**

Li-Hua Du, Rui-Jie Long, Miao Xue, Ping-Feng Chen, Meng-Jie Yang and Xi-Ping Luo
Continuous-Flow Synthesis of β-Amino Acid Esters by Lipase-Catalyzed Michael Addition of Aromatic Amines
Reprinted from: *Catalysts* 2020, 10, 432, doi:10.3390/catal10040432 . . . . **93**

Andrey S. Aksenov, Irina V. Tyshkunova, Daria N. Poshina, Anastasia A. Guryanova, Dmitry G. Chukhchin, Igor G. Sinelnikov, Konstantin Y. Terentyev, Yury A. Skorik, Evgeniy V. Novozhilov and Arkady P. Synitsyn
Biocatalysis of Industrial Kraft Pulps: Similarities and Differences between Hardwood and Softwood Pulps in Hydrolysis by Enzyme Complex of *Penicillium verruculosum*
Reprinted from: *Catalysts* 2020, 10, 536, doi:10.3390/catal10050536 . . . . **107**

Chia-Hung Kuo, Chun-Yung Huang, Chien-Liang Lee, Wen-Cheng Kuo, Shu-Ling Hsieh and Chwen-Jen Shieh
Synthesis of DHA/EPA Ethyl Esters via Lipase-Catalyzed Acidolysis Using Novozym® 435: A Kinetic Study
Reprinted from: *Catalysts* **2020**, *10*, 565, doi:10.3390/catal10050565 133

Daniel Eggerichs, Carolin Mügge, Julia Mayweg, Ulf-Peter Apfel and Dirk Tischler
Enantioselective Epoxidation by Flavoprotein Monooxygenases Supported by Organic Solvents
Reprinted from: *Catalysts* **2020**, *10*, 568, doi:10.3390/catal10050568 147

Ilona Sadauskiene, Arunas Liekis, Inga Staneviciene, Rima Naginiene and Leonid Ivanov
Effects of Long-Term Supplementation with Aluminum or Selenium on the Activities of Antioxidant Enzymes in Mouse Brain and Liver
Reprinted from: *Catalysts* **2020**, *10*, 585, doi:10.3390/catal10050585 161

Magdalena Rychlicka, Natalia Niezgoda and Anna Gliszczyńska
Development and Optimization of Lipase-Catalyzed Synthesis of Phospholipids Containing 3,4-Dimethoxycinnamic Acid by Response Surface Methodology
Reprinted from: *Catalysts* **2020**, *10*, 588, doi:10.3390/catal10050588 173

Liwei Zhang, Yuxiao Lu, Xiaobin Feng, Qinghong Liu, Yuanhui Li, Jiamin Hao, Yanqiong Wang, Yongqiang Dong and Huimin David Wang
Hepatoprotective Effects of *Pleurotus ostreatus* Protein Hydrolysates Yielded by Pepsin Hydrolysis
Reprinted from: *Catalysts* **2020**, *10*, 595, doi:10.3390/catal10060595 187

Andreea Veronica Botezatu (Dediu), Georgiana Horincar, Ioana Otilia Ghinea, Bianca Furdui, Gabriela-Elena Bahrim, Vasilica Barbu, Fanica Balanescu, Lidia Favier and Rodica-Mihaela Dinica
Whole-Cells of *Yarrowia lipolytica* Applied in "One Pot" Indolizine Biosynthesis
Reprinted from: *Catalysts* **2020**, *10*, 629, doi:10.3390/catal10060629 203

José G. Sampedro, Miguel A. Rivera-Moran and Salvador Uribe-Carvajal
Kramers' Theory and the Dependence of Enzyme Dynamics on Trehalose-Mediated Viscosity
Reprinted from: *Catalysts* **2020**, *10*, 659, doi:10.3390/catal10060659 219

Shang-Ming Huang, Hsin-Yi Huang, Yu-Min Chen, Chia-Hung Kuo and Chwen-Jen Shieh
Continuous Production of 2-Phenylethyl Acetate in a Solvent-Free System Using a Packed-Bed Reactor with Novozym® 435
Reprinted from: *Catalysts* **2020**, *10*, 714, doi:10.3390/catal10060714 239

Adama A. Bojang and Ho Shing Wu
Characterization of Electrode Performance in Enzymatic Biofuel Cells Using Cyclic Voltammetry and Electrochemical Impedance Spectroscopy
Reprinted from: *Catalysts* **2020**, *10*, 782, doi:10.3390/catal10070782 253

Thi Huong Ha Nguyen, Su-Min Woo, Ngoc Anh Nguyen, Gun-Su Cha, Soo-Jin Yeom, Hyung-Sik Kang and Chul-Ho Yun
Regioselective Hydroxylation of Naringin Dihydrochalcone to Produce Neoeriocitrin Dihydrochalcone by CYP102A1 (BM3) Mutants
Reprinted from: *Catalysts* **2020**, *10*, 823, doi:10.3390/catal10080823 273

About the Editors

Chia-Hung Kuo (Ph.D.) received his MS in Food Science and Technology from the National Taiwan University, Taiwan and a Ph.D. in Chemical Engineering from the National Taiwan University of Science and Technology, Taiwan. He is currently an associate professor of Seafood Science at the National Kaohsiung University of Science and Technology (NKUST), Taiwan. He has received several awards including the Outstanding New Teacher Award (2016), Outstanding Research Award (2018), and Special Outstanding Research Talent Award (2018–2019) from NKUST, the second largest university in Taiwan. He has also served as a Director of the Center for Aquatic Products Inspection Service at NKUST. He has more than 50 papers published in international journals and indexed on Scopus. His main research interests focus on process biochemistry, food engineering, extraction, oil and fat processing, and fermentation biotechnology.

Chwen-Jen Shieh (Ph.D.) holds a Ph.D. in Food Science and Technology from the University of Georgia, USA. He is currently a distinguished professor of Biotechnology Center at the National Chung-Hsing University, Taiwan. He was also an outstanding research professor at Dayeh University, and a Director of R&D at Taisun Enterprise Company, Taiwan. His main research interests focus on biodiesel, lipid biocatalysis, enzyme technology, bioprocess optimization, supercritical fluid technology, and Chinese herb medicine biotechnology.

Editorial

Biocatalytic Process Optimization

Chia-Hung Kuo [1,*] and Chwen-Jen Shieh [2,*]

1. Department of Seafood Science, National Kaohsiung University of Science and Technology, Kaohsiung 811, Taiwan
2. Biotechnology Center, National Chung Hsing University, Taichung 402, Taiwan
* Correspondence: kuoch@nkust.edu.tw (C.-H.K.); cjshieh@nchu.edu.tw (C.-J.S.); Tel.: +886-7-361-7141 (ext. 23646) (C.-H.K.); +886-4-2284-0450 (5121) (C.-J.S.)

Received: 10 October 2020; Accepted: 25 October 2020; Published: 12 November 2020

Biocatalysis refers to the use of microorganisms and enzymes in chemical reactions, has become increasingly popular and is frequently used in industrial applications due to the high efficiency and selectivity of biocatalysts. Enzymes are effective and precise biocatalysts as they are enantioselective, with mild reaction conditions, and are important tools in green chemistry. Biocatalysis is widely used in the pharmaceutical, food, cosmetic, and textile industries. Biocatalytic processes include enzyme production, biocatalytic process development, biotransformation, enzyme engineering, immobilization, and the recycling of biocatalysts. Factors affecting biocatalytic reactions include substrate concentration, product concentration, enzyme or microorganism stability, inhibitors, temperature, and pH. As such, the optimization of biocatalytic processes is an important issue.

Active compounds in natural products usually contain glycosides, which can be cleaved by glycoside hydrolases to increase biological activity. The cocktail enzyme cellulase has been used in the deglycosylation of piceid to produce resveratrol [1]. Rha et al. [2] tested several commercial food-grade enzymes for producing flavonol aglycones from green tea extracts via deglycosylation. Tannin acyl hydrolase and glycoside hydrolase activities in Plantase-CF (a multi-functional food-grade enzyme from *Aspergillus niger* with the ability of cellulolytic hydrolysis of carbohydrates) trigger the degalloylation of catechins and produce flavonol aglycones from green tea extracts. Optimal conditions for producing flavonol aglycones are pH 4.0 and 50 °C. Biocatalysis can prevent catechin degradation, unlike hydrochloride treatment where 70% (*w/w*) of catechins disappear.

Glycoside hydrolases are enzymes that catalyze the hydrolysis of the glycosidic bonds in glycosides and have many applications, including the production of agaro-oligosaccharides, glucose, xylose, and xylobiose. Jiang et al. [3] reported a novel technique for producing agarotriose and agaropentaose from agaro-oligosaccharide using β-agarase. The results showed that 1.950 U/mL β-agarase AgWH50B was optimal for the preparation of agarotriose from the hydrolysis of agaroheptaose or agarononose, while 0.1900 U/mL β-agarase DagA was optimal for the preparation of agarotriose and agaropentaose from the hydrolysis of agaroheptaose or agarononose. In this study, the authors obtained value-added oligosaccharides from agarose by using different agarolytic enzymes or varying the enzyme amount. Aksenov et al. [4] used cellulolytic enzymes (mainly cellulases and xylanase) from recombinant *Penicillium verruculosum* to hydrolyze hardwood and softwood pulp, though the lignin content and drying process decreased the bioconversion of pulp to glucose. It was determined that fiber morphology, differing xylan and mannan content, and hemicellulose localization in kraft fibers deeply affected the enzymatic hydrolysis of bleached pulp. At a concentration of 10%, never-dried bleached kraft pulp demonstrated highly efficient bioconversion, resulting in a concentration of more than 50 g/L sugar.

Enzymatic biofuel cells rely on enzymes rather than conventional noble metal catalysts. Commonly used redox enzymes include glucose dehydrogenase, glucose oxidase (GOx), lacase (LAc), fructose dehydrogenase, and alcohol dehydrogenase [5]. Bojang and Wu [6] established the use of

GOx/LAc modified electrodes as bioanodes and biocathodes for biofuel cells. Electrochemical analysis methods including cyclic voltammetry, the Nicholson method, the Randles–Sevcik equation, and electrochemical impedance spectroscopy were used to evaluate the performance of prepared electrodes. Following testing, the optimal bioanode and biocathode were determined to be a carbon paper–GOx–mediator–carbon nanotube with a current density of 800 $\mu A/cm^2$ and a carbon paper–Lac–mediator–carbon nanotube with a current density of 600 $\mu A/cm^2$, respectively. The construction and use of enzyme electrodes can be applied to biofuel cells, bioreactors, biosensors, and micro-reactors.

The disaccharide trehalose, a natural biostructure stabilizer that accumulates in the cytoplasm under stress conditions, is present in a wide variety of organisms, including bacteria, yeast, fungi, insects, invertebrates, and lower and higher plants [7]. Sampedro et al. [8] studied the effect of trehalose on enzyme reactions using Kramers' theory. The role of trehalose was reviewed and the molecular interactions of trehalose–water–enzymes/proteins were described in detail, supported by recent in vitro and in silico experimental results. Importantly, the concept of coupling the enzyme's structural dynamics to medium viscosity, as described by Kramers' theory, is the central thesis of this paper and the focus is on enzyme catalysis. As such, the application of Kramers´ theory is reinforced by relating the rate of inactivation, unfolding, and folding of enzymes to trehalose viscosity. The recently observed effects of trehalose viscosity on DNA and RNA folding is mentioned as a corollary.

Proteases belong to the hydrolase class of enzymes and hydrolyze proteins into smaller polypeptides or single amino acids. Protein hydrolysates have many biological functions and demonstrate antioxidant activity [9,10]. Zhang et al. [11] used pepsin, trypsin, dispase, papain, and bromelin to digest *Pleurotus ostreatus* protein extract (POPE). The antioxidant activity of the protein hydrolysates resulting from five different proteases were compared. The results showed that POPEP (POPE hydrolyzed by pepsin), with a molecular weight of 3–5 kDa, had the strongest antioxidant activity. Excessive free radicals or reactive oxygen species (ROS) are harmful to the human body since these components may destroy the normal functions of cells, tissues, and organs [12]. Superoxide dismutase, glutathione peroxide, and catalase can remove free radicals to reduce the risk of oxidative damage during periods of increased ROS. Mice pretreated with POPEP (3–5 kDa) showed significantly increased superoxide dismutase and glutathione peroxide enzyme activity in the liver, demonstrating that POPEP could protect the liver from oxidative damage.

Sadauskiene et al. [13] reported on the effects of long-term supplementation with aluminum (Al) or selenium (Se) on antioxidant enzyme activity in the brains and livers of mice. The results showed that 8 weeks of exposure to Se caused a statistically significant increase in superoxide dismutase, catalase, and glutathione reductase activities in the brain and/or liver, but the changes were dose-dependent. Exposure to Al caused a statistically significant increase in glutathione reductase activity in both organs.

Protein hydrolysates containing bioactive peptides can be used to formulate nutraceuticals or functional ingredients in food. Cheng et al. [14] used three commercial proteases (alcalase, bromelain, and papain) to obtain eel protein hydrolysates (EPHs) from whole eels (*Anguilla marmorata*). The emulsion activity index (EAI) and emulsion stability index (ESI) of each EPH was determined to test the product stability. The EPH obtained from the treatment with alcalase showed optimal EAI and ESI and demonstrated antioxidant activity. The results indicated that alcalase-hydrolyzed EPH had good emulsifying properties and solubility, making it useful in food processing.

The use of biocatalysis or biotransformation to produce pharmaceutical components has become a hot topic in biotechnology research. There are two main types of biocatalysts: whole cells and free enzymes. Both use enzymes to complete the reaction but in the former, the enzyme remains within the microorganism whereas for the latter, the enzyme has been separated and purified. When producing drugs and their intermediates, the most significant difference between biocatalysis and traditional chemical methods is that the former is very effective in the asymmetric synthesis of chiral compounds. In this Special Issue, several studies used biocatalysis to synthesize special compounds. Calycosin-7-O-β-D-glucoside is an isoflavonoid glucoside and one of the principal components of

Radix astragali, a well known medicinal and edible herb cited in European, Japanese, and Chinese literature. Hu et al. [15] used uridine diphosphate-dependent glucosyltransferase to glucosylate the C7 hydroxyl group of calycosin and synthesize calycosin-7-O-β-D-glucoside. Optimal conditions for batch production were determined, including the temperature, pH, and the concentrations of dimethyl sulfoxide, uridine diphosphate, sucrose, and calycosin. Eggerichs et al. [16] used styrene and indole monooxygenase to activate double bonds via chiral epoxidation. The reaction conditions were successfully optimized for two flavins containing two-component monooxygenases during the conversion of large hydrophobic styrene derivatives in the presence of organic cosolvents.

Nguyen et al. [17] used *Bacillus megaterium* CYP102A1 monooxygenase for the regioselective hydroxylation of naringin dihydrochalcone to produce neoeriocitrindihydrochalcone. Kinetic parameters were used to compare the efficiency of dihydrochalcone hydroxylation by different CYP102A1 mutations. The indolizine core is present in many biologically active compounds and can be considered the scaffolding in the preparation of new pharmaceuticals. Indolizines have been synthesized by lipases from *Candida antarctica* [18]. Botezatu et al. [19] used whole cells to catalyze a multicomponent reaction of activated alkynes, α-bromo-carbonyl reagents and 4,4′-bipyridine in the synthesis of bis-indolizines. Several yeast strains were tested to evaluate the effect of the reactants on their physiological activity. The optimal strain was *Yarrowia lipolytica*, which is an effective biocatalyst in cycloaddition reactions and can be used to synthesize indolizines. In recent years, there has been a widespread use of immobilized lipase to catalyze specific reactions in the production of valuable molecules, such as nutraceutical and pharmaceutical compounds [20,21]. Chung et al. [22] used a surface-display system for the expression of lipase A in an *E. coli* expression system. It was reported that lipase A activity was low at lower shaking rates due to the limited amount of dissolved oxygen, while higher shaking rates increased shear stress, leading to a decrease in the specific activity. This phenomenon was confirmed using kinetic studies and it was established that cultivating lipase A at a moderate shaking speed optimized hydrolysis.

Docosahexaenoic acid (DHA) and eicosapentaenoic acid (EPA) ethyl esters are medicines used in the treatment of arteriosclerosis and hyperlipidemia. Kuo et al. [23] studied the lipase-catalyzed synthesis of DHA + EPA ethyl esters via the acidolysis of ethyl acetate with DHA + EPA concentrates. Lipase-catalyzed acidolysis has the advantage of not only synthesizing DHA + EPA ethyl ester efficiently, but also allowing for the easy recovery of the product. Moreover, a response surface methodology (RSM) approach for the evaluation of the kinetic model was successful integrated with the rate equation to simulate the performance of the batch reactor. The integral equation showed a good predictive relationship between the simulated and experimental results. Conversion yields of 88%–94% were obtained for 100–400 mM DHA + EPA concentrate at a constant enzyme activity of 200 U, substrate ratio of 1:1 (DHA + EPA: EA), and reaction time of 300 min. Rychlicka et al. [24] developed a biotechnological method of synthesizing 3,4-dimethoxycinnamoylated phospholipids via the interesterification of egg-yolk phosphatidylcholine with the ethyl ester of 3,4-dimethoxycinnamic acid. RSM and a Box–Behnken design were used to evaluate reaction conditions. The optimal incorporation of 3,4-dimethoxycinnamic acid into phospholipids reached 21 mol%. Moreover, 3,4-dimethoxycinnamoylated lysophosphatidylcholine and 3,4-dimethoxycinnamoylated phosphatidylcholine were obtained in isolated yields of 27.5% and 3.5% (w/w), respectively. Huang et al. [25] developed a biocatalytic process for synthesizing rose-flavored ester-2-phenylethyl acetate using a packed-bed bioreactor system. The synthesis process was performed in a solvent free system, which is an environmentally friendly process. The optimization of the synthesis reaction was carried out by a three-level-three-factor Box–Behnken design and RSM. This continuous process can be applied to the environmentally friendly production of natural flavor compounds, such as rose aromatic esters. Grabner et al. [26] developed a continuous process for the synthesis of a statin side chain precursor using a deoxyribose-5-phosphate aldolase-catalyzed stereoselective aldol addition reaction. A series of substrates was tested but only acetaldehyde and chloroacetaldehyde gave reasonable results. An experimental design was used to optimize pH value, temperature, and flow conditions. The immobilization alginate was chosen and the reaction rates of both alginate beads and the alginate–luffa matrix were tested. The optimized flow process (0.1 mL/min, 0.25 M of chloroacetaldehyde, and 0.5 M of acetaldehyde) produced 4.5 g of product per day in a bench-top reactor. Du et al. [27] developed a continuous-flow procedure for the synthesis of β-amino acid esters via the lipase-catalyzed Michael reactions of various aromatic amines with acrylates. Seventeen β-amino acid esters were rapidly synthesized by lipase TL IM from *Thermomyces lanuginosus* in continuous-flow microreactors. Optimal reaction parameters were determined, including the reaction medium, temperature, enzyme, substrate molar ratio, residence time/flow rate, and substrate structure. The salient features of this study are the green reaction conditions (using methanol as reaction medium), short residence time (30 min), and high yield. Several articles in this Special Issue have demonstrated that a bioreactor with immobilized enzymes is suitable for use in biocatalysis to synthesize active pharmaceutical ingredients, drug precursors, and value-added chemicals. The benefits of continuous flow biocatalysis, including improved reaction rates, in-line product removal and purification, better mixing, improved control, and improved enzyme stability, ultimately minimize labor and reduce production costs. The increased demand for greener and more cost-effective processes will drive the rapid expansion of continuous flow biocatalysis in the next few years.

In conclusion, this Special Issue shows that an optimized biocatalysis process can provide an environmentally friendly, clean, highly efficient, low cost, and renewable process for the synthesis and production of valuable products. With further development and improvements, more biocatalysis processes may be applied in the future.

Author Contributions: C.-H.K. and C.-J.S. prepared the article. All authors have read and agreed to the published version of the manuscript.

Funding: This research received no external funding.

Conflicts of Interest: The authors declare no conflict of interest.

References

1. Kuo, C.-H.; Chen, B.-Y.; Liu, Y.-C.; Chen, J.-H.; Shieh, C.-J. Production of resveratrol by piceid deglycosylation using cellulase. *Catalysts* **2016**, *6*, 32. [CrossRef]
2. Rha, C.-S.; Kim, S.-W.; Byoun, K.H.; Hong, Y.D.; Kim, D.-O. Simultaneous optimal production of flavonol aglycones and degalloylated catechins from green tea using a multi-function food-grade enzyme. *Catalysts* **2019**, *9*, 861. [CrossRef]
3. Jiang, C.; Liu, Z.; Sun, J.; Xue, C.; Mao, X. A novel route for agarooligosaccharide production with the neoagarooligosaccharide-producing β-agarase as catalyst. *Catalysts* **2020**, *10*, 214. [CrossRef]
4. Aksenov, A.S.; Tyshkunova, I.V.; Poshina, D.N.; Guryanova, A.A.; Chukhchin, D.G.; Sinelnikov, I.G.; Terentyev, K.Y.; Skorik, Y.A.; Novozhilov, E.V.; Synitsyn, A.P. Biocatalysis of industrial kraft pulps: Similarities and differences between hardwood and softwood pulps in hydrolysis by enzyme complex of *Penicillium verruculosum*. *Catalysts* **2020**, *10*, 536. [CrossRef]
5. Kuo, C.-H.; Huang, W.-H.; Lee, C.-K.; Liu, Y.-C.; Chang, C.-M.J.; Yang, H.; Shieh, C.-J. Biofuel cells composed by using glucose oxidase on chitosan coated carbon fiber cloth. *Int. J. Electrochem. Sci.* **2013**, *8*, 9242–9255.
6. Bojang, A.A.; Wu, H.S. Characterization of electrode performance in enzymatic biofuel cells using cyclic voltammetry and electrochemical impedance spectroscopy. *Catalysts* **2020**, *10*, 782. [CrossRef]
7. Elbein, A.D.; Pan, Y.; Pastuszak, I.; Carroll, D. New insights on trehalose: A multifunctional molecule. *Glycobiology* **2003**, *13*, 17R–27R. [CrossRef]
8. Sampedro, J.G.; Rivera-Moran, M.A.; Uribe-Carvajal, S. Kramers' theory and the dependence of enzyme dynamics on trehalose-mediated viscosity. *Catalysts* **2020**, *10*, 659. [CrossRef]
9. Yu, Y.; Fan, F.; Wu, D.; Yu, C.; Wang, Z.; Du, M. Antioxidant and ACE inhibitory activity of enzymatic hydrolysates from *Ruditapes philippinarum*. *Molecules* **2018**, *23*, 1189. [CrossRef]
10. Fan, J.; He, J.; Zhuang, Y.; Sun, L. Purification and identification of antioxidant peptides from enzymatic hydrolysates of tilapia (*Oreochromis niloticus*) frame protein. *Molecules* **2012**, *17*, 12836–12850. [CrossRef]

11. Zhang, L.; Lu, Y.; Feng, X.; Liu, Q.; Li, Y.; Hao, J.; Wang, Y.; Dong, Y.; Wang, H.D. Hepatoprotective effects of *Pleurotus ostreatus* protein hydrolysates yielded by pepsin hydrolysis. *Catalysts* **2020**, *10*, 595. [CrossRef]
12. Yang, K.-R.; Yu, H.-C.; Huang, C.-Y.; Kuo, J.-M.; Chang, C.; Shieh, C.-J.; Kuo, C.-H. Bioprocessed production of resveratrol-enriched rice wine: Simultaneous rice wine fermentation, extraction, and transformation of piceid to resveratrol from *Polygonum cuspidatum* roots. *Foods* **2019**, *8*, 258. [CrossRef]
13. Sadauskiene, I.; Liekis, A.; Staneviciene, I.; Naginiene, R.; Ivanov, L. Effects of long-term supplementation with aluminum or selenium on the activities of antioxidant enzymes in mouse brain and liver. *Catalysts* **2020**, *10*, 585. [CrossRef]
14. Cheng, I.; Liao, J.-X.; Ciou, J.-Y.; Huang, L.-T.; Chen, Y.-W.; Hou, C.-Y. Characterization of protein hydrolysates from eel (*Anguilla marmorata*) and their application in herbal eel extracts. *Catalysts* **2020**, *10*, 205. [CrossRef]
15. Hu, Y.; Min, J.; Qu, Y.; Zhang, X.; Zhang, J.; Yu, X.; Dai, L. Biocatalytic synthesis of calycosin-7-O-β-D-glucoside with uridine diphosphate–glucose regeneration system. *Catalysts* **2020**, *10*, 258. [CrossRef]
16. Eggerichs, D.; Mügge, C.; Mayweg, J.; Apfel, U.-P.; Tischler, D. Enantioselective epoxidation by flavoprotein monooxygenases supported by organic solvents. *Catalysts* **2020**, *10*, 568. [CrossRef]
17. Nguyen, T.H.H.; Woo, S.-M.; Nguyen, N.A.; Cha, G.-S.; Yeom, S.-J.; Kang, H.-S.; Yun, C.-H. Regioselective hydroxylation of naringin dihydrochalcone to produce neoeriocitrin dihydrochalcone by CYP102A1 (BM3) mutants. *Catalysts* **2020**, *10*, 823. [CrossRef]
18. Dinica, R.M.; Furdui, B.; Ghinea, I.O.; Bahrim, G.; Bonte, S.; Demeunynck, M. Novel one-pot green synthesis of indolizines biocatalysed by *Candida antarctica* lipases. *Marine Drugs* **2013**, *11*, 431–439. [CrossRef]
19. Botezatu (Dediu), A.V.; Horincar, G.; Ghinea, I.O.; Furdui, B.; Bahrim, G.-E.; Barbu, V.; Balanescu, F.; Favier, L.; Dinica, R.-M. Whole-cells of *Yarrowia lipolytica* applied in "one pot" indolizine biosynthesis. *Catalysts* **2020**, *10*, 629. [CrossRef]
20. Huang, S.-M.; Hung, T.-H.; Liu, Y.-C.; Kuo, C.-H.; Shieh, C.-J. Green synthesis of ultraviolet absorber 2-ethylhexyl salicylate: Experimental design and artificial neural network modeling. *Catalysts* **2017**, *7*, 342. [CrossRef]
21. Kuo, C.-H.; Chen, H.-H.; Chen, J.-H.; Liu, Y.-C.; Shieh, C.-J. High yield of wax ester synthesized from cetyl alcohol and octanoic acid by lipozyme RMIM and Novozym 435. *Int. J. Mol. Sci.* **2012**, *13*, 11694–11704. [CrossRef] [PubMed]
22. Chung, C.-F.; Lin, S.-C.; Juang, T.-Y.; Liu, Y.-C. Shaking rate during production affects the activity of *Escherichia coli* surface-displayed *Candida antarctica* lipase A. *Catalysts* **2020**, *10*, 382. [CrossRef]
23. Kuo, C.-H.; Huang, C.-Y.; Lee, C.-L.; Kuo, W.-C.; Hsieh, S.-L.; Shieh, C.-J. Synthesis of DHA/EPA ethyl esters via lipase-catalyzed acidolysis using Novozym® 435: A kinetic study. *Catalysts* **2020**, *10*, 565. [CrossRef]
24. Rychlicka, M.; Niezgoda, N.; Gliszczyńska, A. Development and optimization of lipase-catalyzed synthesis of phospholipids containing 3, 4-dimethoxycinnamic acid by response surface methodology. *Catalysts* **2020**, *10*, 588. [CrossRef]
25. Huang, S.-M.; Huang, H.-Y.; Chen, Y.-M.; Kuo, C.-H.; Shieh, C.-J. Continuous production of 2-phenylethyl acetate in a solvent-free system using a packed-bed reactor with Novozym® 435. *Catalysts* **2020**, *10*, 714. [CrossRef]
26. Grabner, B.; Pokhilchuk, Y.; Gruber-Woelfler, H. DERA in flow: Synthesis of a statin side chain precursor in continuous flow employing deoxyribose-5-phosphate aldolase immobilized in alginate-luffa matrix. *Catalysts* **2020**, *10*, 137. [CrossRef]
27. Du, L.-H.; Long, R.-J.; Xue, M.; Chen, P.-F.; Yang, M.-J.; Luo, X.-P. Continuous-flow synthesis of β-amino acid esters by lipase-catalyzed michael addition of aromatic amines. *Catalysts* **2020**, *10*, 432. [CrossRef]

Publisher's Note: MDPI stays neutral with regard to jurisdictional claims in published maps and institutional affiliations.

© 2020 by the authors. Licensee MDPI, Basel, Switzerland. This article is an open access article distributed under the terms and conditions of the Creative Commons Attribution (CC BY) license (http://creativecommons.org/licenses/by/4.0/).

Article

Simultaneous Optimal Production of Flavonol Aglycones and Degalloylated Catechins from Green Tea Using a Multi-Function Food-Grade Enzyme

Chan-Su Rha [1,*], Shin-Woo Kim [2], Kyoung Hee Byoun [3], Yong Deog Hong [1] and Dae-Ok Kim [4,5]

1. Basic Research and Innovation Institute, Amorepacific R&D Center, Yongin 17074, Korea; hydhong@amorepacific.com
2. Research and Development Division, Bision Corp., Seoul 05854, Korea; swkim@bision.co.kr
3. Safety and Regulatory Research Institute, Amorepacific Corporation R&D Center, Yongin 17074, Korea; silling@amorepacific.com
4. Department of Food Science and Biotechnology, Kyung Hee University, Yongin 17104, Korea; DOKIM05@khu.ac.kr
5. Graduate School of Biotechnology, Kyung Hee University, Yongin 17104, Korea
* Correspondence: teaman@amorepacific.com; Tel: +82-31-280-5981

Received: 14 September 2019; Accepted: 13 October 2019; Published: 16 October 2019

Abstract: (1) Background: Green tea (GT) contains well-known phytochemical compounds; namely, it is rich in flavan-3-ols (catechins) and flavonols comprising all glycoside forms. These compounds in GT might show better biological activities after a feasible enzymatic process, and the process on an industrial scale should consider enzyme specificity and cost-effectiveness. (2) Methods: In this study, we evaluated the most effective method for the enzymatic conversion of flavonoids from GT extract. One enzyme derived from *Aspergillus niger* (molecular weight 80–90 kDa) was ultimately selected, showing two distinct but simultaneous activities: intense glycoside hydrolase activity via deglycosylation and weak tannin acyl hydrolase activity via degalloylation. (3) Results: The optimum conditions for producing flavonol aglycones were pH 4.0 and 50 °C. Myricetin glycosides were cleaved 3.7–7.0 times faster than kaempferol glycosides. Flavonol aglycones were produced effectively by both enzymatic and hydrochloride treatment in a time-course reaction. Enzymatic treatment retained 80% (*w/w*) catechins, whereas 70% (*w/w*) of catechins disappeared by hydrochloride treatment. (4) Conclusions: This enzymatic process offers an effective method of conditionally producing flavonol aglycones and de-galloylated catechins from conversion of food-grade enzyme.

Keywords: catechin; degalloylation; flavonol; glycoside hydrolase; optimization; tannase

1. Introduction

Green tea (GT) is well-known to be enriched in catechins with flavonols/flavones as the second-most dominant flavonoids; these include myricetin, quercetin, apigenin, and kaempferol [1]. GT generally contains approximately 15% catechins and 0.4% flavonols on a dry weight basis [2]. Many studies have demonstrated that catechins are major sources of the vast diversity of GT bioactivities [3,4]. However, in human nutrition, GT flavonols are generally considered to be less crucial for the utility and functionality of GT. The content and compositions of flavonol and flavone glycosides vary according to GT cultivar [5,6]. Similarly, glycosylated flavonols and flavones have different sugar bonds and compositions according to the plants [7]. Flavonol glycosides consist of various sugar units with –O- or –C-conjugation on flavonol molecules. The glycosidic structure of flavonols affects their biological and physiological properties, such as digestive stability and bioaccessibility [8–10]. For example, quercetin glycosides could be hydrolyzed to aglycone in the intestine [8], and quercetin 3-glucoside and quercetin 4-glucoside can be completely digested in humans [9]. However, flavonol glycosides with more complex structures undergo less hydrolysis in digestive conditions [10]. To date, the majority of research on the functionalities of flavonols has focused on the aglycone forms. Xiao et al. [11] proposed that flavonol aglycones have a higher affinity for proteins due to their hydrophobic characteristics, allowing for easy absorption by cells. Moreover, flavone aglycones exhibited more potent anti-inflammatory effects than their corresponding glycosides [12]. Flavonols, including myricetin, quercetin, and kaempferol, are known to possess numerous beneficial activities, such as antioxidative, anticancer, and antihyperlipidemic effects [13–15]. Furthermore, flavonol supplementation was proven to potentially reduce the risk of cardiometabolic disease in a clinical trial [16].

Plumb et al. [17] presented that the antioxidant activities of GT flavonol glycosides were lower than those of their corresponding aglycones. For example, quercetin rhamnoside, a GT flavonol glycoside, was found to have an affinity to bovine serum albumin 5600-fold lower than its corresponding aglycone, quercetin [11]. Owing to such limitations in the utilization of flavonols in human nutrition, a feasible enzymatic process was developed to break down plant-based flavonoid glycosides to their aglycones [18]. For example, the antioxidant capacity of soybean flour was enhanced by the enzymatic hydrolysis of phenolic glucoside in solid-state fungi fermentation [19]. In particular, tannase (EC 3.1.1.20; tannin acyl hydrolase) shows good ability in the bioconversion of green tea extract (GTE), resulting in gallic acid (GA) and degalloylated catechins [20,21].

Increasing the content of GA through the conversion of GTE by tannase has been shown to improve the radical scavenging activities of GTE [22]. Moreover, conversion of (−)-epigallocatechin gallate (EGCG) by tannase attenuated its toxicity without affecting the antiproliferative effects [21]. Another study demonstrated that tannase-treated catechins influenced the expression of genes involved in the sodium-glucose transport proteins [23]. For their effective absorption and utilization in the human body, some flavonol glycosides must be converted to their aglycone forms by digestive enzymes. Only 2% of the dietary flavonols that are digested and absorbed in the duodenum ultimately reach the plasma in intact form [24]. Flavonol glycosides are hydrolyzed by mammalian glucosidase in the small intestine before being absorbed in aglycone forms [25]. Some of the flavonol glycosides that are not hydrolyzed in the small intestine then move to the large intestine, where they undergo further metabolic reactions through bioconversions mediated by intestinal microorganisms [26]. Approximately 65% of human adults are estimated to have downregulated production of intestinal lactase (lactase phlorizin hydrolase; LPH) [27]. Lactase catalyzes the hydrolysis of β-glucosides, including phlorizin and flavonoid glucosides [28]. Furthermore, it was previously reported that flavonol-enriched fractions of GTE enhanced the bioavailability of the catechin epimers of GTE by downregulating the expression of the catechol-O-methyltransferase gene [29]. To date, research on the bioconversion of flavonol glycosides in GT to verify whether flavonol glycosides or aglycones are more beneficial to human nutrition and health is scarce.

Other than tannase-based research, there are limited studies on the bioconversion of GT flavonoids by food-grade enzymes that show high specificity to flavonol glycosides. Since the enzymes for hydrolyzing glycosides are very diverse and specific to the type of glycosidic bond, multiple enzymes generally need to be used to accomplish complex food treatments. We hypothesized that a few multi-activity enzymes could be used to improve the efficiency of conversion for obtaining functional compounds with nutritional benefits.

To test this possibility, we screened multi-functional food-grade enzymes among eight kinds of enzymes from a broth of fungi. We then selected an enzyme and further evaluated its kinetic characteristics and compared the effectiveness of enzymatic treatment for producing flavonol aglycones from flavonol glycosides in GTE to provide an optimized method.

2. Results

2.1. Composition Changes of GTE by Various Food–Grade Enzymes

The tested commercial food-grade enzymes exhibited various multi-functional activities (Table 1). Although all of the enzymes were derived from the broth of *Aspergillus* spp., the significant activities differed depending on the culture conditions. Nine types of glycosidase activities of the enzymes used in this study were determined (Table 1). The enzyme Peclyve ARA-NS (NS) showed the highest activity for arabinase (175 units/mL), Plantase-CF (CF) showed the highest activity for α-arabinofuranosidase (3406 units/g), Sumizyme-AC (AC) activity was highest for α-arabinopyranosidase (567 units/g) and β-mannosidase (130 units/g), Glucosidase-BT (BT) showed the greatest β-glucosidase (280 units/mL) activity, and Cellulase-KN (KN) had the highest α-rhamnosidase activity (245 units/g) (Table 1). Among the enzymes with multi-functional characteristics, only CF and Tannase KTFHR (TN) produced GA from GTE; these enzymes also produced decreased amounts of the gallated catechins (–)-epicatechin gallate (ECG) and EGCG (Figure 1B, Figure 2A, and Figure S1 in Supplementary Materials). GA is an indicator of tannin acyl hydrolase activity, which promotes the degalloylation of gallated catechins. Thus, except for CF and TN, none of the other enzymes cleaved the galloyl moiety of gallated catechins. The tannin acyl hydrolase activity of CF was lower than that of TN. However, among the eight enzymes, CF showed the best results for producing flavonol aglycone (Figure 1D, Figure 2B, and Table S1).

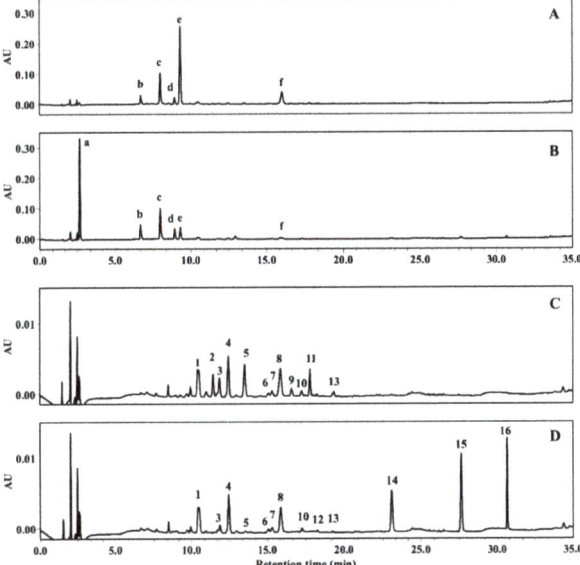

Figure 1. High-performance liquid chromatography traces of green tea extract (GTE) and enzyme-treated GTE. (**A**) Catechin chromatogram of GTE at 275 nm, (**B**) catechin chromatogram of enzyme-treated GTE at 275 nm, (**C**) flavonol chromatogram of GTE at 365 nm, and (**D**) flavonol chromatogram of enzyme-treated GTE at 365 nm. Peaks were identified by mass analysis: **a**, gallic acid (GA); **b**, (–)-epigallocatechin (EGC); **c**, caffeine; **d**, (–)-epicatechin (EC); **e**, (–)-epigallocatechin gallate (EGCG); **f**, (–)-epicatechin gallate (ECG). The numbers 1–13, 14, 15, and 16 indicate flavonol glycosides, myricetin, quercetin, and kaempferol, respectively. Refer to the exact mass and structural information of numbered peaks in previous research [30].

Table 1. Characteristics of enzymes and test conditions of screening for hydrolyzing flavonol glycosides in green tea extract.

Name of Enzyme [a] (abbreviation)		Cellulyve-AN (AN)	Cellulase-KN (KN)	Glucosidase-BT (BT)	Plantase-CF (CF)	Plantase-UF (UF)	Peclyve ARA-NS (NS)	Sumizyme-AC (AC)	Tannase-KTFHR (TN)
EC number		3.2.1.4	3.2.1.4	3.2.1.21	3.2.1.15	3.2.1.6.; 3.2.1.99	3.2.1.99	3.2.1.4	3.1.1.20
Class of glycosidase [b]		C	C	B	P	G; A	A	C	T
Activity (U [c])	arabinase	2	0.2	N/D [d]	62	55	175	N/D	N/D
	α-arabinofuranosidase	36	124	N/D	3406	1058	473	69	N/A [e]
	α-arabinopyranosidase	313	36	N/D	260	235	51	567	N/A
	β-glucosidase	32	6	280	28	40	12	36	N/A
	β-galactosidase	0	17	2	10	9	14	3	N/A
	β-mannosidase	40	2	3	58	1	1	130	N/A
	α-rhamnosidase	4	245	N/D	3	1	6	4	N/A
	tannin acyl hydrolase	N/D	3.9	0.7	14.8	N/D	N/D	3.4	520.4
	β-xylosidase	12	14	40	74	3	1	55	N/A
Test pH (optimum [f])		5.0 (3.0–5.0)	5.0 (3.0–5.0)	5.0 (3.0–6.0)	5.0 (3.0–5.5)	5.0 (4.0–6.5)	5.0 (4.0–6.5)	5.0 (3.5–5.0)	5.0 (3.0–6.0)
Test temp. (°C; optimum [f])		50 (30–50)	50 (40–50)	50 (35–65)	50 (50–60)	50 (40–60)	50 (40–60)	60 (40–65)	40 (40)
Applied amount (% w/v) of enzyme		10	10	10	10	10	10	10	10

[a] The name of the enzyme is followed by the trade name of the manufacturer; all enzymes were collected from the culture broth of *Aspergillus niger*, except for tannase (*A. oryzae*). Multi-functional activity comparison was performed among enzymes with the ability of cellulolytic hydrolysis of carbohydrates. [b] Nomenclature Committee of the International Union of Biochemistry and Molecular Biology, and the abbreviations indicate; C, cellulase; B, β-glucosidase; P, polygalacturonase; G, endo-1,3(4)-β-glucanase; A, arabinan endo-1,5-α-arabinanase; T, tannase. [c] U: unit/mL or unit/g (single-point test). [d] N/D: not detected. [e] N/A: not available (no test). [f] Citrate phosphate buffer (50 mM) was used. The applied GTE concentration was set at 10% (w/v) in the reaction.

All enzymes exhibited glycoside hydrolase (GH; EC 3.2.1) activity. Although BT had the highest β-glucosidase activity, it could not produce any aglycones from the flavonol glycosides of GTE (Figure 2B and Table S2). CF and TN showed the most potent and second most potent activity of glycoside hydrolase to the flavonols of GTE, respectively. The other enzymes (i.e., AC, AN, BT, KN, and NS) showed weak glycoside hydrolase activity (Table S2) and did not produce aglycones of kaempferol glycosides.

Multivariate analysis indicated that only α-arabinofuranosidase activity had a significant main effect on the output of flavonol aglycones (Table S3). Activities of arabinofuranosidase and xylosidase were positively correlated (>0.9; Table S3) to the glycoside hydrolase activity on the cleavage of flavonol glycosides. Thus, reactions were performed with GTE solution by standard enzymes of arabinofuranosidase (40 °C, 50 mM of citrate phosphate buffer, and pH 4.0) and xylosidase (35 °C, 50 mM of Tris-HCl, and pH 7.5). However, no flavonol aglycone-producing activity was observed for the two standard enzymes. This result indirectly implies that these two activities do not co-exist in the same protein molecule and that the microorganism shows correlated expression of glycoside hydrolase protein with arabinofuranosidase.

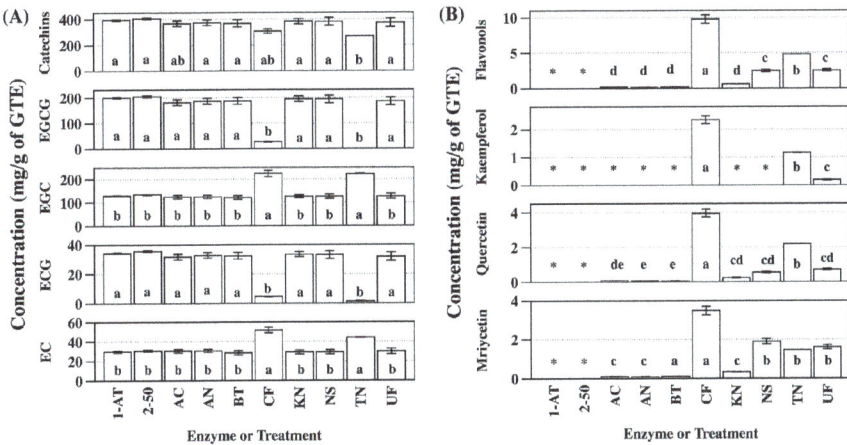

Figure 2. Changes of (**A**) catechins and (**B**) flavonols by the hydrolysis of cellulolytic enzymes. Data shown are the mean ± standard error of the mean (SEM) (mg/g GTE). Different letters on the bars indicate significant differences according to Tukey-Kramer's honestly significant difference test ($p < 0.05$). * means not detected. Flavonols represent the sum of myricetin, quercetin, and kaempferol. Catechins in Y-axis of (**A**) represent the sum of EC, ECG, EGC, and EGCG. 1-AT, control ambient temperature; 2–50, control 50 °C; refer to the abbreviations of enzymes in Table 1. The numeric data are presented in Tables S1 and S2.

2.2. Optimum Condition of Enzyme CF for Producing Flavonol Aglycones and GA

Based on the above screen, enzyme CF showed the best reaction result among the eight enzymes evaluated. The optimum condition for the simultaneous production of flavonol aglycones and GA by the enzyme CF was determined to be 50 °C and pH 4.0 (Figure 3). The increment in flavonol and GA production with increasing temperature could be expressed by a first-order and second-order relationship, respectively. The amount of flavonol aglycones and GA almost doubled at 50 °C compared to that determined at the lower temperature condition (35 °C) (Figure 3A). The linear range of α-arabinofuranosidase activity in CF was determined at 2–15 units/mL of the enzyme (Figure 3B). The sensitivity to variations in pH was more drastic than that to temperature, and no significant reactions to gallated catechins and flavonols were obtained at pH 7.0 (Figure 3C).

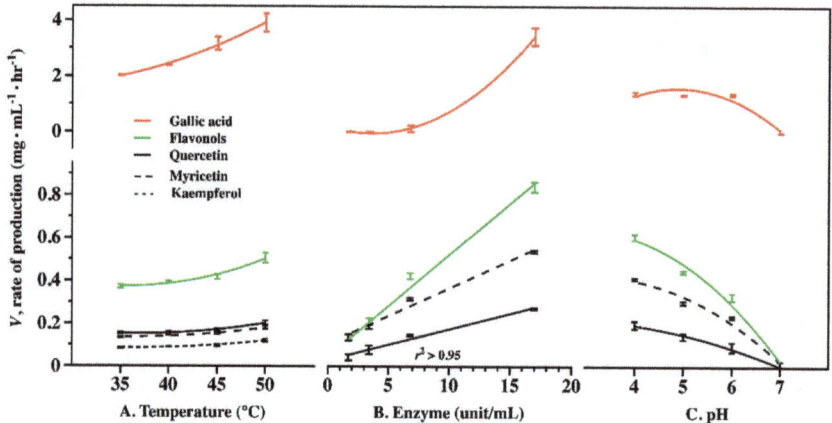

Figure 3. Effects of (**A**) temperature, (**B**) enzyme amount, and (**C**) pH on production of flavonol aglycones and gallic acid from green tea extract in enzyme CF reaction.

2.3. Investigation of Enzyme Activities by Fractionation

CF appears to harbor at least seven kinds of proteins, including A1 of approximately 80–100 kDa, A2 of 60–70 kDa, and A3 of 20–50 kDa (Figure 4A) are the major proteins. Fraction chromatography showed the approximate molecular weight of the CF. Four distinctive peaks were apparent, matching the sodium dodecyl sulfate–polyacrylamide gel electrophoresis (SDS-PAGE) result (Figure 4A). The size distribution of CF from the fast protein liquid chromatography (FPLC) was in the range of 1–100 kDa according to the size-exclusion column based on the manufacturer's reference. The fractions No. 6 and No. 7 had strong simultaneous glycoside hydrolase and tannin acyl hydrolase activities, which sharply decreased in fraction Nos. 8–12, and these two activities were not detected in fraction Nos. 13–20 (Figure 4C). The first major peak of size-exclusion chromatography (SEC; Figure 4B), corresponding to fraction No. 6 and No. 7, revealed a protein of approximately 100 kDa and the second major peak of SEC, corresponding to fraction No. 9 and No. 10, revealed a protein of multiple sizes (approximately 30 and 50 kDa). The third major peak of SEC (t_R = 11.56, corresponding to fraction Nos. 14–16 of Figure 4C) was determined to be a protein or peptides smaller than 1 kDa (Figure 4B). The enzyme fraction (A1 band of Figure 4A or fraction No. 6 and No. 7 of Figure 4C) of CF was considered to have multi-functional activities in a single protein.

2.4. Kinetic Analysis of Enzyme CF for Producing Flavonol Aglycones

The tannin acyl hydrolase activity showed substrate inhibition profiles with a high concentration (> 20 mg/mL) of GTE (Figure 5A5), based on GA production. The amounts of ECG and EGCG in GTE decreased along an exponential decay pattern, with an increase in GTE concentration (Figure 5B2,B4, respectively); (−)-epicatechin (EC) and (−)-epigallocatechin (EGC) were produced or accumulated as a result of the cleavage of GA from EGCG and ECG (Figure 5B3,B5, respectively). Notably, the GH activity of CF to GTE conformed with a typical Michaelis-Menten plot of enzyme reaction.

Kinetic values were calculated from linear regression (Lineweaver–Burk (LB) and Hanes–Woof (HW) plots) and non-linear regression. LB and HW plots revealed that the proper range of substrate concentrations for calculating the kinetic values was 5–100 mg/mL and 2–180 mg/mL, respectively, with an acceptable fit (r^2 > 0.95 and r^2 > 0.97, respectively) for each flavonol aglycone. However, the fitting result of the LB plot (data not shown) was not as suitable as that of the HW plot. Furthermore, over-estimated kinetic values of HW were observed, in which the kinetic value of myricetin was considerably high. Therefore, we compared the kinetic values of non-linear regression. Total flavonol aglycones were produced at a 3.7-fold faster rate at 50 °C than at 35 °C (Table 2 and Figure S2). There were noticeable differences in the maximum velocity (V_m) for the production of myricetin glycosides compared to the two other flavonol aglycones from quercetin and kaempferol glycosides. Myricetin was produced with a V_m of 83 µg·mL^{-1}·h^{-1}, while quercetin and kaempferol were produced with V_m values of 33 and 12 µg·mL^{-1}·h^{-1} at 50 °C, respectively. GA was produced with a V_m of 153 µg·mL^{-1}·h^{-1} at 50 °C and that of 95 µg·mL^{-1}·h^{-1} at 35 °C.

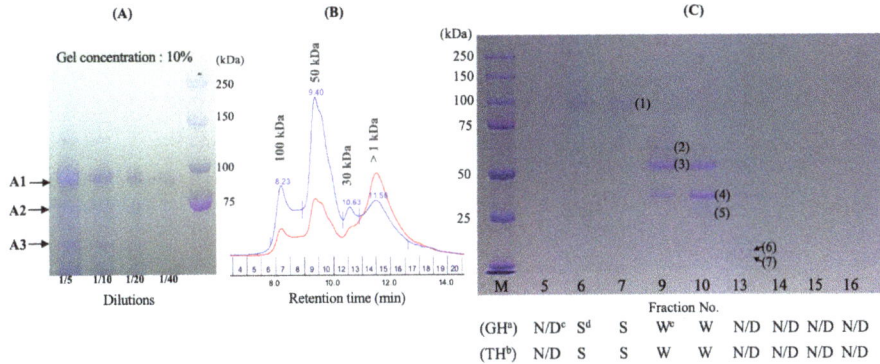

Figure 4. Sodium dodecyl sulfate–polyacrylamide gel electrophoresis (SDS-PAGE) analysis and determination of activities by the fraction of enzyme. (**A**) Size distribution of enzyme CF by SDS-PAGE. (**B**) Size exclusion chromatography of enzyme CF. Blue and red chromatograms were detected at 280 nm and 245 nm, respectively. The numbers on the x-axis of chromatography indicate the fraction of the separation by 0.5 mL, (**C**) Size distribution of the fraction which was obtained by SDS-PAGE. The numbers within the parentheses in (**C**) indicate distinguished protein bands. The fraction No. of (**C**) corresponds to the number on the x-axis of (**B**). Fraction No. 11 was omitted owing to the elution method of the fast protein liquid chromatography (FPLC) system. Abbreviations mean; [a] GH, glycoside hydrolase; [b] TH, tannin acyl-hydrolase; [c] N/D, not detected; [d] S, strong activity; [e] W, weak activity. The activities were classified by the occurrence of flavonol aglycones or GA on the chromatography.

Myricetin was released five-fold faster with an increase in temperature from 35 °C to 50 °C, while quercetin and kaempferol were produced 2.6- and 2.6-fold faster at the higher temperature, respectively (Table 2). These results show that myricetin glycosides are more prone to breaking down to aglycone compared with quercetin and kaempferol glycosides, which may indicate a relatively simpler structure of myricetin glycosides so that they are more readily hydrolyzed by the enzyme. The K'_m values for producing myricetin, quercetin, and kaempferol were 46, 20, and 11 mg/mL at 50 °C, respectively. The higher value of K'_m was correlated to the V_m of corresponding flavonol aglycone. Similar to the differences of V_m values, the K'_m value of myricetin, quercetin, and kaempferol were 5.6-, 3.2-, and 2.7-fold higher at 50 °C than that of 35 °C, respectively. The K'_m values for GA production were lower than those for flavonol aglycone production, which indicated substantially faster hydrolysis activity at low GTE concentrations.

Table 2. Kinetic parameters of the enzyme CF.

Regression[a]	Gallic acid			Myricetin			Quercetin			Kaempferol			Sum of flavonols		
	Temperature (°C)			Temperature (°C)			Temperature (°C)			Temperature (°C)			Temperature (°C)		
	35	50	50/35[b]	35	50	50/35	35	50	50/35	35	50	50/35	35	50	50/35
							V_m (µg·mL^{-1}·h^{-1})								
HW	N.A.[c]	169.5[d]	N.A.	21.9	279.6	12.8	14.4	42.7	3.0	5.7	13.7	2.4	34.9	195.7	5.6
NL[e]	95.1	153.3	1.6	17.0	82.9	4.9	12.5	32.8	2.6	4.6	11.8	2.6	35.5	131.6	3.7
							K'_m (mg·mL^{-1})								
HW	N.A.	2.9	N.A.	14.3	286.3	20.0	7.4	31.9	4.3	5.1	15.0	2.9	6.3	68.2	10.8
NL	2.8	3.8	1.3	8.3	46.3	5.6	6.4	20.2	3.2	4.2	11.4	2.7	7.4	87.5	11.8

[a] The regression method is the Hanes–Woof plot (HW) (first-order regression of HW; [S] 5–100 mg/mL) and non-linear regression (NL; [S] 2–200 mg/mL). [b] 50/35 means the velocity enhancement (fold change) from 35 °C to 50 °C. [c] N.A. means not available due to the negative value of the intercept. [d] The values in the Table were calculated from the regressions plotted from three replicates. [e] NL of GA and other substrates was conducted with biexponential 4P and exponential 3P, respectively, using JMP 12.

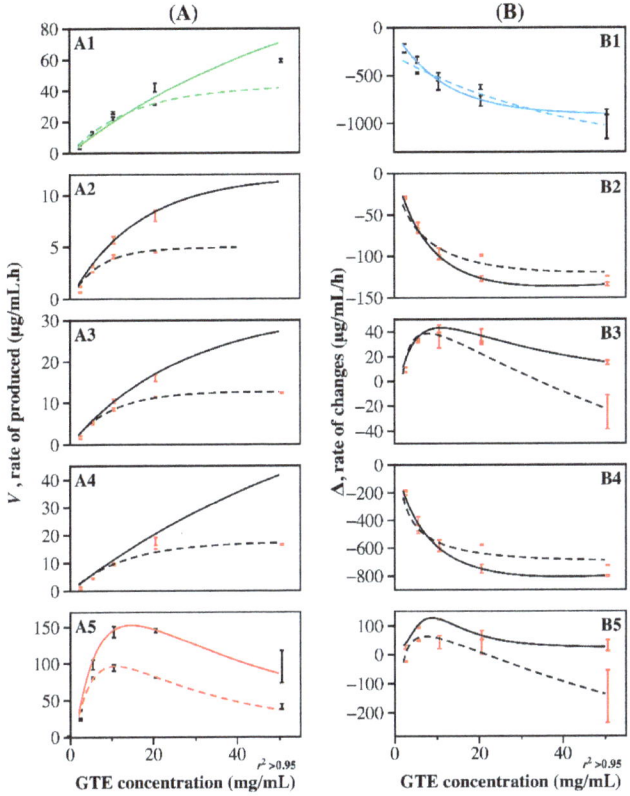

Figure 5. Effects of substrate concentration for producing flavonol aglycones and gallic acid (GA) in enzymatic reactions. (**A1**) sum of flavonols, (**A2**) kaempferol, (**A3**) quercetin, (**A4**) myricetin, (**A5**) GA, (**B1**) sum of catechins, (**B2**) ECG, (**B3**) EC, (**B4**) EGCG, (**B5**) EGC. Lines indicate: dotted line, 35 °C; solid line, 50 °C. All graphs were applied fitted regression curves ($r^2 > 0.95$) from an equation of JMP12 (SAS Institute Inc., Cary, NC, USA).

2.5. Hydrolysis of GTE in Enzymatic and Acid Hydrolysis Conditions

To demonstrate the effectiveness of the enzymatic hydrolysis of the flavonol glycosides and gallated catechins of GTE with enzyme CF, the enzymatic and acid hydrolysis reactions were compared regarding to the GH and tannin acyl hydrolase activities, respectively, in a time-course experiment. GA and flavonol aglycones were simultaneously produced with the enzymatic treatment of GTE, following a first-order reaction and second-order saturation plot, respectively (Figure 6A1). However, acid hydrolysis of GTE did not result in any GA production, whereas the production of flavonol aglycones followed a second-order saturation plot as observed with enzymatic treatment (Figure 6B1). The estimated maximum amounts of flavonol aglycones over time were 11.0 mg/g GTE and 18.9 mg/g GTE by enzymatic and acid hydrolysis, respectively. Furthermore, the half-time of producing the maximum flavonol aglycone concentration was estimated to be 11.2 h and 20.7 h by enzymatic and acid hydrolysis based on a double reciprocal plot as shown in Figure 6A1,B1, respectively. Enzymatic and acid hydrolysis of the flavonol glycosides of GTE resulted in a similar slope in the double-reciprocal plots (Figure 6A2,B2). These results suggest that the rates of producing flavonols with the enzyme are at least comparable, and probably considerably more effective, than those of acid hydrolysis. However, the degalloylation effect on the gallated catechins was quite different between the two treatments. EGC and EC accumulated, while EGCG and ECG decreased over time with the enzyme treatment, resulting in only a 20% loss of total initial catechins (Figure 6C1,C2). By contrast, the acid treatment caused overall losses of the total catechins, with a fairly large amount of loss for each catechin (>70% of the initial values). Taken together, these results demonstrate that acid hydrolysis is a simple way to obtain flavonol aglycones from GTE; however, it has drawbacks in maintaining the stability of catechins.

These findings provide useful information for manipulating GTE when a food-grade enzyme is utilized for the benefit of multi-functionalities. To obtain flavonol aglycones or obtain the hydrolysis benefits from GTE, a higher GTE concentration could deliver more flavonol aglycones with less overall catechin losses during the enzyme reaction (Figures 5 and 6). Furthermore, this inhibition condition (high GTE concentration) could offer efficient molar or weight usage of the enzyme in the reaction, which minimizes the tannin acyl hydrolase activity to gallated catechins.

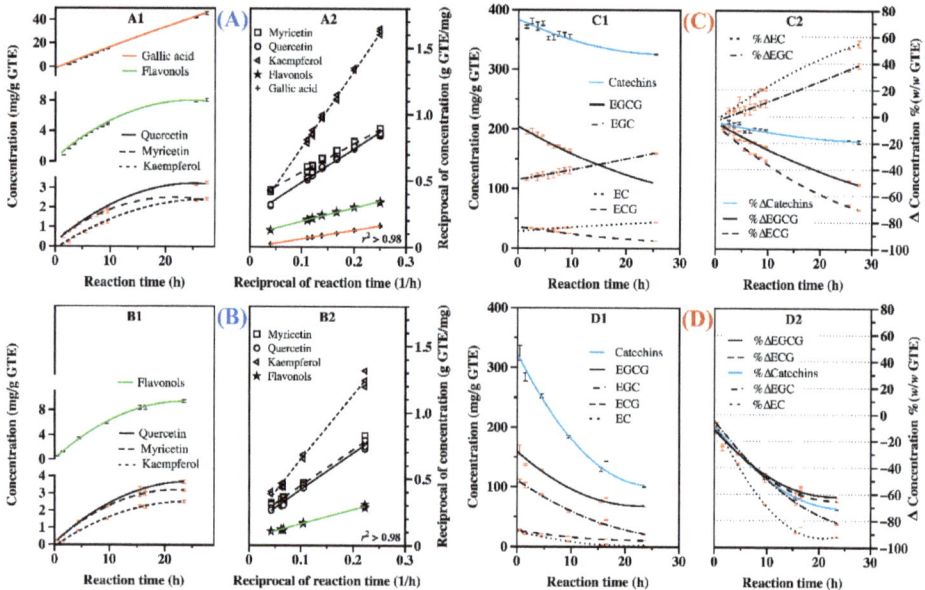

Figure 6. Reaction trends of CF enzyme-based (**A**,**C**) and hydrochloride-based (**B**,**D**) hydrolysis of GTE. The components of GTE were measured (n = 3) by time-course reactions. A2 and B2 were fitted by a first-order regression, while the others were fitted by a second-order regression from an equation of SigmaPlot 10 (Systat Software, Inc., San Jose, CA, USA).

2.6. Rationale of the Enzymatic Reactions from Molecular Docking Simulation

The simultaneous hydrolysis of flavonol glycoside and gallated catechins (ECG and EGCG) was observed in the reaction of the enzyme CF and tannase, respectively. To demonstrate the rationale and possibilities of these phenomena, polygalacturonase, a class of the GH family derived from *A. niger*, and tannase derived from *L. plantarum*, were chosen to evaluate the ligand–protein interaction. Three representative residues (Asp186, Asp207, and Asp208) of the active site in polygalacturonase interacted with isoquercitrin as reported by van Pouderoyen et al. [31] (Figure 7A). Similar to isoquercitrin, EGCG fits into this active site via proper conformation, which provides a clue for the breakage of GA from EGCG by the action of the enzyme. One more simulation was conducted in the same way with tannase, which has six residues in the active sites of Ser163, Lys343, Glu357, Asp419, Asp421, and His451. As shown in Figure 7B, the GA moiety of EGCG interacts with Ser163 and Asp421, as reported by Ren et al. [32] In the case of isoquercitrin with tannase, a similar conformation of isoquercitrin in the active site is obtained, which indirectly indicates the possibility of hydrolysis of the glucose moiety of isoquercitrin. These docking simulations (refer to Table S4 for the result of the docking run) may provide insights into the simultaneous catalytic actions of the two enzymes used in our study, although the exact enzyme structures of enzyme CF and tannase from *A. niger* were not obtained.

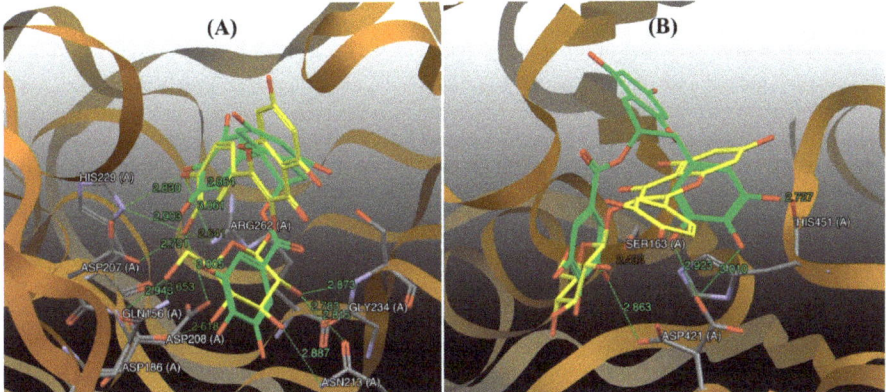

Figure 7. Molecular docking simulation in the active site of (**A**) glycoside hydrolase and (**B**) tannase with the superimposed isoquercitrin (yellow) or EGCG (green). The colors of atoms represent their interaction with ligands as follows: gray, carbon; red, oxygen; and blue, nitrogen. Hydrogen atoms of protein residues and ligands are omitted for improving a graphical comprehension. The green dotted lines represent the hydrogen bonding. Other protein residues are depicted in ribbon, while catalytic residues are illustrated as capped sticks.

3. Discussion

Glycoside hydrolases (GHs; EC 3.2.1) are a group of enzymes that hydrolyze the glycosidic bonds between carbohydrates or between a carbohydrate and a non-carbohydrate [33]. The GH family consists of 13 members (GH1–13), which can be used to obtain flavonol aglycones from the flavonol glycosides of GTE, and to obtain flavonol aglycones from the flavonol glycosides of GTE, such as α-amylase, cellulase, α-glucosidase, β-glucosidase, β-galactosidase, β-mannosidase, β-xylosidase, α-arabinofuranosidase, phlorizin hydrolase, exo-β-1,4-glucanase, glucan 1,4-β-glucosidase, glucan endo-1,6-β-glucosidase, cellulose β-1,4-cellobiosidase, and α-arabinopyranosidase. Hur et al. [18] summarized the multi-functional enzymes of lactic acid bacteria, fungi, and yeasts; they found that fungi are the richest sources of enzyme cocktails for treating food constituents, including amylase, cellulase, esterase, fucoidanase, glucoamylase, β-glycosidase, invertase, lipase, mannanase, pectinases, phytase, protease, tannase, β-xylosidase, and xylanase. Consequently, in our research, we evaluated food-grade enzyme cocktails derived from *A. niger* in order to find which were most effective at producing flavonol aglycones from GTE. We found that the enzyme CF is able to hydrolyze the flavonol glycosides to aglycones more effectively than the other seven GHs. Various methods for the use of GTE have been explored, including purification of EGCG, biotransformation of catechins to degallated catechins, encapsulation of catechins, and stabilization of catechins by adopting carriers [34]. Enzymatic treatments are generally used to obtain specific modifications of food materials, such as clarification of juice, saccharification of starch, removal of unwanted substances, and enhancement of flavor [35]. Accordingly, tannase, which cleaves the ester bonds in tannins and catalyzes the hydrolysis of the ester linkages between galloyl group-containing compounds such as ECG and EGCG, is the enzyme used most commonly to change the potential of GTE [36]. Lu and Chen [37] elucidated that tannase-treated GT contained 29 mg/g of GA (from 13% to 39%), 40 mg/g of EGC (from 30% to 54%), and 5 mg/g of EC (from 1.5% to 16%) after the hydrolysis reaction. In the present study, the content of

GA, EC, and EGC in CF-treated GTE increased from 0 to 78.1 mg/g, 29 to 52 mg/g, and 130 to 224 mg/g, respectively (data not shown).

Furthermore, the enzyme CF displayed characteristics desirable in industrial application of enzymes for food processing, such as optimal reaction conditions in the mesothermal range (40–50 °C) and in moderately acidic conditions (pH 4) (Figure 3). These characteristics are advantageous, as they allow for the use of more cost-effective and environmentally friendly industrial processes.

The two different activities of enzyme CF may be due to a similarly-sized protein producing a single protein band or peak (Figure 4). Comparable results were identified by Matsumoto et al. [38], who found that a multi-functional GH showed α-arabinofuranosidase, β-glucosidase, and β-xylosidase activities with a single protein band determined by SDS-PAGE analysis. However, in our study, the protein band of fraction No. 6 and No. 7 in Figure 4C was relatively broad; thereby, further study is needed to reveal the exact molecular weight or size of proteins that could alternatively be responsible for these different activities.

Some tannases show a substrate inhibition characteristic while others do not; this variation has been largely attributed differences in the sources from which the tannases are derived [39]. Interestingly, enzyme CF showed substrate inhibition kinetics for GA production along with normal Michaelis-Menten kinetics for flavonol aglycones production. This contrasting result may be useful for the industrial processing of GTE, as flavonol aglycone may be obtained with less GA production in settings with high GTE concentration (Figure 5).

Acid hydrolysis is non-specific and cheaper process in raw food material processing. However, the reaction solution must be neutralized with a base, resulting in the production of large quantities of salt byproducts and have high salt removal costs. Consequently, enzymatic treatment is a more preferable option for the production of high-value food supplements. We found that CF treatment can obtain an amount of flavonol aglycones comparable to that obtained via acid hydrolysis. However, GA was not produced in the HCl treatment, while the CF treatment of GTE produced approximately 4% (w/w) of GA (Figure 6). Moreover, during acid hydrolysis, the non-specific and strong hydrolytic action of HCl resulted in the disappearance of approximately 70% (w/w) of the catechins that had been generated (Figure 6). Therefore, for applications in which minimal loss of catechins is desirable, such as the production of GTE food supplements, enzymatic processing with CF is a preferable to acid hydrolysis for the production of flavonol aglycones from GTE. As increased GA content improves the radical scavenging effects of GTE [22], and beneficial dietary flavonols are best absorbed as aglycones [24], our research on the enzyme CF for producing flavonol aglycones from GT might potentially deliver useful information in processing application and beneficial advantage in human health.

A study of the crystal structure of an enzyme is relatively difficult to obtain and requires the enzyme to be isolated and highly purified. Consequently, we focused on food-grade enzymes that were good candidates for potential use on an industrial scale. As illustrated in Figure 7, similar enzymes derived from the same microorganism often have similar active site residues, although some subunits differ. We believe that such an approach provides useful insight into the catalytic action of the enzyme CF. The docking results indicate that the glucose moiety of isoquercitrin and the GA moiety of EGCG are correctly positioned in the active site of enzyme CF (Figure 7A). The hydrolytic action may occur in a similar way as the moieties of the two compounds interacting with the representative residues of active site.

4. Materials and Methods

4.1. Chemicals and Enzymes

Quercetin, kaempferol, and dimethyl sulfoxide (DMSO) were purchased from Sigma-Aldrich Co., LLC (St. Louis, MO, USA). Formic acid, acetonitrile, and methanol were purchased from Fisher Scientific (Waltham, MA, USA). Water for high-performance liquid chromatography (HPLC) was purchased from Burdick & Jackson (Muskegon, MI, USA). Myricetin was purchased from

Extrasynthese (Genay, France). EC, EGC, EGCG, and ECG were purchased from Wako Pure Chemicals Industries, Ltd. (Osaka, Japan). Nitrophenol, 4-nitrophenyl-α-arabinofuranoside, 4-nitrophenyl-α-arabinopyranoside, 2-nitrophenyl-β-D-galactopyranoside, 4-nitrophenyl-β-D-xylopyranoside, 4-nitrophenyl-β-D-mannopyranoside, α-arabinofuranosidase, α-arabinopyranosidase, β-galactosidase, β-xylosidase, and β-mannosidase were purchased from Megazyme (Bray, Ireland). All commercial enzymes were supplied from Bision Corp. (Seoul, Korea). All other chemicals used were of analytical grade or higher.

4.2. Preparation of GTE

GTE was prepared according to the method described by Rha et al. [30]. Briefly, GT (Osulloc Farm Corp., Seogwipo, Republic of Korea) was harvested from May–June, 2017, and then processed into a loose tea. To obtain GTE, the dried GT was soaked in a 10-fold weight of 70% (v/v) aqueous ethanol for 3 h at 60 °C with stirring; the solvent in the extract was then removed under vacuum condition before being pulverized by spray dryer (Seogang Engineering, Cheonan, Korea). The powered GTE was used in all subsequent experiments as the substrate for the test enzymes.

4.3. Assay of Multi-Functional Activities of Food-Grade Enzymes

The enzyme activities were measured according to the corresponding universal method as described in Appendix A.

4.4. Screening of Food-Grade Enzymes for Producing Flavonol Aglycones from GTE

To find the most suitable enzyme for food-grade commercial-scale purposes, we screened the activities of eight cellulolytic enzymes. The reaction conditions are described in Table 1. The reaction was performed in a 1-mL microtube in a thermomixer (500 rpm; Eppendorf AG, Hamburg, Germany) for 2 h with a sufficient amount of enzyme (100 mg/mL) and substrate (100 mg/mL of GTE). After the reaction, 500 µL of 10% (v/v) DMSO in absolute methanol was added to halt the enzyme activity and to fully recover the flavonol aglycones produced therein. The mixture was sonicated for 20 min and properly diluted (1.0 mg/mL) for quantitative analysis. The screening criterion was the total amount of flavonols, including myricetin, quercetin, and kaempferol, produced. The statistical correlation between the amount of flavonol created and individual activity of the multi-functional enzyme was analyzed by JMP 12 for Windows 7 or higher (SAS Institute Inc., Cary, NC, USA).

4.5. Investigation of Optimal Enzymatic Conditions for Producing Flavonol Aglycones and GA from GTE

Enzyme CF was reacted with the GTE solution (100 mg/mL in 50 mM of citrate phosphate buffer; pH 5.0) and 3 units/mL of the enzyme at 35, 40, 45, and 50 °C for 4 h. A linear range of enzyme units was tested with 100 mg/mL of GTE in 50 mM of citrate phosphate buffer (pH 5.0) for 2 h. The optimal pH for the enzyme reaction was investigated using 100 mg/mL of GTE, 10 units/mL of the enzyme (based on the activity of α-arabinofuranosidase) at 50 °C for 2 h by varying the pH in the range of 4.0–7.0 with 50 mM of citrate phosphate buffer.

4.6. Size Distribution, Separation, and Determination of the Core Activities of CF

SDS-PAGE was performed with 10% (w/w) agarose gel to determine the size distribution of CF with reference to a size marker (25–250 kDa). The separation was performed using SEC with an Agilent Bio SEC-5 (300 Å, 5 µm, 7.8 × 300 mm; Santa Clara, CA, USA) column to determine which fraction of CF showed GH (deglycosylation) activity or tannin acyl hydrolase (degalloylation) activity on GT flavonols. The CF solution (10 mg/mL) was prepared with 50 mM of citrate phosphate buffer (pH 5.0), and then, 25 µL of the solution was injected into the column. The elution buffer was 50 mM citrate phosphate (pH 5.0), and the flow rate was 1.0 mL/min. The fractions were collected every 0.5 mL after injection by fast protein liquid chromatography (ÄKTA purifier 10; GE Healthcare, Stockholm, Sweden).

Each fraction was mixed with 0.5 mL of the GTE solution (2.0 mg/mL) in 50 mM of citrate phosphate buffer (pH 5.0) and then reacted for 24 h at 45 °C in a thermoshaker (Eppendorf AG). The reaction results were analyzed by HPLC as described below. To confirm the molecular weight and size distribution of the fraction obtained by SEC, the fractions were loaded on a 10% (*w/w*) SDS-PAGE gel.

4.7. Kinetic Analysis of CF for Producing Flavonol Aglycones

The stock substrate solution (200 mg GTE/mL) was prepared with 50 mM of citrate phosphate buffer (pH 4.0), and the enzymatic reactions were performed with 2–180 mg of GTE/mL buffer and 10 units/mL for 2 h in a thermoshaker (Eppendorf AG). The test temperatures were 35 °C and 50 °C. The same volume of 10% (*v/v*) DMSO in aqueous methanol was mixed in the reacted solution, followed by sonication for 20 min. After that, the mixture was filtered through a 0.45-μm GH Polypro syringe filter (Pall Corp, Port Washington, NY, USA) before HPLC analysis.

4.8. Time-Course of Enzyme CF and Hydrochloride Hydrolysis of GTE for Producing Flavonol Aglycones

To perform hydrochloride hydrolysis, 250 μL of GTE solution (200 mg/mL) was mixed with 250 μL of 6 N hydrochloride, and the reaction was maintained at 30 °C. To neutralize the acidic solution, 250 μL of 6 M sodium hydroxide was mixed in at the end of the reaction and then properly diluted with 10% (*v/v*) DMSO in absolute methanol. The temperature (35 °C) of enzymatic treatment was chosen to minimize GA production with 100 mg/mL of GTE and 100 mg/mL of the enzyme in 50 mM of citrate phosphate buffer at pH 4.0. The results were collected from 0 to 30 h during the reaction and quantified by HPLC.

4.9. HPLC Analysis of Catechins and Flavonols

For analysis of catechins, GA, caffeine, and flavonols, the reaction mixtures were diluted with 10% (*v/v*) DMSO in absolute methanol, sonicated for 20 min, and then filtered through a 0.45-μm GH Polypro syringe filter (Pall Corp.). The analytes were then analyzed by an Alliance HPLC system (Waters, Milford, MA, USA) with a Poroshell 120 SB octadecyl silica column (120 Å, 2.7 μm, 4.6 × 150 mm; Agilent). The column temperature was 30°C, and the injection volume was 5 μL. Peaks were monitored at 275 nm for catechins, GA, and caffeine, and at 365 nm for flavonols. Gradient elution was performed with 0.1% (*v/v*) formic acid in water (solvent A) and 0.1% (*v/v*) formic acid in acetonitrile (solvent B). All solvents were filtered and degassed. The flow rate was 0.8 mL/min. The linear gradient of the binary mobile phases was as follows: 92% A/8% B at 0 min, 92% A/8% B at 2 min, 88% A/12% B at 3 min, 84% A/16% B at 4 min, 84% A/16% B at 15 min, 80% A/20% B at 18 min, 76% A/24% B at 21 min, 70% A/30% B at 22 min, 70% A/30% B at 26 min, 50% A/50% B at 28 min, 50% A/50% B at 30 min, 20% A/80% B at 32 min, 20% A/80% B at 33 min, 92% A/8% B at 34 min, and 92% A/8% B at 35 min.

4.10. Molecular Docking Simulations

Molecular dynamics was conducted according to the method described by Jones et al. [40] with modifications. GOLD v5.7.2 (Genetic Optimisation for Ligand Docking; the Cambridge Crystallographic Data Centre, Cambridge, UK) for Windows 10 was used for modeling the enzymes and ligand interactions. The coordinates of GH (polygalacturonase from *A. niger*; Protein Data Bank (PDB) code: 1NHC) and tannase (tannin acyl hydrolase from *Lactobacillus plantarum*; PDB code: 4J0C) were obtained from the PDB, and the tertiary data for isoquercitrin and EGCG were obtained from the PubChem website (NIH, Bethesda, MD, USA). The PDB files were manipulated by eliminating the water and ligand records using the program GOLD. The coordinates (x × y × z) of the grid were 25 × 38 × −15 and 20 × 95 × 72 for GH and tannase, respectively. The protein residues located in the active site were considered in a space less than 10 Å wide. The ChemPLP fitness function of GOLD was applied to obtain a suitable conformation of ligand-protein binding. The genetic algorithm was employed with the slow searching option. One conformation that provided the highest scoring was chosen among 10 estimations, and then, a graphical calculation was performed for hydrogen bonds between protein residues and ligand. The docking results are listed Table S4.

4.11. Statistical Analysis

Data are expressed as the means ± standard error of the mean of three replicates. One-way analysis of variance was carried out and assessed at a significance level of $\alpha = 0.05$ using JMP 12 to evaluate the significance of differences among the means.

5. Conclusions

In this study, we screened a single food-grade enzyme cocktail to assess its use for increasing the benefits of flavonols during digestion of GTE. Among various food-grade enzymes originating from fungal culture, the enzyme CF with multi-functional activity showed specificity for effectively cleaving various glycosidic linkages such as 3-*O*-glucosides, 3-*O*-galactosides, and 3-*O*-glucosylrutinosides. Moreover, CF showed potent activity on gallated catechins, resulting in GA and degalloylated catechin. The optimum conditions for these two major activities were similar, but tannase activity was inhibited by high substrate (GTE) concentrations. These results may provide useful guidance for controlling the compositions of catechins and flavonol glycosides in GT by enzymatic treatment. We suggest that, compared to acid treatment, this enzymatic treatment for producing flavonol aglycones from GTE is safer, more effective, more convenient, and more environmentally–friendly, and also provides value-added GTE for food fortification or as dietary supplements. Ultimately, this study suggests that a GH enzyme should be selected to best utilize the flavonols in GTE more efficiently and effectively. Such processing could allow; the production of GTE in which beneficial flavonol aglycones can be digested and absorbed regardless of whether or not the individual consuming the product can produce LPH. Further studies for elucidating the specific health-promoting effects of GT flavonols are warranted.

Supplementary Materials: The following are available online at http://www.mdpi.com/2073-4344/9/10/861/s1, Figure S1: Chemical structures of catechins and flavonols in GTE, Tables S1 and S2: Changes of catechins and flavonols by hydrolysis of cellulolytic enzymes (numerical values), Table S3: Correlation result for producing flavonol aglycones by multivariate analysis, Figure S2: Kinetics of enzyme CF, Table S4: Parameters and result of docking run.

Author Contributions: Conceptualization, C.-S.R.; methodology, C.-S.R., S.-W.K., and K.H.B.; software, C.-S.R.; validation, C.-S.R. and S.-W.K.; formal analysis, C.-S.R., S.-W.K., and K.H.B.; investigation, C.-S.R. and S.-W.K.; resources, C.-S.R., S.-W.K., and K.H.B.; data curation, C.-S.R. and D.-O.K.; writing—original draft preparation, C.-S.R.; writing—review and editing, C.-S.R. and D.-O.K.; visualization, C.-S.R. and D.-O.K.; supervision, C.-S.R. and D.-O.K.; project administration, Y.D.H.

Funding: This research received no external funding.

Acknowledgments: We are grateful to Professor Cheon-Seok Park (Kyung Hee University, Yongin, Republic of Korea) for supporting the SDS-PAGE analysis and molecular docking simulation.

Conflicts of Interest: The authors declare no conflict of interest.

Appendix A. Assay of Multi-Functional Activities of Commercial Enzymes

The enzyme activities were measured according to the corresponding universal method as described below. Substrate blanks were prepared by adding buffer instead of the enzyme solution.

The assay of arabinase activity was determined using Arabinazyme Tablets (T-ARZ200; Megazyme International, Bray, Ireland) [41]. The assay of α-rhamnosidase activity followed the method described by Gallego et al. [42] In brief, the mixture contained 250 µL of substrate (3 mM 4-nitrophenyl-α-rhamnopyranoside in 50 mM succinate buffer, pH 5.5) and 250 µL of the enzyme solution, which was incubated at 50 °C. The reaction was stopped by adding 1 mL of stop reagent (2 M Na_2CO_3) after 10 min of incubation, and then the absorbance was measured at 400 nm. One unit of α-rhamnosidase is considered the amount of enzyme required to liberate 1 µmol of nitrophenol per minute at 50 °C.

The assay of β-glucosidase activity was conducted according to a modification of the method reported by Kim and Park [43] In brief, the mixture contained 400 µL of substrate (125 mM 4-nitrophenyl-β-D-glucoside in 20 mM sodium acetate buffer, pH 4.5) and 100 µL of enzyme solution, and incubated at 40 °C. The reaction was stopped by adding 5 mL of stop solution (50 mM Na_2CO_3) after 10 min of incubation, and the absorbance was read at 412 nm. One unit of β-glucosidase is the amount of enzyme that catalyzes the hydrolysis of 1.0 µmol of nitrophenol per minute.

The assay of α-arabinofuranosidase activity followed the method described by McCleary et al. [44]. In brief, the mixture contained 400 µL of substrate (2.5 mM 4-nitrophenyl-α-arabinofuranoside in 50 mM citrate-phosphate buffer, pH 4.0) and 60 µL of enzyme solution, which was incubated at 40 °C. The reaction was stopped by the addition of 460 µL of stop reagent (0.2 M Na_2CO_3) after 10 min of incubation, and the absorbance was measured at 410 nm. One α-arabinofuranosidase unit is the amount of enzyme that liberates 1 µmol of nitrophenol per minute at 40 °C.

The assay of α-arabinopyranosidase activity was applied modifying the method for α-arabinofuranosidase activity described above. The mixture contained 200 µL of substrate (2.0 mM 4-nitrophenyl-α-arabinopyranoside in 20 mM sodium acetate buffer, pH 4.5) and 200 µL of enzyme solution, which was incubated at 50 °C. The reaction was stopped by adding 400 µL of stop reagent (0.2 M Na_2CO_3) after 10 min of incubation and the absorbance was measured at 400 nm. One unit of α-arabinopyranosidase is the amount of enzyme that liberates 1 µmol of nitrophenol per minute at 50 °C.

The β-galactosidase activity was assessed according to a modified method described by the Food Chemicals Codex [45]. A mixture containing 800 µL of substrate (12.5 mM 2-nitrophenyl-β-D-galactopyranoside in 20 mM of sodium acetate buffer, pH 4.5) and 200 µL of enzyme solution was incubated at 50 °C. The reaction was stopped by adding 1 mL of stop reagent [10% (w/v) Na_2CO_3] after 10 min of incubation, followed by the addition of 8 mL of water, and then, the absorbance was measured at 420 nm. One unit of β-galactosidase is the amount of enzyme that liberates 1 µmol of nitrophenol per minute at 50 °C.

The β-xylosidase activity was evaluated by modifying the method used for α-arabinofuranosidase activity describe above. The mixture contained 400 µL of the substrate (2.5 mM 4-nitrophenyl-β-D-xylopyranoside in 20 mM sodium acetate buffer, pH 4.5) and 400 µL of enzyme solution, which was incubated at 50 °C. The reaction was stopped by adding 800 µL of stop reagent (0.2 M Na_2CO_3) after 10 min of incubation and the absorbance was read at 400 nm. One unit of β-xylosidase is the amount of enzyme that liberates 1 µmol of nitrophenol per minute at 50 °C.

The assay of β-mannosidase activity was modified from that used for α-arabinofuranosidase activity. The mixture contained 400 µL of substrate (2.5 mM 4-nitrophenyl-β-D-mannopyranoside in 20 mM sodium acetate buffer, pH 4.5) and 400 µL of enzyme solution, which was incubated at 50 °C. The reaction was stopped by adding 800 µL of stop reagent (0.2 M Na_2CO_3) after 10 min of incubation, and then the absorbance was read at 400 nm. One unit of β-mannosidase is the amount of enzyme that liberates 1 µmol of nitrophenol per minute at 50 °C.

The tannase activity was evaluated according to the method described by Iibuchi et al. [46]. The mixture contained 2 mL of substrate [0.35% (w/v) tannic acid in 50 mM citrate buffer, pH 5.5] and 0.5 mL of enzyme solution, which was incubated at 37 °C. After 10 min of incubation, 20 µL of the reaction mixture and 2 mL of 80% (v/v) ethanol solution were mixed, and the absorbance was measured at 310 nm. The activity was calculated according to the difference in absorbance to the blank without enzyme solution. One unit of tannase activity is the amount which was hydrolyzed 1 µmol of ester in 1 min at 37 °C.

References

1. Monobe, M.; Nomura, S.; Ema, K.; Matsunaga, A.; Nesumi, A.; Yoshida, K.; Maeda-Yamamoto, M.; Horie, H. Quercetin glycosides-rich tea cultivars (*Camellia sinensis* L.) in Japan. *Food Sci. Technol. Res.* **2015**, *21*, 333–340. [CrossRef]
2. Peterson, J.; Dwyer, J.; Bhagwat, S.; Haytowitz, D.; Holden, J.; Eldridge, A.L.; Beecher, G.; Aladesanmi, J. Major flavonoids in dry tea. *J. Food Compos. Anal.* **2005**, *18*, 487–501. [CrossRef]
3. Khan, N.; Mukhtar, H. Tea polyphenols for health promotion. *Life Sci. Adv. Exp. Clin. Endocrinol.* **2007**, *81*, 519–533. [CrossRef]
4. Cabrera, C.; Artacho, R.; Gimenez, R. Beneficial effects of green tea—A review. *J. Am. Coll. Nutr.* **2006**, *25*, 79–99. [CrossRef] [PubMed]
5. Wu, C.; Xu, H.; Heritier, J.; Andlauer, W. Determination of catechins and flavonol glycosides in Chinese tea varieties. *Food Chem.* **2012**, *132*, 144–149. [CrossRef] [PubMed]
6. Jiang, H.; Engelhardt, U.H.; Thräne, C.; Maiwald, B.; Stark, J. Determination of flavonol glycosides in green tea, oolong tea and black tea by UHPLC compared to HPLC. *Food Chem.* **2015**, *183*, 30–35. [CrossRef]
7. Iwashina, T. The structure and distribution of the flavonoids in plants. *J. Plant Res.* **2000**, *113*, 287–299. [CrossRef]
8. Walle, T.; Otake, Y.; Walle, U.K.; Wilson, F.A. Quercetin glucosides are completely hydrolyzed in ileostomy patients before absorption. *J. Nutr.* **2000**, *130*, 2658–2661. [CrossRef]
9. Olthof, M.R.; Hollman, P.C.; Vree, T.B.; Katan, M.B. Bioavailabilities of quercetin-3-glucoside and quercetin-4′-glucoside do not differ in humans. *J. Nutr.* **2000**, *130*, 1200–1203. [CrossRef]
10. Hollman, P.C.H.; Bijsman, M.N.C.P.; van Gameren, Y.; Cnossen, E.P.J.; de Vries, J.H.M.; Katan, M.B. The sugar moiety is a major determinant of the absorption of dietary flavonoid glycosides in man. *Free Radic. Res.* **1999**, *31*, 569–573. [CrossRef]
11. Xiao, J.; Cao, H.; Wang, Y.; Zhao, J.; Wei, X. Glycosylation of dietary flavonoids decreases the affinities for plasma protein. *J. Agric. Food. Chem.* **2009**, *57*, 6642–6648. [CrossRef] [PubMed]
12. Hostetler, G.; Riedl, K.; Cardenas, H.; Diosa-Toro, M.; Arango, D.; Schwartz, S.; Doseff, A.I. Flavone deglycosylation increases their anti-inflammatory activity and absorption. *Mol. Nutr. Food Res.* **2012**, *56*, 558–569. [CrossRef] [PubMed]
13. Semwal, D.K.; Semwal, R.B.; Combrinck, S.; Viljoen, A. Myricetin: A dietary molecule with diverse biological activities. *Nutrients* **2016**, *8*, 90. [CrossRef] [PubMed]
14. Lea, M.A. Flavonol regulation in tumor cells. *J. Cell. Biochem.* **2015**, *116*, 1190–1194. [CrossRef] [PubMed]
15. Nomura, S.; Monobe, M.; Ema, K.; Matsunaga, A.; Maeda-Yamamoto, M.; Horie, H. Effects of flavonol-rich green tea cultivar (*Camellia sinensis* L.) on plasma oxidized LDL levels in hypercholesterolemic mice. *Biosci. Biotechnol. Biochem.* **2016**, *80*, 360–362. [CrossRef]
16. Menezes, R.; Rodriguez-Mateos, A.; Kaltsatou, A.; Gonzalez-Sarrias, A.; Greyling, A.; Giannaki, C.; Andres-Lacueva, C.; Milenkovic, D.; Gibney, E.R.; Dumont, J.; et al. Impact of flavonols on cardiometabolic biomarkers: A meta-analysis of randomized controlled human trials to explore the role of inter-individual variability. *Nutrients* **2017**, *9*, 117. [CrossRef]
17. Plumb, G.W.; Price, K.R.; Williamson, G. Antioxidant properties of flavonol glycosides from tea. *Redox Rep.* **1999**, *4*, 13–16. [CrossRef]
18. Hur, S.J.; Lee, S.Y.; Kim, Y.C.; Choi, I.; Kim, G.B. Effect of fermentation on the antioxidant activity in plant-based foods. *Food Chem.* **2014**, *160*, 346–356. [CrossRef]
19. Georgetti, S.R.; Vicentini, F.T.; Yokoyama, C.Y.; Borin, M.F.; Spadaro, A.C.; Fonseca, M.J. Enhanced *in vitro* and *in vivo* antioxidant activity and mobilization of free phenolic compounds of soybean flour fermented with different β-glucosidase-producing fungi. *J. Appl. Microbiol.* **2009**, *106*, 459–466. [CrossRef]
20. Ni, H.; Chen, F.; Jiang, Z.D.; Cai, M.Y.; Yang, Y.F.; Xiao, A.F.; Cai, H.N. Biotransformation of tea catechins using *Aspergillus niger* tannase prepared by solid state fermentation on tea byproduct. *LWT Food Sci. Technol.* **2015**, *60*, 1206–1213. [CrossRef]
21. Macedo, J.A.; Ferreira, L.R.; Camara, L.E.; Santos, J.C.; Gambero, A.; Macedo, G.A.; Ribeiro, M.L. Chemopreventive potential of the tannase-mediated biotransformation of green tea. *Food Chem.* **2012**, *133*, 358–365. [CrossRef] [PubMed]

22. Baik, J.H.; Shin, K.S.; Park, Y.; Yu, K.W.; Suh, H.J.; Choi, H.S. Biotransformation of catechin and extraction of active polysaccharide from green tea leaves via simultaneous treatment with tannase and pectinase. *J. Sci. Food Agric.* **2015**, *95*, 2337–2344. [CrossRef] [PubMed]
23. Farrell, T.L.; Ellam, S.L.; Forrelli, T.; Williamson, G. Attenuation of glucose transport across Caco-2 cell monolayers by a polyphenol-rich herbal extract: Interactions with SGLT1 and GLUT2 transporters. *BioFactors* **2013**, *39*, 448–456. [CrossRef] [PubMed]
24. Clifford, M.N. Diet-derived phenols in plasma and tissues and their implications for health. *Planta Med.* **2004**, *70*, 1103–1114. [CrossRef] [PubMed]
25. Walle, T. Absorption and metabolism of flavonoids. *Free Radic. Biol. Med.* **2004**, *36*, 829–837. [CrossRef] [PubMed]
26. Van Duynhoven, J.; Vaughan, E.E.; Jacobs, D.M.; Kemperman, R.A.; van Velzen, E.J.J.; Gross, G.; Roger, L.C.; Possemiers, S.; Smilde, A.K.; Dore, J.; et al. Metabolic fate of polyphenols in the human superorganism. *Proc. Natl. Acad. Sci. USA* **2010**, *108*, 4531–4538. [CrossRef] [PubMed]
27. Itan, Y.; Jones, B.L.; Ingram, C.J.; Swallow, D.M.; Thomas, M.G. A worldwide correlation of lactase persistence phenotype and genotypes. *BMC Evol. Biol.* **2010**, *10*, 36. [CrossRef]
28. Nemeth, K.; Plumb, G.W.; Berrin, J.G.; Juge, N.; Jacob, R.; Naim, H.Y.; Williamson, G.; Swallow, D.M.; Kroon, P.A. Deglycosylation by small intestinal epithelial cell β-glucosidases is a critical step in the absorption and metabolism of dietary flavonoid glycosides in humans. *Eur. J. Nutr.* **2003**, *42*, 29–42. [CrossRef]
29. Choi, E.H.; Rha, C.S.; Balusamy, S.R.; Kim, D.O.; Shim, S.M. Impact of bioconversion of gallated catechins and flavonol glycosides on bioaccessibility and intestinal cellular uptake of catechins. *J. Agric. Food. Chem.* **2019**, *67*, 2331–2339. [CrossRef]
30. Rha, C.-S.; Jeong, H.W.; Park, S.; Lee, S.; Jung, Y.S.; Kim, D.-O. Antioxidative, anti-inflammatory, and anticancer effects of purified flavonol glycosides and aglycones in green tea. *Antioxidants* **2019**, *8*, 278. [CrossRef]
31. van Pouderoyen, G.; Snijder, H.J.; Benen, J.A.E.; Dijkstra, B.W. Structural insights into the processivity of endopolygalacturonase I from *Aspergillus niger*. *FEBS Lett.* **2003**, *554*, 462–466. [CrossRef]
32. Ren, B.; Wu, M.; Wang, Q.; Peng, X.; Wen, H.; McKinstry, W.J.; Chen, Q. Crystal structure of tannase from *Lactobacillus plantarum*. *J. Mol. Biol.* **2013**, *425*, 2737–2751. [CrossRef] [PubMed]
33. AFMB. Glycoside Hydrolase Family Classification. Available online: http://www.cazy.org/Glycoside-Hydrolases.html (accessed on 25 November 2018).
34. Dube, A.; Ng, K.; Nicolazzo, J.A.; Larson, I. Effective use of reducing agents and nanoparticle encapsulation in stabilizing catechins in alkaline solution. *Food Chem.* **2010**, *122*, 662–667. [CrossRef]
35. Kirk, O.; Borchert, T.V.; Fuglsang, C.C. Industrial enzyme applications. *Curr. Opin. Biotechnol.* **2002**, *13*, 345–351. [CrossRef]
36. Yao, J.; Fan, X.J.; Lu, Y.; Liu, Y.H. Isolation and characterization of a novel tannase from a metagenomic library. *J. Agric. Food. Chem.* **2011**, *59*, 3812–3818. [CrossRef]
37. Lu, M.-J.; Chen, C. Enzymatic modification by tannase increases the antioxidant activity of green tea. *Food Res. Int.* **2008**, *41*, 130–137. [CrossRef]
38. Matsumoto, T.; Shimada, S.; Hata, Y.; Tanaka, T.; Kondo, A. Multi-functional glycoside hydrolase: Blon_0625 from *Bifidobacterium longum subsp. infantis* ATCC 15697. *Enzym. Microb. Technol.* **2015**, *68*, 10–14. [CrossRef]
39. Baik, J.H.; Suh, H.J.; Cho, S.Y.; Park, Y.; Choi, H.S. Differential activities of fungi-derived tannases on biotransformation and substrate inhibition in green tea extract. *J. Biosci. Bioeng.* **2014**, *118*, 546–553. [CrossRef]
40. Jones, G.; Willett, P.; Glen, R.C.; Leach, A.R.; Taylor, R. Development and validation of a genetic algorithm for flexible docking. *J. Mol. Biol.* **1997**, *267*, 727–748. [CrossRef]
41. Flipphi, M.J.; Visser, J.; van der Veen, P.; de Graaff, L.H. Arabinase gene expression in *Aspergillus niger*: Indications for coordinated regulation. *Microbiology* **1994**, *140*, 2673–2682. [CrossRef]
42. Gallego, M.V.; Piñaga, F.; Ramón, D.; Vallés, S. Purification and characterization of an α-L-rhamnosidase from *Aspergillus terreu*s of interest in winemaking. *J. Food Sci.* **2001**, *66*, 204–209. [CrossRef]
43. Kim, B.H.; Park, S.K. Enhancement of volatile aromatic compounds in black raspberry wines via enzymatic treatment. *J. Inst. Brew.* **2017**, *123*, 277–283. [CrossRef]

44. McCleary, B.V.; McKie, V.A.; Draga, A.; Rooney, E.; Mangan, D.; Larkin, J. Hydrolysis of wheat flour arabinoxylan, acid-debranched wheat flour arabinoxylan and arabino-xylo-oligosaccharides by β-xylanase, α-L-arabinofuranosidase and β-xylosidase. *Carbohydr. Res.* **2015**, *407*, 79–96. [CrossRef] [PubMed]
45. Codex, F.C. *Food Chemicals Codex*, 10th ed.; Available online: http://www.foodchmicalscodex.org (accessed on 11 March 2019).
46. Iibuchi, S.; Minoda, Y.; Yamada, K. Studies on tannin acyl hydrolase of microorganisms: Part II. A new method determining the enzyme activity using the change of ultra violet absorption. *Agric. Biol. Chem.* **1967**, *31*, 513–518. [CrossRef]

© 2019 by the authors. Licensee MDPI, Basel, Switzerland. This article is an open access article distributed under the terms and conditions of the Creative Commons Attribution (CC BY) license (http://creativecommons.org/licenses/by/4.0/).

Article

DERA in Flow: Synthesis of a Statin Side Chain Precursor in Continuous Flow Employing Deoxyribose-5-Phosphate Aldolase Immobilized in Alginate-Luffa Matrix

Bianca Grabner *, Yekaterina Pokhilchuk and Heidrun Gruber-Woelfler *

Institute of Process and Particle Engineering, Graz University of Technology, Inffeldgasse 13/III, 8010 Graz, Austria; y.pokhilchuk@student.tugraz.at
* Correspondence: b.grabner@tugraz.at (B.G.); woelfler@tugraz.at (H.G.-W.)

Received: 16 December 2019; Accepted: 14 January 2020; Published: 18 January 2020

Abstract: Statins, cholesterol-lowering drugs used for the treatment of coronary artery disease (CAD), are among the top 10 prescribed drugs worldwide. However, the synthesis of their characteristic side chain containing two chiral hydroxyl groups can be challenging. The application of deoxyribose-5-phosphate aldolase (DERA) is currently one of the most promising routes for the synthesis of this side chain. Herein, we describe the development of a continuous flow process for the biosynthesis of a side chain precursor. Design of experiments (DoE) was used to optimize the reaction conditions (pH value and temperature) in batch. A pH of 7.5 and a temperature of 32.5 °C were identified to be the optimal process settings within the reaction space considered. Additionally, an immobilization method was developed using the alginate-luffa matrix (ALM), which is a fast, simple, and inexpensive method for enzyme immobilization. Furthermore, it is non-toxic, biodegradable, and from renewable resources. The final continuous process was operated stable for 4 h and can produce up to 4.5 g of product per day.

Keywords: immobilized DERA; statin side chain; continuous flow synthesis; alginate-luffa matrix; design of experiments; optimization

1. Introduction

Cardiovascular diseases (CVD) are the number one cause of death worldwide [1]. In Europe each year, 3.9 million deaths (45% of all deaths) are associated with CVD [2]. The primary reason for death among CVD patients is coronary artery disease (CAD). CAD is characterized by arthrosclerosis—the formation of sedimentation (plaque) in the blood vessel that leads to reduced oxygen supply to the heart [3]. Arterial plaque formation is often attributed to an increased level of low-density lipoprotein (LDL) cholesterol, which in Western culture is often caused by unhealthy foods and little exercise. Beside a change in lifestyle being the first choice for lowering the cholesterol level in blood, one can choose between three medical mechanisms for the treatment of hypercholesterolemia increase bile synthesis, decreased intestinal cholesterol absorption, or inhibition of the 3-hydroxy-3-methylglutary coenzyme A (HMG-CoA) reductase, an essential enzyme in the synthetic route to cholesterol [4,5]. Statins competitively inhibit HMG-CoA reductase and thus approach the reduction of LDL cholesterol concentration by the latter of the above-mentioned methods. One of the most prevalent statins is atorvastatin (Lipitor®), synthesized by Pfizer since 1996. It is known as the best-selling blockbuster in the past two decades [6]. Despite the patent expiration in 2011, Lipitor® was the third most commonly prescribed medication in the U.S. in 2016. An additional statin, simvastatin, was the eighth most prescribed medication; therefore, statins represent a significant share of the pharmaceutical market [6].

Natural and semi-synthetic statins possess side chains in the form of lactones, which are in vivo hydrolyzed to the corresponding and biologically active hydroxyl acid. Synthetic statins, so-called super-statins such as rosuvastatin (Crestor®), are provided in the active form of dihydroxy heptanoic acid with two chiral alcohol groups attached to a heterocyclic core [7,8]. The structures of the three above-mentioned statins are depicted in Figure 1.

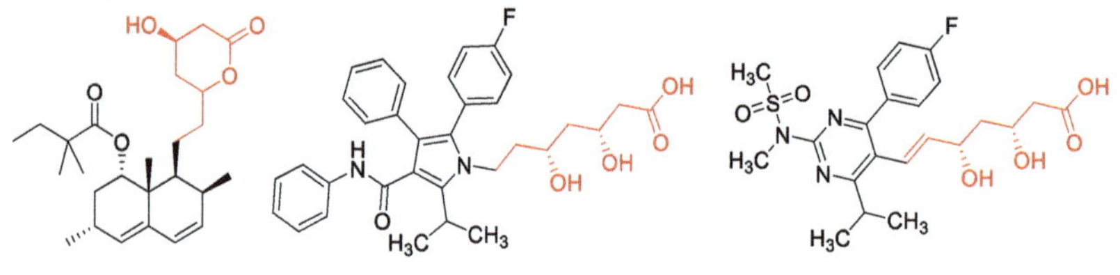

Figure 1. Molecular structures of three statins: Simvastatin, atorvastatin, and rosuvastatin. Dihydroxy heptanoic acid side chain and its cyclic precursor are colored in red.

Only one of the enantiomers of the chiral side chain is active and needs to be provided in high purity for adequate activity. This is a major challenge for manufacturers [9,10]. In the past decade, numerous approaches for the enantiomerically pure synthesis of this side chain were published [11–15]. Chemical routes requiring harsh chemicals and numerous additional steps for the protection/de-protection of sensitive functional groups, by-product formation, and waste generation are an issue. In contrast to this, it was shown by Tao et al. that biocatalysis could be a comparably sustainable approach [16,17]. Beside numerous chemo-enzymatic routes, the one employing deoxyribose-5-phosphate aldolase (DERA, EC 4.1.2.4) is very promising. DERA is a unique enzyme able to catalyze the aldol addition of two aldehydes resulting in an aldehyde product, which can again serve as a substrate for another aldol addition. The product after two sequential addition reactions and spontaneous cyclization is the hemiacetal 2,4,6-trideoxyhexose **1c** (Figure 2). This unique property of DERA was discovered by Gijsen et al. in 1994 and extensively studied around the millennium [18–23]. The product of this biotransformation can be further processed to the statin side chain via the oxidation and subsequent ring-opening of **1e** (Figure 2). The application of DERA on an industrial scale is challenging as the enzyme is sensitive to high concentrations of acetaldehyde, its natural substrate. The active site of the wildtype DERA (DERA$_{WT}$) is irreversibly inhibited by the covalent binding of the side-product, crotonaldehyde. In 2016, the group of Pietruszka at the Research Center Jülich GmbH tackled this issue and developed a mutant (C47M), which is resistant to acetaldehyde to a high degree [24]. This mutant showed outstanding catalytic activity in tests using acetaldehyde as the donor molecule. Further implementation of the mutant was not reported by the group.

In the present work, a continuous process for the synthesis of a statin side chain precursor was developed. Continuous processes go hand in hand with a number of advantages such as reduced reaction time, constant product quality and cost reduction, compared to batch mode. While continuous operations are well implemented in bulk industries such as paper and food, they have barely made their way into pharmaceutical drug synthesis [25]. Fortunately, numerous researchers are passionate about continuous flow synthesis and aim to accelerate the establishment of continuous processes in the industry [26–29]. In this manuscript, we describe the process of developing a continuous biocatalytic synthesis employing the novel DERA mutant. Immobilized freeze-dried whole cells (*Escherichia coli* hosting DERA (C47M)) in an alginate gel matrix on a luffa sponge were used in a packed-bed reactor. This immobilization method was originally developed by Phisalaphong et al. and named alginate-loofa matrix (ALM) [30]. This method is simple, inexpensive, and fast in preparation. Loofa sponge (*Luffa cylindrical*) as support brings a number of advantages. The matrix originates from a renewable source, is a highly porous material, and is fully biodegradable [31–33].

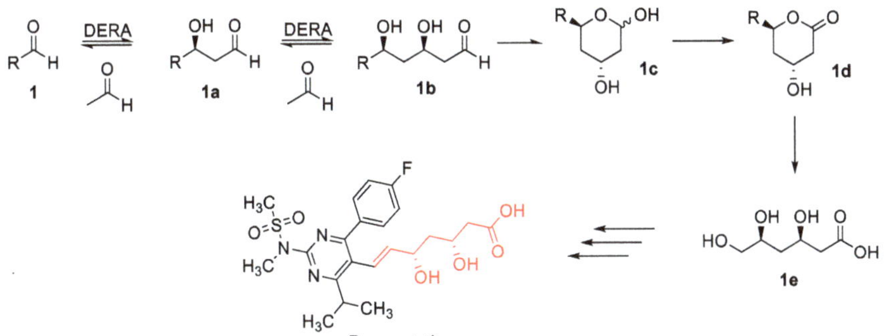

Figure 2. Deoxyribose-5-phosphate aldolase (DERA)-catalyzed stereoselective aldol addition of three aldehydes to produce a lactol (**1c**), which can be further oxidized to a lactone results in the typical statin side chain after ring-opening.

In this work, design of experiments (DoE) was applied for the optimization of pH value and temperature for the enzymatic reaction and optimizing the flow conditions. DoE is a multivariate approach for parameter screening and optimization. In contrast to the original approach, where one factor is changed at a time, this method allows the identification of the interaction between the individual parameters. DoE helps in gaining maximum information from a minimum number of experiments [34–36].

So far, no such development process using DoE for the optimization of a continuous enzymatic process was described in the literature, to the best of our knowledge. Furthermore, this immobilization method and the application of DERA in continuous flow processes are barely found in the literature.

2. Results and Discussion

2.1. Design of Experiments for Optimal Batch Conditions

In order to operate the continuous process under ideal conditions, the optimal parameters for the enzymatic reaction needed to be determined. For that, design of experiments (DoE) was used to evaluate the effect of the two crucial process parameters, temperature, and pH value on intermediate and product formation. In the first circuit of experiments, a rough, full-factorial lattice was designed. The temperature ranged between 28 °C and 37 °C in steps of 4.5 °C and the pH ranged between 6.0 and 8.0 in steps of 0.5 in order to screen a wide range of process settings. For the second circuit, a fine full-factorial lattice was laid in the optimum of the response surface of the first experimental circuit. Both designs are shown in Figure 3.

For each point on the lattice, an experiment was conducted. On a 500 µL scale, 1.5 M of acetaldehyde was dissolved in 0.1 M of TEOA buffer set to the respective pH value via HCl. Samples were collected over time and analyzed by means of GC-FID. The collected data (reaction rate for intermediate and product formation and enzyme stability) was evaluated using MATLAB®. Details on the experimental design and parameters can be found in ESI (Electronic Supplementary Information).

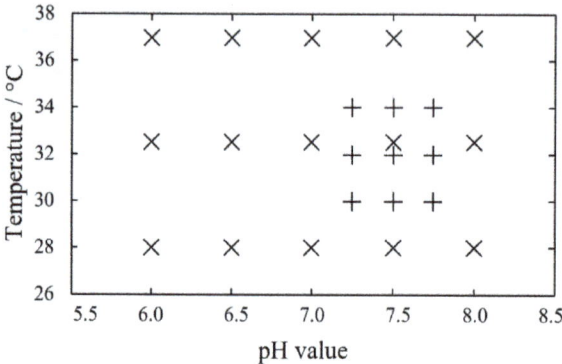

Figure 3. Experimental design chosen for the two circuit of design of experiments (DoE) (1st circuit: x; 2nd circuit: +).

The response surface for the rate for product formation shown in Figures 4 and 5 was obtained from the contour plot in Figure S9. It shows an almost linear incline in enzyme activity with increasing reaction temperature. This correlation between the temperature and reaction rate is also clearly visible in Figure S7 in ESI, which shows the data points without surface. The shape of this surface can be described by the Arrhenius law and the rate of inactivation, which is mathematically a logarithmic function. The productivity of the catalytic system grows until it reaches the point where the inactivation rate is higher than the reaction rate [37]. The highest activity is supposed to be at 34.5 °C. However, the deactivation rate due to denaturation is also high at this temperature. At 34 °C, the activity was reduced by 30% after 1 h of reaction time, while a reaction temperature of 32.5 °C retained more than 95% of the initial activity after 1 h reaction time. Therefore, for all the following experiments 32.5 °C, was the temperature of choice, as it constitutes the ideal compromise between the required reaction rate for a continuous application and stability for a steady state over several hours. The connection between pH value and reaction rate shows the typical Gaussian-like distribution, with an optimum value at pH 7.5 (Figure 5).

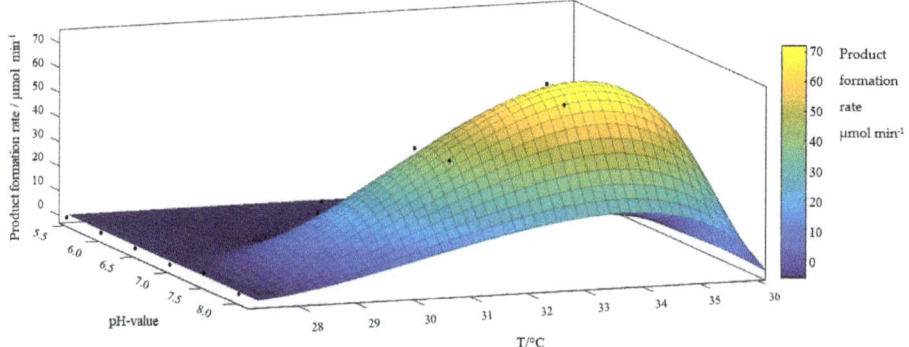

Figure 4. Response surface (view from the perspective over T) for the experimental design over T. An amount of 1.5 M of **1** in a final volume of 500 µL, 0.1 M of buffer (6.0 ≤ pH ≤ 8.5), 7 µL of DMSO, and 10 mg of freeze-dried *E. coli* cells hosting DERA, 28 °C ≤ T ≤ 37.

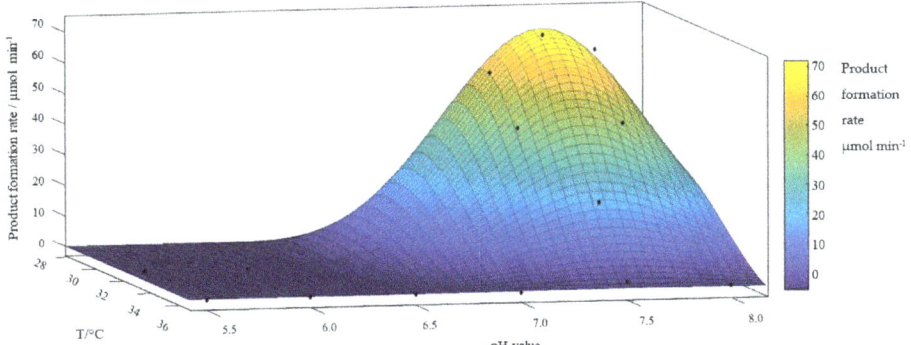

Figure 5. Response surface (view from the perspective over pH) for the experimental design over pH value. An amount of 1.5 M of **1** in a final volume of 500 µL, 0.1 M of buffer (6.0 ≤ pH ≤ 8.5), 7 µL of DMSO, and 10 mg of freeze-dried *E. coli* cells hosting DERA, 28 °C ≤ T ≤ 37.

At pH ≤ 7.0, no lactol was formed at all. Based on the results of the DoE, all the following experiments were conducted at 32.5 °C and pH 7.5.

2.2. Substrate Screening

Five substrates were tested under optimized batch conditions for their potential to serve as an acceptor for aldol addition (Figure 6). Acetaldehyde **1** and its chloro derivative **2** were converted to the desired products, while larger residues (benzaldehyde **4** and cinnamaldehyde **5**) and acrolein **3** were not accepted. The major issue with aromatic substrates is the limited solubility in aqu. buffer systems, even in mixtures with DMSO [38]. These substrates would be especially interesting, as the direct addition of acetaldehyde to the molecule core would reduce the number of process steps drastically. Acrolein was also tested as substrate. Although the mutant was successfully tested for its tolerance toward crotonaldehyde, acrolein still inhibits the enzyme [38,39]. Further investigations were conducted using acetaldehyde **1** and chloroacetaldehyde **2**.

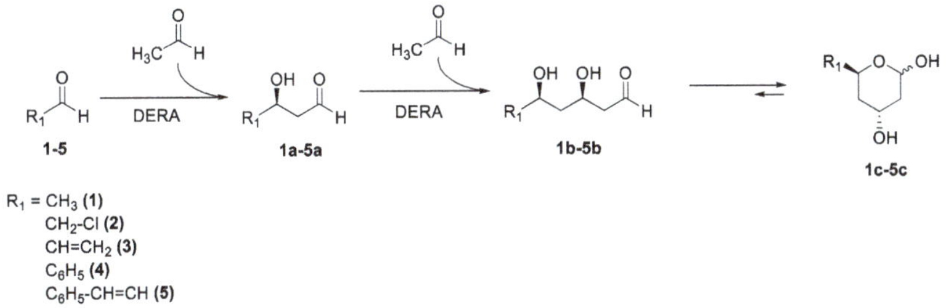

Figure 6. Overview of tested substrates and corresponding products.

2.3. Kinetics

After the optimal reaction conditions and accepted substrates were identified, it was of interest to investigate the kinetic behavior of the addition reaction. This reveals important information required for designing the continuous process. First, a batch experiment in which three molecules of acetaldehyde **1** are linked to the lactol **1c** was conducted. It shows that product formation is initiated as soon as intermediate concentration exceeds 100 mM (Figure 7). After 3 h, the conversion exceeded 95% and the yield (determined by GC-FID) reached 88.5%.

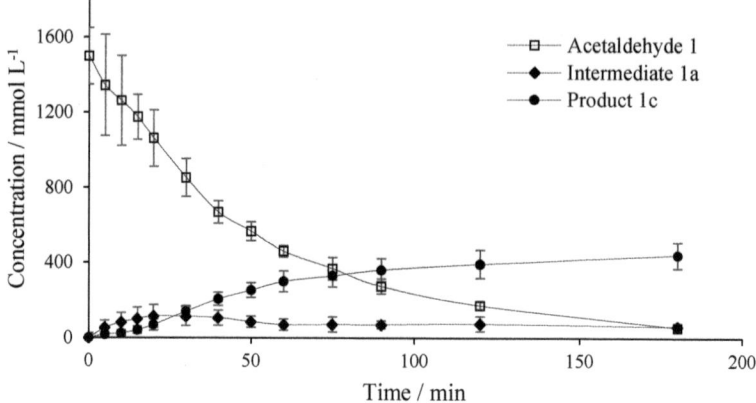

Figure 7. Time course of conversion of acetaldehyde **1** with DERA to the dimer **1a**, and subsequently to the lactol **1c**, at the optimized reaction conditions (32 °C, pH 7.5) based on the DoE, 5 mL of final volume, 0.1 M of TEOA buffer, 1.5 M of substrate concentration, 70 µL of DMSO, and 100 mg of DERA in freeze-dried *E. coli* cells.

As **1** and thus **1c** do not host any functional groups that could serve in further coupling reactions to link the product to the core of the API, the chloro derivative **2** is of greater interest for the industry. Fortunately, **2** proved to be converted faster than acetaldehyde. Most of the reaction progress can be observed within the first 60 min of the reaction, after which a 75% yield was detected (Figure 8). After 3 h, more than 90% was reached. For the continuous flow application, **2** was chosen as the aldehyde acceptor for the reasons mentioned above.

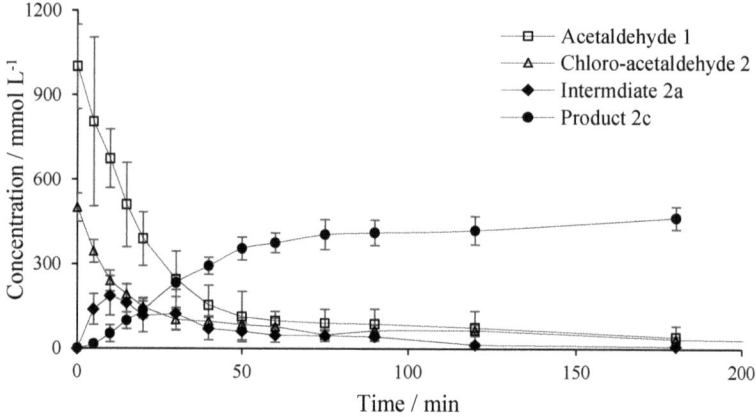

Figure 8. Time course of batch bioconversion of chloroacetaldehyde **2** (0.5 M) and acetaldehyde **1** (1.0 M) to the dimer **2a**, and subsequently to the lactol **2c**, at the optimized reaction conditions (32 °C, pH 7.5) based on the DoE, 5 mL of final volume, 0.1 M of TEOA buffer, 70 µL of DMSO, and 100 mg of DERA in freeze-dried *E. coli* cells.

2.4. Immobilization of DERA in Alginate-Luffa Matrix

Since the use of a packed-bed reactor brings along a number of advantages over the application of homogeneous catalysts, an appropriate immobilization method for the enzyme was required. Covalent binding to a solid support needs an additional purification step of the enzyme prior to linking, which is a labor-intense process. In order to keep catalyst preparation simple, adsorption and encapsulation were the remaining options. Adsorption bears the risk of enzyme leaching due to loose binding. Encapsulation into a matrix is a fast, inexpensive, and simple technique to immobilize isolated biocatalysts or whole cells.

Alginate is non-toxic, biodegradable, and made from a renewable feedstock (bacteria), and thus fulfills all requirements for a "green" matrix for biocatalyst encapsulation. Freeze-dried *E. coli* cells hosting overexpressed DERA were immobilized in two ways in alginate beads and an alginate of luffa sponge. The former of the two was prepared by dissolving 2% (w/v) Na-alginate in 2 mL of a 0.9% (w/v) NaCl solution. After a clear solution was obtained, the biocatalyst was suspended in the viscose liquid. To form spherical beads from the suspension, the mixture was dropwise added to a 2% (w/v) solution of a bivalent cation for cross-linking, Ca^{2+} ($CaCl_2$) or Ba^{2+} ($BaCl_2$). Other tested cations (Zn^{2+}, Fe^{2+}, and Mg^{2+}) did not lead to sufficient cross-linking to form a stable alginate matrix. Since the alginate matrix displays a barrier for the substrate, which results in a reduction of reaction rate, an increase in surface can be beneficial for the reaction rate. For that, a porous material serving as support, which could be coated with the alginate matrix enclosing the enzyme, was desired. A luffa sponge was chosen as support because it is a natural product of high porosity. This immobilization technique is called the alginate-luffa matrix (ALM). In order to immobilize the same volume of an alginate-enzyme mixture on the luffa sponge, a volume of 2.5 cm^3 (245 mg) was required. The sponge was soaked with the cell suspension and cross-linking was induced either by Ba^{2+} or Ca^{2+} (details in ESI). The results of the ALM were compared with cells immobilized in conventional alginate beads (I.D. 2 mm) (Figure 9). The comparison shows that ALM is four times more active than the beads. The type of cross-linking cation has hardly any effect on the performance of the enzyme. ALM enclosing DERA was further employed for application in continuous flow.

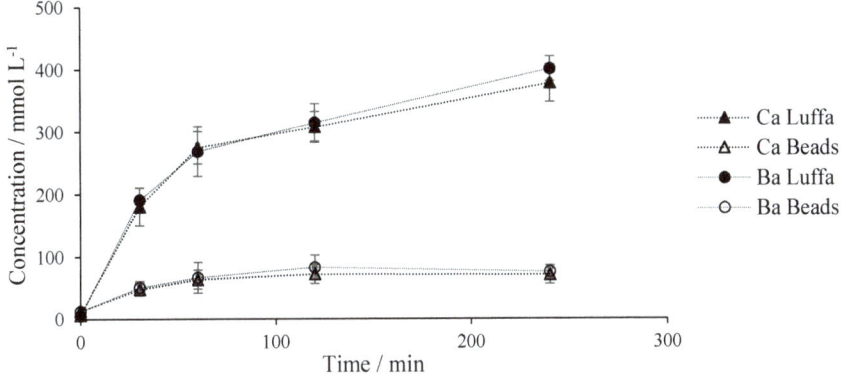

Figure 9. Time course of lactol formation by immobilized DERA on Luffa or in beads, 5 mL of final volume, 0.1 M of TEOA buffer with pH 7.5, 70 mL of DMSO, 0.5 M of **2**, 1.0 M of **1**, 32.5 °C, 100 mg of freeze-dried *E. coli* cells hosting DERA, 2% (w/v) Na-alginate in 0.9% (w/v) NaCl solution, cross-linking induced by Ca^{2+} or Ba^{2+}, 2% (w/v), respectively.

2.5. Flow Application

Flow experiments were carried out in the so-called "Plug and Play" reactor [40]. It is a modular bench-top reactor (20 cm × 15 cm) equipped with heating/cooling shell and openings for reaction modules (commercially available HPLC columns). For catalyst immobilization, a cylindrical piece of the sponge fitting into the column was cut manually to serve as solid support. Eight hundred and fifty milligrams (10 cm^3) of luffa was fitted into the column. Immobilized DERA on the loofa sponge was packed in an HPLC column (20 cm × 8 mm) (Figure 10), which was placed in the reactor. After heating the catalyst to the respective temperature, the bed was flushed with buffer in order to remove residual chemicals from the immobilization process prior to running the reaction.

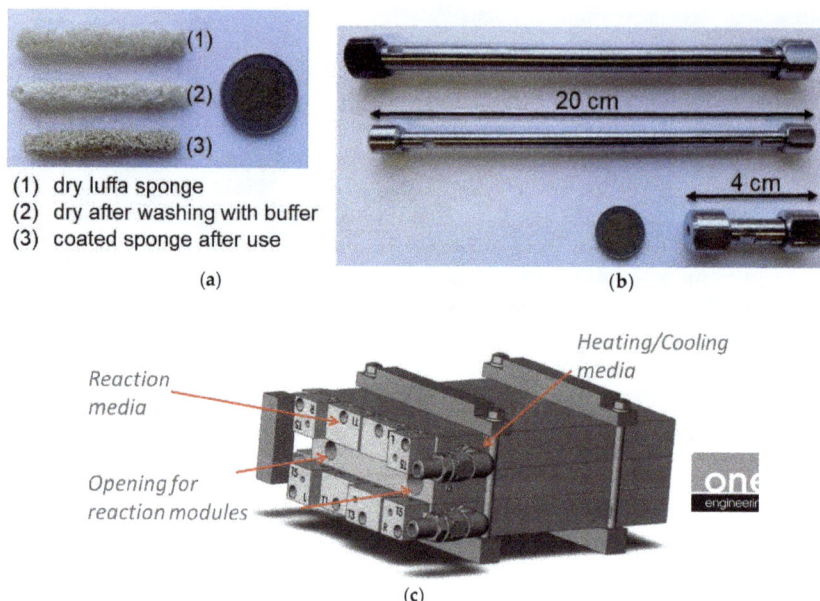

Figure 10. Equipment used for continuous synthesis: luffa sponge (a), stainless steel tube (b), and "Plug and Play" reactor (c).

2.5.1. Residence Time Distribution

Prior to conducting experiments in flow, the mean residence time in the reactor for various flow rates was determined. For that, the column was packed with a luffa sponge and the mean residence time was determined (details in ESI). The results are summed up in Table 1.

Table 1. Theoretically and experimentally determined mean residence times and bodenstein (Bo) numbers for the three test flow velocities going through an HPLC column (20 cm × 0.8 cm) packed with a loofa sponge.

v	$\bar{t}_{th}$ [min]	$\bar{t}_{exp}$ [min]	Bo [–]
0.10	60.5	62.7	15.0
0.25	24.2	26.8	9.5
0.50	12.1	11.4	33.3

$\bar{t}_{th}$ = theoretical calculated mean residence time, $\bar{t}_{exp}$ = experimentally determined mean residence time. See Electronic Supplementary Information (ESI) for details.

2.5.2. Design of Experiments for Continuous Flow Application

For optimizing the flow process, the flow rate and substrate concentration were of interest. Furthermore, a potential effect of the cation used to induce cross-linking of alginate on the catalyst performance should be investigated. A full-factorial design for three parameters (flow rate, concentration, cation) was developed. Three levels were set for flow rate (0.1, 0.25, and 0.5 mL/min) and substrate concentration (0.25, 0.5, and 0.75 M of **2a**). The influence of the chosen cross-coupling ion was investigated on two levels (Ba^{2+} and Ca^{2+}). The performance of the process was followed by collecting samples at the outlet of the reactor and determining the yield by means of GC-FID. The result was evaluated using MODDE® (Figures 11 and 12).

DERA immobilized in alginate cross-linked with Ba^{2+} showed stronger dependence on the flow rate and also on substrate concentration. At high substrate concentration, this catalyst is less active. High flow rate leads to low yield. Details on the results of the DoE can be found in ESI.

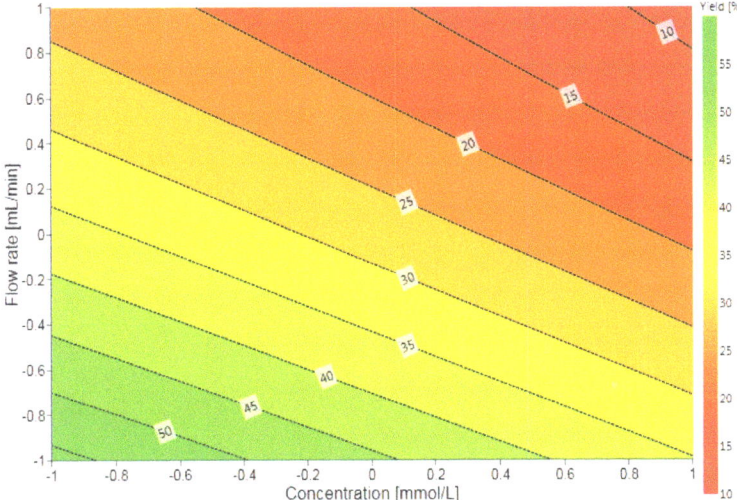

Figure 11. Contour plot for the performance of the continuous flow process for Ba^{2+} cross-linked alginate. Stock: 250–750 mM of **2**, 500–1500 mM (2 mol-eq.) of **1** in 0.1 M of TEOA buffer, pH 7.5, with 1.4% of DMSO. T = 32.5 °C; flow rate: 0.1–0.5 mL/min; 350 mg of freeze-dried *E. coli* hosting DERA on 850 mg of luffa sponge (10 mL of 2% (*w/v*) Na-alginate solution); cross-linking by 2% (*w/v*) $BaCl_2$ solution.

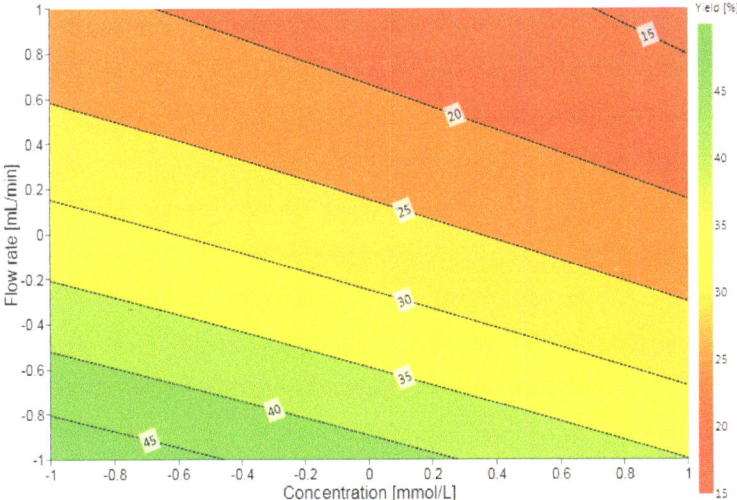

Figure 12. Contour plot for the performance of the continuous flow process for Ca^{2+} cross-linked alginate. Stock: 250–750 mM of 2; 500–1500 mM (2 mol-eq.) of **1** in 0.1 M of TEOA buffer, pH 7.5, with 1.4% of DMSO. T = 32.5 °C; flow rate: 0.1–0.5 mL/min; 350 mg of freeze-dried *E. coli* hosting DERA on 850 mg of luffa sponge (10 mL of 2% (*w/v*) Na-alginate solution); cross-linking by 2% (*w/v*) $CaCl_2$ solution.

Ca²⁺ cross-linked alginate was found to not immobilize the catalyst sufficiently, which led to enzyme leaching and rapid loss of catalyst activity within the reactor. Enzyme leaching was proved by not quenching the samples collected at the outlet of the reactor. The reaction proceeded, indicating the presence of active enzymes in the reaction mixture leaving the column. Ba²⁺ cross-linked alginate did not suffer from enzyme leaching. Therefore, this catalyst could also be applied for an increased period of time in a continuous process.

2.5.3. Continuous Synthesis

As a final step, the ideal reaction conditions obtained from the DoE approach were applied for a continuous flow process for the synthesis of a statin side chain precursor. The time course for the process output is shown in Figure 13. An increase in enzyme loading from 350 mg to 700 mg in the reactor led to almost an 80% yield of **2c**. However, higher cell concentration in the alginate solution resulted in a highly viscous mixture, which made the handling and immobilization process more difficult.

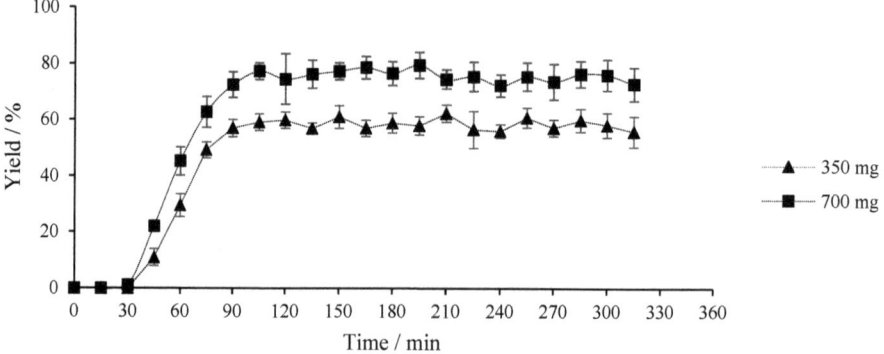

Figure 13. Yield over time for continuous flow process using 350 mg and 700 mg of freeze-dried cells hosting DERA; 32.5 °C, 0.1 M of TEOA buffer, pH 7.5, 0.1 mL/min flow rate, 0.25 M of **2**, 0.5 M of **1**.

3. Materials and Methods

3.1. General

All chemicals used were purchased from Sigma Aldrich in analytical grade and used as received.

3.2. Enzyme Expression and Purification

Two microliters of plasmid isolation and 100 µL of competent *Escherichia coli* cells (strain BL21(DE3)) were combined in a 0.2 cm electroporation cuvette and placed in the BioRad MicroPulser set to "Ec2". The mixture was pulsed once. One milliliter of SOC medium (0.5% (w/v) yeast extract, 2% (w/v) tryptone, 10 mM of NaCl, 2.5 mM of KCl, 20 mM of MgSO₄, pH 7.5) was added and the cells were suspended in the medium. After incubating for 1 h at 37 °C, the cells were applied on agar plates containing the antibiotic Ampicillin and were grown overnight at 37 °C. On the following day, one culture spot was transferred into a shaking flask together with 200 mL of LB medium (5 g/L of yeast extract, 10 g/L of tryptone, 10 g/L of NaCl) and 200 µL of Ampicillin. The culture was incubated overnight at 37 °C. The cells were harvested by centrifugation (15 min, 7000× *g*, 4 °C) and the obtained pellets were freeze-dried.

3.3. Optimization of Reaction Conditions in Batch Using Design of Experiments (DoE)

In the first round, a rough, full-factorial lattice with two variables (temperature and pH) was designed. The temperature ranged between 28 and 36 °C in steps of 4.5 °C, while the pH was set to a value between 6.0 and 8.0 in steps of 0.5. Each reaction was carried out with 10 mg of freeze-dried cells hosting DERA suspended in 500 µL of buffer with 1.5 M of acetaldehyde as substrate and 7 µL of DMSO [24]. The mixture was stirred with 200 rpm at the respective temperature. Product formation was followed by taking samples after 30 and 60 min. For the samples, 100 µL of reaction suspension were mixed with 400 µL of acetonitrile to stop the reaction by precipitating the enzyme. After separating the inactivated enzyme by centrifugation (10 min, 15,000 rpm), the clear solution was transferred into a GC vial and analyzed by means of GC-FID. The results (yield of intermediate and product) were evaluated by surface fitting in MATLAB® using the least square method. A Gaussian function was the result for the description of the activity over pH, while a combination of a polynomial function (increasing activity with rising temperature) and a logarithmic function (describing inactivation at elevated temperature) fitted the surface over temperature. From the response surface, a new, narrow, and full-factorial lattice was planned. This time, the temperature ranged between 30 and 34 °C (steps of 2 °C) and the pH was adjusted to 7.25, 7.5, or 7.75. The reaction and analysis were carried out the same way as for the first round.

3.4. GC Analysis

Samples were analyzed by means of gas chromatography (GC) using a Perkin Elmer (USA) Clarus 500 equipped with an Optima 5-MS 0.25 µm, 30 m × 0.32 mm ID capillary column, and a flame ionization detector (FID) run on H_2 and synthetic air. N_2 was used as carrier gas. The heating program was set as follows: initial temperature of 50 °C (5 min), and gradient of 10 °C min^{-1} to 250 °C (5 min). Injection volume: 1 µL. This method was adapted from the literature [41]. Retention times: **1**, 1.6 min; **1a**, 3.3 min; **1b**, 10.2 min; **1c**, 11.9 min; **2**, 2.0 min; **2a**, 7.9 min; **2c**, 15.4 min.

3.5. Substrate Screening

One hundred milligrams of freeze-dried cells were suspended in 5 mL of 0.1 M of triethanol amine (TEOA) buffer, pH 7.5. After heating to 32.5 °C, 70 µL of DMSO, the substrates (500 mM of acceptor 1-5, respectively, and 1 M of acetaldehyde 1) were added. The reaction mixture was stirred with 600 rpm. Samples were taken over time. For each sample, 200 µL of the reaction mixture were quenched with 800 µL of acetonitrile. Centrifugation (10 min, 15,000 rpm) led to good separation of the precipitated catalyst. The supernatant was transferred into a GC vial and analyzed by means of GC-FID.

3.6. Immobilization in Beads

A 2% *w/v* sodium alginate solution was formulated by dissolving 40 mg of Na-alginate in 2 mL of 0.9% *w/v* aqueous NaCl solution. One hundred milligrams of lyophilized *E. coli* cells were added and the mixture was stirred to homogeneity. In order to form beads, the mixture was then added dropwise to a 2% (*w/v*) solution of the cross-linking cation ($CaCl_2$, $BaCl_2$, $ZnCl_2$, $MgCl_2$, or $FeCl_2$) using a syringe and a needle. The beads were stirred in the cation solution for 60 min to let them solidify. The size of the beads could be varied by the diameter of the needle on the syringe. After filtering, the beads were washed with a 0.9% (*w/v*) NaCl solution and kept under ambient conditions for 30 min to let the surface solidify and become more resistant to mechanical abrasion.

3.7. Immobilization in ALM for Batch Reactions

A 2% *w/v* sodium alginate solution was formulated by dissolving 40 mg of Na-alginate in 2 mL of 0.9% *w/v* aqueous NaCl solution. One hundred milligrams of lyophilized *E. coli* cells were added and the mixture was stirred to homogeneity. Afterward, the carrier (luffa sponge, 245 mg cut into pieces 2.5 cm^3 in volume) was soaked in the mixture and transferred to a 2% (*w/v*) cation solution ($CaCl_2$, $BaCl_2$) where it was gently stirred for 60 min to solidify. The loaded carrier was then washed with 0.9% NaCl solution, kept at ambient conditions for 30 min to solidify, and stored in purified water at 4 °C until usage.

3.8. Immobilization in ALM for Flow Application

The loofa sponge was cut to completely be packed into the 20 cm × 0.8 cm HPLC column. The immobilization process is illustrated in Figure S11 in ESI. Three hundred and fifty milligrams or 700 mg of freeze-dried *E. coli* cells were suspended in a 2% (w/v) (200 mg) Na-alginate in 10 mL of 0.9% (w/v) NaCl solution (Figure S10a). Eight hundred and fifty milligrams (10 cm^3) of cylindrically cut luffa sponge was used to soak up the mixture. After completely soaking up the entire DERA-alginate solution, the loofa sponge was submerged into 2% w/v $CaCl_2$ or $BaCl_2$ for 1 h with stirring at room temperature for cross-linking, followed by 30 min of air-drying (Figure S10b). The resultant loofa sponge carrying immobilized freeze-dried whole cells entrapped by alginate was used to pack the HPLC column for further use in continuous flow experiments.

3.9. Design of Experiments for Optimizing Flow Process

Continuous experiments were carried out in the so-called "Plug and Play" reactor [40]. The 850 mg loofa sponge with the immobilized enzyme (350 mg of cells) was packed in the stainless steel column (20 cm × 8 mm). The reaction medium consisting of 0.25, 0.5, or 0.75 M of chloroacetaldehyde **2** and 2 mol-eq. of acetaldehyde **1**, with respect to **2**, in 0.1 M of TEOA buffer, pH 7.5, was pumped through the column using an HPLC pump (Knauer, Azura P4.1 S) set to 0.10, 0.25, and 0.50 mL/min, respectively. The reaction temperature was 32.5 °C. A sample was taken every 15 min by collecting an aliquot of 200 µL of the product stream and diluting it with 800 µL of acetonitrile. After centrifugation (10 min, 15,000 rpm), the supernatant was transferred to a GC vial and analyzed by means of GC-FID. The results were evaluated by means of MODDE®. Details (table of experiments and results) are available in ESI.

3.10. Continuous Synthesis

The 850 mg loofa sponge with the immobilized enzyme (350 mg of cells) was packed in the stainless steel column (20 cm × 8 mm). The reaction medium, consisting of 0.25 M of **2** and 0.5 M of **1** in 0.1 M of TEOA buffer pH, 7.5, was pumped through the column using an HPLC pump (Knauer, Azura P4.1 S) set to 0.10 mL/min. The reaction temperature was 32.5 °C. A sample was taken every 15 min by collecting an aliquot of 200 µL of the product stream and diluting it with 800 µL of acetonitrile. After centrifugation (10 min, 15,000 rpm), the supernatant was transferred to a GC vial and analyzed by means of GC-FID.

4. Conclusions

We were able to develop an optimized continuous flow process for the synthesis of a statin side chain precursor by using the DoE approach. Herein, 32.5 °C and pH 7.5 turned out to be the ideal process parameters for DERA (C47M). A series of substrates was tested for its applicability as substrate, but only acetaldehyde **1** and chloroacetaldehyde **2** gave reasonable results. For immobilization alginate was chosen and tested as both alginate beads and alginate-luffa matrix (ALM), of which the latter of them showed a fourfold higher reaction rate, most likely due to an increased surface area. After identifying ALM as a "green" technique for immobilizing biocatalysts, the enzyme was applied in continuous flow. While ALM cross-linked by Ca^{2+} suffered from enzyme leaching, Ba^{2+} led to the sufficiently strong enclosing of the enzyme into the matrix. The usage of freeze-dried cells benefits from the size of the biocatalyst because it can be enclosed into the network more sufficiently. ALM has the major advantage that it can be used for almost all biocatalysts to immobilize them in non-covalent encapsulation. The application is only restricted by the limited stability of alginate against harsh chemicals. However, these chemicals are in most cases not in the application field of biocatalysis. Finally, the optimized flow process (0.1 mL/min, 0.25 M of chloroacetaldehyde, and 0.5 M of acetaldehyde) produced 4.5 g of product per day in a bench-top reactor not bigger than a sheet of paper in area and 15 cm in height. The heterogeneous biocatalyst performed stable for 4 h and convinced by its simple, inexpensive, and fast preparation. The mutant proved stability over the whole course of the continuous process and in the batch processes (homogeneous and heterogeneous). In addition, the whole catalytic system is biodegradable and made from renewable resources. In order to further increase the yield of the process, a longer reaction time could help. An alternative immobilization method, which is not limited to a certain enzyme loading due to rapidly increasing viscosity, can also be a solution.

Supplementary Materials: The following information available online at http://www.mdpi.com/2073-4344/10/1/137/s1, **1.** Figure S1: SDS-PAGE of harvested *E. coli* cells overexpressing DERA; **2.** List of experiments (DoE) for optimization of pH and temperature in batch; **3.** Figure S2: GC-FID spectrum of aldol addition with acetaldehyde-substrates and products (**1**, **1a**, **1b**, and **1c**); Figure S3: GC-FID spectrum of aldol addition with chloroacetaldehyde-substrates and products ((**1**, **2**, **2a**, **2b**, and **2c**); **4.** GC-FID spectra of DoE runs, result of DoE without surface-product formation rate over pH and temperature and response surface view from top; **5.** Coating procedure; **6.** Determination of the residence time distribution in the flow reactor incl. scheme of setup; **7.** Product synthesis in semi-batch for reference; **8.** Product isolation procedure; **9.** NMR of intermediate **2a** and product **2c**; **10.** List of experiments (DoE) for optimization of flow process; **11.** Details on data evaluation in MODDE® for optimization of flow process.

Author Contributions: Conceptualization, B.G.; Investigation, B.G. and Y.P.; Methodology, Y.P.; Resources, H.G.-W.; Supervision H.G.-W.; Writing—original draft, B.G.; Writing—review & editing, B.G.; and H.G.-W. All authors have read and agreed to the published version of the manuscript.

Funding: This research received no external funding.

Acknowledgments: The authors B.G. and Y.P. would like to thank Anna K. Schweiger from the Institute for Molecular Biotechnology at Graz University of Technology for expressing the enzyme and her support when it came to questions regarding biochemistry. Furthermore, all authors thank the research group of Pietruszka at the Research Center Jülich GmbH for providing the plasmid of their DERA mutant.

Conflicts of Interest: The authors declare no conflicts of interest

References

1. World Health Organization. The top 10 Causes of Death. Available online: https://www.who.int/news-room/fact-sheets/detail/the-top-10-causes-of-death (accessed on 24 January 2019).
2. Wilkins, E.; Wilson, L.; Wickramasinghe, K.; Bhatnagar, P.; Leal, J.; Luengo-Fernandez, R.; Burns, R.; Rayner, M.; Townsend, N. *European Cardiovascular Disease Statistics 2017*; European Heart Network: Brussels, Belgium, 2017.
3. National Heart, Lung, and Blood Institute (NHLBI). Ischemic Heart Disease. Available online: https://www.nhlbi.nih.gov/health-topics/ischemic-heart-disease (accessed on 13 September 2019).
4. Wald, E.J.; Law, M.R. A strategy to reduce cardiovascular disease by more than 80%. *BMJ* **2003**, *326*, 1419. [CrossRef] [PubMed]
5. Feingold, K.R.; Grunfeld, C. *Endotext. Cholesterol Lowering Drugs*; MDText. com, Inc.: South Dartmouth, MA, USA, 2000.
6. Istvan, E.S.; Deisenhofer, J. Structural mechanism for statin inhibition of HMG-CoA reductase. *Science* **2001**, *292*, 1160–1164. [CrossRef] [PubMed]
7. Gazzerro, P.; Proto, M.C.; Gangemi, G.; Malfitano, A.M.; Ciaglia, E.; Pisanti, S.; Santoro, A.; Laezza, C.; Bifulco, M. Pharmacological actions of statins: A critical appraisal in the management of cancer. *Pharmacol. Rev.* **2012**, *64*, 102–146. [CrossRef] [PubMed]
8. Stancu, C.; Sima, A. Statins: Mechanism of action and effects. *J. Cell. Mol. Med.* **2001**, *5*, 378–387. [CrossRef] [PubMed]
9. Müller, M. Chemoenzymatic synthesis of building blocks for statin side chains. *Angew. Chem. Int. Ed.* **2005**, *44*, 362–365. [CrossRef] [PubMed]
10. Liljeblad, A.; Kallinen, A.; Kanerva, L. Biocatalysis in the Preparation of the Statin Side Chain. *Curr. Org. Synth.* **2009**, *6*, 362–379. [CrossRef]
11. Ručigaj, A.; Krajnc, M. Optimization of a Crude Deoxyribose-5-phosphate Aldolase Lyzate-Catalyzed Process in Synthesis of Statin Intermediates. *Org. Process Res. Dev.* **2013**, *17*, 854–862. [CrossRef]
12. Ručigaj, A.; Krajnc, M. Kinetic modeling of a crude DERA lysate-catalyzed process in synthesis of statin intermediates. *Chem. Eng. J.* **2015**, *259*, 11–24. [CrossRef]

13. Wolberg, M.; Dassen, B.H.N.; Schürmann, M.; Jennewein, S.; Wubbolts, M.G.; Schoemaker, H.E.; Mink, D. Large-Scale Synthesis of New Pyranoid Building Blocks Based on Aldolase-Catalysed Carbon-Carbon Bond Formation. *Adv. Synth. Catal.* **2008**, *350*, 1751–1759. [CrossRef]
14. Jennewein, S.; Schürmann, M.; Wolberg, M.; Hilker, I.; Luiten, R.; Wubbolts, M.; Mink, D. Directed evolution of an industrial biocatalyst: 2-deoxy-D-ribose 5-phosphate aldolase. *Biotechnol. J.* **2006**, *1*, 537–548. [CrossRef]
15. Fei, H.; Zheng, C.-C.; Liu, X.-Y.; Li, Q. An industrially applied biocatalyst: 2-Deoxy-d-ribose-5-phosphate aldolase. *Process Biochem.* **2017**, *63*, 55–59. [CrossRef]
16. Tao, J.; Xu, J.-H. Biocatalysis in development of green pharmaceutical processes. *Curr. Opin. Chem. Biol.* **2009**, *13*, 43–50. [CrossRef] [PubMed]
17. Britton, J.; Majumdar, S.; Weiss, G.A. Continuous flow biocatalysis. *Chem. Soc. Rev.* **2018**, *47*, 5891–5918. [CrossRef] [PubMed]
18. Barbas, C.F.; Wang, Y.-F.; Wong, C.-H. Deoxyribose-5-phosphate Aldolase as a Synthetic Catalyst. *J. Am. Chem. Soc.* **1990**, *112*, 2013–2014. [CrossRef]
19. Gijsen, H.J.M.; Wong, C.-H. Unprecedented Asymmetric Aldol Reactions with Three Aldehyde Substrates Catalyzed by 2-Deoxyribose-5-phosphate Aldolase. *J. Am. Chem. Soc.* **1994**, *116*, 8422–8423. [CrossRef]
20. Gijsen, H.J.M.; Wong, C.-H. Sequential One-Pot Aldol Reactions Catalyzed by 2-Deoxyribose-5-phosphate Aldolase and Fructose-1,6-diphosphate Aldolase. *J. Am. Chem. Soc.* **1995**, *117*, 2947–2948. [CrossRef]
21. Gijsen, H.J.M.; Wong, C.-H. Sequential Three- and Four-Substrate Aldol Reactions Catalyzed by Aldolases. *J. Am. Chem. Soc.* **1995**, *117*, 7585–7591. [CrossRef]
22. Sakuraba, H.; Yoneda, K.; Yoshihara, K.; Satoh, K.; Kawakami, R.; Uto, Y.; Tsuge, H.; Takahashi, K.; Hori, H.; Ohshima, T. Sequential aldol condensation catalyzed by hyperthermophilic 2-deoxy-d-ribose-5-phosphate aldolase. *Appl. Environ. Microbiol.* **2007**, *73*, 7427–7434. [CrossRef]
23. Wong, C.-H.; Garcia-Junceda, E.; Chen, L.; Blanco, O.; Gijsen, H.J.M.; Steensma, D.H. Recombinant 2-Deoxyribose-5-phosphate Aldolase in Organic Synthesis: Use of Sequential Two-Substrate and Three-Substrate Aldol Reactions. *J. Am. Chem. Soc.* **1995**, *117*, 3333–3339. [CrossRef]
24. Dick, M.; Hartmann, R.; Weiergräber, O.H.; Bisterfeld, C.; Classen, T.; Schwarten, M.; Neudecker, P.; Willbold, D.; Pietruszka, J. Mechanism-based inhibition of an aldolase at high concentrations of its natural substrate acetaldehyde: Structural insights and protective strategies. *Chem. Sci.* **2016**, 4493–4502. [CrossRef]
25. Gutmann, B.; Cantillo, D.; Kappe, C.O. Continuous-Flow Technology—A Tool for the Safe Manufacturing of Active Pharmaceutical Ingredients. *Angew. Chem. Int. Ed.* **2015**, *54*, 6688–6728. [CrossRef] [PubMed]
26. Plutschack, M.B.; Pieber, B.; Gilmore, K.; Seeberger, P.H. The Hitchhiker's Guide to Flow Chemistry. *Chem. Rev.* **2017**, *117*, 11796–11893. [CrossRef] [PubMed]
27. Steinreiber, J.; Schurmann, M.; Wolberg, M.; van Assema, F.; Reisinger, C.; Fesko, K.; Mink, D.; Griengl, H. Overcoming thermodynamic and kinetic limitations of aldolase-catalyzed reactions by applying multienzymatic dynamic kinetic asymmetric transformations. *Angew. Chem. Int. Ed.* **2007**, *46*, 1624–1626. [CrossRef] [PubMed]
28. Colella, M.; Carlucci, C.; Luisi, R. Supported Catalysts for Continuous Flow Synthesis. *Top. Curr. Chem.* **2018**, *376*, 46. [CrossRef] [PubMed]
29. Britton, J.; Raston, C.L. Multi-step continuous-flow synthesis. *Chem. Soc. Rev.* **2017**, *46*, 1250–1271. [CrossRef] [PubMed]
30. Phisalaphong, M.; Budiraharjo, R.; Bangrak, P.; Mongkolkajit, J.; Limtong, S. Alginate-loofa as carrier matrix for ethanol production. *J. Biosci. Bioeng.* **2007**, *104*, 214–217. [CrossRef]
31. Dzionek, A.; Wojcieszyńska, D.; Guzik, U. Natural carriers in bioremediation: A review. *Electron. J. Biotechnol.* **2016**, *23*, 28–36. [CrossRef]
32. Shen, J.; Min Xie, Y.; Huang, X.; Zhou, S.; Ruan, D. Mechanical properties of luffa sponge. *J. Mech. Behav. Biomed.* **2012**, *15*, 141–152. [CrossRef]
33. Ogbonna, J.C.; Liu, Y.-C.; Liu, Y.-K.; Tanaka, H. Loofa (*Luffa cylindrica*) sponge as a carrier for microbial cell immobilization. *J. Biosci. Bioeng.* **1994**, *78*, 437–442. [CrossRef]
34. Murray, P.M.; Bellany, F.; Benhamou, L.; Bucar, D.-K.; Tabor, A.B.; Sheppard, T.D. The application of design of experiments (DoE) reaction optimisation and solvent selection in the development of new synthetic chemistry. *Org. Biomol. Chem.* **2016**, *14*, 2373–2384. [CrossRef]
35. Weissman, S.A.; Anderson, N.G. Design of Experiments (DoE) and Process Optimization. A Review of Recent Publications. *Org. Process Res. Dev.* **2015**, *19*, 1605–1633. [CrossRef]

36. Lendrem, D.; Owen, M.; Godbert, S. DOE (Design of Experiments) in Development Chemistry: Potential Obstacles. *Org. Process Res. Dev.* **2001**, *5*, 324–327. [CrossRef]
37. Robinson, P.K. Enzymes: Principles and biotechnological applications. *Essays Biochem.* **2015**, *59*, 1–41. [CrossRef] [PubMed]
38. Dick, M.; Weiergräber, O.H.; Classen, T.; Bisterfeld, C.; Bramski, J.; Gohlke, H.; Pietruszka, J. Trading off stability against activity in extremophilic aldolases. *Sci. Rep.* **2016**, *6*. [CrossRef] [PubMed]
39. Wilton, D.C. Acrolein, an Irreversible Active-Site-Directed Inhibitor of Deoxyribose 5-phosphate Aldolase? *Biochem. J.* **1976**, *153*, 495–497. [CrossRef] [PubMed]
40. Lichtenegger, G.J.; Tursic, V.; Kitzler, H.; Obermaier, K.; Khinast, J.G.; Gruber-Wölfler, H. The Plug & Play Reactor: A Highly Flexible Device for Heterogeneous Reactions in Continuous Flow. *Chem. Ing. Tech.* **2016**, *88*, 1518–1523. [CrossRef]
41. Ošlaj, M.; Cluzeau, J.; Orkić, D.; Kopitar, G.; Mrak, P.; Casar, Z. A highly productive, whole-cell DERA chemoenzymatic process for production of key lactonized side-chain intermediates in statin synthesis. *PLoS ONE* **2013**, *8*, e62250. [CrossRef]

© 2020 by the authors. Licensee MDPI, Basel, Switzerland. This article is an open access article distributed under the terms and conditions of the Creative Commons Attribution (CC BY) license (http://creativecommons.org/licenses/by/4.0/).

Article

Characterization of Protein Hydrolysates from Eel (*Anguilla marmorata*) and Their Application in Herbal Eel Extracts

I-Chun Cheng [1], Jin-Xian Liao [1], Jhih-Ying Ciou [2], Li-Tung Huang [3,4], Yu-Wei Chen [5] and Chih-Yao Hou [1,*]

1. Department of Seafood Science, National Kaohsiung University of Science and Technology, No. 142, Haijhuan Rd., Nanzih Dist., Kaohsiung City 81157, Taiwan; s0928522928@gmail.com (I-C.C.); j0920181@gmail.com (J.-X.L.)
2. Department of Food Science, Tunghai University, No. 1727, Section 4, Taiwan Boulevard, Xitun District, Taichung City 40704, Taiwan; jyciou@thu.edu.tw
3. Institute for Translational Research in Biomedicine, College of Medicine, Kaohsiung Chang Gung Memorial Hospital and Chang Gung University, Kaohsiung 833, Taiwan; huangli@cgmh.org.tw
4. Department of Traditional Medicine, Chang Gung University, Linkow 333, Taiwan
5. Department of Medicine, Chang Gung University, No. 259, Wenhua 1st Rd., Guishan Dist., Taoyuan City 33302, Taiwan; naosa720928@gmail.com
* Correspondence: chihyaohou@gmail.com; Tel.: +886-985300345; Fax: +886-7-3640634

Received: 17 December 2019; Accepted: 7 February 2020; Published: 8 February 2020

Abstract: The enzymatic hydrolysis of fish proteins is the principle method for converting under-utilized fish into valuable products for the pharmaceutical and health food industries. In this study, three commercial enzymes (alcalase, bromelain, and papain) were tested for their ability to create eel protein hydrolysates (EPHs) from whole eel (*Anguilla marmorata*). Freeze-dried EPHs had almost more than 80% solubility ($p < 0.05$) in solutions ranging from pH 2–10. The amino acid profiles of the EPHs showed a high percentage of essential amino acids, including histidine, threonine, valine, isoleucine, and leucine. The emulsion activity index (EAI) of EPH resulted as follows: alcalase group (36.8 ± 2.00) > bromelain group (21.3 ± 1.30) > papain group (16.2 ± 1.22), and the emulsion stability index (ESI) of EPH was: alcalase group (4.00 ± 0.34) > bromelain group (2.62 ± 0.44) > papain group (1.44 ± 0.09). As such, EPH has a high nutritional value and could be used as a supplement to diets lacking protein. EPH showed excellent solubility and processed interfacial properties, which are governed by its concentration. Among of them the alcalase group had the best antioxidant effect at 1,1-diphenyl-2-pyridinohydrazinyl (DPPH) radical method, determination of reducing power and ABTS test compared with other groups. EPH may be useful in developing commercial products like herbal eel extracts that are beneficial to human health.

Keywords: *Anguilla marmorata*; eel protein hydrolysates; functional properties; herbal eel extracts

1. Introduction

As the world's population continues its expansion towards nearly 10 billion people by 2050, increasing wealth in developing countries adds to the demand for protein [1], an essential component of the human diet. This is due to changing food preferences and a growing recognition of the importance of protein as a key dietary ingredient. Dietary protein supplements are becoming popular, especially for people on restricted diets, like athletes and the elderly [2]. Dairy and soy are the main sources of protein in nutritional beverages and herbal extract products, with a whey protein concentration of 80% the most widely used. However, fish is an excellent source of protein, with proven satiating effects and higher protein content than most terrestrial animals. The proper utilization of limited aquatic resources has been a topic of great interest for many decades. The use of proteases in fish processing leads to the hydrolysis of proteins in the source material, which can then be separated from the muscle [3,4] and used in beverage or herbal products.

Anguillid eels are one of the main high value species of fish used in aquaculture, with the Japanese eel (*Anguilla japonica*) and European eel (*Anguilla anguilla*) the most popular [5,6]. World production and consumption of *Anguilla* sp. eels in 1987 was ~100,000–110,000 tons and ~70–80% of it was produced and consumed in Japan and Taiwan [7]. Eel has been an important aquatic export in these countries for nearly 30 years. In recent years, there has been a severe shortage of *Anguilla japonica* glass eels, as artificial propagation of freshwater eels had not yet been commercially successful. Most eel seedlings come from natural fishing, which has prompted a search for other *Anguilla* species, such as *Anguilla marmorata*. These adult eels are greyish-yellow with a white belly and brownish-black marbling on their backs that can fade over time. Mass production of *Anguilla marmorata* has been cultivated gradually and its breeding habits are still being explored, but little research exists on its functionality [8]. *Anguilla marmorata* are not only economically valuable, but are also rich in nutritional value due to their high levels of proteins, Carnosine, vitamins, and minerals [9,10]. However, research into *Anguilla marmorata* is relatively rare. Recent research into fish protein hydrolysates (FPHs) and their antioxidant activity, for instance skin gelatin hydrolysates from brownstripe red snapper or meat protein hydrolysates from yellowstripe trevally (*Selaroides leptolepis*) [11,12], show that hydrolysis can release functional peptides which may be used in numerous food products. R. Hartmann and H. Meisel (2007) pointed out that many peptides released in vitro or in vivo from animal or plant proteins are bioactive and have regulatory functions in humans beyond normal adequate nutrition. Different health effects have been attributed to food-derived peptides [13]. Water-holding and fat-binding capacities are functional properties that are closely related to texture based on the interactions between water, oil, and other components [14]. Protein hydrolysates rich in bioactive compounds represent promising ingredients for food and industrial applications. Some recent studies have gone beyond producing and characterizing EPH and have tested their ability to fortify foods [15,16]. Because the protein hydrolysates characterization and process application properties research of *Anguilla marmorata* is relatively few, and protein hydrolysates are often used in nutritional supplements. Therefore, the objectives of this study are to compare the use of three commercial enzymes (alcalase, bromelain and papain) in the production of eel protein hydrolysates (EPHs) and to characterize these EPHs, compare their chemical, functional properties, and sensory properties, and application to herbal eel extracts.

2. Results and Discussion

2.1. Enzyme Activity and Degree of Hydrolysis

Before hydrolysis, an enzyme activity assay was performed to estimate the quality of the respective enzyme. Proteolytic activity for each enzyme was: alcalase 2.4 L at 2500 U/mg, papain at 10,000 U/mg, and bromelain at 1250 U/mg. Alcalase is produced from microbes while the other two enzymes are produced from plants. Of the three, bromelain had the lowest activity level. Degree of hydrolysis (DH) is defined as the percentage of peptide bonds cleaved and it is the standard parameter commonly used to monitor and compare the level of protein proteolysis, which also affects protein solubility, emulsification, and foaming. Protein hydrolysates with high solubility can be easily mixed into liquid and have excellent wettability. Different DH values affect the structure and amino acid composition. Figure 1a shows the DH for alcalase, bromelain, and papain at optimal pH and temperature (pH 7.5 and 55 °C, pH 4.5 and 45 °C, and pH 6.0 and at 50 °C, respectively) for different reaction times (from 1 h to 16 h). Figure 1b shows DH depending on the amount of enzyme added (from 0.1 to 2.0 g/g). In Figure 1a, there are significant differences in DH between the enzymes, as alcalase shows greater hydrolytic ability than bromelain or papain. Similar results are seen in Figure 1b, with alcalase still showing the greatest degree of hydrolysis. The functionality of protein hydrolysates is a major factor in their success as functional supplements in food. The physicochemical properties of protein hydrolysates depend on the protein substrate, the specificity of the enzyme used for proteolysis, and the hydrolysis conditions [14].

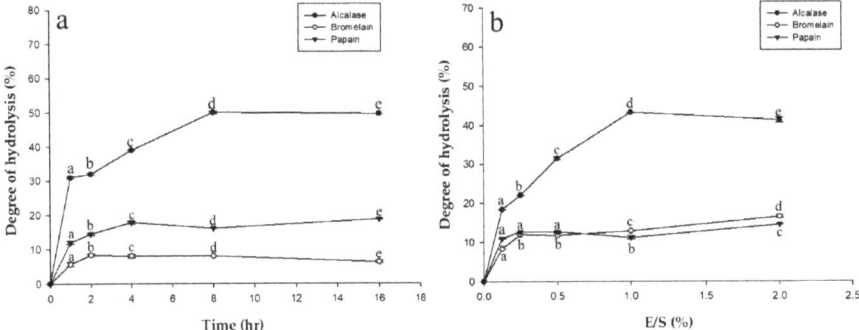

Figure 1. (a) Degree of hydrolysis of eel (*Anguilla marmorata*) at different times with three different commercial proteases (alcalase, bromelain, and papain) at optimal pH and temperature; (b) Degree of hydrolysis of eel (*Anguilla marmorata*) with different amounts of commercial proteases (alcalase, bromelain, and papain) at optimal pH and temperature. Means in the same form with various characters have significant differences ($p < 0.05$). Data are expressed as mean ± SD, from experiments performed in triplicate.

2.2. Molecular Mass Distribution Profile

Prior to enzymatic hydrolysis, the peptide molecular mass distribution of eel (*Anguilla marmorata*) was analyzed via high performance liquid chromatography (HPLC). All EPHs were reduced to smaller peptides with molecular mass between 1056.55 Da and 1554.53 Da (Table 1). Small molecule proteins and peptides are easier for the human body to digest and absorb which mainly include dipeptide or tripeptide [17].

Table 1. Average molecular mass of peptides in eel (*Anguilla marmorata*) following hydrolysis with alcalase, bromelain, and papain.

	Enzyme	Molecular mass (Da)
	Alcalase	1056.55 ± 2.60 [a]
EPH *	Bromelain	1482.24 ± 3.36 [b]
	Papain	1554.53 ± 19.22 [c]

* Prior to enzymatic hydrolysis, eel (*Anguilla marmorata*) were analyzed by HPLC. [a-c] Different letters in the same column denote significant differences between hydrolysates ($p < 0.05$).

2.3. Amino Acid Composition

Amino acid levels in the hydrolyzed EPHs were higher than in unhydrolyzed eels. Among the 20 types of standard amino acids, there are 9 essential amino acids (EAAs) that adults cannot produce and must obtain from their diet; arginine (Arg), in particular, is essential for babies. Among the EPHs, the levels of histidine (His), threonine (Thr), valine (Val), isoleucine (Ile), and leucine (Leu) were higher with alcalase than for bromelain or papain (Table 2). In fact, the alcalase EPH provided the equivalent of the World Health Organization's recommended daily intake of EAAs needed to meet the protein requirement for adults. Protein hydrolysates are mainly used in health products and energy drinks, but hydrolysis results in protein structural changes, causing bitterness. Since bitterness was produced by proline (Pro), Leu, Val, and phenylalanine (Phe), an herbal eel extract recipe was used to remove the unpleasant taste. The amino acid composition of the hydrolyzed eel protein solution could help supply adults with EAAs and be used to develop nutritional supplements in the future.

Table 2. Total amino acids of the three EPHs (A = alcalase, B = bromelain, P = papain) and recommended amino acid requirements of adults (WHO, 2002). Branched chain amino acids are highlighted in bold font. WHO = World Health Organization.

Amino Acid	Eel	A	B	P	WHO
His *	0.81	1.01	0.76	0.95	1.50
Thr *	0.63	2.98	2.75	0.71	2.30
Val *	0.31	1.66	3.07	2.23	3.90
Ile *	1.87	1.93	1.27	2.15	3.00
Leu *	1.74	2.85	1.48	0.77	5.90
Asp	4.51	5.14	5.33	6.44	-
Glu	7.18	7.10	6.96	6.95	-
Ser	0.80	1.48	2.31	0.92	-
Gly	0.30	6.38	5.58	3.09	-
Ala	5.83	5.03	5.02	6.37	-
Pro	1.06	1.37	0.70	1.02	-
Arg	0.41	2.04	1.95	0.55	-
Phe *	1.12	1.19	0.29	0.98	-

* required for adults. Results are expressed as the mean ± SD from triplicate determinations. Values in the same column with different superscript letters are significantly different ($p < 0.05$).

2.4. Functional Properties of EPHs

The functional properties of the hydrolyzed protein solution are affected by amino acid composition and the molecular weight of the peptides. The solubility, emulsifying properties, stability, turbidity, color, oil binding capacity, and flavor are important in food processing. Functional properties can also affect the sensory evaluation, especially for chewy and smooth textures.

2.4.1. Emulsifying Properties

EAI is a function of oil volume fraction, protein concentration, and the type of equipment used to produce the emulsion [14] EAI (m^2g^{-1}) and ESI (min) of EPHs at different concentrations (0.5%, 1%, and 2% w/v) are shown in Table 3. All EPHs had reduced EAI as concentration increased. The most significant difference among the EPHs was at 0.5% (w/v) concentration, with Alcalsae having the highest EAI (36.8 ± 2.00 m^2g^{-1}) and papain having the lowest (16.2 ± 1.22 m^2g^{-1}). ESI was also significantly impacted by concentration. Similar to EAI, differences in ESI were largest at a concentration of 0.5% (w/v). Alcalase had the highest ESI (4.00 ± 0.34 min) at a concentration of 2% (w/v). Emulsification occurs during homogenization when proteins are absorbed at the surface of the oil droplets as they form, creating a membrane which prevents them from consolidating [18].

Table 3. Emulsifying activity index (EAI, $m^2 g^{-1}$) and emulsion stability index (ESI, min) at different EPH concentrations (0.5%, 1%, and 2% w/v) and oil binding capacities (OBC, g/g).

	(EAI m^2g^{-1})			ESI (min)			OBC (g/g)
EPH conc.	0.50%	1%	2%	0.50%	1%	2%	
Alcalase	36.8 ± 2.00 [a]	19.7 ± 1.00 [a]	11.3 ± 0.28 [a]	1.11 ± 0.31 [a]	2.70 ± 0.20 [a]	4.00 ± 0.34 [a]	1.58 ± 0.07 [a]
Bromelain	21.3 ± 1.30 [b]	10.2 ± 1.99 [b]	8.59 ± 0.87 [b]	0.92 ± 0.89 [b]	1.92 ± 0.29 [b]	2.62 ± 0.44 [b]	1.21 ± 0.01 [b]
Papain	16.2 ± 1.22 [c]	8.63 ± 0.58 [c]	3.25 ± 0.10 [c]	0.81 ± 0.96 [c]	1.37 ± 0.75 [c]	1.44 ± 0.09 [c]	1.12 ± 0.01 [b]

Values are given as mean ± SD from triplicate determinations ($n = 3$). [a-c] Different letters in the same column denote significant differences between hydrolysates ($p < 0.05$).

2.4.2. Oil Binding Capacity (OBC)

Oil binding capacity was determined by physical methods, the density of oil in the bulk, and use the ratio of the mass of bound liquid oil to the solid fat content. In this study, alcalase has the highest OBC value of the three EPHs (Table 3). The oil binding capacity of a protein affects its functional properties and the taste of the end product. Water-holding and fat-binding capacities are functional properties that are closely related to texture, due to the interactions between water, oil, and other components.

2.5. Antioxidant Properties of EPHs

Antioxidants are very important in food preservation, and also play an important role in improving immunity and delaying human aging. In Figure 2a, DPPH free radical scavenging activity was compared for the three enzymes (alcalase, bromelain, and papain) at different times. Bromelain had significantly higher scavenging activity than the other enzymes. Figure 2b shows the DPPH free radical scavenging activity of EPHs at different concentration. The results for alcalase are significantly higher. Alcalase addition ratios of 1.0 and 2.0 resulted in significantly higher scavenging ability. Figure 2c,d used the reducing power assay to compare antioxidant activity at different times and with different amounts of the enzymes. Of the three, the alcalase hydrolysate had the highest reducing power. Figure 2e,f compare the impact hydrolysis time and the amount of enzyme have on ABTS total antioxidant capacity. Of the three, bromelain has a significantly higher ABTS total antioxidant capacity. Antioxidant tests were performed for three different mechanisms. As such, the antioxidant capacities of the EPHs are clearly different, since they were hydrolyzed by different enzymes, with different hydrolysis times, and with differing amounts of enzymes. However, based on all three assays, the alcalase hydrolysate appears to have the greatest total antioxidant capacity.

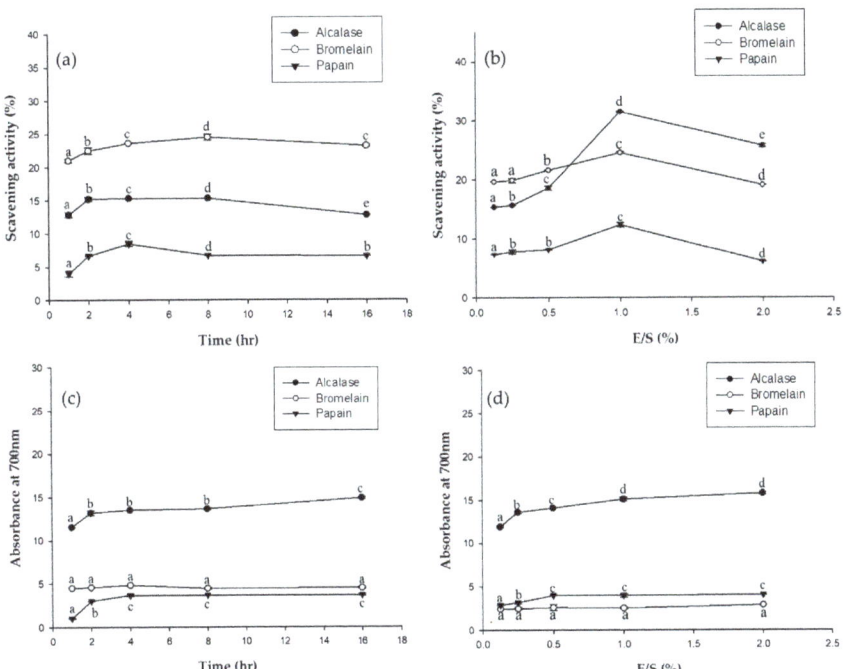

Figure 2. Cont.

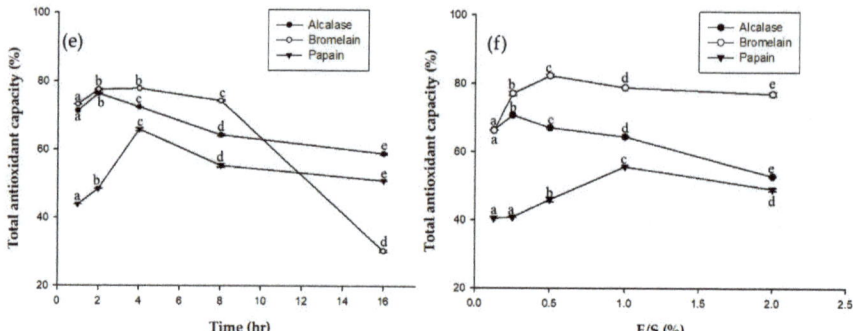

Figure 2. Antioxidant effect of the three EPHs (alcalase, bromelain, and papain) using three antioxidant assays: (**a**) Antioxidant effect at different times based on DPPH free radical scavenging assay; (**b**) Antioxidant effect at different enzyme concentrations based on DPPH free radical scavenging assay; (**c**) Antioxidant effect at different times based on reducing power assay; (**d**) Antioxidant effect at different enzyme concentrations based on reducing power assay; (**e**) Antioxidant effect at different time based on ABTS total antioxidant capacity; (**f**) Antioxidant effect at different enzyme concentrations based on ABTS total antioxidant capacity. Means in the same form with various characters denote significant differences ($p < 0.05$). Data are expressed as mean±SD from triplicate determinations.

2.6. Nitrogen Solubility of EPH and Herbal Eel Extracts

Solubility is often considered the most important physicochemical property in protein hydrolysates, especially when considered in terms of fortifying herbal extracts. Many other functional properties, such as emulsification and foaming, are affected by solubility. Figure 3a shows the nitrogen solubility profile as a function of pH for EPHs produced from different enzymes. All three EPHs had high solubility (more than 80%) in different pH conditions and 85% solubility in alkaline conditions. Among them, EPH produced from alcalase performed the best. At low pH, the charges on the weakly acidic and basic side-chains of the amino acids become less soluble, and resulted in precipitation [18]. The reduced solubility of the alcalase hydrolysate may be due to the greater proportion of small molecule amino acids due to their hydrophilic relationship to the water molecules prior to nitrogen solubility analysis. As shown in Figure 3b, the pH of the herbal eel extract was 7.0 ± 1, and it showed good solubility at higher pH. The beverage matrix will contain other compounds that promote or hinder solubility. Higher solubility gives food and beverage products an appealing appearance and a smooth mouthfeel.

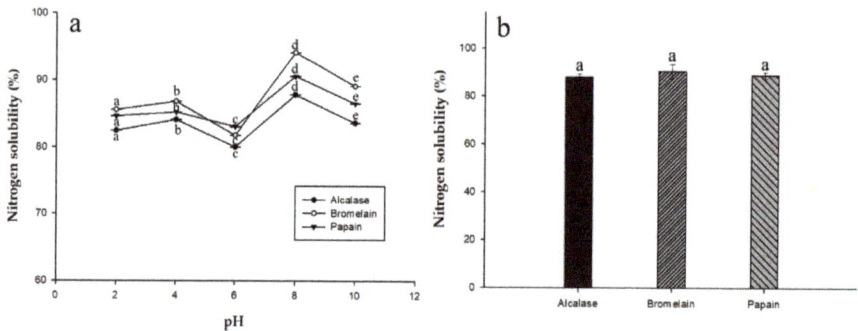

Figure 3. Nitrogen solubility (%) of the three EPHs in (**a**) water at different pH levels (2, 4, 6, 8, and 10) and (**b**) herbal eel extracts. Means in the same form with various characters have significant differences ($p < 0.05$). Data are expressed as mean ± SD from triplicate determinations.

2.7. Color of EPH and Herbal Eel Extracts

In the food industry, food color is the deciding factor in consumers' overall acceptance of a product. The color of eel protein hydrolysates is based on the ingredients, enzymes, and hydrolysis conditions. At a concentration of 15% (w/v), there were significant differences between the EPHs. As summarized in Table 4, L^* represents brightness, while a^* and b^* represent the color space. Both the EPHs and the herbal eel extract were brownish-yellow in color. Specifically, all three EPHs were yellowish in color, although alcalase (3.96 ± 0.04) was darker than both bromelain (6.37 ± 0.43) and papain (8.10 ± 0.18). Based on a^* value, alcalase (0.42±0.03) was preferable to bromelain (0.78 ± 0.02) and papain (0.28 ± 0.01). The same trend was seen with the b^* value, with alcalase (0.74 ± 0.07) again preferable to both bromelain (2.39 ± 0.02) and papain (5.75 ± 0.10). The color differences were mainly due to the ingredient sources and enzymes. Alcalase had the lowest brightness and therefore showed the darkest color following hydrolysis.

Table 4. Color analysis of the three EPHs (alcalase, bromelain, and papain) at concentration of 15% (1.5 g/10 mL).

		Alcalase	Bromelain	Papain
EPH	L^*	3.96 ± 0.04 [c]	6.37 ± 0.43 [b]	8.10 ± 0.18 [a]
	a^*	0.42 ± 0.03 [b]	0.78 ± 0.02 [a]	0.28 ± 0.01 [c]
	b^*	0.74 ± 0.07 [c]	2.39 ± 0.02 [b]	5.75 ± 0.10 [a]
Herbal extract with 15% EPH (1.5 g/10 mL)	L^*	2.17 ± 0.00 [c]	2.59 ± 0.04 [b]	4.52 ± 0.07 [a]
	a^*	0.98 ± 0.00 [a]	0.62 ± 0.00 [c]	0.78 ± 0.11 [b]
	b^*	2.68 ± 0.00 [c]	3.75 ± 0.06 [b]	5.84 ± 0.05 [a]

Values are given as mean ± SD from triplicate determinations (n = 3). [a-c] Different letters in the same column denote significant differences between hydrolysates ($p < 0.05$).

3. Materials and Methods

3.1. Materials

2,2-diphenyl-1-picrylhydrazyl, phthaldialdehyde, and trifluoroacetic acid were purchased from Alfa Aesar (Haverhill, MA, USA). Glycerol and ascorbic acid were purchased from R&D Systems (Minneapolis, MN, USA). Glacial acetic acid was purchased from ThermoFisher Scientific (Waltham, MA, USA). Anhydrous iron (III) chloride, citric acid monohydrate, and sodium tetraborate decahydrate were purchased from Showa Chemical Industry Company (Tokyo, Japan). Methanol, hydrochloric acid, sodium bicarbonate, and anhydrous sodium acetate were purchased from Aencore Chemical Co. (Melbourne, VIC, Australia). Disodium (hydrogen) phosphate was purchased from J.T. Baker (Phillipsburg, NJ, USA). Hydrogen peroxide was purchased from SHOWA (Japan). Alcalase®2.5 L was purchased from Strem Chemicals (Newburyport, MA, USA, CAS number 9014-01-1), and papain (CAS number 9001-73-4) and bromelain (CAS number 9001-00-7) were purchased from ChappionBio (Taiwan). Olive oil was purchased from Carrefour (Taiwan).

3.2. Methods

3.2.1. Anguilla Marmorata Raw Material Preparation

Anguilla marmorata were provided by the Honya eel farm (Dongshi Township, Cjiayi County, Taiwan). The heads of the eels were excised, and the skin and bones were removed. The remaining fish was sterilized at 121 °C for 30 min before storage at −20 °C for subsequent experiments and analysis.

3.2.2. Anguilla Marmorata Meat Hydrolysate Preparation

1 g eel meat was added to 0.1 M phosphate-citrate buffer solution (pH 7.5) at a 1:50 ratio then homogenized, before the addition of different concentrations of the three enzymes, as shown in Table 5.

The enzyme was inactivated by heating the solution to 95 °C for 20 min before it was freeze-dried to obtain EPH. The sample was then stored at −20 °C until further analysis. The scheme of the hydrolysis process of *Anguilla marmorata* is shown in Figure 4.

Table 5. Parameters used to obtain the optimal hydrolysis conditions for eel, using three different enzymes.

Factors	Unit	Symbol	Enzyme		
			Alcalase	Bromelain	Papain
Temperature	°C	T	55	55	45
Time	h	t		1.0, 2.0, 4.0, 8.0, 16.0	
Enzyme concentration	%	E/S		0.125, 0.25, 0.5, 1.0, 2.0	

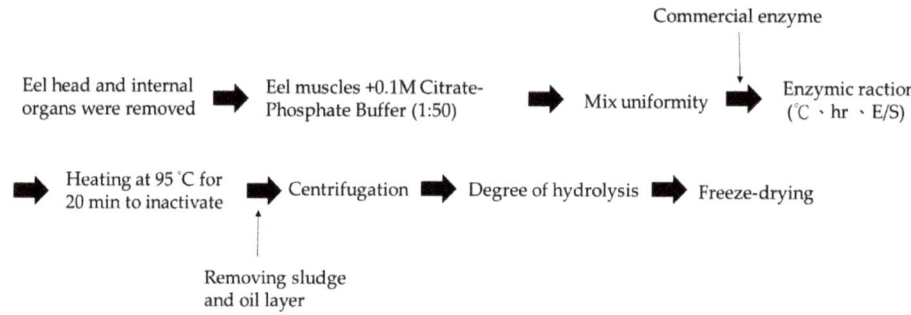

Figure 4. Scheme of the hydrolysis process of *Anguilla marmorata*.

3.2.3. Herbal Eel Extracts

The recipe of herbal eel extracts is to add 23 g of wolfberry, 20 g of red dates, 8 g of *Astragalus propinquus*, 8 g *Angelica sinensis*, 6 g *Cinnamomum cassia*, 6 g Licorice to the medicine bag, and put into the 1000 mL water contains 49 g sugar and 6 g salt for boil 30 min together. After cooling down remove the medicine bag and mix with 500 mL of 15% EPH as herbal eel extract for assay.

3.2.4. Determination of Degree of Hydrolysis (DH)

The degree of hydrolysis (DH), defined as the percentage ratio of the number of peptide bonds cleaved to the total number of peptide bonds available in the substrate, was monitored throughout the reaction. 100 ppm serine (Seine) standard and a sample containing 0.08% protein content were added to 150 µL OPA colorant (1 mL 4% o-phthalaldehyde in ethanol added to 40 mL 1250 ppm sodium dodecyl sulfate in 0.125 M sodium tetraborate buffer solution, further dissolved in 44 mg of dithiothreitol to 50 mL). The reaction was allowed to stand for 2 min, before absorbance was measured by an ELISA reader at 340 nm [19].

DH (%) = [(OD(Sample) − OD(Blank))/(OD(Standard) − OD(Blank)) × 0.9516 meqv/L × 0.1/(X × P) − β]/α/h_{tot} × 100%

where OD (Standard) = standard absorbance, OD (Sample) = sample absorbance, OD(Blank) = blank absorbance, X = g sample, and P = protein % in sample.

The values of the constants α, β, and h_{tot} (total number of peptide bonds per protein equivalent, which is dependent on the amino acid composition of the raw material) for fish protein are estimated to be 1.00, 0.40, and 8.6, respectively.

3.2.5. Molecular Mass Identification

Each EPH was filtered and its molecular mass determined by HPLC (Hitachi, Chromaster, Japan) and BioSep-SEC-S 2000 (600 × 7.8 mm). The mobile phase used was 45% (v/v) acetonitrile with 1% trifluoracetic acid buffer at a rate of 1 mL/min by gradient. 25 µL of the sample was injected and determined at 214 nm. The molecular mass was analyzed by the standard curve with standard proteins maltose (360 Da), apoinin (6500 Da), ribonuclease (13,700 Da), and carbonic anhydrase (29,000 Da).

3.2.6. Analysis of Amino Acid Composition

The sample or standard was dissolved in 0.1 N HCl, before 50 µL was added to 250 µL of 50 mM NaHCO$_3$ (pH 8.1) and 200 µL of DABS-Cl solution (1.3 mg/mL) was dissolved in acetonitrile. The reaction was shaken for 5 min, then placed in a water bath at 70 °C for 12 min. After cooling to room temperature (about 5 min), 300 µL of 70% ethanol was added to stop the reaction and the sample was filtered through a 0.45 µm PVDF filter, using methods previously described [20].

Mobile phase:
(1) Solution A: 25 mM sodium acetate solution (pH 6.5) containing 4% dimethylformamide
(2) Solution B: acetonitrile
Flow rate setting: 1 mL/min
Gradient setting: initial concentration (solution A: solution B = 85: 15)
0~5 min (solution A: solution B = 78.7: 21.3)
5~60 min (solution A: solution B = 70: 30)
60~70 min (solution A: solution B = 30: 70)
70~75 min (solution A: solution B = 30: 70)
75~76 min (solution A: solution B = 30: 70)
76~80 min (solution B: solution C = 15: 85)
80~86 min (solution A: solution B = 85: 15)
86~90 min (solution A: solution B = 85: 15)
Column: Inspire 5 µm C18, 250 × 4.6 mm
Detection wavelength: 436 nm

3.2.7. Functional Properties

Nitrogen Solubility of EPH and Herbal Eel Extracts

Nitrogen solubility was initially determined in distilled water over a range of pH values (2, 4, 6, 8, and 10), as described elsewhere [14]. 200 mg of EPH was added to 30 mL distilled water and mixed, then centrifuged at room temperature (5000 rpm, 20 min) and filtered (ADVANTEC paper no. 1). The nitrogen content of the resulting supernatant was determined using the Kjeldahl method and the nitrogen solubility was calculated as follows:

$$\text{Nitrogen solubility}(\%) = \frac{N_1}{N_0} \times 100$$

where N_1 = supernatant nitrogen concentration and N_0 = sample nitrogen concentration. Solubility analysis was carried out in triplicate. Subsequently, nitrogen solubility of the herbal eel extracts was determined following the method above.

Emulsifying Properties

The emulsion activity index (EAI) and emulsion stability index (ESI) of each EPH was determined with some modifications [14]. EPHs were reconstituted in distilled water (15 mL) at concentrations of 0.5%, 1%, and 2% (w/v). Olive oil (5 mL) was homogenized with the EPH solution for 1 min at room temperature. 50 µL aliquots of the emulsion were taken from the bottom of the conical flask directly after homogenization and again 10 min later, then diluted 100-fold in 0.1% (*w/v*) SDS solution. The new solution was mixed for 10 s and the absorbance measured at 500 nm

$$ESI\ m^2g^{-1} = \frac{2 \times 2.303 \times A_0}{0.25 \times Protein\ weight\ (g)}$$

$$ESI = \frac{\Delta A}{A_0} \times t$$

where A = absorbance, ΔA = (A0 − A10) and t = 10 min (A_0 = Absorbance at 0 min, A_{10} = absorbance at 10 min).

3.2.8. Antioxidant Capacity Analysis

Analysis of Reducing Power

Ascorbic acid was used as a standard. 200 µL of the standard and sample were mixed with 200 µL of 200 mM phosphate buffer solution (pH 6.6) and 200 µL of 1% potassium ferricyanide, then allowed to react in a 50 °C water bath for 20 min. 200 µL of 10% trichloroacetic acid solution was then added, and 200 µL of the supernatant was combined with 200 µL of distilled water and 40 µL of a 0.1% ferric chloride solution dissolved in 10% HCl. After standing for 10 min, the absorbance of the solution was measured using an ELISA plate reader at a wavelength of 700 nm [21].

Analysis of DPPH Free Radical Scavenging Ability

Using ascorbic acid as a standard, 160 µL of the standard and sample were added to 40 µL of 1 mM DPPH dissolved in methanol, then protected from the light for 30 min. The absorbance was measured using a reader at a wavelength of 517 nm [22].

ABTS Radical-Scavenging Activity Assay

The stock solution of 2,2'-azino-bis-(3-ethylbenzothiazoline-6-sulfonic acid (ABTS) radical was produced by reacting 7mM ABTS with 2.45mM $K_2S_2O_8$ (final concentration) and allowing the mixture to stand in dark incubation for 12 h at room temperature condition. The ABTS radical working solution was diluted with phosphate buffered saline (PBS) (pH7.4) at 730 nm to absorbance of 0.70 (±0.02). 1 mL of diluted solution is added with 10 µL of different sample and reacted for a minute, then tested under 730 nm [23].

3.2.9. Color of EPH and Herbal Eel Extracts

Aliquots of EPH were diluted to 15% (*w/v*) and the herbal eel extracts to measured three times by a color difference meter (Nippon Denshoku Industries Co., Ltd., SA2000, Japan). L^* denoted brightness, a^* indicated red green and b^* indicated yellow-blue values [24].

3.2.10. Statistical Analysis

Data are expressed as the mean±standard deviation of three independent replicates. Statistical analysis was performed using the software SPSS statistics v24. All data were analyzed using one-way analysis of variance (ANOVA). Post hoc analysis of group differences was performed using Duncan's new multiple range tests. The criterion for significance was set at $p < 0.05$.

4. Conclusions

This study has shown that eel (*Anguilla marmorata*) are a good source of highly nutritious protein. Through hydrolysis, three commercial enzymes (alcalase, bromelain, and papain) produced low molecular mass peptides that are easy to digest. Among the three, the EAA content in the alcalase-hydrolyzed EPH was the most similar to the daily dose recommended by WHO. EAI was highest with the alcalase EPH (36.8 ± 2.00), while the bromelain and papain EPHs had EAI values of 21.3 ± 1.30 and 16.2 ± 1.22, respectively. Alcalase EPH also had a higher solubility (>80%) in a pH range of 2 to 10, as well as demonstrating the greatest antioxidant capacity. Finally, the alcalase-hydrolyzed EPH had a brighter color, good emulsifying properties, and showed good solubility, making it useful in food processing. In future studies of our research will investigate the fractionation, purification, structure identification of EPH then evaluate human nutrition efficacy by animal model assessment potential application of EPH.

Author Contributions: Conceptualization, L.-T.H., Y.-W.C., and C.-Y.H.; Data curation, I-C.C., J.-X.L., Y.-W.C., and J.-Y.C.; Formal analysis, I-C.C., J.-X.L., and Y.-W.C.; Funding acquisition, Y.-W.C., J.-Y.C., and C.-Y.H.; Investigation, I-C.C. and J.-X.L.; Methodology, I-C.C., J.-X.L., and C.-Y.H.; Project administration, C.-Y.H.; Resources, J.-Y.C.; Software, I-C.C., Y.-W.C., and C.-Y.H.; Supervision, L.-T.H. and J.-Y.C.; Validation, J.-X.L. and L.-T.H.; Visualization, J.-X.L. and L.-T.H.; Writing—original draft, L.-T.H. and C.-Y.H.; Writing—review and editing, J.-Y.C. and C.-Y.H. All authors have read and agreed to the published version of the manuscript.

Funding: This research was funded by the National Kaohsiung University of Science and Technology, grant number NKUST-108D04 and The APC was funded by National Kaohsiung University of Science and Technology, grant number NKUST-108D04.

Conflicts of Interest: The authors declare no conflict of interest. The authors alone are responsible for the content and writing of the manuscript.

References

1. Rees Clayton, E.M.; Specht, E.A.; Welch, D.R.; Berke, A.P. *Addressing Global Protein Demand Through Diversification and Innovation: An Introduction to Plant-Based and Clean Meat*; Melton, L., Shahidi, F., Varelis, P.B.T.-E.o.F.C., Eds.; Academic Press: Oxford, UK, 2019; pp. 209–217. [CrossRef]
2. Galaz, G.A. *Chapter 20—An Overview on the History of Sports Nutrition Beverages*; Bagchi, D., Nair, S., Sen, C.K.B.T.-N.a.E.S.P., Eds.; Academic Press: San Diego, CA, USA, 2013.
3. Kristinsson, H.G. 10—Aquatic food protein hydrolysates. In *Woodhead Publishing Series in Food Science, Technology and Nutrition*; Shahidi, F.B.T.-M.t.V.o.M.B.-P., Ed.; Woodhead Publishing: Sawston, UK, 2007; 229p.
4. Egerton, S.; Culloty, S.; Whooley, J.; Stanton, C.; Ross, R.P. Characterization of protein hydrolysates from blue whiting (*Micromesistius poutassou*) and their application in beverage fortification. *Food Chem.* **2018**, *245*, 698–706. [CrossRef] [PubMed]
5. Aoyama, J.; Watanabe, S.; Miyai, T.; Sasai, S.; Nishida, M.; Tsukamoto, K. The European eel, *Anguilla anguilla* (L.), in Japanese waters. *Dana* **2000**, *12*, 1–5.
6. Schlegel, T. Anguilla japonica. *FAO* **1847**, *79*, 198–205.
7. Heinsbroek, L.T.N. A review of eel culture in Japan and Europe. *Aquac. Res.* **1991**, *22*, 57–72. [CrossRef]
8. Minegishi, Y.; Aoyama, J.; Tsukamoto, K.J.M.E. Multiple population structure of the giant mottled eel, *Anguilla marmorata*. *Mol. Ecol.* **2008**, *17*, 3109–3122. [CrossRef] [PubMed]
9. Jamaluddin, J.; Agustinus, W. Comparative study of mineral content of sidat fish meat (anguilla marmorata quoy gaimard) on yellow eel phase from palu river and lake poso. *J. Islamic Pharm.* **2018**, *3*, 8–14. [CrossRef]
10. Everaert, I.; Stegen, S.; Vanheel, B.; Taes, Y.; Derave, W. Effect of beta-alanine and carnosine supplementation on muscle contractility in mice. *Med. Sci. Sports Exerc.* **2013**. [CrossRef]
11. Khantaphant, S.; Benjakul, S. Comparative study on the proteases from fish pyloric caeca and the use for production of gelatin hydrolysate with antioxidative activity. *Comp. Biochem. Physiol. Part B Biochem. Mol. Biol.* **2008**, *151*, 410–419. [CrossRef]
12. Klompong, V.; Benjakul, S.; Yachai, M.; Visessanguan, W.; Shahidi, F.; Hayes, K.D. Amino Acid Composition and Antioxidative Peptides from Protein Hydrolysates of Yellow Stripe Trevally (*Selaroides leptolepis*). *J. Food Sci.* **2009**, *74*, C126–C133. [CrossRef]
13. Hartmann, R.; Meisel, H. Food-derived peptides with biological activity: From research to food applications. *Curr. Opin. Biotechnol.* **2007**, *18*, 163–169. [CrossRef]
14. Sila, A.; Sayari, N.; Balti, R.; Martinez-Alvarez, O.; Nedjar-Arroume, N.; Moncef, N.; Bougatef, A. Biochemical and antioxidant properties of peptidic fraction of carotenoproteins generated from shrimp by-products by enzymatic hydrolysis. *Food Chem.* **2014**, *148*, 445–452. [CrossRef] [PubMed]

15. Mendis, E.; Rajapakse, N.; Kim, S.-K. Antioxidant properties of a radical-scavenging peptide purified from enzymatically prepared fish skin gelatin hydrolysate. *J. Agric. Food Chem.* **2005**, *53*, 581–587. [CrossRef] [PubMed]
16. Kristinsson, H.G.; Rasco, B.A. Fish protein hydrolysates: Production, biochemical, and functional properties. *Crit. Rev. Food Sci. Nutr.* **2000**, *40*, 43–81. [CrossRef] [PubMed]
17. Nikoo, M.; Benjakul, S.; Ahmadi Gavlighi, H.; Xu, X.; Regenstein, J.M. Hydrolysates from rainbow trout (*Oncorhynchus mykiss*) processing by-products: Properties when added to fish mince with different freeze-thaw cycles. *Food Biosci.* **2019**. [CrossRef]
18. Gbogouri, G.A.; Linder, M.; Fanni, J.; Parmentier, M. Influence of Hydrolysis Degree on the Functional Properties of Salmon Byproducts Hydrolysates. *J. Food Sci.* **2004**, *69*, C615–C622. [CrossRef]
19. Nielsen, P.M.; Petersen, D.; Dambmann, C. Improved method for determining food protein degree of hydrolysis. *J. Food Sci.* **2001**, *66*, 642–646. [CrossRef]
20. Yang, X.; Wang, G.; Gong, X.; Huang, C.; Mao, Q.; Zeng, L.; Zheng, P.; Qin, Y.; Ye, F.; Lian, B.; et al. Effects of chronic stress on intestinal amino acid pathways. *Physiol. Behav.* **2019**. [CrossRef]
21. Wang, T.; Zhou, Y.; Cao, S.; Lu, J.; Zhou, Y. Degradation of sulfanilamide by Fenton-like reaction and optimization using response surface methodology. *Ecotoxicol. Env. Saf.* **2019**. [CrossRef]
22. Wang, C.-Y.; Chen, Y.-W.; Hou, C.-Y. Antioxidant and antibacterial activity of seven predominant terpenoids. *Int. J. Food Prop.* **2019**. [CrossRef]
23. Kim, S.S.; Ahn, C.B.; Moon, S.W.; Je, J.Y. Purification and antioxidant activities of peptides from sea squirt (*Halocynthia roretzi*) protein hydrolysates using pepsin hydrolysis. *Food Biosci.* **2018**. [CrossRef]
24. Zakaria, N.A.; Sarbon, N.M. Physicochemical properties and oxidative stability of fish emulsion sausage as influenced by snakehead (*Channa striata*) protein hydrolysate. *LWT* **2018**. [CrossRef]

© 2020 by the authors. Licensee MDPI, Basel, Switzerland. This article is an open access article distributed under the terms and conditions of the Creative Commons Attribution (CC BY) license (http://creativecommons.org/licenses/by/4.0/).

Article

A Novel Route for Agarooligosaccharide Production with the Neoagarooligosaccharide-Producing β-Agarase as Catalyst

Chengcheng Jiang [1], Zhen Liu [1,*], Jianan Sun [1], Changhu Xue [1,2] and Xiangzhao Mao [1,2,*]

[1] College of Food Science and Engineering, Ocean University of China, Qingdao 266003, China; cz941220@163.com (C.J.); sunjianan@ouc.edu.cn (J.S.); xuech@ouc.edu.cn (C.X.)
[2] Laboratory for Marine Drugs and Bioproducts of Qingdao National Laboratory for Marine Science and Technology, Qingdao 266237, China
* Correspondence: liuzhenyq@ouc.edu.cn (Z.L.); xzhmao@ouc.edu.cn (X.M.); Tel.: +86-532-8203-1360 (Z.L.); +86-532-8203-2660 (X.M.); Fax: +86-532-8203-1789 (X.M.)

Received: 28 January 2020; Accepted: 9 February 2020; Published: 10 February 2020

Abstract: Enzymes are catalysts with high specificity. Different compounds could be produced by different enzymes. In case of agaro-oligosaccharides, agarooligosaccharide (AOS) can be produced by α-agarase through cleaving the α-1,3-glycosidic linkages of agarose, while neoagarooligosaccharide (NAOS) can be produced by β-agarase through cleaving the β-1,4-glycosidic linkages of agarose. However, in this study, we showed that β-agarase could also be used to produce AOSs with high purity and yield. The feasibility of our route was confirmed by agarotriose (A3) and agaropentaose (A5) formation from agaroheptaose (A7) and agarononoses (A9) catalyzed by β-agarase. Agarose was firstly liquesced by citric acid into a mixture of AOSs. The AOSs mixture was further catalyzed by β-agarase. When using the neoagarotetraose-forming β-agarase AgWH50B, agarotriose could be produced with the yield of 48%. When using neoagarotetraose, neoagarohexaose-forming β-agarase DagA, both agarotriose and agaropentaose could be produced with the yield of 14% and 13%, respectively. Our method can be used to produce other value-added agaro-oligosaccharides from agarose by different agarolytic enzymes.

Keywords: agarose; agarase; agarotriose; agaropentaose; expression

1. Introduction

Agarose is a polysaccharide which is widely restricted to the red algae, and it is composed of D-galactose (D-gal) and 3,6-anhydro-L-galactose (L-AHG) with alternate β-1,4- and α-1,3-glycosidic linkages [1,2]. Agaro-oligosaccharides are originated from agarose and can be divided into agarooligosaccharide (AOS) and neoagarooligosaccharide (NAOS); the non-reducing end of AOS is D-gal while the NAOS taking the L-AHG as non-reducing end. In recent years, considerable studies demonstrated that AOSs possess a variety of physiological activities, such as prebiotic properties, antioxidant activity, α-glucosidase inhibition activity, anti-inflammatory effects, hepatoprotective effects and potential protective effects against neurotoxicity [3–6], which suggested that AOSs have the potential to be used for food, cosmetic and pharmaceutical industries.

Agaro-oligosaccharides can be produced by agarase, which is divided into α-agarase (EC 3.2.1.158) and β-agarase (EC 3.2.1.81) based on their cleavage actions [7,8]. α-Agarase could act on the α-1,3-glycosidic linkages of agarose to prepare even-numbered AOSs [9–11], while β-agarase could act on the β-1,4-glycosidic linkages of agarose to prepare even-numbered NAOSs, such as the β-agarase Aga16B, originating from *Saccharophagus degradans* 2-40T and can be used for obtaining neoagarotetraose (NA4) and neoagarohexaose (NA6) [12].

In this study, a novel route was established to produce AOSs by using the NAOS-producing β-agarase. More specifically, the citric acid hydrolysis was used to produce AOSs with high degrees of polymerization (DP). Furthermore, NA4-forming β-agarase AgWH50B from *Agarivorans gilvus* WH0801 (GenBank accession no. CP013021.1) or NA4, NA6-forming β-agarase DagA from *Streptomyces coelicolor* A3(2) was used to hydrolyse the AOSs with high DP for preparation of agarotriose (A3) or A3 and agaropentaose (A5) [13–15], respectively. More importantly, our route is also suitable for preparing odd-numbered AOSs with higher DP, such as A7, A9, and so on.

2. Results and Discussion

2.1. Feasibility of the Chemical-Biological Route

Our research route was shown in Figure 1. Under the mild acid condition, AOSs were predominantly produced because the α-1,3-glycosidic linkages of agarose were preferentially cleaved [8,16]. In previous reports, sulfuric acid, hydrochloric acid, and acetic acid were often used for obtaining AOSs from agarose or agar [17,18]. Citric acid, an organic acid that can be used in food industry [19], also can be used for AOSs production [3,20]. However, these produced AOSs have different DPs. As shown in step II of Figure 1, if β-agarase can act on the β-1,4-glycosidic linkages of the AOSs produced after liquefaction of agarose, an odd-numbered AOS could be cut off from the non-reducing end. Furthermore, because of the specificity of enzyme, the cut-off AOSs would have the same DP. Therefore, plenty of AOSs with special DP would be produced by a two-step chemical-biological reaction. To verify the feasibility of this route, the AgWH50B and DagA were used to act on A7, A9, respectively. Results showed that A3, NA4 and neoagarobiose (NA2) were generated from hydrolysis of A7 or A9 by AgWH50B; A3, A5, NA2 and NA4 were generated from hydrolysis of A7 or A9 by DagA (Figures 2a and 3). According to the composition of A7 reaction products, we could speculate the reaction mechanism of AgWH50B towards A7. Obviously, AgWH50B could cleave β2 to create A3 and NA4, but it could cleave β3 as well in a small degree to create A5 and NA2, and A5 could be further degraded into A3 and NA2. The degradation of A9 has similar results. AgWH50B first cleaves β2 and β4 to create A3, NA6 and A7, NA2 separately. In addition, a slight cleavage towards β3 could create A5 and NA4. In the process of further reaction, A5 was degraded into A3 and NA2 while NA6 was degraded into NA4 and NA2. DagA could help us prepare A3 and A5 simultaneously. There exist two cutting sites of β2 and β3 to the same extent when using A7 as reaction substrate. Therefore, we could get A3, A5, NA2 and NA4 from the products. However, when we change the substrate to A9, DagA shows a preference towards β3 to create A5 and NA4; after that, DagA could cleave β2 in a relative weak way, which could create a small quantity of A3 and NA6. Furtherly, NA6 could be catalyzed to create NA4 and NA2 (Figure 2b). These results demonstrated that there is an easy and feasible method for odd-numbered AOSs' production.

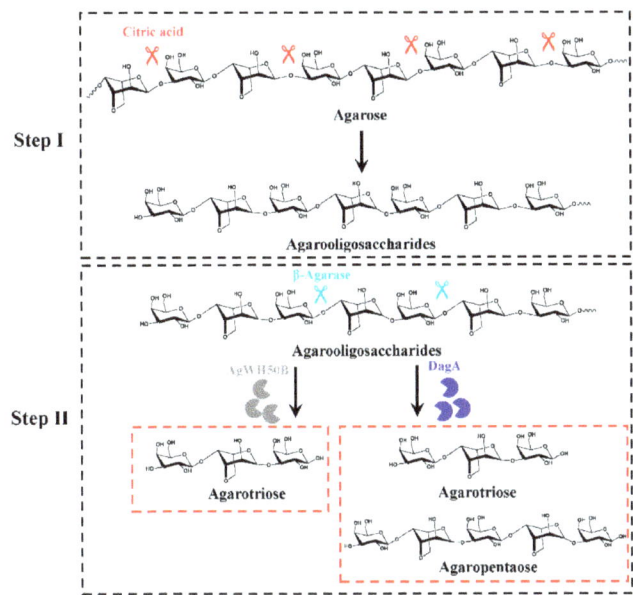

Figure 1. Technical route for preparation of A3 or both A3 and A5 by acid hydrolysis followed by enzymatic hydrolysis.

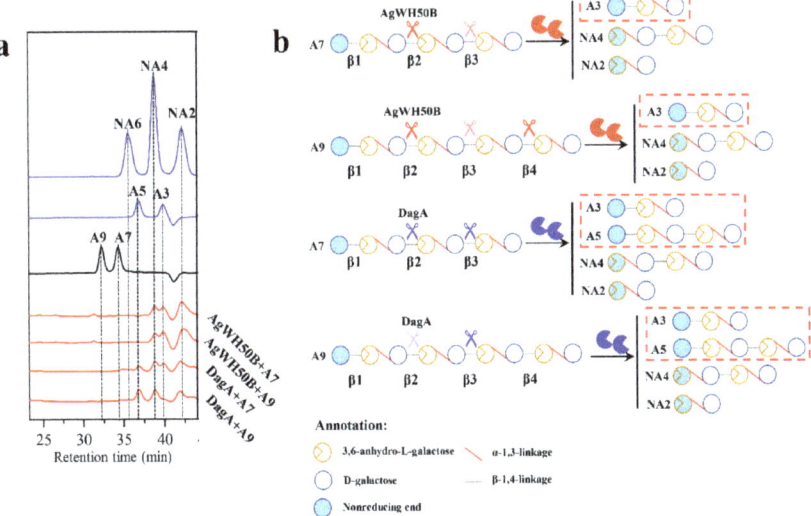

Figure 2. HPLC analysis of enzymatic production of agaroheptaose (A7) and agarononoses (A9) by AgWH50B, DagA (**a**). The cleavage sites of agarase AgWH50B (✂) and DagA (✂) act on the A7 or A9 (**b**). The shades of scissors mean the priority of the hydrolyzed AOSs.

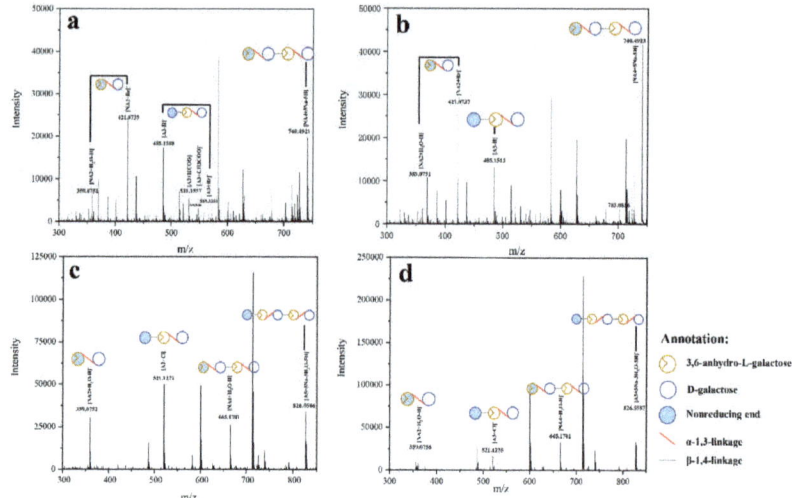

Figure 3. ESI-MS analysis of enzymatic production of agaroheptaose (A7), by AgWH50B (**a**) or DagA (**c**) and agarononoses (A9) by AgWH50B (**b**) or DagA (**d**).

2.2. Preparation of A3, A5 by the Chemical-Biological Route

As shown in Figure 4, in the Step II of the chemical-biological route, the yield of A3 was increased with the enzyme amount raising from 0.244 U/mL to 1.950 U/mL. However, there is no significant increase of the yield of A3 when AgWH50B amount was improved to 2.925 U/mL. Therefore, 1.950 U/mL AgWH50B was chosen as the optimum enzyme amount for the preparation of A3. In addition, the yield of A3 and A5 from combining acidolysis with DagA catalysis was increased with the increase of enzyme amount during 0.0238–0.1900 U/mL DagA, and the yield increase of A3 and A5 was not obvious when using 0.2850 U/mL DagA. Therefore, 0.1900 U/mL DagA was chosen as the optimum enzyme amount for the preparation of A3 and A5. Furthermore, the processes of A3 or both A3 and A5 production were shown in Figure 5b,c, respectively. A3 was gradually produced during the enzymatic hydrolysis by AgWH50B. A volume of 85.50 ± 0.02 g/L A3 was obtained after enzymatic hydrolysis for 24 h in 10 mL reaction mixture, which means 0.855 ± 0.020 g A3 was produced from 1.5 g agarose. Moreover, A3 and A5 were gradually produced during enzymatic hydrolysis by DagA. A volume of 23.30 ± 2.01 g/L A3 and 22.80 ± 0.22 g/L A5 were obtained after enzymatic hydrolysis in 10 mL reaction mixture, which means 0.233 ± 0.020 g A3 and 0.228 ± 0.002 g A5 were produced from 1.5 g agarose. As shown in Figure 5a, there are no AOSs smaller than DP4 were produced after acidolysis by citric acid. After further enzymatic hydrolysis of the neutralized solution, a large amount of A3 was produced by AgWH50B, and A3 and A5 became the major components after catalysis by DagA. However, it showed that there still exist byproducts; therefore, removing these byproducts is important in the following step for the purification of A3 or A5.

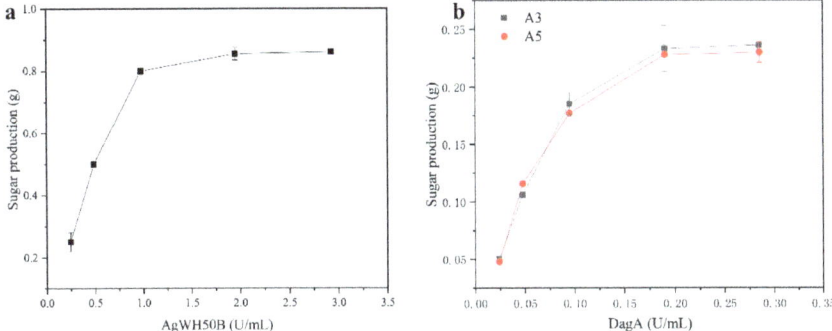

Figure 4. The optimization of enzyme amount for producing A3 (**a**) or both A3 and A5 (**b**). The value of each point was the mean of three experiments (n = 3). Error bars represent one standard deviation of triplicate measurements.

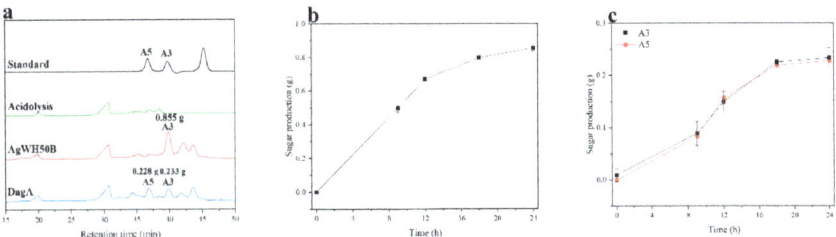

Figure 5. HPLC analysis of AOSs production during the chemical-biological route (**a**). Time-course of the product formation during the processes of enzymatic hydrolysis by AgWH50B (**b**) or DagA (**c**). The value of each point was the mean of three experiments (n = 3). Error bars represent one standard deviation of triplicate measurements.

2.3. Purification of A3, A5

To remove the byproducts of crude A3 solution prepared by AgWH50B or A3 and A5 solution prepared by DagA, the Bio-Gel P2 chromatography was used for purification [2,21]. Purified A3 prepared by AgWH50B and A3, A5 prepared by DagA were collected after purification process. The aliquots containing target oligosaccharides were identified by TLC and then freeze-dried to obtain oligosaccharides powder. In this way, 0.727 g A3 powder could be prepared by AgWH50B and 0.212 g A3, 0.206 g A5 powder could be prepared by DagA in total. This means that the yield of A3 prepared by AgWH50B was 48%, the yields of A3 and A5 prepared by DagA were 14% and 13%, respectively. Purified powders were dissolved with certain volume of ultrapure water, then analyzed by HPLC and ESI-MS. These results suggested that the purified A3 and A5 were acquired successfully (Figure 6). According to the results of HPLC, the amount of A3 prepared by AgWH50B was 0.700 g with a purity of 96.23%, and the amounts of A3 and A5 prepared by DagA were 0.203 g and 0.200 g with the purity of 95.89% and 97.09%, respectively.

AgWH50B is a NA4-forming β-agarase, which can act on the first β-1,4 bond from the non-reducing end of agarose. NA4 would be produced when the substrate was agarose; however, the A3 of the agarose's non-reducing end was retained (Figure 7a). In our study, the agarose was liquefied by citric acid into AOSs because the α-1,3-glycosidic linkages of agarose were preferentially cleaved. Therefore, when AgWH50B acts on these AOSs, A3 would be cut off from the non-reducing end of each AOS with different PD (Figure 7b). This explains why A3 is dominant in the product mixture of the NA4-producing β-agarase AgWH50B, with using the liquefied agarose as substrate. Our work indicates that for an enzyme-catalyzing reaction, the chemical bond acted on by the enzyme is more essential than the usually regarded product of the enzyme.

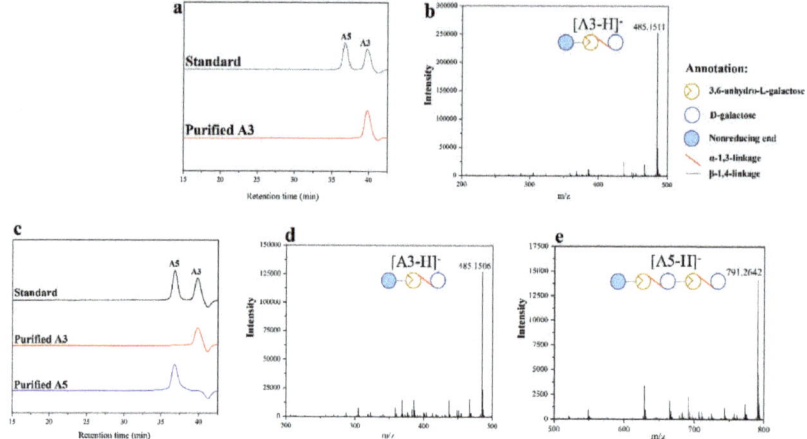

Figure 6. HPLC analysis of the obtained pure A3 produced by AgWH50B (**a**) and pure A3, A5 produced by DagA (**c**). ESI-MS analysis of the results of pure A3 produced by AgWH50B (**b**) and pure A3, A5 produced by DagA (**d**,**e**).

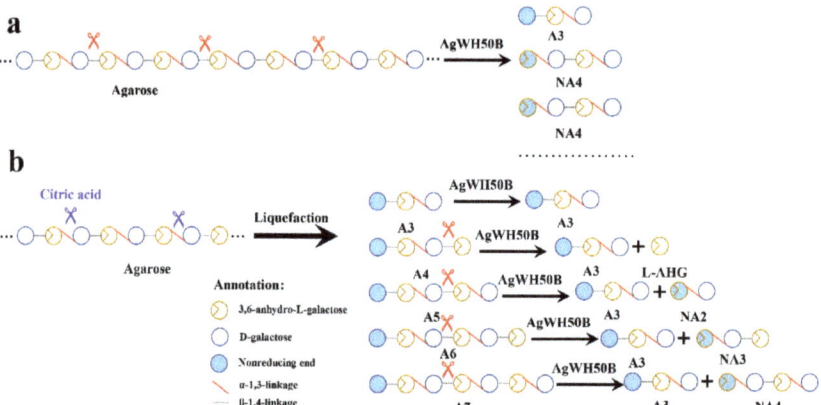

Figure 7. Comparison of simple enzymatic hydrolysis of agarose with our chemical-biological route, to show the reason of A3 accumulation. NA4 production from agarose by enzymatic hydrolysis with NA-forming β-agarase AgWH50B (**a**); A3 production by acid hydrolysis followed by enzymatic hydrolysis with NA-forming β-agarase AgWH50B (**b**).

In a previous study, a 33.2% yield of AOSs from agar acidolysis was reported [20]. Another report indicated that acidolysis with higher acid concentration generated smaller molecular weight products with higher yield. The product DP ranges and reaction yields for different treatments were: DP 2–22, 14.5% (0.1 M HCl); DP 2–16, 28.3% (0.2 M HCl); DP 1–14, 45, 6% (0.4 M HCl); and DP 1–6, 47.0% (0.8 M HCl). According to a recent report, 20.51% AOSs were produced through cellulase hydrolysis from 0.25% agarose; the AOSs produced by acidolysis were an extremely complex mixture [22]. Moreover, in our previous study, the β-agarase AgWH50B and a α-neoagarobiose hydrolase (NABH) AgWH117A were combined for obtaining pure A3 from 0.3% low-gelling-temperature agarose [23,24]. These results indicated the low yield of AOSs production and the complexity of separating products from acidolysis agarose or agar. In addition, preparation of pure AOS through single enzymatic hydrolysis is efficient only when the substrate concentration is very low, which then contributes to a low output. Compared with these studies, for the above problems, our study combined these two ways and then provided a specific and efficient method for preparation of pure AOSs, which is not only suitable for A3 and A5 but also can be used for preparation of odd-numbered AOSs with higher DP, as long as the suitable enzyme was used.

3. Materials and Methods

3.1. Materials

Agarose for A3, A5 production was purchased from TsingKe (Beijing, China). *Escherichia coli* BL21(DE3) was used for expression of AgWH50B (GenBank accession no. KY417136) and DagA (GenBank accession no. CAA29257.1) with plasmid pET21a (+). The citric acid monohydrate for pretreating the agarose was ordered from Chinese Medicine Ltd (Qingdao, China). Yeast extract and tryptone used for medium were purchased from Oxoid (Basingstoke, England). A3, A5, A7, and A9 were obtained from Bz Oligo Biotech (Tsingtao, China).

3.2. Acidolysis

Agarose (15% w/v) was dissolved in 2.5% (w/v) citric acid monohydrate solutions with a total reaction mixture volume of 200 mL. The acidolysis process was performed at 90 °C and 0.5 MPa for 50 min using a Ldzx-50kbs vertical pressure steam sterilizer from Shenan Medical Apparatus Plant (Shanghai, China). After cooling down to room temperature, the reaction mixture was neutralized to pH 7.0 with 20% (w/v) sodium hydroxide.

3.3. Preparation of AgWH50B and DagA

The *E. coli* BL21(DE3) harboring pET21a (+)-*agWH50B* or pET21a (+)-*dagA* were cultured at Luria–Bertani (LB) medium (1% peptone, 0.5% yeast extract, and 1% NaCl), shaking at 37 °C and 200 rpm for 12 h [25]. Then, transfer to fermentation medium ZYP-5052 (1% tryptone, 0.5% yeast extract, 0.2% MgSO4, 1.25% glycerol, 0.125% glucose, and 10% α-galactose), shaking (220 rpm) for 48 h at 20 °C. The cell was collected by centrifugation at 8000× g for 15 min at 4 °C, per 100 mL of fermentation broth was resuspended in 20 mL acid-base buffer (20 mM PBS buffer, pH 7.0) and subsequently disrupted for at a cycle for 30 minutes by Ultrasonic cell wall breaking instrument from Xinzhi Biotechnology Co., Ltd (Ningbo, China); the cycle program was on for 0.3 seconds and off for 0.3 seconds lasting for 30 min. The crude extracts were obtained by centrifugation at 8500× g for 15 min at 4 °C, then freeze-dried to get the crude enzyme powder. Powder (3 g) could be obtained per litre fermentation medium.

3.4. Enzyme Assay

The concentrations of reducing sugars were determined by 3,5-dinitrosalicylic acid (DNS) method [26]. One unit of enzymatic activity (U) was defined as the amount of enzyme that produced 1 μmol of reducing sugar per min by hydrolyzing agarose. The crude enzyme activity of AgWH50B and DagA were 0.195 U/mg and 0.019 U/mg, respectively. The amount of AgWH50B and DagA was optimized for the last reaction to prepare A3, A5. 0.244, 0.488, 0.975, 1.950, 2.925 U/mL AgWH50B, and 0.0238, 0.0475, 0.0950, 0.1900, 0.2850 U/mL DagA were used for the optimization with a 10 mL reaction volume at 40 °C for 12 h. All measurements were performed in triplicate.

3.5. Preparation of A3, A5

Taking the neutralized oligosaccharides solution obtained by acidolysis as the substrate, we prepared A3 and A3, A5 respectively from AgWH50B and DagA enzymolysis with a total volume of 10 mL at 40 °C for 24 h. At the same time, we optimized the enzyme amount of AgWH50B and DagA to realize producing target oligosaccharides with relative lower cost. After reaction, the reaction mixture was boiled for 10 min followed by centrifuging at 8500× g for 20 min to obtain crude sugar solution.

3.6. Purification of A3, A5

In order to obtain pure A3 and A5. The crude sugar solution from the part 2.4 was separated by Bio-Gel P2 chromatography column eluting by hyperpure water with the rate of 1 mL/min. Aliquots were sampled every 5 min; ingredient of every sample was determined by thin layer chromatography (TLC), the plates were eluted in a developing solvent composed of n-butanol/acetic acid/water (2:1:1, v:v:v). The spots were visualized by soaking in an ethanol solution containing 10% (v/v) H_2SO_4, then coloration by heating at 100 °C for 5 min. The purified samples were further analyzed through high performance liquid chromatography (HPLC) with a Superdex 30 increese 10/300 gel filtration column (GE Health, Marlborough, MA, USA) with 5 mM ammonium formate as the mobile phase at a flow rate of 0.4 mL/min; the detector was a refractive index detector (RID) (Agilent, USA) [27]. Furthermore, the molecular weight of each sample was determined using the ESI-MS method on microTOF-Q II equipment (Agilent, USA) in a negative mode with ion spray voltage of 4 kV and source temperature of 350 °C.

4. Conclusions

In summary, this study successfully employed a specific and efficient method for production of pure A3 and A5 with NAOS-producing β-agarase as catalyst. By using citric acid to produce AOSs with high DP and then NA4-forming β-agarase AgWH50B from *A. gilvus* WH0801 and NA4, NA6-forming β-agarase DagA from *S. coelicolor* A3(2) was used to hydrolyze the AOSs with high DP for preparation of A3 or both A3 and A5, respectively. Moreover, it is indicated that our method has potential for the production of odd-numbered AOSs with higher DP.

Author Contributions: Conceptualization, C.J. and X.M.; methodology, C.J. and Z.L.; software, J.S. and C.X.; validation, C.J. and C.X.; data curation, C.J. and J.S.; writing—original draft preparation, C.J.; writing—review and editing, Z.L.; visualization, J.S.; supervision, Z.L.; project administration, X.M.; funding acquisition, X.M. All authors have read and agreed to the published version of the manuscript.

Funding: This work was supported by the National Key R&D Program of China (2018YFC0311200), the Taishan Scholar Project of Shandong Province (tsqn201812020), and Fundamental Research Funds for the Central Universities (201941002).

Conflicts of Interest: The authors declare no conflict of interest.

References

1. Lahaye, M.; Yaphe, W.; Viet, M.T.P.; Rochas, C. 13 C-n.m.r. spectroscopic investigation of methylated and charged agarose oligosaccharides and polysaccharides. *Carbohydr. Res.* **1989**, *190*, 249–265. [CrossRef]
2. Li, J.; Han, F.; Lu, X.; Fu, X.; Ma, C.; Chu, Y.; Yu, W. A simple method of preparing diverse neoagaro-oligosaccharides with β-agarase. *Carbohydr. Res.* **2007**, *342*, 1030–1033. [CrossRef]
3. Chen, H.M.; Zheng, L.; Yan, X.J. The preparation and bioactivity research of agaro-oligosaccharides. *Food Technol. Biotechnol.* **2005**, *43*, 29–36.
4. Jin, M.; Liu, H.; Hou, Y.; Chan, Z.; Di, W.; Li, L.; Zeng, R. Preparation, characterization and alcoholic liver injury protective effects of algal oligosaccharides from *Gracilaria lemaneiformis*. *Food Res. Int.* **2017**, *100*, 186–195. [CrossRef]
5. Ramnani, P.; Chitarrari, R.; Tuohy, K.; Grant, J.; Hotchkiss, S.; Philp, K.; Campbell, R.; Gill, C.; Rowland, I. In vitro fermentation and prebiotic potential of novel low molecular weight polysaccharides derived from agar and alginate seaweeds. *Anaerobe* **2012**, *18*, 1–6. [CrossRef]
6. Ye, Q.; Wang, W.; Hao, C.; Mao, X. Agaropentaose protects SH-SY5Y cells against 6-hydroxydopamine-induced neurotoxicity through modulating NF-κB and p38MAPK signaling pathways. *J. Funct. Foods* **2019**, *57*, 222–232. [CrossRef]

7. Ekborg, N.A.; Taylor, L.E.; Longmire, A.G.; Henrissat, B.; Weiner, R.M.; Hutcheson, S.W. Genomic and proteomic analyses of the agarolytic system expressed by *Saccharophagus degradans* 2-40. *Appl. Environ. Microb.* **2006**, *72*, 3396–3405. [CrossRef]
8. Yun, E.J.; Kim, H.T.; Cho, K.M.; Yu, S.; Kim, S.; Choi, I.G.; Kim, K.H. Pretreatment and saccharification of red macroalgae to produce fermentable sugars. *Bioresour. Technol.* **2016**, *199*, 311–318. [CrossRef]
9. Flament, D.; Barbeyron, T.; Jam, M.; Potin, P.; Czjzek, M.; Kloareg, B.; Michel, G. Alpha-agarases define a new family of glycoside hydrolases, distinct from beta-agarase families. *Appl. Environ. Microb.* **2007**, *73*, 4691–4694. [CrossRef]
10. Potin, P. Purification and characterization of the alpha-agarase from *Alteromonas agarlyticus* (Cataldi) comb. nov. strain GJ1B. *Eur. J. Biochem.* **1993**, *214*, 599–607. [CrossRef]
11. Liu, J.; Liu, Z.; Jiang, C.; Mao, X. Biochemical characterization and substrate degradation mode of a novel alpha-agarase from *Catenovulum agarivorans*. *J. Agric. Food Chem.* **2019**, *67*, 10373–10379. [CrossRef]
12. Kim, J.H.; Yun, E.J.; Seo, N.; Yu, S.; Kim, D.H.; Cho, K.M.; An, H.J.; Kim, J.H.; Choi, I.G.; Kim, K.H. Enzymatic liquefaction of agarose above the sol–gel transition temperature using a thermostable endo-type β-agarase, Aga16B. *Appl. Microbiol. Biot.* **2017**, *101*, 1111–1120. [CrossRef]
13. Liang, Y.; Ma, X.; Zhang, L.; Li, F.; Liu, Z.; Mao, X. Biochemical characterization and substrate degradation mode of a novel exotype beta-agarase from *Agarivorans gilvus* WH0801. *J. Agric. Food Chem.* **2017**, *65*, 7982–7988. [CrossRef]
14. Bibb, M.J.; Jones, G.H.; Joseph, R.; Buttner, M.J.; Ward, J.M. The agarase gene (*dag A*) of *Streptomyces coelicolor* A3(2): Affinity purification and characterization of the cloned gene product. *J. Gen. Appl. Microbiol.* **1987**, *133*, 2089–2096. [CrossRef]
15. Zhang, P.; Rui, J.; Du, Z.; Xue, C.; Li, X.; Mao, X. Complete genome sequence of *Agarivorans gilvus* WH0801(T), an agarase-producing bacterium isolated from seaweed. *J. Biotechnol.* **2016**, *219*, 22–23. [CrossRef]
16. Yang, B.; Yu, G.; Zhao, X.; Jiao, G.; Ren, S.; Chai, W. Mechanism of mild acid hydrolysis of galactan polysaccharides with highly ordered disaccharide repeats leading to a complete series of exclusively odd-numbered oligosaccharides. *FEBS J.* **2009**, *276*, 2125–2137. [CrossRef]
17. Kazlowski, B.; Pan, C.L.; Ko, Y.T. Separation and quantification of neoagaro- and agaro-oligosaccharide products generated from agarose digestion by beta-agarase and HCl in liquid chromatography systems. *Carbohydr. Res.* **2008**, *343*, 2443–2450. [CrossRef]
18. Kazlowski, B.; Pan, C.L.; Ko, Y.T. Monitoring and preparation of neoagaro- and agaro-oligosaccharide products by high performance anion exchange chromatography systems. *Carbohydr. Polym.* **2015**, *122*, 351–358. [CrossRef]
19. Henry, R.W.; Pickard, D.W.; Hughes, P.E. Citric acid and fumaric acid as food additives for early-weaned piglets. *Anim. Sci.* **2010**, *40*, 505–509. [CrossRef]
20. Chen, H.M.; Zheng, L.; Lin, W.; Yan, X.J. Product monitoring and quantitation of oligosaccharides composition in agar hydrolysates by precolumn labeling HPLC. *Talanta* **2004**, *64*, 773–777. [CrossRef]
21. Jang, M.-K.; Lee, D.-G.; Kim, N.; Yu, K.-H.; Jang, H.-J.; Lee, W.S.; Jang, J.H.; Lee, J.Y.; Lee, S.-H. Purification and characterization of neoagarotetraose from hydrolyzed agar. *J. Microbiol. Biotechnol.* **2009**, *19*, 1197–1200. [CrossRef]
22. Kang, O.L.; Ghani, M.; Hassan, O.; Rahmati, S.; Ramli, N. Novel agaro-oligosaccharide production through enzymatic hydrolysis: Physicochemical properties and antioxidant activities. *Food Hydrocoll.* **2014**, *42*, 304–308. [CrossRef]
23. Liu, N.; Yang, M.; Mao, X.; Mu, B.; Wei, D. Molecular cloning and expression of a new alpha-neoagarobiose hydrolase from *Agarivorans gilvus* WH0801 and enzymatic production of 3,6-anhydro-l-galactose. *Biotechnol. Appl. Biochem.* **2016**, *63*, 230–237. [CrossRef]
24. Wang, Q.; Sun, J.; Liu, Z.; Huang, W.; Xue, C.; Mao, X. Coimmobilization of beta-agarase and alpha-neoagarobiose hydrolase for enhancing the production of 3,6-anhydro-l-galactose. *J. Agric. Food Chem.* **2018**, *66*, 7087–7095. [CrossRef]
25. Bertani, G. Studies on lysogenesis.1. The mode of phage liberation by lysogenic *Escherichia-coli*. *J. Bacteriol.* **1951**, *62*, 293–300. [CrossRef]

26. Miller, G.L. Use of dinitrosalicylic acid reagent for determination of reducing sugar. *Anal. Biochem.* **1959**, *31*, 426–428. [CrossRef]
27. Zhu, B.; Tan, H.; Qin, Y.; Xu, Q.; Du, Y.; Yin, H. Characterization of a new endo-type alginate lyase from *Vibrio* sp. W13. *Int. J. Biol. Macromol.* **2015**, *75*, 330–337. [CrossRef]

© 2020 by the authors. Licensee MDPI, Basel, Switzerland. This article is an open access article distributed under the terms and conditions of the Creative Commons Attribution (CC BY) license (http://creativecommons.org/licenses/by/4.0/).

Article

Biocatalytic Synthesis of Calycosin-7-O-β-D-Glucoside with Uridine Diphosphate–Glucose Regeneration System

Yumei Hu [1,2,†], **Jian Min** [2,†], **Yingying Qu** [2], **Xiao Zhang** [2], **Juankun Zhang** [1,*], **Xuejing Yu** [2,*] **and Longhai Dai** [2,*]

[1] Tianjin Key Laboratory of Industrial Microbiology, College of Biotechnology, Tianjin University of Science and Technology, Tianjin 300457, China; hu_ym@tib.cas.cn
[2] State Key Laboratory of Biocatalysis and Enzyme Engineering, Hubei Collaborative Innovation Center for Green Transformation of Bio-Resources, Hubei Key Laboratory of Industrial Biotechnology, School of Life Sciences, Hubei University, Wuhan 430062, China; Jianmin@hubu.edu.cn (J.M.); chiuyy@126.com (Y.Q.); zhangxiao_HUBU@126.com (X.Z.)
* Correspondence: dailonghai@hubu.edu.cn (L.D.); zhangjk@tust.edu.cn (J.Z.); xjy51@hotmail.com (X.Y.)
† These authors contributed equally to this work.

Received: 3 February 2020; Accepted: 16 February 2020; Published: 20 February 2020

Abstract: Calycosin-7-O-β-D-glucoside (Cy7G) is one of the principal components of *Radix astragali*. This isoflavonoid glucoside is regarded as an indicator to assess the quality of *R. astragali* and exhibits diverse pharmacological activities. In this study, uridine diphosphate-dependent glucosyltransferase (UGT) UGT88E18 was isolated from *Glycine max* and expressed in *Escherichia coli*. Recombinant UGT88E18 could selectively and effectively glucosylate the C7 hydroxyl group of calycosin to synthesize Cy7G. a one-pot reaction by coupling UGT88E18 to sucrose synthase (SuSy) from *G. max* was developed. The UGT88E18–SuSy cascade reaction could recycle the costly uridine diphosphate glucose (UDPG) from cheap sucrose and catalytic amounts of uridine diphosphate (UDP). The important factors for UGT88E18–SuSy cascade reaction, including UGT88E18/SuSy ratios, different temperatures, and pH values, different concentrations of dimethyl sulfoxide (DMSO), UDP, sucrose, and calycosin, were optimized. We produced 10.5 g L^{-1} Cy7G in the optimal reaction conditions by the stepwise addition of calycosin. The molar conversion of calycosin was 97.5%, with a space–time yield of 747 mg L^{-1} h^{-1} and a UDPG recycle of 78 times. The present study provides a new avenue for the efficient and cost-effective semisynthesis of Cy7G and other valuable isoflavonoid glucosides by UGT–SuSy cascade reaction.

Keywords: calycosin; calycosin-7-O-β-D-glucoside; glucosyltransferase; sucrose synthase; UDP-glucose recycle; UGT–SuSy cascade reaction

1. Introduction

Flavonoids are a group of structurally diverse, plant-derived polyphenols widely present in fruits, vegetables, and beverages [1,2]. These compounds play an important role in human diet related to health [3]. More than 10,000 flavonoids have been isolated from the plant kingdom, which can be categorized into flavonols, flavones, isoflavonoids, flavanones, chalcone, catechins, and anthocyanins according to the variability in their chemical structures [4,5]. Glucosylation mediated by uridine diphosphate-dependent glucosyltransferase (UGT) is one of the most common modifications of flavonoids, which presents immense structural diversity after this process [6]. The attachment of glucosyl moiety to flavonoids not only increases their solubility and stability, but also improves their bioavailability and pharmacokinetic properties [7,8].

Calycosin-7-O-β-D-glucoside (Cy7G) is an isoflavonoid glucoside (Figure 1). This compound is one of the principal components of *Radix astragali*—a famous medicinal and edible herbal plant recorded in European, Japanese, and Chinese pharmacopoeias [9]. The contents of Cy7G are generally regarded as an indicator to assess the quality of R. astragali and R. astragali-based drugs or foods [9,10]. Cy7G exhibits various health-beneficial bioactivities, including antioxidant, anti-inflammatory, antiapoptotic, and bone-regeneration-enhancing activities [11,12]. Commercially available Cy7G is mainly extracted from plants; however, the natural scarcity and high cost of production are major obstacles that limit its applications in pharmaceutical and healthcare industries [13].

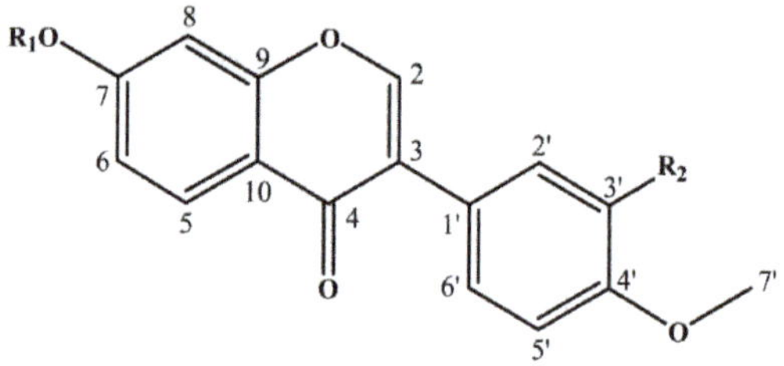

Compounds	R1	R2
Calycosin (1)	-H	-OH
Cy7G (1a)	-Glc	-OH
Formononetin (2)	H	H
Ononin (2a)	-Glc	H

Figure 1. Chemical structures of calycosin, formononetin, and glucosylated products synthesized by UGT88E18.

Chemical glucosylation is notoriously complicated because of various disadvantages such as low efficiency, poor stereospecificity, a series of protection and deprotection steps, and environmental pollution [14]. In this respect, enzymatic glucosylation using regio- and stereoselective UGTs can alleviate these disadvantages [15,16]. Extensive studies indicate that sucrose synthase (SuSy) is a versatile biocatalyst that can synthesize costly uridine diphosphate glucose (UDPG) from abundant and cheap sucrose and catalytic amounts of uridine diphosphate (UDP) [17,18]. Furthermore, SuSy can be coupled with a UGT in a one-pot reaction that could recycle UDPG from UDP and sucrose, thereby providing a cost-effective approach for the glucosylation of natural products [19,20].

A considerable number of UGTs involved in the glucosylation of flavonoids have been isolated and functionally characterized from plants [21]. These UGTs typically tolerate a broad range of aglycones with high regiospecificity in vitro [22,23]. In the present study, the regioselective and effective glucosylation of calycosin was achieved using UGT88E18 isolated from *Glycine max*. In addition, a UGT88E18–SuSy cascade reaction was developed to synthesize Cy7G using cheap sucrose as the expedient glucosyl donor.

2. Results and Discussion

2.1. Regioselective Glucosylation of Calycosin and Formononetin by UGT88E18

UGT88E18 from *G. max* was specific for isoflavones, including genistein and daidzein [23]. Thus, N-terminal His$_6$-tagged UGT88E18 was heterologously expressed in *Escherichia coli* BL21 (DE3), and purified by one-step nickel chelate affinity chromatography. SDS-PAGE analysis showed a clear band at around 72.2 kDa that corresponded to the calculated molecular weight of recombinant UGT88E18 (Figure S1). The purified UGT88E18 was applied to the in vitro reaction with calycosin (Compound 1) or formononetin (Compound 2, another important isoflavonoid present in *R. astragali*) as the acceptor and UDPG as the glucosyl donor, respectively (Figure 1). High-performance liquid chromatography (HPLC) analysis of the reactants revealed that over 97% of calycosin and formononetin could be transformed into their corresponding glucosides (Figure 2A), suggesting that UGT88E18 could effectively glucosylate calycosin and formononetin. These results were in agreement with a previous study that reported that UGT88E18 is an isoflavone UGT [23]. Only one new product with higher polarity was produced using calycosin or formononetin as the aglycone. The retention time (RT) of new Products **1a** and **2a** was 9.8 and 13.1 min, respectively. HPLC electrospray-ionization mass-spectrometry (HPLC-ESI-MS) analysis of Products **1a** ($[2M+H]^+$ m/z$^+$ ~893.2509]$^+$, $[M+H]^+$ m/z$^+$ ~447.1304, and $[M-Glc+H]^+$ m/z$^+$ ~285.0747) and **2a** ($[2M+H]^+$ m/z$^+$ ~861.2614]$^+$, $[M+H]^+$ m/z$^+$ ~431.1365, and $[M-Glc+H]^+$ m/z$^+$ ~269.0808) indicated the addition of one glucosyl moiety from calycosin ($C_{16}H_{12}O_5$, $[M+H]^+$ m/z$^+$ ~285.0743) and formononetin ($C_{16}H_{12}O_4$, $[M+H]^+$ m/z$^+$ ~269.0809), respectively (Figure 2B).

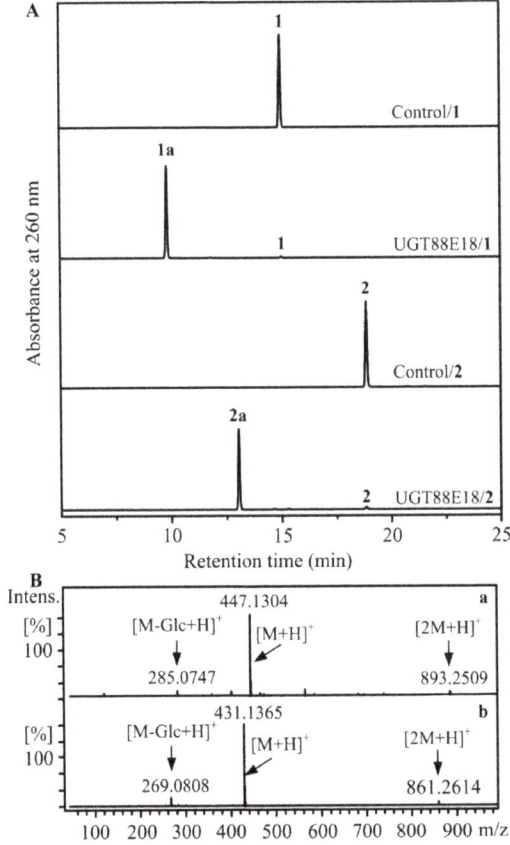

Figure 2. HPLC electrospray-ionization mass-spectrometry (HPLC-ESI-MS) analysis of glucosylated products of calycosin and formononetin. (**A**) HPLC chromatograms of calycosin, formononetin, and glucosylated products synthesized by UGT88E18. (**B**) MS spectra for Products (**a**) **1** and (**b**) **2**.

Products **1a** and **2a** were purified by preparative HPLC system, and their chemical structures were determined by 1D NMR (^{1}H NMR and ^{13}C NMR) and 2D NMR spectra (homonuclear correlation spectroscopy (COSY), heteronuclear multiple-bond correlation spectroscopy (HMBC), and heteronuclear singular quantum correlation (HSQC)) (Figures S2–S11). For Product **1a** (Table S1), glucosyl proton signals (δ_H 3.1–5.2 ppm) in the ^{1}H NMR spectra and carbon signal (δ_C 60–101 ppm) in the ^{13}C NMR spectra suggested the presence of a glucosyl moiety [24]. The HMBC correlations of the glucosyl moiety anomeric signal H1' (δ_H 5.11 ppm, d, J = 7.3 Hz) with C7 (δ_C 161.32 ppm) indicated the attachment of a glucosyl moiety to the C7 hydroxyl group of calycosin. The ^{1}H- and ^{13}C-NMR spectra of Product **2a** were highly similar to those of Product **1a** [24]. The additional signals of glucosyl proton in the δ_H 3.1–5.2 ppm region and anomeric proton H1' at δ_H 5.12 ppm (d, J=7.4 Hz) in the ^{1}H NMR spectra indicated the presence of a glucosyl moiety. In ^{13}C-NMR analysis, the observation of the anomeric carbon signal C1' (δ_C 100.47 ppm) and five other new carbon signals (δ_C 61.12, 70.11, 73.61, 76.96, and 77.69 ppm) further justified the presence of a glucosyl moiety. The long-range correlations between glucosyl anomeric signal H1' (δ_H 5.12 ppm, d, J = 7.4 Hz) and C7 (δ_H 161.93 ppm) suggested that the glucosyl moiety was attached to the C7 hydroxyl group of formononetin. In addition, the large anomeric proton-coupling constants (J = 7.3–7.4 Hz) of Products **1** and **2** suggested the formation of the β-anomers, which agreed with the inverting mechanism for UGTs [25]. The present results revealed that UGT88E18 could selectively glucosylate the C7 hydroxyl group of calycosin and formononetin to synthesize Cy7G and formononetin-7-O-β-D-glucoside (ononin), respectively.

2.2. Kinetic Analysis of UGT88E18 Toward Formononetin and Calycosin

The K_m values of UGT88E18 toward calycosin and formononetin were 18.67 and 30.64 µM, respectively (Table 1 and Figure S12). These results suggested that UGT88E18 had a higher substrate preference for calycosin than for formononetin. In addition, these K_m values were comparable to those of other UGTs involved in glycosylation of isoflavonoids [22,26]. The kcat values of UGT88E18 toward calycosin and formononetin were 7.67 and 5.39 s^{-1}, respectively. The turnover rates of UGT88E18 toward formononetin and calycosin were considerably high compared with other UGTs involved in flavonoid biosynthesis [1,22]. Thus, the catalytic efficiency of UGT88E18 toward calycosin and formononetin was 4.11 × 10^5 and 1.76 × 10^5 M^{-1} s^{-1}, respectively. Kinetic analysis of UGT88E18 suggested that it was a powerful catalyst for the in vitro glucosylation of calycosin and formononetin.

Table 1. Kinetic parameters of UGT88E18 toward calycosin and formononetin.

Substrate	K_m (µM)	kcat (s^{-1})	kcat/ K_m (s^{-1} M^{-1})
Calycosin	18.67 ± 1.36	7.67 ± 0.13	4.11 × 10^5
Formononetin	30.64 ± 2.30	5.39 ± 0.11	1.76 × 10^5

2.3. Optimization of UGT88E18–SuSy Cascade Reaction for Cy7G Synthesis

The UGT88E18-catalyzed synthesis of Cy7G requires the costly UDPG as the glucosyl donor, which is one of the major bottlenecks for in vitro enzymatic reactions [27]. Thus, SuSy from G. max was also heterologously expressed in E. coli and purified to homogeneity (Figure S1). A one-pot reaction was developed by coupling UGT88E18 to SuSy to regenerate costly UDPG from UDP and cheap sucrose. As expected, calycosin could be transformed into Cy7G by a UGT88E18–SuSy cascade reaction, similar to in vitro reactions using UDPG as the glucosyl donor (Figure S13), suggesting that the UGT88E18–SuSy cascade reaction was applicable to semisynthesize Cy7G from calycosin.

The effects of different ratios of UGT88E18 and SuSy on Cy7G production were primarily determined (Table 2). Only 1.30 mM Cy7G was produced in the reaction mixtures containing 50 mU mL^{-1} UGT88E18. The content of Cy7G doubled when the amount of UGT88E18 was raised from 50 to 100 mU mL^{-1}. The concentration of Cy7G continued to increase from 3.35 to 3.78 mM when the amount of UGT88E18 was increased from 150 to 200 mU mL^{-1}. Thus, 200 mU mL^{-1} UGT88E18 was selected for the following UGT88E18–SuSy cascade reactions. The titer of Cy7G increased by 106% when the amount of SuSy was increased from 50 to 150 mU mL^{-1}, suggesting that glucosylation catalyzed by UGT88E18 was the rate-limiting step in the UGT88E18–SuSy cascade reaction system. Approximately 95% of calycosin (3.80 mM) could be transformed into Cy7G in the presence of 150 and 200 mU mL^{-1} SuSy. Thus, 200 mU mL^{-1} UGT88E18 and 150 mU mL^{-1} SuSy were selected as the optimal ratios for the subsequent optimization assays.

Table 2. Effect of different enzyme ratios on product yields of calycosin-7-O-β-D-glucoside (Cy7G).

Entry	UGT88E18 (mU mL^{-1})	SuSy (mU mL^{-1})	Cy7G (mM)
1	50	150	1.30
2	100	150	2.56
3	150	150	3.35
4	200	150	3.78
5	200	50	1.82
6	200	100	3.31
7	200	150	3.75
8	200	200	3.80

Different temperature and pH values and different concentrations of dimethyl sulfoxide (DMSO), UDP, sucrose, and calycosin were further optimized (Figure 3). Cy7G production in UGT88E18–SuSy cascade reaction was investigated in different temperature ranges (25–45 °C). Only 1.25 mM Cy7G was produced at 25 °C. The concentration of Cy7G increased to 3.28 mM at 30°C. Over 93% calycosin could be transformed into Cy7G (3.72–3.84 mM) at 35 and 40 °C. However, the concentration of Cy7G decreased to 3.09 mM at 45°C. Considering a lower temperature could be more conducive to maintaining the thermostability of plant-derived UGT88E18 and SuSy, 35 °C was selected as the optimal temperature. The pH values were in the range of 6.5–8.0 and were also optimized. The UGT88E18–SuSy cascade reaction performed well in the selected pH values. The highest Cy7G titers (~3.75 mM) were obtained at pH 7.0 and 7.5. Thus, pH 7.5 in the Tris-HCl buffer was selected in the following experiments. The poor solubility of aglycones could inhibit the in vitro UGT–SuSy cascade reactions. DMSO is a versatile organic solvent to promote the dissolution of hydrophilic aglycones [28]. The Cy7G titer increased from 2.60 to 4.70 mM when the DMSO concentration increased to 10% (v/v). The UGT–SuSy cascade system could tolerate 10%–15% DMSO when the conversion of calycosin reached over 98%. However, Cy7G production (2.32 mM) decreased rapidly when the concentration of DMSO increased to 20%. Considering the concentration of DMSO would be further increased in the subsequent stepwise addition of calycosin (stocked in DMSO), 10% DMSO (v/v) was selected as the optimal condition. UDP is a costly cofactor in the UGT–SuSy cascade system. UDP is involved in the formation of UDPG in the SuSy-catalyzed hydrolysis of sucrose. In addition, as a byproduct of the UGT-catalyzed reaction, a high concentration of UDP could inhibit UGT activity [29]. Only 1.30 mM Cy7G was formed in the presence of 0.1 mM UDP. The production of Cy7G increased to 3.74 mM when the concentration of UDP exceeded 0.3 mM. Thus, 0.3 mM UDP was adequate for the UGT88E18–SuSy cascade reaction. Cy7G titers increased with increasing concentration of sucrose. Maximal Cy7G production reached 3.74 when the concentration of sucrose increased to 400–500 mM. Then, 2–3 mM calycosin could be completely transformed into Cy7G by the UGT88E18–SuSy cascade reaction. The maximal titer of Cy7G reached 3.75 mM in the presence of 4 mM calycosin. Collectively, 35 °C, pH 7.5, 10% DMSO (v/v), 0.3 mM UDP, 400 mM sucrose, and 4 mM calycosin were selected as the optimal parameters for the subsequent fed-batch reaction.

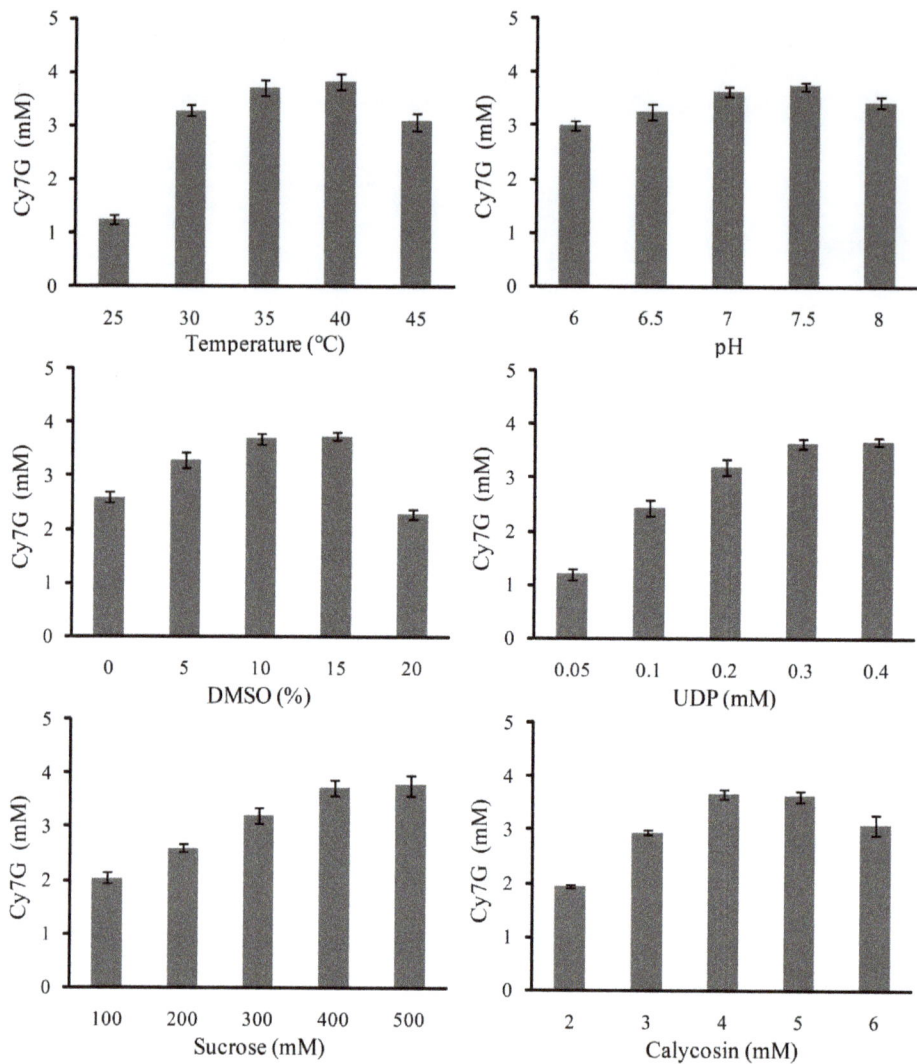

Figure 3. Optimization of reaction conditions for UGT88E18–SuSy cascade reactions.

2.4. Fed-Batch Synthesis of Cy7G by UGT88E18–SuSy Cascade Reaction

UGT88E18–SuSy cascade reaction by the periodical addition of calycosin was conducted to increase the final titer of Cy7G and to avoid the inhibition of a high concentration of calycosin to the reaction system. As shown in Figure 4, 7.80 mM Cy7G was produced in the first hour with a space–time yield (STY) of 3487 mg L^{-1} h^{-1}. Specific productivity gradually decreased with the stepwise addition of calycosin. STYs were 849 mg L^{-1} h^{-1} over a reaction time of 1–5 h, 581 mg L^{-1} h^{-1} over a reaction time of 5–8 h, and 305 mg L^{-1} h^{-1} over a reaction time of 8–14 h. The main cause of the decrease in STY might be the product inhibition by a high concentration of fructose and Cy7G in the system [30–33]. Eventually, 10.46 g L^{-1} Cy7G was produced with the conversion rate of calycosin being 97.5%. The overall STY of Cy7G was 747 mg L^{-1} h^{-1}, with a maximal number of UDPG regeneration cycles of 78 times (= 23.4/0.3). These results provide a new approach for the efficient and cost-effective semisynthesis of Cy7G and other valuable isoflavonoid glucosides by UGT–SuSy cascade reaction.

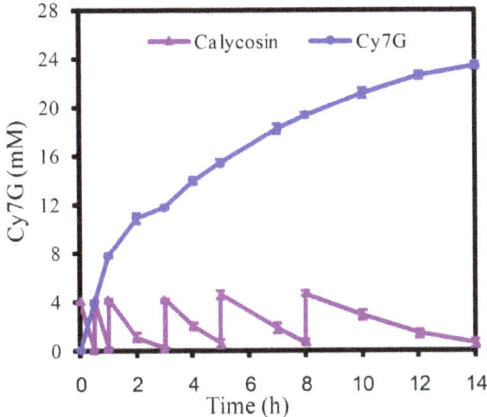

Figure 4. Fed-batch synthesis of Cy7G by UGT88E18–SuSy cascade reaction. Stepwise addition of 100 μL of fresh calycosin, stocked in dimethyl sulfoxide (DMSO, 400 mM), to reactants at 0.5, 1, 3, 5, and 8 h.

3. Materials and Methods

3.1. Chemicals and Reagents

Calycosin was provided by Chengdu Biopurify Phytochemicals Ltd. (Chengdu, China). Formononetin, UDP, and DMSO were purchased from J&K Scientific Ltd. (Beijing, China). UDPG, deuterium dimethyl sulfoxide (DMSO-d_6), and chromatography-grade acetonitrile were obtained from Sigma-Aldrich (St. Louis, MO, USA). All other chemicals and reagents were of the highest chemical grade.

3.2. Expression and Purification of Recombinant Proteins

UGT88E18 (GenBank accession number, AB904893) and SuSy (GenBank accession number, NP_001237525.1) from *G. max* were codon-optimized and synthesized by Shanghai Sangon Biotechnology Co., Ltd. (Shanghai, China). UGT88E18 and SuSy were subcloned into a pET32a vector to construct recombinant vectors pET32-UGT88E18 and pET32-SuSy, respectively. Expression vectors pET32-UGT88E18 and pET32-SuSy were transformed into *E. coli* BL21 (DE3) cells for heterologous expression, respectively. The recombinant *E. coli* cells were cultured in a Luria–Bertani medium containing 100 mg L^{-1} ampicillin at 37 °C and 200 rpm. After optical density (OD$_{600}$) reached 0.6–0.8, the recombinant *E. coli* cells were induced with 0.1 mM isopropyl-β-D-thiogalactopyranoside for 20 h at 16 °C and 200 rpm. Subsequently, induced cells were harvested by centrifugation at 5000 g for 10 min at 4 °C and resuspended in lysis buffer (50 mM Tris-HCl, 25 mM imidazole, and 150 mM NaCl, pH 7.5). After disruption by high-press homogenization, the resultant cell debris was removed by centrifugation at 17,000 g for 60 min at 4 °C. The supernatant was applied to an AKTA Purifier system equipped with a nickel nitrilotriacetic acid agarose affinity column (GE Healthcare, Piscataway, NJ, USA). The purity of the purified proteins was analyzed by 4%–12% SurePAGE Bis-Tris gels (GenScript, Nanjing, China). Protein concentration was determined by bicinchoninic acid assay using bovine serum albumin as a reference according to the manufacturer's instructions (Solarbio, Beijing, China). Purified UGT88E18 and SuSy were kept in 50 mM Tis-HCl (pH 7.5) at −80 °C.

3.3. Enzyme Activity

The reaction mixtures (3 μg UGT88E18, 1 mM calycosin or formononetin, 5 mM UDPG, 10% DMSO (v/v), 50 mM Tris-HCl buffer, pH 7.5) were carried out in a total reaction volume of 500 μL.

The reactions were performed at 35 °C for 1 h, and were stopped by adding twofold volumes of methanol. After centrifugation at 13,000 g for 10 min, the reactants were filtered with a 0.22 µm nylon filter prior to analysis by a Shimadzu LC-20AD HPLC system coupled with a photodiode array detector (Shimadzu Corporation, Kyoto, Japan). The reverse-phase InertSustain C18 column (4.6 × 250 mm, 5 µm; GL Science, Kyoto, Japan) was eluted with double-distilled water and chromatography-grade acetonitrile using a gradient of 15%–85% chromatography-grade acetonitrile in 0–25 min. HPLC-ESI-MS was performed in the same positive-ion mode as in our previous study [24]. One unit of enzyme activity was defined as the amount of UGT88E18 that generated 1 µmole of Cy7G per min under the assay conditions. Enzyme activity of SuSy was measured by the bicinchoninic acid method as described previously [31].

3.4. Kinetic Parameters of UGT88E18

Kinetic analysis of UGT88E18 was carried out in 300 µL volumes containing 50 mM Tris-HCl (pH 7.5), 5 mM of UDPG, 10% DMSO (v/v), and 10–400 µM calycosin or formononetin. The reaction mixtures were precultured at 35 °C for 5 min and initiated by adding 0.5 µg UGT88E18. Subsequently, reactants were incubated at 35 °C for 10 min and quenched by an equal volume of methanol. Finally, the aliquots were centrifuged at 13,000 g for 10 min, filtered with a 0.22 µm nylon filter, and analyzed by HPLC, as described above. Data were obtained from three parallel experiments. Kinetic parameters were determined by nonlinear regression analysis using GraphPad Prism 5.0 software (GraphPad Software, San Diego, CA, USA).

3.5. Optimization of UGT88E18–SuSy Cascade Reaction

UGT88E18–SuSy cascade reactions (4 mM calycosin, 0.5 mM UDP, 10% DMSO (v/v), 500 mM sucrose, 50 mM Tris-HCl (pH 7.5), 200 mU mL^{-1} UGT88E18, and 150 mU mL^{-1} SuSy) were performed at 35 °C for 0.5 h. The optimal ratio of UGT88E18 and SuSy was determined by adding different ratios of UGT88E18 and SuSy (50–200 mU mL^{-1}). Subsequently, different temperature values (25, 30, 35, 40, and 45 °C) and pH values (NaH$_2$PO$_4$–Na$_2$HPO$_4$ buffer, 6.0, 6.5, and 7.0; Tris-HCl buffer, 7.5 and 8.0) were further determined. Finally, the effects of different concentrations of DMSO (0–20%, v/v), UDP (0.05–0.4 mM), sucrose (100–500 mM), and calycosin (2–6 mM) on the UGT88E18–SuSy cascade reaction were optimized.

3.6. Fed-Batch Synthesis of Cy7G

UGT88E18–SuSy cascade reaction (10 mL) was conducted in optimal conditions (50 mM Tris-HCl (pH 7.5), 4 mM calycosin, 0.3 mM UDP, 10% DMSO (v/v), 400 mM sucrose, 200 mU mL^{-1} UGT88E18, and 150 mU mL^{-1} SuSy) at 35 °C and 150 rpm. Samples (100 µL) were collected at different time intervals (0.5, 1, 2, 3, 4, 5, 7, 8, 10, 12, and 14 h). Samples were quenched and diluted using 2.9 mL of methanol. After filtration with 0.22 µm nylon filters, the reaction mixtures were analyzed by HPLC, as described above. In addition, 100 µL of fresh calycosin stocked in DMSO (400 mM) was stepwise added to the reaction mixtures at 0.5, 1, 3, 5, and 8 h.

3.7. Purification and Structural Elucidation of the Glucosylated Products

A scaled-up UGT88E18–SuSy cascade reaction (30 mL) was performed with calycosin or formononetin as the substrate, respectively. After quenching using twofold volumes of methanol, the reaction mixtures were concentrated by reduced-pressure distillation and resuspended in 10 mL volumes of methanol. The synthesized compounds were purified using a preparative HPLC system coupled with a reverse-phase Ultimate C18 column (21.2 × 250 mm, 5 µm particles, Welch, Shanghai, China). The purified glucosylated products were dissolved in DMSO-d_6, respectively. The chemical structures of the purified products were determined by 1D NMR (^{1}H NMR and ^{13}C NMR) and 2D NMR spectra (COSY, HMBC and HSQC) using a 400 MHz NMR spectrometer (Bruker, Karlsruhe,

Germany). The chemical-shift values were quoted in parts per million (ppm). All raw data were analyzed using the MestReNova 9 program.

4. Conclusions

In summary, UGT88E18 from *G. max* was demonstrated to selectively and effectively glucosylate the C7 hydroxyl group of calycosin to synthesize Cy7G. Our study revealed that the UGT88E18–SuSy cascade reaction is a powerful approach for the biocatalytic synthesis of Cy7G. UGT88E18–SuSy cascade reaction provides a new avenue for the cost-effective and scaled-up production of Cy7G and other isoflavonoids.

Supplementary Materials: The following are available online at http://www.mdpi.com/2073-4344/10/2/258/s1. Table S1. ^{1}H- and ^{13}C-NMR spectra for Products **1** and **2**. Figure S1. SDS-PAGE analysis of purified UGT88E18 and SuSy. Figure S2. ^{1}H NMR of Product **1**. Figure S3. ^{13}C NMR of Product **1**. Figure S4. HMBC spectra of Product **1**. Figure S5. HSQC spectra of Product **1**. Figure S6. COSY spectra of Product **1**. Figure S7. ^{1}H NMR of Product **2**. Figure S8. ^{13}C NMR of Product **2**. Figure S9. HMBC spectra of Product **2**. Figure S10. HSQC spectra of Product **2**. Figure S11. COSY spectra of Product **2**. Figure S12. Kinetic analysis of UGT88E18 toward formononetin and calycosin. Figure S13. Fed-batch synthesis of Cy7G by UGT88E18–SuSy cascade reaction.

Author Contributions: Conceptualization, Y.H. and J.M.; methodology, Y.Q. and X.Z.; data curation, Y.H., J.M., and Y.Q.; writing—review and editing, Y.H. and X.Y.; supervision, J.Z. and L.D.; funding acquisition, L.D. All authors have read and agreed to the published version of the manuscript.

Funding: This research was supported by the National Natural Science Foundation of China (no. 21702226), the China Postdoctoral Science Foundation (no. 2019M662575), and the Hubei Postdoctoral Sustentation Foundation.

Conflicts of Interest: The authors declare no conflict of interest.

References

1. Chiang, C.M.; Wang, T.Y.; Yang, Z.Y.; Wu, J.Y.; Chang, T.S. Production of new isoflavone glucosides from glycosylation of 8-hydroxydaidzein by glycosyltransferase from *Bacillus subtilis* ATCC 6633. *Catalysts* **2018**, *9*, 387. [CrossRef]
2. Hertog, M.G.; Hollman, P.C.; Putte, B.V. Content of potentially anticarcinogenic flavonoids of tea infusions, wines, and fruit juices. *J. Agric. Food Chem.* **1993**, *41*, 1242–1246. [CrossRef]
3. Hertog, M.G.; Hollman, P.C.; Katan, M.B. Content of potentially anticarcinogenic flavonoids of 28 vegetables and 9 fruits commonly consumed in the Netherlands. *J. Agric. Food Chem.* **1992**, *40*, 2379–2383. [CrossRef]
4. Saraei, R.; Marofi, F.; Naimi, A.; Talebi, M.; Ghaebi, M.; Javan, N.; Salimi, O.; Hassanzadeh, A. Leukemia therapy by flavonoids: Future and involved mechanisms. *J. Cell. Physiol.* **2019**, *234*, 8203–8220. [CrossRef] [PubMed]
5. Pandey, R.P.; Gurung, R.B.; Parajuli, P.; Koirala, N.; Sohng, J.K. Assessing acceptor substrate promiscuity of YjiC-mediated glycosylation toward flavonoids. *Carbohyd. Res.* **2014**, *393*, 26–31. [CrossRef] [PubMed]
6. Chen, K.; Hu, Z.M.; Song, W.; Wang, Z.L.; He, J.B.; Shi, X.M.; Cui, Q.H.; Qiao, X.; Ye, M. Diversity of O-glycosyltransferases contributes to the biosynthesis of flavonoid and triterpenoid glycosides in *Glycyrrhiza uralensis*. *ACS Synth. Biol.* **2019**, *8*, 1858–1866. [CrossRef]
7. Rha, C.S.; Kim, E.R.; Kim, Y.J.; Jung, Y.S.; Kim, D.O.; Park, C.S. Simple and efficient production of highly soluble daidzin glycosides by amylosucrase from *Deinococcus geothermalis*. *J. Agric. Food Chem.* **2019**, *67*, 12824–12832. [CrossRef]
8. Sangeetha, K.S.; Umamaheswari, S.; Reddy, C.U.M.; Kalkura, S.N. Flavonoids: Therapeutic potential of natural pharmacological agents. *Int. J. Pharm. Sci. Res.* **2016**, *7*, 3924.
9. Li, L.; Zheng, S.H.; Brinckmann, J.A.; Fu, J.; Zeng, R.; Huang, L.F.; Chen, S.L. Chemical and genetic diversity of *Astragalus mongholicus* grown in different eco-climatic regions. *PLoS ONE* **2017**, *12*, e0184791. [CrossRef]
10. Qi, L.W.; Yu, Q.T.; Li, P.; Li, S.L.; Wang, Y.X.; Sheng, L.H.; Yi, L. Quality evaluation of *Radix Astragali* through a simultaneous determination of six major active isoflavonoids and four main saponins by high-performance liquid chromatography coupled with diode array and evaporative light scattering detectors. *J. Chromatogr. A* **2006**, *1134*, 162–169. [CrossRef]

11. Jian, J.; Sun, L.J.; Cheng, X.; Hu, X.F.; Liang, J.C.; Chen, Y. Calycosin-7-O-β-d-glucopyranoside stimulates osteoblast differentiation through regulating the BMP/WNT signaling pathways. *Acta Pharm. Sin. B* **2015**, *5*, 454–460. [CrossRef] [PubMed]
12. Zhang, J.Q.; Xue, X.L.; Yang, Y.; Ma, W.; Han, Y.; Qin, X.M. Multiple biological defects caused by calycosin-7-O-β-D-glucoside in the nematode *Caenorhabditis elegans* are associated with the activation of oxidative damage. *J. Appl. Toxicol.* **2018**, *38*, 801–809. [CrossRef]
13. Xiao, W.H.; Han, L.J.; Shi, B. Isolation and purification of flavonoid glucosides from *Radix Astragali* by high-speed counter-current chromatography. *J. Chromatogr. B* **2009**, *877*, 697–702. [CrossRef] [PubMed]
14. Dai, L.H.; Li, J.; Yang, J.G.; Zhu, Y.M.; Men, Y.; Zeng, Y.; Cai, Y.; Dong, C.X.; Dai, Z.B.; Zhang, X.L.; et al. Use of a promiscuous glycosyltransferase from *Bacillus subtilis* 168 for the enzymatic synthesis of novel protopanaxatriol-type ginsenosides. *J. Agric. Food Chem.* **2018**, *66*, 943–949. [CrossRef]
15. Xie, K.B.; Dou, X.X.; Chen, R.D.; Chen, D.W.; Fang, C.; Xiao, Z.Y.; Dai, J.G. Two novel fungal phenolic UDP glycosyltransferases from *Absidia coerulea* and *Rhizopus japonicus*. *Appl. Environ. Microbiol.* **2017**, *83*, e03103–e03116. [CrossRef]
16. Chen, D.W.; Fan, S.; Chen, R.D.; Xie, K.B.; Yin, S.; Sun, L.L.; Liu, J.M.; Yang, L.; Kong, J.Q.; Yang, Z.Y.; et al. Probing and engineering key residues for bis-C-glycosylation and promiscuity of a C-glycosyltransferase. *ACS Catal.* **2018**, *8*, 4917–4927. [CrossRef]
17. Zhang, L.; Gao, Y.N.; Liu, X.F.; Guo, F.; Ma, C.X.; Liang, J.H.; Feng, X.D.; Li, C. Mining of sucrose synthases from *Glycyrrhiza uralensis* and their application in the construction of an efficient UDP-recycling system. *J. Agric. Food Chem.* **2019**, *67*, 11694–11702. [CrossRef]
18. Gutmann, A.; Nidetzky, B. Unlocking the potential of leloir glycosyltransferases for applied biocatalysis: Efficient synthesis of uridine 5′-diphosphate-glucose by sucrose synthase. *Adv. Synth. Catal.* **2016**, *358*, 3600–3609. [CrossRef]
19. Gutmann, A.; Lepak, A.; Diricks, M.; Desmet, T.; Nidetzky, B. Glycosyltransferase cascades for natural product glycosylation: Use of plant instead of bacterial sucrose synthases improves the UDP-glucose recycling from sucrose and UDP. *Biotechnol. J.* **2017**, *12*, 1600557. [CrossRef]
20. Nidetzky, B.; Gutmann, A.; Zhong, C. Leloir glycosyltransferases as biocatalysts for chemical production. *ACS Catal.* **2018**, *8*, 6283–6300. [CrossRef]
21. Wang, X.; Fan, R.Y.; Li, J.; Li, C.F.; Zhang, Y.S. Molecular cloning and functional characterization of a novel (iso) flavone 4′, 7-O-diglucoside glucosyltransferase from *Pueraria lobata*. *Front. Plant Sci.* **2016**, *7*, 387. [CrossRef] [PubMed]
22. Li, J.; Li, Z.B.; Li, C.F.; Gou, J.B.; Zhang, Y.S. Molecular cloning and characterization of an isoflavone 7-O-glucosyltransferase from *Pueraria lobata*. *Plant Cell Rep.* **2014**, *33*, 1173–1185. [CrossRef] [PubMed]
23. Funaki, A.; Waki, T.; Noguchi, A.; Kawai, Y.; Yamashita, S.; Takahashi, S.; Nakayama, T. Identification of a highly specific isoflavone 7-O-glucosyltransferase in the soybean (*Glycine max* (L.) Merr.). *Plant Cell Physiol.* **2015**, *56*, 1512–1520. [CrossRef] [PubMed]
24. Yu, D.H.; Bao, Y.M.; Wei, C.L.; Li, J.A. Studies of chemical constituents and their antioxidant activities from *Astragalus mongholicus* Bunge. *Biomed. Environ. Sci.* **2005**, *18*, 297.
25. Dai, L.H.; Li, J.; Yang, J.G.; Men, Y.; Zeng, Y.; Cai, Y.; Sun, Y.X. Enzymatic synthesis of novel glycyrrhizic acid glucosides using a promiscuous *Bacillus* glycosyltransferase. *Catalysts* **2018**, *8*, 615. [CrossRef]
26. Achnine, L.; Huhman, D.V.; Farag, M.A.; Sumner, L.W.; Blount, J.W.; Dixon, R.A. Genomics-based selection and functional characterization of triterpene glycosyltransferases from the model legume *Medicago truncatula*. *Plant J.* **2005**, *41*, 875–887. [CrossRef]
27. Bungaruang, L.; Gutmann, A.; Nidetzky, B. Leloir glycosyltransferases and natural product glycosylation: Biocatalytic synthesis of the C-glucoside nothofagin, a major antioxidant of redbush herbal tea. *Adv. Synth. Catal.* **2013**, *355*, 2757–2763. [CrossRef]
28. Liu, F.; Ding, F.Y.; Shao, W.M.; He, B.F.; Wang, G.J. Regulated preparation of crocin-1 or crocin-2′ triggered by the cosolvent DMSO using Bs-GT/At-SuSy one-pot reaction. *J. Agric. Food Chem.* **2019**, *67*, 12496–12501. [CrossRef]
29. Dai, L.H.; Liu, C.; Li, J.; Dong, C.X.; Yang, J.G.; Dai, Z.B.; Zhang, X.L.; Sun, Y.X. One-pot synthesis of ginsenoside Rh2 and bioactive unnatural ginsenoside by coupling promiscuous glycosyltransferase from *Bacillus subtilis* 168 to sucrose synthase. *J. Agric. Food Chem.* **2018**, *66*, 2830–2837. [CrossRef]

30. Dai, L.H.; Li, J.; Yao, P.Y.; Zhu, Y.M.; Men, Y.; Zeng, Y.; Yang, J.G.; Sun, Y.X. Exploiting the aglycon promiscuity of glycosyltransferase Bs-YjiC from *Bacillus subtilis* and its application in synthesis of glycosides. *J. Biotechnol.* **2017**, *248*, 69–76. [CrossRef]
31. Diricks, M.; Bruyn, F.D.; Daele, P.V.; Walmagh, M.; Desmet, T. Identification of sucrose synthase in nonphotosynthetic bacteria and characterization of the recombinant enzymes. *Appl. Microbiol. Biotechnol.* **2015**, *99*, 8465–8474. [CrossRef] [PubMed]
32. Pei, J.J.; Chen, A.N.; Zhao, L.G.; Cao, F.L.; Ding, G.; Xiao, W. One-pot synthesis of hyperoside by a three-enzyme cascade using a UDP-galactose regeneration system. *J. Agric. Food Chem.* **2017**, *65*, 6042–6048. [CrossRef] [PubMed]
33. Hu, Y.M.; Xue, J.; Min, J.; Qin, L.J.; Zhang, J.K.; Dai, L.H. Biocatalytic synthesis of ginsenoside Rh2 using *Arabidopsis thaliana* glucosyltransferase-catalyzed coupled reactions. *J. Biotechnol.* **2020**, *309*, 107–112. [CrossRef] [PubMed]

© 2020 by the authors. Licensee MDPI, Basel, Switzerland. This article is an open access article distributed under the terms and conditions of the Creative Commons Attribution (CC BY) license (http://creativecommons.org/licenses/by/4.0/).

Article

Shaking Rate during Production Affects the Activity of *Escherichia coli* Surface-Displayed *Candida antarctica* Lipase A

Chen-Fu Chung [1], Shih-Che Lin [1], Tzong-Yuan Juang [2,*] and Yung-Chuan Liu [1,*]

1. Department of Chemical Engineering, National Chung Hsing University, Taichung 402, Taiwan; hahahaandy@gmail.com (C.-F.C.); leo83873@yahoo.com.tw (S.-C.L.)
2. Department of Cosmeceutics, China Medical University, Taichung 40402, Taiwan
* Correspondence: tyjuang@mail.cmu.edu.tw (T.-Y.J.); ycliu@dragon.nchu.edu.tw (Y.-C.L.)

Received: 5 March 2020; Accepted: 30 March 2020; Published: 1 April 2020

Abstract: In this study, a surface-display system was applied for the expression of lipase A in an *E. coli* expression system. Since the target protein was exposed on the cell membrane, the shaking rate during culturing might have increased the oxygen mass transfer rate and the shear stress, both of which would be detrimental to the surface-displayed protein. The shaking rate did indeed have an effect on the properties of the surface-displayed lipase A from *Candida antarctica* (sdCALA). When cultivated at a shaking rate of less than 50 rpm, the specific activity of sdCALA was low, which was due to the limited amount of dissolved oxygen. When the shaking rate was greater than 100 rpm, the specific activity decreased as a result of shear stress. When cultivating CALA and sdCALA at various temperatures and values of pH, both proteins displayed the same activity profile, with the optimum conditions being 60 °C and pH 6. A kinetic study revealed that the sdCALA cultivated at 100 rpm gave a higher value of v_m (0.074 µmol/mL/min) and a lower value of K_m (0.360 µmol/mL) relative to those obtained at 200 rpm and relative to those of the free CALA. sdCALA retained over 80% of its activity after treatment at 70 °C for 30 min, but its activity decreased rapidly when the temperature was above 80 °C. The specific activity of sdCALA decreased in the presence of acetonitrile and acetone relative to that of the control (50% ethanol), regardless of the solvent concentration. The highest activity (0.67 U/mL) was obtained when the ethanol concentration was 30%.

Keywords: *Candida antarctica* lipase A; surface-display system; shear rate; mass transfer rate; enzymatic kinetic study

1. Introduction

Candida antarctica is a psychrophilic Basidiomycetous yeast originally selected from a strain found in Antarctic habitats. This strain expresses two lipolytic enzymes displaying different catalytic properties: *Candida antarctica* lipase A (CALA) and lipase B (CALB) [1]. CALB, first described in 1994 [2], has a molecular weight of approximately 34–43 kDa. As a result of its high selectivity toward secondary alcohols and its endurance in organic solvents and at high temperatures [2,3], it has been widely applied industrially and academically. The structure of CALA was solved using X-ray crystallography [2,4], with a molecular weight of approximately 45–53 kDa. Compared with CALB, the properties of CALA have been less well explored. CALA displays specific hydrolytic properties; for example, it retains its activity at high temperature (>70 °C) and in acidic environments. In addition, CALA preferentially hydrolyzes triglycerides at position sn-2, which is located at the center of the carbon chain; therefore, it has been applied in the preparation of fat substitutes and to the development of new modes of drug delivery [5]. At present, other microorganisms known to produce CALA include *Aspergillus sp.* [6] and *Pichia sp.* [2,7] fungi and prokaryotic *E. coli* [8]. Studies of CALA have mainly been focused on its use in different host cells [9], food processing [10], and transesterification [10,11]. To render CALA more applicable, new processes should be developed for its rapid and efficient production.

In 1985, Smith inserted the recombinant antigens to the N-terminus of the filamentous bacteriophage coat protein pIII and expressed them on the phage surface [12]. Many carrier proteins have been demonstrated to carry the protein outside the membranes of cells. Common carrier proteins include S-layer protein, lipoprotein, bacterial fimbriae, and ice nucleation protein (INP) [10,13,14]. A variety of biotechnological processes and materials—for example, antibody production, bioconversion, biosensors, bioadsorbents, and oral vaccines—have adopted this technique for process improvement [13,14].

INP is a membrane-bound protein that is capable of binding to the phospholipids of cell membranes [15,16]. The main structure of INP can be divided into three parts: the N-terminus, a hydrophobic region that binds to the phospholipids of the outer membrane; an intermediate region, comprising mainly AlaGlyTyrGlySerThrLeuThr (AGYGSTLT) repeat fragments that form an ice core; and the C-terminus, which is a hydrophilic region that can link with the target protein and expose it to the outer membrane surface. INP-based systems can suitably express recombinant protein molecules after modulating the length of the intermediate repeat region, thereby minimizing the issue of steric disorder [17]. According to its genotypes, INP can be classified into four systems: *inaK* [18], *inaQ* [19], *inaV* [20], and *inaX* [21]. The expression of INP is most stable during the stationary phase of culturing. INP-based systems are among the most useful for Gram-negative bacteria surface-display applications [18].

Many genetic engineering technologies require cell surface carriers for the rapid expression of foreign proteins. Through the surface expression of lipases, for example, various surface-display lipases have been developed with the goal of applying them directly to lipolytic reactions. Matsumoto et al. constructed a yeast cell surface-display system for lipase expression, using a flocculation functional domain of Flo1p as the carrier [22]. Moura et al. assessed the biochemical features of lipase B when immobilized on the cell surface of the methylotrophic yeast *Pichia pastoris*, using the yeast surface-display approach [23]. Liu et al. displayed lipase Lip2 from *Yarrowia lipolytica* on the cell surface of *Saccharomyces cerevisiae*, using Cwp2 as an anchor protein; the thermostability of the displayed lipase was superior to that of free lipase Lip2 [24].

The use of recombinant *E. coli* to express intracellular target proteins is common. The main issue hindering its application has been the separation and purification of the target protein from the harvested cultural broth—a tedious and labor-intensive process that increases the complexity of production. Surface-display systems could provide an alternative means of obtaining enzymes for direct industrial applications. Wilhelm et al. demonstrated that the cell surface-display protein LipH from *P. aeruginosa* could properly express foldase–lipase complexes on the surface of *E. coli* cells [25]. Jo et al. displayed the lipase from *Bacillus licheniformis* ATCC14580 on the cell surface of *E. coli* by using Lpp'OmpA as the anchoring protein; this development suggested that *E. coli*, through the direct display of lipase on its cell membrane, could be applied as a whole cell biocatalyst [26].

In this present study, we constructed an INP surface-display system to express lipase A in the *E. coli* outer membrane. Although some articles describe the expression of lipase anchoring on the *E. coli* cell membrane [25,26], those studies did not focus on the effects of cultivation on the biochemical features of surface-display CALA (sdCALA). For example, mass transfer and shear stress during cultivation might directly affect the properties of the enzyme through its exposure to the outside environment. To test this concept, here we performed a preliminary comparison of the biochemical properties of CALA produced intracellularly with those of sdCALA anchored on the membrane. In addition, we also evaluated the kinetic properties of sdCALA prepared under various cultivation conditions.

2. Results and Discussion

2.1. Gene Construction and Protein Expression

The host strains *E. coli* JM109 (DE3)/pET22b-calawt and *E. coli* JM109 (DE3)/pINP-CALA were constructed according to the methods listed in Section 3.2. Both types of cells were cultivated and induced according to the methods listed in Section 3.3. To obtain the intracellular CALA, the cells were disrupted through sonication, followed by centrifugation to collect the supernatant containing CALA. To obtain sdCALA, the collected cells were dissolved in the same volume of sodium phosphate buffer (pH 6.0) for further use. Sodium dodecyl sulfate–polyacrylamide gel electrophoresis (SDS-PAGE) was performed to check for the correct expression of intracellular and surface-display CALA (Figure 1). In the PAGE gel, a band in Lane 1 revealed the successful expression of CALA (53 kDa) and that the expression was mainly intracellular as the soluble protein (Lane 2). In addition, no clear band at 53 kDA appeared in the pellets (Lane 3), implying that no inclusion body was formed. For the surface expression system, a band for INP-CALA (80 kDa) appeared in Lane 4, confirming the accurate expression of the fusion protein INP-CALA (Lane 4), with the main expression located at the pellet (Lane 6) with only a very unclear band in Lane 5. Thus, the expression of sdCALA occurred mainly on the cell pellet.

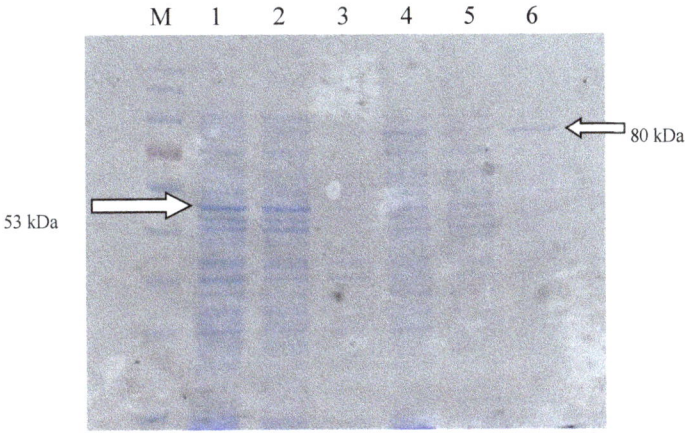

Figure 1. SDS-PAGE of intracellular *Candida antarctica* lipase A (CALA) and surface-display CALA. Lane M: Marker; Lanes 1–3: Whole cells, supernatant, and pellet of intracellular CALA, respectively; Lanes 4–6: Whole cells, supernatant, and pellet for surface-display CALA (sdCALA), respectively.

The activities of the CALA enzymes obtained from the two systems were assayed (Table 1). The activity of CALA obtained from the intracellular system was approximately 0.326 U/mL, while that from the surface expression system was 0.285 U/mL. The specific activity of sdCALA (0.285 U/mg) was larger than that of CALA (0.251 U/mg). We suspected that sdCALA on the cell membrane might have had a larger space for its proper extension. Although the volumetric activity of the lipase A produced through surface expression was slightly lower than that of the intracellular system, the higher specific activity and avoidance of cell disruption suggest that it might be convenient in certain applications. In addition, we found, interestingly, that a small amount of soluble fusion protein INP-CALA existed in the periplasm, with a relative activity of 0.057 ± 0.04 U/mL. The total expression of sdCALA with the supernatant plus pellet would compete with that of the intracellular CALA.

Table 1. Expression and activities of intracellular CALA and surface-displayed CALA.

Expression System	Biomass (mg/mL)	Activity (U/mL)	Specific Activity (U/mg)
Intracellular [a] (pET22b-calawt)	1.298 ± 0.232	0.326 ± 0.11	0.251 ± 0.047
Surface display [b] (pINP-CALA)	0.999 ± 0.052	0.285 ± 0.07	0.285 ± 0.013

[a] To measure the volumetric activity of CALA, the supernatant obtained as stated in Section 3.3 was used for the assay. [b] To measure the volumetric activity of sdCALA, the harvested cells were resuspended in the same volume of phosphate buffer (pH 6.0) for the assay.

In our laboratory, we have frequently applied INP to expose outer membrane proteins; some papers concerning this approach have been published [27–30]. The most typical use has been in the expression of INP-INT-EGFP (INP-intein-enhanced green fluorescent protein) outside the cell membrane and then, through the self-cleavage of INT, to relieve the EGFP to the supernatant [30]. In such a case, the expression of EGFP as the outer membrane protein would result in a green-colored cell pellet. In addition, this approach has also been used for the expression of D-hydantoinase outside the cell membrane and to simplify the enzyme purification process [28]. Those previous studies demonstrated that INP could be used not only to correctly express the target protein as an outer membrane protein but also to maintain the correct conformation of the protein and its usual activity (measured through a whole-cell activity assay). In this study, we used INP to express lipase A as an outer membrane enzyme and applied a tributyrin agar plate test to demonstrate the existence of outer membrane INP-CALA, as described in Section 3.6. Figure 2 displays the results. In the case of sdCALA, a clear zone formed around the cell colony, which was due to direct hydrolysis of the tributyrin. In contrast, the cells prepared with the intracellular CALA did not form any such clear zone. Thus, the JM109/pINP-CALA system could indeed express sdCALA properly outside the cell membranes.

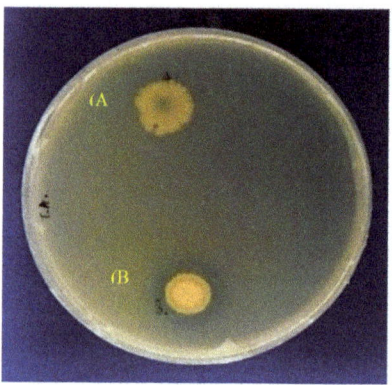

Figure 2. Tributyrin plate tests for intracellular and surface-display CALA. (A: Intracellular CALA; B: sdCALA).

2.2. Effect of Shaking Rate on sdCALA Activity and Biomass

Since the expressed sdCALA was anchored outside the cell membrane, the cultivation conditions—especially the shaking rate—would presumably induce various mass transfer rates and shear rates on the produced enzyme. To examine this concept, sdCALA production was performed with cultivation at various shaking rates (Figure 3). The shaking rate did indeed affect the enzyme specific hydrolysis activity and biomass production. The biomass increased remarkably upon increasing the shaking rate from 50 to 100 rpm, but the increases were less dramatic thereafter. The specific activity also greatly increased upon increasing the shaking rate from 50 to 100 rpm. However, when the shaking rate was at or above 150 rpm, the specific activity decreased significantly. To tests the effects of mass transfer and shear on the biomass production and lipase activity, relevant data were calculated according to Equations (1)–(5), as described in Section 3.4, and these are listed in Table 2. At the low shaking rate of 50 rpm, a low rate of oxygen mass transfer would occur, limiting cell growth and resulting in very low biomass production. This limited oxygen transfer rate would also block the proper expression of sdCALA. When the shaking rate was 100 rpm, the higher rate of oxygen mass transfer would benefit cell growth, resulting in a high expression of sdCALA. When the shaking rate was 150–200 rpm, the higher rates of oxygen mass transfer would slightly increase the biomass production, but high shear rates would have a detrimental effect on the expression of sdCALA, due to its direct exposure to the broth. To minimize the damage caused at high shear rates, we selected a shaking rate of 100 rpm for subsequent expressions of sdCALA.

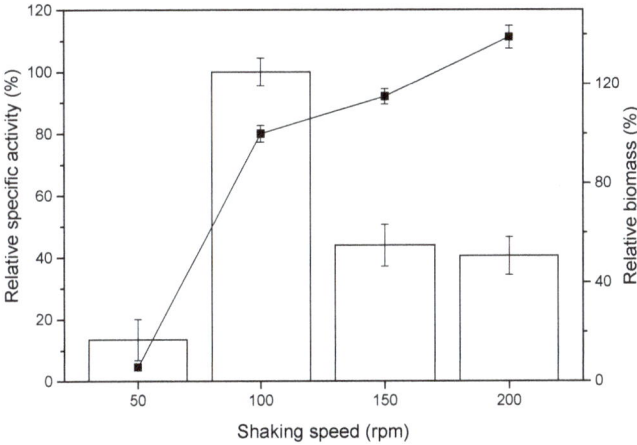

Figure 3. Effect of shaking rate on the specific activity (bars) and biomass production (black squares) of sdCALA. To calculate relative values, the value measured at 100 rpm was set as 100%.

Table 2. Effect of mass transfer and shear rate on the specific activity and biomass production of sdCALA.

Shaking Rate (rpm)	K_{La} (h^{-1})	γ (h^{-1})	Relative Specific Activity (%)	Relative Biomass (%)
50	9.6	13	13.5 ± 6.6	5.7 ± 1.2
100	21.4	89	100 ± 4.5	100 ± 3.4
150	34.2	276	43.9 ± 6.7	115 ± 3.1
200	47.8	617	40.5 ± 6.1	139 ± 4.6

2.3. Optimal Temperature and pH for Production of CALA and sdCALA

In reactions involving biocatalysts, the enzyme activity would increase upon increasing the temperature within a certain range. However, the activity would decrease if the temperature was too high, due to the effect of enzyme denaturing. For example, most lipolytic enzymes lose their activity when the temperature exceeds 40 °C [31]. For CALA, a thermophilic lipolytic enzyme [32,33], the optimal temperature has been reported to be in the range from 50 to 70 °C [33,34] and possibly even as high as 90 °C [33]. We prepared two types of sdCALA through the cultivation of JM109/pINP-CALA at 100 and 200 rpm (denoted herein as sdCALA100 and sdCALA200, respectively). As references, we also prepared two samples of intracellular CALA enzymes from cultures of JM109/pET22b-calawt under the same conditions (CALA100 and CALA200, respectively). The activities of all of these tested enzymes were adjusted to 0.35 U/mL using 50 mM phosphate buffer (pH 6).

Figure 4 displays the effect of temperature on the activities of CALA and sdCALA. All of the enzymes exhibited the same profile in response to the temperature. The optimal temperature was 60

°C in each case. Thus, even though they had been subjected to different culture conditions and various expression systems, the resulting enzymes exhibited the same dependency on temperature.

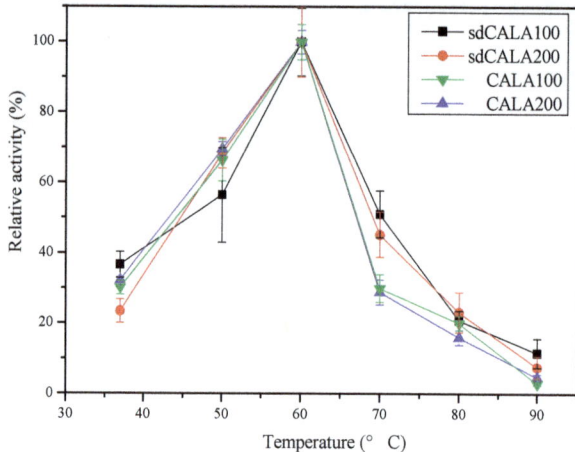

Figure 4. Effect of temperature on the relative activities of samples of CALA and sdCALA prepared at different shaking rates.

Various lipases have their own operating pH ranges, from as low as pH 3 to as high as pH 9 [35,36]. Since the structure of the active site might be affected by the pH, a lipase might display acid- or alkali-resistant properties. Figure 5 displays the effect of pH on the activities of the CALA and sdCALA samples prepared at the two spinning rates. All of these enzymes had the same pH-dependency, with the optimal activities occurring at pH 6. The activities decreased sharply when the pH was 7, and the enzymes were almost inactive at pH 9. Thus, we conclude that CALA is not an alkaline enzyme. Furthermore, the use of different shaking rates for the expression of CALA and sdCALA did not have any effect on their pH dependency.

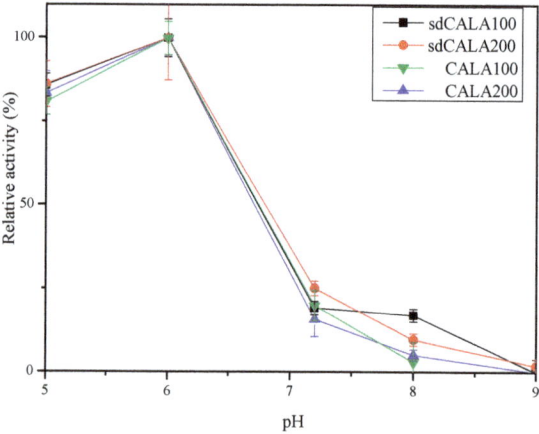

Figure 5. Effect of pH on the relative activities of CALA and sdCALA samples prepared at different shaking rates.

2.4. Kinetic Analyses of Various CALA Enzymes

We adopted the Michaelis–Menten kinetics model in Equation (6) to test the kinetic behavior of our various lipase A enzymes. Figure 6 presents a Hanes–Woolf plot of the substrate concentration divided by the activity with respect to the substrate concentration (Equation (7)). The kinetic parameters for each enzyme were collected through linear fitting of the data. Table 3 lists the calculated values of K_m and v_m. All of the enzymes displayed values of v_m in range of 0.059–0.072 (µmol/mL/min) and K_m in range of 1.032–1.821 (µmol/mL), except for sdCALA100, which had a slightly higher value of v_m (0.074 µmol/mL/min) and a much lower value of K_m (0.360 µmol/mL). Thus, preparing sdCALA at 100 rpm significantly improved both its reaction rate (high v_m) and substrate affinity (low K_m). The sdCALA obtained at 200 rpm gave a value of K_m close to those of the intracellular enzymes. Notably, the collection of intracellular enzymes requires harsh sonication or high pressures for cell disruption, potentially influencing the properties of the enzymes and decreasing the affinity to the substrate. Furthermore, we suspect that sdCALA200, which had been subjected to a higher shaking rate and a higher shear rate, had poorer affinity than sdCALA100 toward the substrate. Therefore, such kinetic analysis can provide indirect evidence for the shear rate affecting the performance of CALA prepared under various cultivation conditions.

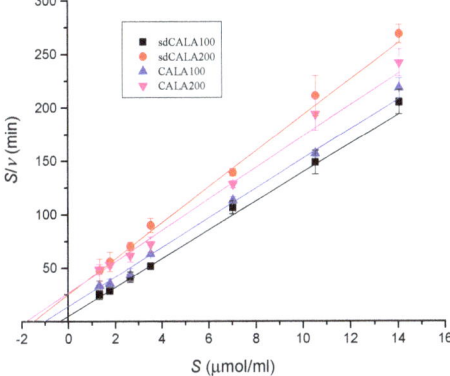

Figure 6. Hanes–Woolf diagram for CALA and sdCALA.

Table 3. Kinetics parameters of CALA and sdCALA.

Lipase	v_m (µmol/mL/min)	K_m (µmol/mL)
sdCALA100	0.074	0.360
sdCALA200	0.059	1.510
CALA100	0.072	1.032
CALA200	0.068	1.821

2.5. Thermal Stability of sdCALA

We tested the thermal stability of sdCALA100 at various temperatures. Samples were taken every 10 min to obtain the activity assay at the optimal temperature of 60 °C. Figure 7 reveals that sdCALA retained greater than 80% of its activity when heated at a temperature lower than or equal to 70 °C for 30 min; however, at 80 °C, its activity dropped remarkably, to less than 20% after 30 min, and it was almost completely lost after being heated at 90 °C for 10 min. We found in our earlier study of the optimal temperature that the relative activity of sdCALA decreased to less than 50% when it was in contact with 50% ethanol (in the activity assay) at 70 °C for 5 min; however, in this stability test, greater than 80% of the residual activity remained after heating at 70 °C for 30 min in the absence of contact with ethanol. Thus, a high concentration of ethanol presumably had a negative effect on the activity of sdCALA. Accordingly, we tested the influence of various organic solvents at various concentrations on the hydrolysis reactions of sdCALA.

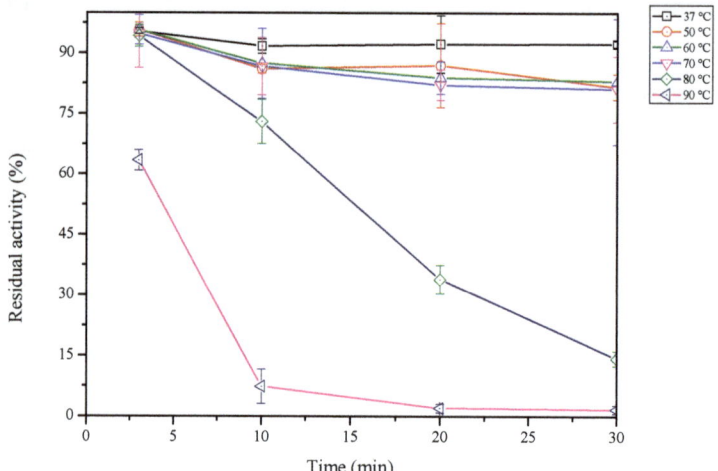

Figure 7. Thermal stability tests for sdCALA, where the initial activity was set as the control.

2.6. Effect of Organic Solvents on sdCALA Activity

Yang et al. studied the effects of various organic solvents on the activity of purified CALA produced by *Pichia pastoris* [37]. They found that when CALA was treated with acetonitrile or acetone, its activities toward esterification and hydrolysis increased. However, when treated with ethanol, only 20% of the lipolytic activity remained; when treated with isopropanol, 20% of the hydrolysis and esterification activities were lost.

To explore the effects of organic solvents on sdCALA, we applied ethanol, acetonitrile, acetone, and isopropanol at various concentrations as replacements for 50% ethanol and analyzed the activities (Figure 8). The activity obtained in the presence of 50% ethanol was set as the control (100%). When the concentration of ethanol decreased to 30%, the activity of sdCALA increased significantly, but it decreased upon lowering the concentration thereafter, which was possibly because of the low solubility of the substrate [*p*-nitrophenyl palmitate (p-NPP)] at such low concentrations of ethanol. In the case of isopropanol, the activity increased upon decreasing the concentration of the organic solvent, but all of the activities were lower than that of the control (50% ethanol), except when the concentration of isopropanol was at or below 30%. The results obtained when using acetonitrile or acetone were quite different. Increasing the concentration of either of these organic solvents from 20% to 40% enhanced the activity, but a concentration of 50% had a negative effect. Nevertheless, all of the activities in the presence of these two solvents were lower than that of the control. Thus, the results of this study were quite different from those reported by Yang et al. [37]. Presumably, the organic solvents affected the conformation of CALA, exposing its active site more to the substrate. On the other hand, the organic solvents might also have decreased the activities through denaturing of the structure of CALA. In the case of sdCALA, the host *E. coli* might also have been a factor, which was subjected to its own solvent effect—the organic solvents might have varied the cell membrane structure and indirectly affected the structure of the anchored CALA [38,39]. A more detailed study of the effect of solvents on hosts would be worthy in future.

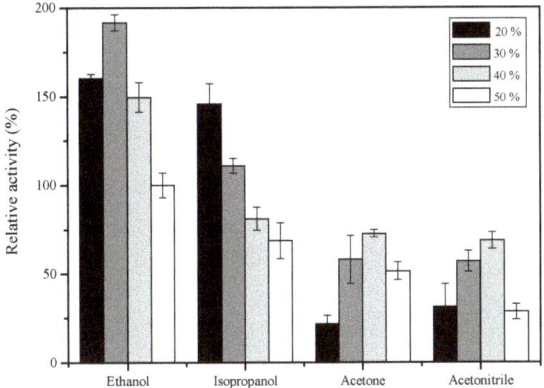

Figure 8. Effect of the concentrations of various organic solvents on the activity of sdCALA with 50% ethanol set as the control.

3. Materials and Methods

3.1. Strains, Plasmids, and Source

E. coli DH5α and E. coli JM109 (DE3) were used as host cells for gene transfer and recombinant protein expression, respectively. The plasmid pET22b-calawt containing the CALA(*cala*) gene sequence was a gift from Professor Bornscheuer at Greifswald University, Germany; the pET28a-inaX having the INP(*inaX*) gene was provided by Professor Wen-Teng Wu of National Cen Kong University, Taiwan. The E. coli host strains and vectors used in this study are listed in Table 4. All other chemicals were of analytical grade and obtained from a local supplier.

Table 4. E. coli strains and plasmids used in this study. INP: ice nucleation protein.

Strain or Plasmid	Genotype and Relevant Characteristics	Source
DH5α	F- endA1 glnV44 thi-1 recA1 relA1 gyrA96 deoR nupG Φ80dlacZΔM15 Δ(lacZYA-argF)U169, hsdR17(rk − mk +), λ−	Novagen
JM109 (DE3)	endA1 glnV44 thi-1 relA1 gyrA96 recA1 mcrB+ Δ(lac-proAB) e14- (F′ traD36 proAB+ lacIq lacZΔM15) hsdR17(rk − mk +) λ(DE3)	NEB
pET22b-calawt	CALA (calA) gene fragment in pET22b	Professor Bornscheuer (Greifswald University, Germany)
pET26b-inaX	INP (inaX) gene fragment in pET26b	Professor Wu (NCKU, Taiwan)
yT&A	pBluescript IISK(−) with modified MCS	Yeastern Biotech

3.2. Construction of Expression Systems

pET22b-calawt was used as the template to amplify the *cala* gene with restriction sites of *EcoRI* and *XhoI*. The primers are listed in Table 5. The obtained inserts (PCR products) were further ligated to vector yT&A (Yeastern Biotech) to build the constructs of pT-CALA. The *cala* gene was cut from pT-CALA via the restriction enzyme sites of *EcoRI* and *XhoI* and was ligated into pET26b-INP to form the pINP-CALA construct (Figure 9). The production strains were prepared via heat-shock, transforming the respective pET22b-calawt and pINP-CALA into the host E. coli JM109 (DE3). All recombinant DNA manipulations were performed using standard procedures [40].

Table 5. Primers used in this study.

Primer Name	Primer Sequence (5′3′)	Template
CALA-F	GAATTCATGCGTGTGAGCCTGCGTAG (*EcoRI*)	pET22b-calawt
CALA-R	CACTCGAGCACCACCACCA (*XhoI*)	pET22b-calawt

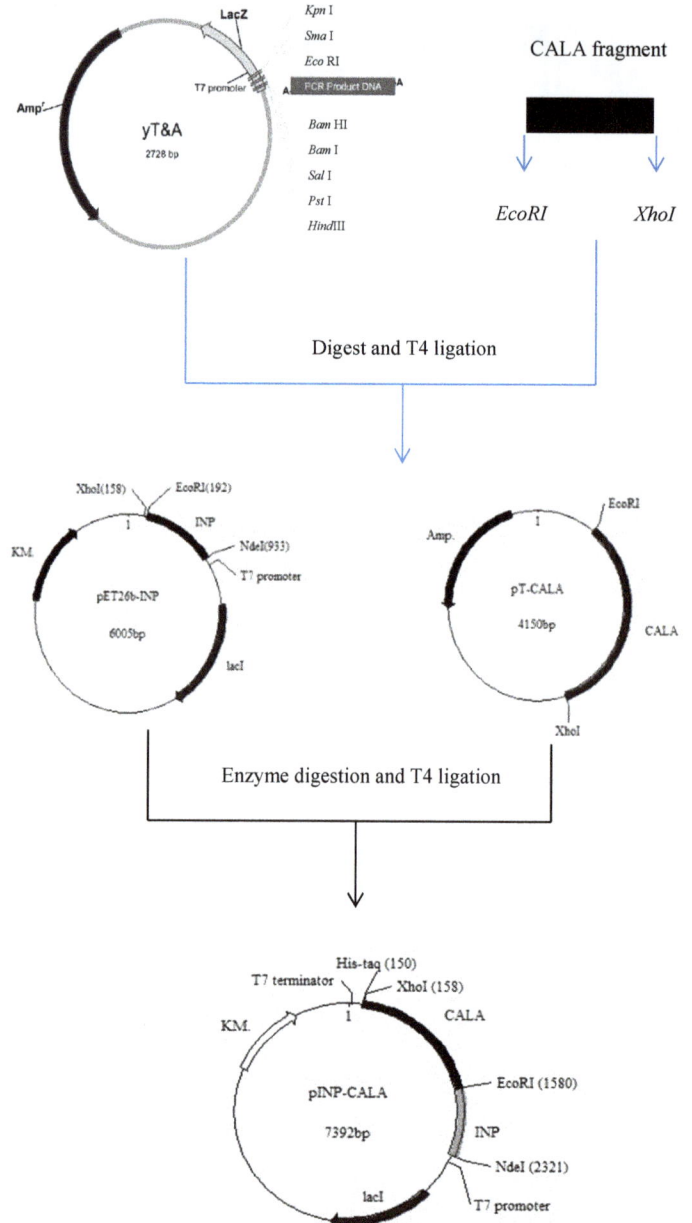

Figure 9. Construction of pINP-CALA.

3.3. Cultivation Conditions

E. coli JM109 strains carrying *cala* and *inaX-cala* genes were activated from the preservation freezer to a Petri dish with Luria-Bertani (LB) agar plus 10 µL (50 mg/mL) ampicillin and cultured at 37 °C for 12 h. A single colony was picked to inoculate 10 mL LB medium plus 10 µL (50 mg/mL) ampicillin; the culture was operated at 37 °C and a shaking rate of 200 rpm for 12 h. A volume of 1% seed culture was used to inoculate 100 mL of LB medium plus 100 µL (50 mg/mL) ampicillin. Cultivation was conducted at 37 °C and 200 rpm until the cell optical density (OD_{600}) reached 0.8, followed by the addition of isopropyl β-D-1-thiogalactopyranoside (IPTG) to a final concentration of 0.1 mM. The cultivation temperature and time were set at 15 °C and 24 h, respectively, while the shaking rate was varied from 50 to 200 rpm. To prepare CALA, the broth was centrifuged at 3000 g for 10 min to collect the cells. The cells were subjected to sonication disruption (ultrasonic processing at 20 W for 10 cycles: 30 s working, 30 s free) and centrifugation (10,000 g, 10 min) to separate the supernatant from the cell debris pellet. Both samples were subjected to a dyeing process prior to performing the SDS-PAGE assay. The dyeing solution comprised 10% SDS, 50% glycerol, 1.5 M Tris-Cl (pH 6.8), 2% β-mercaptoethanol, and 0.5% bromophenol blue. The samples were mixed with the dyeing solution at a ratio of 5:1, followed by heating at 100 °C for 10 min, to give the loading samples for SDS-PAGE. To prepare the whole-cell sdCALA, centrifugation (3000 g, 10 min) was applied to separate the cells from the broth; the cells were resuspended in phosphate buffer (pH 6.0) for use in all of the enzyme tests.

3.4. Estimation of Shear Stress and Mass Transfer Coefficient

The oxygen mass transfer coefficient (K_{La}) for an Erlenmeyer flask can be estimated according to the following equation described by Sanchez et al. [41]:

$$K_{La} = 0.024 n^{1.16} V_L^{-0.83} d_0^{0.38} d^{1.92} \tag{1}$$

where n is the rate (rpm); V_L is the liquid volume in the shake flask (ml); d_0 is the incubator shaking diameter (cm); and d is the maximal diameter of the shake flask (cm). The units of K_{La} are (h^{-1}).

The average shear rate (γ) and shear stress (τ_t) of a liquid in a shake flask can be estimated using the following formulas [42].

$$\gamma = 3600 \frac{\tau_t}{\mu_L} \tag{2}$$

$$\tau_t = 0.0676 \left(\frac{d_p}{\lambda}\right)^2 (\rho_L \mu_L \varepsilon)^{0.5} \tag{3}$$

where ρ_L is the density of the liquid (kg/m³), μ_L is the viscosity of the liquid (Pas), d_p is the average cell diameter (m), λ is the Kolmogorov length scale (m), and ε is the energy dissipation rate (m²/s³), which are calculated from the following equation:

$$\lambda = \left(\frac{\mu_L}{\rho_L}\right)^{3/4} \varepsilon^{-1/4} \tag{4}$$

$$\varepsilon = \frac{1.94 n^{2.8} d_s^{3.6} \mu_L^{0.2}}{V_L^{2/3} \rho_L^{0.2}} \tag{5}$$

where n is the rate (1/s), d_s is the flask diameter (m), and V_L is the working volume of the medium (m³). The units of γ are (h^{-1}); the units of τ_t are (N/m²).

3.5. Kinetics Models

The kinetics were derived from the Michaelis–Menten equation as follows:

$$v = \frac{v_m S}{K_m + S} \tag{6}$$

Rearranging Equation (6) by inversion and multiplication by S yields the Hanes–Woolf equation

$$\frac{S}{v} = \frac{1}{v_m} S + \frac{K_m}{v_m} \tag{7}$$

where S is the initial substrate concentration, v is the reaction rate, K_m is the Michaelis–Menten constant, and v_m is the maximum reaction rate. The Hanes–Woolf equation was applied by plotting S/v with respect to S, to yield a straight line of slope $1/v_m$ and an intercept of K_m/v_m.

Lipase hydrolysis reactions were performed to estimate the reaction rate constants. In the tests, the initial activities of CALA and sdCALA were set at 0.35 U/mL, while the concentration of p-NPP was varied from 0.1 to 10 mg/mL (equivalent to 0.265–26.5 μmol/mL). The activity measured at 60 °C and pH 6 was used to represent the sample reaction rate [43].

3.6. Assays

Protein analysis was performed using 15% SDS-PAGE and Coomassie Blue staining [44]. The lipase activity was assayed using the modified method described by Chiou et al. [43]. A 50% ethanol solution of the reaction mixture was prepared as follows: 0.1 mL of free lipase was added to a mixture of 1 mL of p-NPP (5 mg/mL in 99% ethanol) and 1 mL of 50 mM phosphate buffer (pH 6.0), followed by incubating at 60 °C and 100 rpm for 5 min. The reaction was terminated by adding 0.9 mL of 0.5 N Na_2CO_3 to a 0.1 mL sample; all of the samples were centrifuged at 10,000 g for 15 min. The absorbance of the supernatant at 410 nm was measured using a UV–Vis spectrophotometer (Beckman DU-530, Fullerton, CA, USA). A molar extinction coefficient for p-nitrophenol of 15,000 M^{-1} cm^{-1} was used. Unless stated otherwise, one unit (1U) of enzyme activity was defined as the amount of enzyme required to produce 1 μmol of p-nitrophenol per min at 60 °C and pH 6.0.

For the thermal stability tests, 1 mL of sdCALA (0.35 U/mL) was heated on an incubator set at 37–90 °C for 30 min, followed by cooling in an ice bath to remove the residual heat. The sdCALA activity was measured at 60 °C and pH 6 for 5 min. For the organic solvent tests, the reaction substrate (p-NPP) was dissolved in various concentrations of ethanol, isopropanol, acetone, and acetonitrile, and then the activity was measured at 60 °C and pH 6 for 5 min. Tributyrin agar plate tests were based on the method described by Jung et al. [45]. A drop (10 μL) of the harvested broth was placed on a prepared tributyrin agar plate and left at room temperature for 3 days; then, the clear zone around the tested sample was scrutinized.

3.7. Statistical Analysis

To obtain statistical results, all data were analyzed using Origin software (v. 9.0). Experiments were performed in triplicate and data are expressed as means ± SD; significant differences between means were identified through analysis of variance (Origin software, v.9.0).

4. Conclusions

We have observed that the shaking rate has an effect on the production of surface-displayed lipase A. Since sdCALA was exposed outside the cell membrane, an increase in the shaking rate during cultivation had a negative effect on this surface-displayed enzyme. Indeed, the increased shaking rate lowered the specific activity of sdCALA as a result of shear stress. In contrast, the cultivation conditions did not affect the temperature and pH dependency of CALA and sdCALA. A kinetic study revealed that the cultivation of sdCALA at a moderate shaking rate optimized the hydrolysis performance.

Furthermore, the observed decrease in the specific activity of sdCALA in the presence of various organic solvents was presumably due to the effect of decomposition of the host cells.

Author Contributions: Data curation and methodology, C.-F.C. and S.-C.L., conceptualization, writing—original draft preparation; writing—review and editing, project administration, T.-Y.J. and Y.-C.L. All authors have read and agreed to the published version of the manuscript.

Funding: This study was supported by research grants from the National Science Council of Taiwan (grant nos. MOST 103-2221-E-005-071–MY3 and MOST 106-2313-B-005-029) and China Medical University (grant nos. CMU108-AWARD-01 and CMU-108-MF-122).

Conflicts of Interest: The authors declare no conflicts of interest.

References

1. Kirk, O.; Christensen, M.W. Lipases from Candida antarctica: Unique biocatalysts from a unique origin. *Org. Process Res. Dev.* **2002**, *6*, 446–451. [CrossRef]
2. Uppenberg, J.; Parkar, S.; Bergfors, T.; Jones, T.A. Crystallization and preliminary X-ray studies of lipase B from Candida antarctica. *J. Mol. Biol.* **1994**, *235*, 790–792. [CrossRef] [PubMed]
3. Blank, K.; Morfill, J.; Gumpp, H.; Gaub, H.E. Functional expression of Candida antarctica lipase B in Eschericha coli. *J. Biotechnol.* **2006**, *125*, 474–483. [CrossRef] [PubMed]
4. Widmann, M.; Juhl, P.B.; Pleiss, J. Structural classification by the lipase engineering database: A case study of Candida antarctica lipase A. *BMC Genom.* **2010**, *11*, 123–131. [CrossRef]
5. de María, P.D.; Carboni-Oerlemans, C.; Tuin, B.; Bargeman, G.; van der Meer, A.; van Gemert, R. Biotechnological applications of Candida antarctica lipase A: State-of-the-art. *J. Mol. Catal. B-Enzym.* **2005**, *37*, 36–46. [CrossRef]
6. Hoegh, I.; Patkar, S.; Halkier, T.; Hansen, M.T. Two lipases from Candida antarctica: Cloning and expression in Aspergillus oryzae. *Can. J. Bot.* **1995**, *73*, 869–875. [CrossRef]
7. Pfeffer, J.; Richter, S.; Nieveler, J.; Hansen, C.E.; Rhlid, R.B.; Schmid, R.D.; Rusnak, M. High yield expression of lipase A from Candida antarctica in the methylotrophic yeast Pichia pastoris and its purification and characterisation. *Appl. Microbiol. Biotechnol.* **2006**, *72*, 931–938. [CrossRef]
8. Pfeffer, J.; Rusnak, M.; Hansen, C.-E.; Rhlid, R.B.; Schmid, R.D.; Maurer, S.C. Functional expression of lipase A from Candida antarctica in Escherichia coli—A prerequisite for high-throughput screening and directed evolution. *J. Mol. Catal. B-Enzym.* **2007**, *45*, 62–67. [CrossRef]
9. Larsen, M.W.; Bornscheuer, U.T.; Hult, K. Expression of Candida antarctica lipase B in Pichia pastoris and various Escherichia coli systems. *Protein Expr. Purif.* **2008**, *62*, 90–97. [CrossRef]
10. Larios, A.; Garcia, H.S.; Oliart, R.M.; Valerio-Alfaro, G. Synthesis of flavor and fragrance esters using Candida antarctica lipase. *Appl. Microbiol. Biotechnol.* **2004**, *65*, 373–376. [CrossRef]
11. Ognjanovic, N.; Bezbradica, D.; Knezevic-Jugovic, Z. Enzymatic conversion of sunflower oil to biodiesel in a solvent-free system: Process optimization and the immobilized system stability. *Bioresour. Technol.* **2009**, *100*, 5146–5154. [CrossRef]
12. Smith, G.P. Filamentous fusion phage-novel expression vectors that display cloned antigens on the virion surface. *Science* **1985**, *228*, 1315–1317. [CrossRef] [PubMed]
13. Lee, S.Y.; Choi, J.H.; Xu, Z. Microbial cell-surface display. *Trends Biotechnol.* **2003**, *21*, 45–52. [CrossRef]
14. Wernerus, H.; Stahl, S. Biotechnological applications for surface engineered bacteria. *Biotechnol. Appl. Biochem.* **2004**, *40*, 209–228. [PubMed]
15. Margaritis, A.; Bassi, A.S. Principles and biotechnological applications of bacterial ice nucleation. *Crit. Rev. Biotechnol.* **1991**, *11*, 277–295. [CrossRef] [PubMed]
16. Edwards, A.R.; Van Den Bussche, R.A.; Wichman, H.A.; Orser, C.S. Unusual pattern of bacterial ice nucleation gene evolution. *Mol. Biol. Evol.* **1994**, *11*, 911–920.
17. van Bloois, E.; Winter, R.T.; Kolmar, H.; Fraaije, M.W. Decorating microbes: Surface display of proteins on Escherichia coli. *Trends Biotechnol.* **2011**, *29*, 79–86. [CrossRef]
18. Samuelson, P.; Gunneriusson, E.; Nygren, P.A.; Stahl, S. Display of proteins on bacteria. *J. Biotechnol.* **2002**, *96*, 129–154. [CrossRef]
19. Li, Q.; Yu, Z.; Shao, X.; He, J.; Li, L. Improved phosphate biosorption by bacterial surface display of phosphate binding protein utilizing ice nucleation protein. *FEMS Microbiol. Lett.* **2009**, *299*, 44–52. [CrossRef]

20. Shimazu, M.; Mulchandani, A.; Chen, W. Cell surface display of organophosphorus hydrolase using ice nucleation protein. *Biotechnol. Prog.* **2001**, *17*, 76–80. [CrossRef]
21. Wu, P.H.; Giridhar, R.; Wu, W.T. Surface display of transglucosidase on *Escherichia coli* by using the ice nucleation protein of *Xanthomonas campestris* and its application in glucosylation of hydroquinone. *Biotechnol. Bioeng.* **2006**, *95*, 1138–1147. [CrossRef] [PubMed]
22. Matsumoto, T.; Fukuda, H.; Ueda, M.; Tanaka, A.; Kondo, A. Construction of yeast strains with high cell surface lipase activity by using novel display systems based on the Flo1p flocculation functional domain. *Appl. Environ. Microbiol.* **2002**, *68*, 4517–4522. [CrossRef] [PubMed]
23. Moura, M.V.H.; Silva, G.P.D.; Machado, A.C.D.O.; Torres, F.A.G.; Freire, D.M.G.; Almeida, R.V. Displaying lipase B from Candida antarctica in Pichia pastoris using the yeast surface display approach: Prospection of a new anchor and characterization of the whole cell biocatalyst. *PLoS ONE* **2015**, *12*, e0141454. [CrossRef] [PubMed]
24. Liu, W.; Zhao, H.; Jia, B.; Xu, L.; Yan, Y. Surface display of active lipase in Saccharomyces cerevisiae using Cwp2 as an anchor protein. *Biotechnol. Lett.* **2010**, *32*, 255–260. [CrossRef]
25. Wilhelm, S.; Rosenau, F.; Becker, S.; Buest, S.; Hausmann, S.; Kolmar, H.; Jaeger, K.-E. Functional cell-surface display of a lipase-specific chaperone. *Chembiochem* **2007**, *8*, 55–60. [CrossRef]
26. Jo, J.-H.; Han, C.-W.; Kim, S.-H.; Kwon, H.-J.; Lee, H.-H. Surface display expression of Bacillus licheniformis lipase in Escherichia coli using Lpp'OmpA Chimera. *J. Microbiol.* **2014**, *52*, 856–862. [CrossRef]
27. Kan, S.-C.; Chen, C.-M.; Lin, C.-C.; Wu, J.-Y.; Shieh, C.-J.; Liu, Y.-C. Deciphering EGFP production via surface display and self-cleavage intein system in different hosts. *J. Taiwan Inst. Chem. Eng.* **2015**, *55*, 1–6. [CrossRef]
28. Lin, C.-C.; Liu, T.-T.; Kan, S.-C.; Zang, C.-Z.; Yeh, C.-W.; Wu, J.-Y.; Chen, J.-H.; Shieh, C.-J.; Liu, Y.-C. Production of D-Hydantoinase via surface display and self-cleavage system. *J. Biosci. Bioeng.* **2013**, *116*, 562–566. [CrossRef]
29. Wu, J.-Y.; Chen, C.-I.; Chen, C.-M.; Lin, C.-C.; Kan, S.-C.; Shieh, C.-J.; Liu, Y.-C. Cell disruption enhanced the pure EGFP recovery from an EGFP-intein- surface protein production system in recombinant *E. coli*. *Biochem. Eng. J.* **2012**, *68*, 12–18. [CrossRef]
30. Wu, J.-Y.; Tsai, T.-Y.; Liu, T.-T.; Lin, C.-C.; Chen, J.-H.; Yang, S.-C.; Shieh, C.-J.; Liu, Y.-C. Production of recombinant EGFP via surface display of ice nucleation protein and self-cleavage intein. *Biochem. Eng. J.* **2011**, *54*, 158–163. [CrossRef]
31. Shamel, M.M.; Ramachandran, K.B.; Hasan, M. Operational stability of lipase enzyme: Effect of temperature and shear. *Dev. Chem. Eng. Min. Process.* **2005**, *13*, 599–604. [CrossRef]
32. Dimitrijevic, A.; Velickovic, D.; Bihelovic, F.; Bezbradica, D.; Jankov, R.; Milosavic, N. One-step, inexpensive high yield strategy for Candida antarctica lipase A isolation using hydroxyapatite. *Bioresour. Technol.* **2012**, *107*, 358–362. [CrossRef] [PubMed]
33. Zamost, B.L.; Nielsen, H.K.; Starnes, R.L. Thermostable enzymes for industrial applications. *J. Ind. Microbiol.* **1991**, *8*, 71–81. [CrossRef]
34. Panpipat, W.; Xu, X.; Guo, Z. Improved acylation of phytosterols catalyzed by Candida antarctica lipase A with superior catalytic activity. *Biochem. Eng. J.* **2013**, *70*, 55–62. [CrossRef]
35. Boran, R.; Ugur, A. Partial purification and characterization of the organic solvent-tolerant lipase produced by Pseudomonas fluorescens RB02-3 isolated from milk. *Prep. Biochem. Biotechnol.* **2010**, *40*, 229–241. [CrossRef] [PubMed]
36. El-Batal, A.I.; Farrag, A.A.; Elsayed, M.A.; El-Khawaga, A.M. Effect of environmental and nutritional parameters on the extracellular lipase production by Aspergillus niger. *Int. Lett. Nat. Sci.* **2016**, *60*, 18–29. [CrossRef]
37. Yang, C.; Wang, F.; Lan, D.; Whiteley, C.; Yang, B.; Wang, Y. Effects of organic solvents on activity and conformation of recombinant *Candida antarctica* lipase A produced by *Pichia pastoris*. *Process Biochem.* **2012**, *47*, 533–537. [CrossRef]
38. Ramos, J.L.; Duque, E.; Rodríguez-Herva, J.J.; Godoy, P.; Haïdour, A.; Reyes, F.; Fernández-Barrero, A. Mechanisms for solvent tolerance in bacteria. *J. Biol. Chem.* **1997**, *272*, 3887–3890. [CrossRef]
39. Segura, A.; Duque, E.; Mosqueda, G.; Ramos, J.L.; Junker, F. Multiple responses of Gram-negative bacteria to organic solvents. *Environ. Microbiol.* **1999**, *1*, 191–198. [CrossRef]
40. Sambrook, J.; Russel, D.W. *Molecular Cloning: A Laboratory Manual*, 3rd ed.; Cold Spring Harbor Laboratory Press: New York, NY, USA, 2001.

41. Sanchez, C.E.G.; Martınez-Trujillo, A.; Osorio, G.A. Oxygen transfer coefficient and the kinetic parameters of exo-polygalacturonase production by Aspergillus flavipes FP-500 in shake flasks and bioreactor. *Lett. Appl. Microbiol.* **2012**, *55*, 444–452. [CrossRef]
42. Camacho, F.G.; Rodrıguez, J.J.G.; Miron, A.S.; Garcıa, M.C.C.; Belarbi, E.H.; Grima, E.M. Determination of shear stress thresholds in toxic dinoflagellates cultured in shaken flasks Implications in bioprocess engineering. *Process Biochem.* **2007**, *42*, 1506–1515. [CrossRef]
43. Chiou, S.H.; Wu, W.T. Immobilization of Candida rugosa lipase on chitosan with activation of the hydroxyl groups. *Biomaterials* **2004**, *25*, 197–204. [CrossRef]
44. Laemmli, U.K. Cleavage of structural proteins during the assembly of the head of bacteriophage T4. *Nature* **1970**, *227*, 680–685. [CrossRef] [PubMed]
45. Jung, H.C.; Ko, S.; Ju, S.J.; Kim, E.J.; Kim, M.K.; Pan, J.G. Bacterial cell surface display of lipase and its randomly mutated library facilitates high-throughput screening of mutants showing higher specific activities. *J. Mol. Catal. B-Enzym.* **2003**, *26*, 177–184. [CrossRef]

© 2020 by the authors. Licensee MDPI, Basel, Switzerland. This article is an open access article distributed under the terms and conditions of the Creative Commons Attribution (CC BY) license (http://creativecommons.org/licenses/by/4.0/).

Communication

Continuous-Flow Synthesis of β-Amino Acid Esters by Lipase-Catalyzed Michael Addition of Aromatic Amines

Li-Hua Du [1,*]**, Rui-Jie Long** [1]**, Miao Xue** [1]**, Ping-Feng Chen** [1]**, Meng-Jie Yang** [1] **and Xi-Ping Luo** [2,*]

[1] College of Pharmaceutical Science, ZheJiang University of Technology, Hangzhou 310014, China; longruijie@zjut.edu.cn (R.-J.L.); xuemiao@zjut.edu.cn (M.X.); chenpingfeng@zjut.edu.cn (P.-F.C.); yangmj1002@126.com (M.-J.Y.)
[2] Zhejiang Provincial Key Laboratory of Chemical Utilization of Forestry Biomass, Zhejiang A&F University, Hangzhou, 311300, China
* Correspondence: orgdlh@zjut.edu.cn (L.-H.D.); luoxiping@zafu.edu.cn (X.-P.L.); Tel.: +86-571-8832-0903 (X.-P.L.)

Received: 16 March 2020; Accepted: 14 April 2020; Published: 16 April 2020

Abstract: A continuous-flow procedure for the synthesis of β-amino acid esters has been developed via lipase-catalyzed Michael reaction of various aromatic amines with acrylates. Lipase TL IM from *Thermomyces lanuginosus* was first used to catalyze Michael addition reaction of aromatic amines. Compared with other methods, the salient features of this work include green reaction conditions (methanol as reaction medium), short residence time (30 min), readily available catalyst and a reaction process that is easy to control. This enzymatic synthesis of β-amino acid esters performed in continuous-flow microreactors is an innovation that provides a new strategy for the fast biotransformation of β-amino acid esters.

Keywords: enzymatic synthesis; β-amino acid esters; microreactor; aromatic amines; Michael addition

---

1. Introduction

β-Amino acid ester derivatives have been widely found in biologically active molecules, many of which possess useful biological activities (e.g., anticancer, antiviral, antibacterial, antifungal, antipsychotics) [1–6]. β-Amino acid esters have also been used to construct peptidomimetic oligomers, which are of high interest in medicinal chemistry [7,8]. In particular, β-amino acid esters based on aromatic amines are important intermediates for the effective synthesis of urease inhibitors, which can be used in the treatment of *Helicobacter pylori* infection [9]. Due to these widespread applications in pharmacy and materials science, the development of synthesis methods of β-amino acid esters has become a hot topic in the chemical field.

Michael addition is one of the simplest and most effective strategies to prepare β-amino acid esters. Recently, Ahmed has reported the Michael addition of aromatic amines mediated by KOtBu [10]; HOTf (trifluoromethylsulfonic acid) was also used to catalyze the Michael addition of aromatic amines, and good yields were obtained within 4 h [11]. However, the use of acids and alkalis has brought challenges to the environment. In order to avoid such problems, various transition-metal catalysts have been developed and used to efficiently catalyze the Michael addition to synthesize β-amino acid esters [12–17]. Although desired results were obtained, the use of metal catalysts and expensive ligands are still shortcomings for the practical utility.

Enzymes are versatile catalysts of biological origin that can catalyze reactions with great specificity and high efficiency under the benign environmental conditions [18–20]. In the past few years, several works about the synthesis of β-amino esters by enzymatic Michael addition reaction were reported [21–23]. However, there are few reports on the preparation of β-amino acid esters by the enzymatic Michael addition of aromatic amines. Bhanage reported that immobilized HMC:PFL (lipase from *Pseudomonas fluorescence* immobilized on hydroxypropyl methyl cellulose) could be used to catalyze the Michael addition of aromatic amines in toluene, although only 33%–40% yields could be obtained in 3 h [24]. Lipase CAL-B (lipase B from *Candida antarctica*) has also been used to catalyze the Michael addition of aromatic amines, but it requires a long reaction time (72 h) to achieve desired results [25]. Our attention is focused on finding a more efficient and green synthesis method of β-amino acid esters based on aromatic amines. In recent years, more and more reports have been published on enzymatic reactions in continuous-flow microreactors due to their advantages of improving process efficiency and material transfer [26–28]. The small size of the continuous-flow microreactors facilitates control of the reaction parameters, which can reduce waste generation and increase productivities [29–31].

In the interest of developing an efficient and green synthesis method for β-amino acid esters based on aromatic amines, we employed a continuous flow technology for the synthesis of β-amino acid esters. Lipase TL IM from *Thermomyces lanuginosus* was first used to catalyze the Michael addition reaction of aromatic amines with acrylates in continuous-flow microreactors, and 17 β-amino acid esters (3a–3r) were obtained with excellent yields in 30 min (Scheme 1). Effects of different reaction parameters including reaction solvent, temperature, enzyme, substrate ratio, residence time and aromatic amine structure on the reaction were studied.

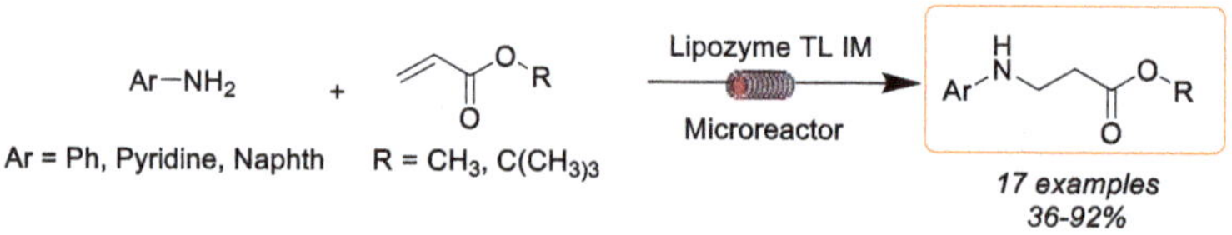

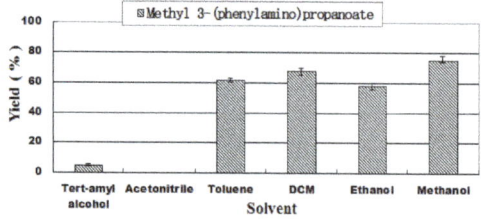

Scheme 1. Synthesis of β-amino acid esters by enzymatic Michael reaction of aromatic amines in continuous-flow microreactors.

2. Results and Discussion

2.1. Effect of Reaction Media

The solvent plays an important role in enzymatic reaction by promoting the dissolution of the substrates and controlling the reaction chemo-selectivity [32]. Therefore, we first studied the effect of reaction media on β-amino acid ester synthesis in continuous-flow microreactors (Figure 1). We chose aniline (**1a**) and methyl acrylate (**2a**) as the model reaction; several solvents, including tert-amyl alcohol, acetonitrile, toluene, DCM (dichloromethane), ethanol and methanol, were tested. As shown in Figure 1, the reaction did not proceed in acetonitrile, and only 5.2% of the desired product was obtained in tert-amyl alcohol. Further solvent screening disclosed that methanol was the best solvent. Aromatic amines are a class of poor nucleophiles whose nucleophilicity is highly related to solvents. Research has shown that polar solvents can promote Michael addition of aromatic amines [33].

Figure 1. The effect of reaction media on enzymatic synthesis of β-amino acid esters in continuous-flow microreactors.

2.2. Effect of Reaction Temperature

Temperature plays significant functions, such as improving the interaction between the substrate and catalyst and affecting the stability of enzyme [34]. Therefore, we investigated the effect of reaction temperature on the synthesis of β-amino acid esters; the results are shown in Figure 2. We performed the reactions from 30 to 55 °C and found that the best yield (80.3%) can be achieved at 35 °C. After that, as the reaction temperature continued to increase, the yield was not improved. The higher temperature may cause a decrease in the selectivity of enzymatic reaction, which favors the double Michael product formation. Therefore, the optimal reaction temperature for the synthesis of β-amino acid esters is 35 °C.

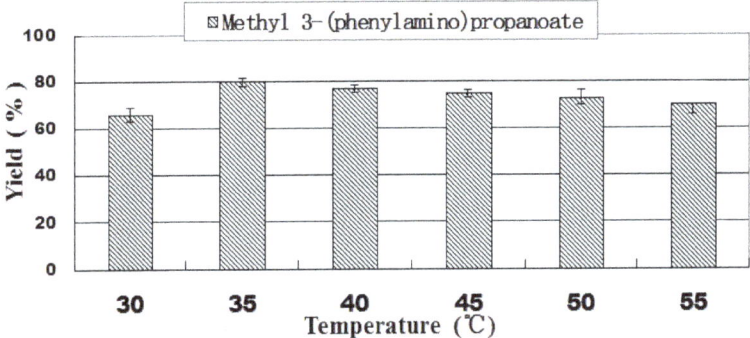

Figure 2. The effect of reaction temperature on enzymatic synthesis of β-amino acid esters in continuous-flow microreactors.

2.3. Catalyst Screening

We tested the efficiency of three enzymes in catalyzing β-amino acid ester synthesis (Figure 3). Further screening of enzymes showed that lipozyme TL IM was the best catalyst. In the absence of lipozyme TL IM, a group of blank tests was performed, and it was found that the reaction did not occur. This indicated that lipozyme TL IM played a key role in the synthesis of β-amino acid esters.

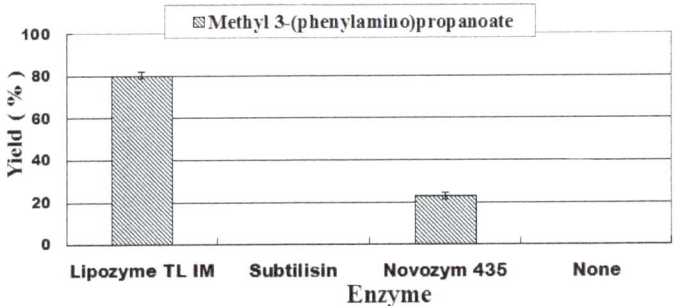

Figure 3. The catalyst screening on enzymatic synthesis of β-amino acid esters in continuous-flow microreactors.

2.4. Effect of Substrate Ratio

Another important factor that affects the synthesis of β-amino acid ester is the molar ratio of aniline to methyl acrylate. The influence of molar ratio (aniline/methyl acrylate) was investigated from 1:1 to 1:6. According to the Figure 4, the best reaction yield 80.3% was obtained when the molecular ratio of aniline (**1a**) and methyl acrylate (**2a**) reached 1:4. As the content of methyl acrylate continued to increase, the yield of target product **3a** was decreased. This occurred since a higher amount of methyl acrylate favored the double Michael product, which reduced the yield of the target product **3a**. Therefore, we decided to choose aniline/methyl acrylate = 1:4 as the optimum substrate ratio.

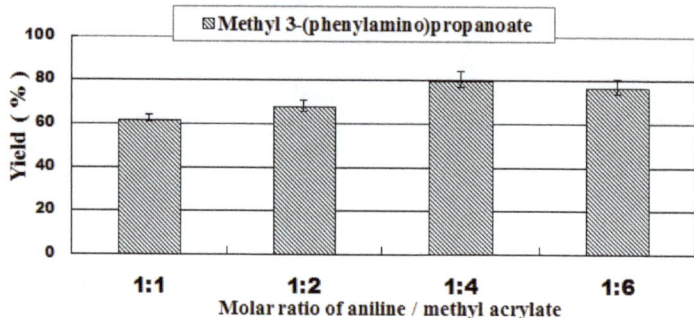

Figure 4. The effect of substrate ratio on enzymatic synthesis of β-amino acid esters in continuous-flow microreactors.

2.5. Effect of Residence Time

It is well known that prolonged reactions can promote the formation of impurities and reduce the selectivity of products. In addition, in the case of Michael reaction, the long-term reaction facilitates the accumulation of double Michael products [33]. In view of this, we performed the reaction from 20 to 40 min to investigate the effect of residence time/flow rate on the reaction; the results are shown in Figure 5. The best yield was reached in 30 min with a flow rate of 20.8 µL min^{-1}. As we continued to extend residence time, the yield of **3a** was decreased slightly. Detection found that with the increase of time, the target product underwent a second Michael addition reaction, which reduced the yield of the target product. Thus, we chose 30 min as the optimum residence time for the following experiment.

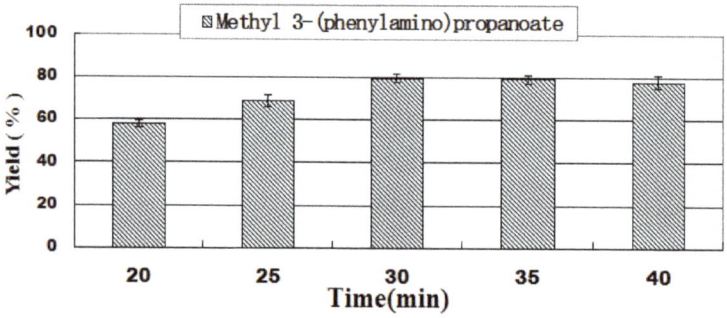

Figure 5. The effect of residence time on enzymatic synthesis of β-amino acid esters in continuous-flow microreactors.

2.6. The Effect of Aromatic Amine Structure on the Reaction

After identifying the optimum reaction conditions, we continued to investigate the effect of aromatic amine structure on the enzymatic β-amino acid ester synthesis reaction under continuous-flow microreactors. As shown in Figure 6, the reaction yields were 80.3%, 92.4%, 43.7% and 36.3% for aniline (**1a**), 4-toluidine (**1b**), 2-aminopyridine (**1g**) and 4-chloroaniline (**1e**), respectively. Experimental results showed that aromatic amines with electron-donating functional groups exhibited high reactivity in the enzymatic Michael addition reaction. The electron-donating groups increased the nucleophilicity of aniline, making the conjugated addition easier. In contrast, aromatic amines containing electron-withdrawing groups had lower yields in the enzymatic Michael addition reaction.

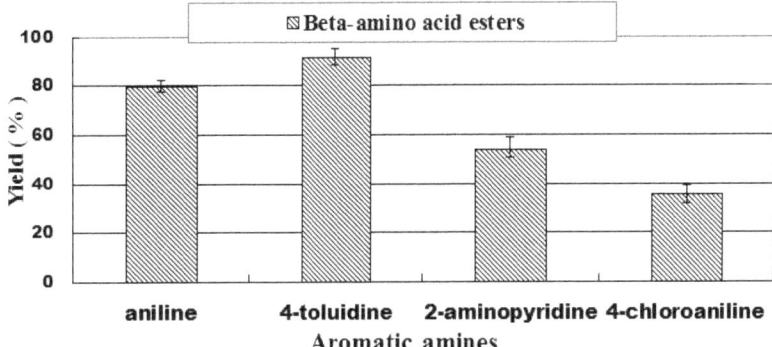

Figure 6. The effect of aromatic amine structure on enzymatic synthesis of β-amino acid esters in continuous-flow microreactors.

2.7. Enzymatic Synthesis of β-Amino Acid Ester in Continuous-Flow Microreactor and Batch Bioreactor

Due to high surface-to-volume ratio, better heat exchange and efficient mixing, continuous-flow microreactor technology has become increasingly popular as an alternative to conventional batch chemistry synthesis. In order to investigate the effects of enzymatic reactions in different reactors, we performed the enzymatic Michael addition of aniline to synthesize β-amino acid ester in the continuous-flow microreactor and in a batch bioreactor (Table 1). We detected the reaction process in the batch bioreactor and found that it took about 24 h to achieve the ideal yield (method B). However, in the continuous-flow microreactor, we could get better yield in 30 min (method A). Experiments showed that the continuous-flow microreactor can improve the efficiency of enzymatic synthesis of β-amino acid esters.

Table 1. Enzymatic Michael addition of aniline in the continuous-flow reactor or batch bioreactor. [a]

Entry	Method	Yield [b] (%)
1	A	80.3 ± 1.5
2	B	62.3 ± 2.5

[a] General reaction conditions: Method A: Continuous-flow microreactor, feed A. (5.0 mmol aniline in 10 mL methanol), feed B (20.0 mmol methyl acrylate in 10 mL methanol), residence time: 30 min, flow rate: 20.8 µL min−1, 35 °C, lipozyme TL IM 870 mg. Method B: Batch bioreactor, 5.0 mmol aniline, 20.0 mmol methyl acrylate and 20 mL methanol in a 50 mL conical flask, 200 r·min−1, lipozyme TL IM 870 mg, 35 °C, 24 h. [b] Yield of 3a. Yield: 100× (actual received quality/ideal calculated quality). The data are presented as average ± SD of triplicate experiments.

2.8. Combined Effects of Process Parameters

Lipase-catalyzed synthesis of β-amino acid ester by Michael addition between aniline and methyl acrylate in continuous-flow microreactors was studied, and the optimum conditions for the Michael addition were investigated. The results are shown in Tables 2 and 3.

Table 2. Solvent, temperature and enzyme screening for enzymatic synthesis of β-amino acid esters. [a]

Entry	Catalyst	Solvent	T (°C)	Yield [b] (%)
1	Lipozyme TL IM	Tert-amyl alcohol	45	5.2 ± 0.7
2	Lipozyme TL IM	Acetonitrile	45	n.d.
3	Lipozyme TL IM	Toluene	45	62.0 ± 1.2
4	Lipozyme TL IM	DCM	45	68.3 ± 1.9
5	Lipozyme TL IM	Ethanol	45	58.2 ± 2.3
6	Lipozyme TL IM	Methanol	45	75.1 ± 0.9
7	Lipozyme TL IM	Methanol	30	66.5 ± 2.1
8	Lipozyme TL IM	Methanol	35	80.3 ± 1.5
9	Lipozyme TL IM	Methanol	40	77.4 ± 2.5
10	Lipozyme TL IM	Methanol	50	73.1 ± 2.6
11	Lipozyme TL IM	Methanol	55	70.0 ± 1.7
12	Subtilisin	Methanol	35	n.d.
13	Novozym 435	Methanol	35	23.2 ± 0.9
14	None	Methanol	35	n.d.

[a] Experimental conditions: feed A, 5.0 mmol aniline (**1a**) was dissolved in 10 mL solvent, feed B, 20.0 mmol methyl acrylate (**2a**) was dissolved in 10 mL solvent, enzyme 870 mg, 30 min. [b] Yield of **3a**. n.d. means no reaction was found. Yield: 100× (actual received quality/ideal calculated quality). The data are presented as average ± SD of triplicate experiments.

Table 3. Effects of substrate ratio and residence time on enzymatic synthesis of β-amino acid esters. [a]

Entry	$n_{1a}:n_{2a}$	Time/Min	Flow Rate/μL min^{-1}	Yield [b] (%)
1	1:1	30	20.8	62.6 ± 2.3
2	1:2	30	20.8	68.1 ± 1.6
3	1:4	30	20.8	80.3 ± 1.5
4	1:6	30	20.8	77.4 ± 2.8
5	1:4	20	31.2	58.7 ± 2.2
6	1:4	25	25.0	69.0 ± 0.9
7	1:4	35	17.8	80.1 ± 1.6
8	1:4	40	15.6	78.0 ± 1.2

[a] Experimental conditions: feed A, 5.0 mmol aniline (**1a**) was dissolved in 10 mL methanol, feed B, methyl acrylate (**2a**) was dissolved in 10 mL methanol, lipozyme TL IM 870 mg, 35 °C. [b] Yield of **3a**. Yield: 100× (actual received quality/ideal calculated quality). The data are presented as average ± SD of triplicate experiments.

For enzymatic reactions, the reaction medium and temperature are important influencing factors which affect the enzyme activity and process of the reaction. Preliminary experiments showed that the best reaction yield can be achieved when the reaction was catalyzed by lipozyme TL IM in methanol at 35 °C (Entry 8, Table 2). As we continued to increase the reaction temperature, the yield was not improved. Higher temperature may decrease the selectivity of enzymatic reactions, which increases the by-product formation. Residence time and substrate ratio are also important factors affecting the synthesis of β-amino acid esters. With the increase of residence time or methyl acrylate, the target product **3a** underwent a second Michael addition reaction, which reduced the yield of the target product.

2.9. The Scope and Limitation for the β-Amino Acid Ester Synthesis Methodology

Finally, we explored the scope and limitation of this continuous flow methodology for β-amino acid ester synthesis catalyzed by lipase TL IM from *Thermomyces lanuginosus*. Nine aromatic amines (aniline (**1a**), 4-toluidine (**1b**), 4-tert-butylaniline (**1c**), 4-methoxyaniline (**1d**), 4-chloroaniline (**1e**), 4-bromoaniline (**1f**), 2-aminopyridine (**1g**), 3,4-(methylenedioxy)aniline (**1h**) and 2-naphthylamine (**1i**)) and two acrylates (methyl acrylate (**2a**) and tert-butyl acrylate (**2b**)) were subjected to the optimal reaction conditions, using continuous-flow microreactors. Seventeen β-amino acid esters (**3a–3r**) were synthesized after 30 min residence time in excellent yields (Table 4).

Table 4. Continuous-flow synthesis of β-amino acid esters by enzymatic Michael addition of aromatic amines. [a]

Entry	Aromatic Amine	Product	Yield [b] (%)
1	aniline (**1a**)	**3a**	80.3 ± 1.5
2	aniline (**1a**)	**3b**	75.1 ± 1.9
3	4-toluidine (**1b**)	**3c**	92.4 ± 0.8
4	4-toluidine (**1b**)	**3d**	81.2 ± 1.7
5	4-tert-butylaniline (**1c**)	**3e**	83.6 ± 2.1
6	4-tert-butylaniline (**1c**)	**3f**	87.7 ± 2.9
7	4-methoxyaniline (**1d**)	**3g**	85.0 ± 2.7
8	4-methoxyaniline (**1d**)	**3h**	82.1 ± 2.6
9	4-chloroaniline (**1e**)	**3i**	36.3 ± 1.2
10	4-chloroaniline (**1e**)	**3j**	48.5 ± 3.1
11	4-bromoaniline (**1f**)	**3k**	42.4 ± 1.5
12	4-bromoaniline (**1f**)	**3l**	47.3 ± 1.1
13	2-aminopyridine (**1g**)	**3m**	43.7 ± 2.4
14	2-aminopyridine (**1g**)	**3n**	trace

Table 4. Cont.

Entry	Aromatic Amine	Product	Yield [b] (%)
15	3,4-(methylenedioxy)aniline (1h)	3o	88.1 ± 2.2
16	3,4-(methylenedioxy)aniline (1h)	3p	85.4 ± 1.7
17	2-naphthylamine (1i)	3q	74.0 ± 1.4
18	2-naphthylamine (1i)	3r	72.1 ± 1.2

[a] Experimental conditions: feed A, 5.0 mmol aromatic amine was dissolved in 10 mL methanol, feed B, 20.0 mmol acrylic ester was dissolved in 10 mL methanol, lipozyme TL IM 870 mg, 30 min, 35 °C. [b] Yield of β-amino acid esters. Yield: 100× (actual received quality/ideal calculated quality). The data are presented as average ± SD of triplicate experiments.

The results showed that this continuous-flow enzymatic method has a broad applicability. Most aromatic amines showed good reactivity, except the halogen-substituted anilines and 2-aminopyridine (Entries 9–14, Table 4). The electronic effects of substrates have some impacts on additions. The introduction of electron-donating groups increased the nucleophilicity of aniline, making the conjugated addition easier, which was consistent with previous reports [25]. In addition, our studies have shown that Michael addition of aromatic amines catalyzed by lipase in continuous-flow microreactors was markedly influenced by the reaction parameters. The lipozyme TL IM catalyzed the Michael addition of aromatic amines with acrylates in methanol, which could effectively achieve the synthesis of β-amino acid esters. The selectivity of enzymatic reaction decreased with high reaction temperature or long residence time, which led to the increase of by-products. Compared with conventional batch bioreactor, continuous-flow microreactor improved the efficiency of enzymatic synthesis, which realized the rapid conversion of β-amino acid esters.

3. Materials and Methods

3.1. Materials

Unless otherwise stated, all chemicals were obtained from commercial sources and used without further purification. Lipozyme TL IM from *Thermomyces lanuginosus* was purchased from Novo Nordisk (Copenhagen, Denmark), and 4-tert-butylaniline, 4-methoxyaniline, 3,4-(methylenedioxy)aniline, 2-aminopyridine, 2-naphthylamine were purchased from Energy Chemical (Shanghai, China). Aniline, 4-toluidine, 4-chloroaniline, 4-bromoaniline, methyl acrylate and tert-butyl acrylate were purchased from Aladdin (Shanghai, China). Harvard Apparatus PHD 2000 syringe pumps were purchased from Harvard (Holliston, MA, USA). Bruker-ADNANCE III 500 MHz NMR spectrometer (Billerica, MA, USA) and Liquid Chromatography/Mass Spectrometer Detector (Agilent LC1290-MS6530, Santa Clara, CA, USA) were used in this study.

3.2. β-Amino Acid Ester Synthesis Operating Conditions

3.2.1. Experimental Setup

The equipment configuration that was used for synthesis of β-amino acid esters via lipase-catalyzed Michael reaction of aromatic amines with acrylates is described in Figure 7. The continuous-flow microreactor was composed of a syringe pump (Harvard Apparatus PHD 2000), reactant injectors, Y-shaped mixer (Φ = 1.8 mm; M), microchannel reactor and product collector. Syringe pump was used to deliver reagents from reactant injectors to the Y-shaped mixer and then to the microchannel reactor. Reagent feed A (10 mL) with aromatic amines in methanol solution and reagent feed B (10 mL) with acrylates in methanol solution were fully mixed in the Y-shaped mixer. The microchannel reactor consists of a PFA reactor coil (inner diameter ID = 2.0 mm, length = 100 cm) filled with lipozyme TL IM (catalyst reactivity: 250 IUN g^{-1}). The microchannel reactor was immersed into a thermostatic apparatus (water bath) to control the temperature. The final solution was collected in a product collector connected to the microchannel reactor. The residence time was controlled by setting the flow rates of feed A and feed B.

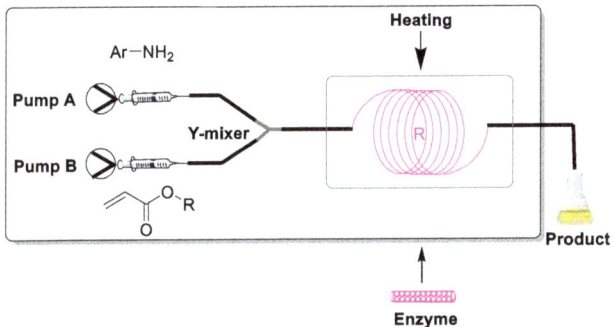

Figure 7. Experimental setup for the synthesis of β-amino acid esters.

3.2.2. General Procedure for β-Amino Acid Ester Synthesis in Continuous Flow Microreactor

Here, 5.0 mmol of aromatic amines were dissolved in 10 mL methanol (feed A, ~0.5 M), and 20.0 mmol acrylates were dissolved in 10 mL methanol (feed B, ~2.0 M). Lipozyme TL IM (870 mg) was added to the PFA reactor coil (inner diameter ID = 2.0 mm, length = 100 cm). Streams A and B were mixed together at a flow rate of 10.4 µL min^{-1} in a Y-mixer at 35 °C, and the resulting stream (20.8 µL min^{-1}) was connected to a sample vial which was used to collect the final mixture. The final mixture was then evaporated, and the residue was submitted to column chromatography on silica gel (200–300 mesh). The crude product was purified by silica gel flash chromatography with a petroleum ether/ethyl acetate gradient from 20:1 to 5:1. The purification was monitored by TLC. The fractions containing the main products were pooled, and the solvent was evaporated. The residue was analyzed by ^{1}H NMR, ^{13}C NMR and ESI-MS.

3.3. Analytical Methods

3.3.1. Thin-Layer Chromatography

Thin-layer chromatography analysis was conducted on silica gel plates with ethyl acetate/petroleum ether (1:5, by vol) as the eluent. Spots were detected by ultraviolet irradiation at 254 nm.

3.3.2. High-Performance Liquid Chromatography (HPLC)

The reaction was monitored by HPLC analysis using a Shim-Pack VP-ODS column (4.6 × 150 mm) and a UV detector (285 nm). Hexane/2-propanol solution 90/10 (v/v) was used as the mobile phase (flow rate: 1.0 mL min^{-1}).

3.3.3. Nuclear Magnetic Resonance (NMR) and Electrospray Ionization Mass Spectrometry (ESI/MS) Analysis

After purification of the synthesized products by column chromatography, the chemical structures of β-amino acid esters were determined by ^{1}H NMR, ^{13}C NMR and ESI-MS. ^{1}H (500 MHz) and ^{13}C (126 MHz) NMR spectra were acquired on a Bruker-ADNANCE III 500 MHz NMR spectrometer. The sample temperature was 22 °C and using CDCl$_3$ or DMSO-d$_6$ as solvent. ESI-MS was measured on a Liquid Chromatography/Mass Spectrometer Detector (Agilent LC1290-MS6530). Evaporated samples were dissolved in methanol. The injection volume was 10 μL. Reaction mixtures were injected and infused into the electrospray ion source at 0.2 mL min^{-1}. The spectrometer was operated in the positive ionization mode with the capillary voltage set to +3.5 kV. The sheath gas flow rate was 11 L min^{-1}, and the sheath gas temperature was 300 °C.

Methyl 3-(phenylamino)propanoate (**3a**). White solid. ^{1}H NMR (500 MHz, CDCl$_3$) δ 7.23 (dd, J = 8.6, 7.3 Hz, 2H), 6.77 (tt, J = 7.3, 1.1 Hz, 1H), 6.67 (dd, J = 8.7, 1.1 Hz, 2H), 3.74 (s, 3H), 3.49 (t, J = 6.4 Hz, 2H), 2.66 (t, J = 6.4 Hz, 2H); ^{13}C NMR (126 MHz, CDCl$_3$) δ 172.83, 147.59, 129.35, 117.77, 113.09, 51.76, 39.47, 33.73. HRMS (ESI): calcd for C$_{10}$H$_{14}$NO$_2$ [M + H]$^+$: 180.1016, found: 180.1018.

Tert-butyl 3-(phenylamino)propanoate (**3b**). White powder. ^{1}H NMR (500 MHz, CDCl$_3$) δ 7.21 (dd, J = 8.6, 7.3 Hz, 2H), 6.75 (tt, J = 7.3, 1.1 Hz, 1H), 6.65 (dd, J = 8.6, 1.2 Hz, 2H), 3.43 (t, J = 6.4 Hz, 2H), 2.55 (t, J = 6.4 Hz, 2H), 1.49 (s, 9H); ^{13}C NMR (126 MHz, CDCl$_3$) δ 171.72, 147.73, 129.25, 117.60, 113.06, 80.83, 39.65, 35.08, 28.10. HRMS (ESI): calcd for C$_{14}$H$_{16}$NO$_2$ [M + H]$^+$: 222.1489, found: 222.1491.

Methyl 3-(p-tolylamino)propanoate (**3c**). White solid. ^{1}H NMR (500 MHz, DMSO-d$_6$) δ 6.84–6.96 (m, 2H), 6.48 (d, J = 8.4 Hz, 2H), 5.37 (t, J = 6.0 Hz, 1H), 3.61 (s, 3H), 3.24 (q, J = 6.6 Hz, 2H), 2.55 (t, J = 6.8 Hz, 2H), 2.14 (s, 3H); ^{13}C NMR (126 MHz, DMSO) δ 172.18, 146.17, 129.39, 124.31, 112.29, 51.36, 39.77, 33.56, 20.09. HRMS (ESI): calcd for C$_{11}$H$_{16}$NO$_2$ [M + H]$^+$: 194.1176, found: 194.1177.

Tert-butyl 3-(p-tolylamino)propanoate (**3d**). White solid. ^{1}H NMR (500 MHz, CDCl$_3$) δ 7.07 (d, J = 7.8 Hz, 2H), 6.63 (d, J = 8.4 Hz, 2H), 3.44 (t, J = 6.4 Hz, 2H), 2.58 (t, J = 6.4 Hz, 2H), 2.33 (s, 3H), 1.54 (s, 9H); ^{13}C NMR (126 MHz, CDCl$_3$) δ 171.59, 145.38, 129.58, 126.51, 113.12, 80.49, 39.85, 34.92, 27.93, 20.21. HRMS (ESI): calcd for C$_{14}$H$_{22}$NO$_2$ [M + H]$^+$: 236.1643, found: 236.1646.

Methyl 3-((p-(tert-butyl)phenyl)amino)propanoate (**3e**). White powder. ^{1}H NMR (500 MHz, DMSO-d$_6$) δ 6.91–7.24 (m, 2H), 6.66 – 6.18 (m, 2H), 5.41 (t, J = 6.0 Hz, 1H), 3.61 (s, 3H), 3.25 (q, J = 6.6 Hz, 2H), 2.56 (t, J = 6.8 Hz, 2H), 1.21 (s, 9H); ^{13}C NMR (126 MHz, DMSO) δ 172.19, 146.03, 138.17, 125.53, 111.90, 51.36, 38.98, 33.61, 33.46, 31.49. HRMS (ESI): calcd for C$_{14}$H$_{22}$NO$_2$ [M + H]$^+$: 236.1643, found: 236.1645.

Tert-butyl 3-((p-(tert-butyl)phenyl)amino)propanoate (**3f**). Yellow oil. ^{1}H NMR (500 MHz, CDCl$_3$) δ 7.17–7.38 (m, 2H), 6.48–6.77 (m, 2H), 3.45 (t, J = 6.4 Hz, 2H), 2.58 (t, J = 6.4 Hz, 2H), 1.52 (s, 9H), 1.35 (s, 9H); ^{13}C NMR (126 MHz, CDCl$_3$) δ 171.72, 145.36, 140.25, 125.95, 112.76, 80.66, 39.85, 35.18, 33.77, 31.50, 28.06. HRMS (ESI): calcd for C$_{17}$H$_{28}$NO$_2$ [M + H]$^+$: 278.2112, found: 278.2113.

Methyl 3-((p-methoxyphenyl)amino)propanoate (**3g**). Light yellow solid. ^{1}H NMR (500 MHz, DMSO-d$_6$) δ 6.68–6.80 (m, 2H), 6.65 – 6.45 (m, 2H), 5.14 (t, J = 6.1 Hz, 1H), 3.65 (s, 3H), 3.63 (s, 3H), 3.26 (q, J = 6.6 Hz, 2H), 2.58 (t, J = 6.8 Hz, 2H); ^{13}C NMR (126 MHz, DMSO) δ 172.10, 150.86, 142.53, 114.47, 113.21, 55.09, 51.13, 39.54, 33.50. HRMS (ESI): calcd for C$_{11}$H$_{16}$NO$_3$ [M + H]$^+$: 210.1124, found: 210.1127.

Tert-butyl 3-((p-methoxyphenyl)amino)propanoate (**3h**). Yellow oil. ^{1}H NMR (500 MHz, CDCl$_3$) δ 6.71–6.96 (m, 2H), 6.38–6.73 (m, 2H), 3.76 (s, 3H), 3.36 (t, J = 6.4 Hz, 2H), 2.52 (t, J = 6.4 Hz, 2H), 1.47 (s, 9H); ^{13}C NMR (126 MHz, CDCl$_3$) δ 171.76, 152.29, 141.90, 114.84, 114.57, 55.69, 40.74, 35.06, 28.06. HRMS (ESI): calcd for C$_{14}$H$_{22}$NO$_3$ [M + H]$^+$: 252.1596, found: 252.1595.

Methyl 3-((4-chlorophenyl)amino)propanoate (**3i**). White powder. ^{1}H NMR (500 MHz, CDCl$_3$) δ 7.06–7.19 (m, 2H), 6.50–6.58 (m, 2H), 3.71 (s, 3H), 3.42 (t, J = 6.4 Hz, 2H), 2.62 (t, J = 6.3 Hz, 2H); ^{13}C NMR

(126 MHz, CDCl$_3$) δ 172.63, 146.10, 129.05, 122.16, 114.03, 51.75, 39.44, 33.46. HRMS (ESI): calcd for C$_{10}$H$_{13}$ClNO$_2$ [M + H]$^+$: 214.0628, found: 214.0630.

Tert-butyl 3-((4-chlorophenyl)amino)propanoate (**3j**). Yellow oil. ^{1}H NMR (500 MHz, CDCl$_3$) δ 6.98–7.20 (m, 2H), 6.35–6.66 (m, 2H), 3.38 (t, *J* = 6.3 Hz, 2H), 2.53 (t, *J* = 6.3 Hz, 2H), 1.47 (s, 9H); ^{13}C NMR (126 MHz, CDCl$_3$) δ 171.61, 146.20, 129.12, 122.37, 114.28, 81.06, 39.91, 34.85, 28.12. HRMS (ESI): calcd for C$_{13}$H$_{19}$ClNO$_2$ [M + H]$^+$: 256.1096, found: 256.1098.

Methyl 3-((4-bromophenyl)amino)propanoate (**3k**). White powder. ^{1}H NMR (500 MHz, CDCl$_3$) δ 7.02–7.64 (m, 2H), 6.00–6.82 (m, 2H), 3.71 (s, 3H), 3.42 (t, *J* = 6.3 Hz, 2H), 2.62 (t, *J* = 6.3 Hz, 2H); ^{13}C NMR (126 MHz, CDCl$_3$) δ 171.61, 146.20, 129.12, 122.37, 114.28, 81.06, 39.91, 34.85, 28.12. HRMS (ESI): calcd for C$_{10}$H$_{13}$BrNO$_2$ [M + H]$^+$: 258.0124, found:258.0126.

Tert-butyl 3-((4-bromophenyl)amino)propanoate (**3l**). Clear oil. ^{1}H NMR (500 MHz, CDCl$_3$) δ 7.06–7.42 (m, 2H), 6.36–6.65 (m, 2H), 3.37 (t, *J* = 6.3 Hz, 2H), 2.53 (t, *J* = 6.3 Hz, 2H), 1.46 (s, 9H); ^{13}C NMR (126 MHz, CDCl$_3$) δ 171.60, 146.63, 131.98, 114.74, 109.35, 81.07, 39.77, 34.82, 28.12. HRMS (ESI): calcd for C$_{13}$H$_{19}$BrNO$_2$ [M + H]$^+$: 300.0593, found:300.0595.

Methyl 3-(pyridin-2-ylamino)propanoate (**3m**). White solid, ^{1}H NMR (500 MHz, CDCl$_3$) δ 8.04 (ddd, *J* = 5.1, 1.9, 0.9 Hz, 1H), 7.34 (ddd, *J* = 8.4, 7.1, 1.9 Hz, 1H), 6.51 (ddd, *J* = 7.1, 5.1, 1.0 Hz, 1H), 6.35 (dt, *J* = 8.4, 1.0 Hz, 1H), 5.04 (t, *J* = 6.3 Hz, 1H), 3.64 (s, 3H), 3.60 (q, *J* = 6.3 Hz, 2H), 2.61 (t, *J* = 6.3 Hz, 2H); ^{13}C NMR (126 MHz, CDCl$_3$) δ 172.85, 158.12, 147.85, 137.11, 112.76, 107.50, 51.52, 37.19, 33.87. HRMS (ESI): calcd for C$_9$H$_{13}$N$_2$O$_2$ [M + H]$^+$: 181.0972, found: 181.0975.

Methyl 3-(benzo[d][1,3]dioxol-5-ylamino)propanoate (**3o**). Light yellow solid. ^{1}H NMR (500 MHz, CDCl$_3$) δ 6.64 (d, *J* = 8.3 Hz, 1H), 6.25 (d, *J* = 2.3 Hz, 1H), 6.04 (dd, *J* = 8.3, 2.3 Hz, 1H), 5.82 (d, *J* = 0.8 Hz, 2H), 3.68 (d, *J* = 0.7 Hz, 3H), 3.35 (t, *J* = 6.4 Hz, 2H), 2.58 (t, *J* = 6.4 Hz, 2H); ^{13}C NMR (126 MHz, CDCl$_3$) δ 172.59, 148.19, 143.15, 139.62, 108.40, 104.51, 100.38, 96.10, 51.48, 40.24, 33.41. HRMS (ESI): calcd for C$_{11}$H$_{14}$NO$_4$ [M + H]$^+$: 224.0918, found: 224.0919.

Tert-butyl 3-(benzo[d][1,3]dioxol-5-ylamino)propanoate (**3p**). Light yellow oil. ^{1}H NMR (500 MHz, CDCl$_3$) δ 6.66 (d, *J* = 8.3 Hz, 1H), 6.27 (d, *J* = 2.4 Hz, 1H), 6.07 (dd, *J* = 8.3, 2.3 Hz, 1H), 5.86 (s, 2H), 3.33 (t, *J* = 6.3 Hz, 2H), 2.51 (t, *J* = 6.3 Hz, 2H), 1.47 (s, 9H); ^{13}C NMR (126 MHz, CDCl$_3$) δ 171.73, 148.34, 143.48, 139.81, 108.59, 104.90, 100.55, 96.38, 80.84, 40.76, 34.96, 28.09. HRMS (ESI): calcd for C$_{14}$H$_{20}$NO$_4$ [M + H]$^+$: 266.1385, found: 266.1386.

Methyl 3-(naphthalen-2-ylamino)propanoate (**3q**). Pink powder. ^{1}H NMR (500 MHz, CDCl$_3$) δ 7.70 (d, *J* = 8.6 Hz, 1H), 7.66 (d, *J* = 5.8 Hz, 1H), 7.64 (d, *J* = 5.5 Hz, 1H), 7.39 (ddd, *J* = 8.2, 6.8, 1.3 Hz, 1H), 7.23 (ddd, *J* = 8.1, 6.8, 1.2 Hz, 1H), 6.90 (dd, *J* = 8.8, 2.4 Hz, 1H), 6.85 (d, *J* = 2.3 Hz, 1H), 3.74 (s, 3H), 3.59 (t, *J* = 6.4 Hz, 2H), 2.72 (t, *J* = 6.3 Hz, 2H); ^{13}C NMR (126 MHz, CDCl$_3$) δ 172.85, 145.18, 135.11, 129.06, 127.67, 127.63, 126.37, 125.94, 122.16, 118.11, 104.66, 51.81, 39.41, 33.52. HRMS (ESI): calcd for C$_{14}$H$_{16}$NO$_2$ [M + H]$^+$: 230.1176, found: 230.1181.

Tert-butyl 3-(naphthalen-2-ylamino)propanoate (**3r**). Pale violet powder. ^{1}H NMR (500 MHz, CDCl$_3$) δ 7.69 (d, *J* =8.10 Hz, 1H), 7.66 (d, *J* =6.25 Hz, 1H), 7.64 (d, *J* =5..45 Hz, 1H), 7.37–7.40 (m, 1H), 7.21–7.24 (m, 1H), 6.92 (dd, *J* =8.8, 2.3 Hz, 1H), 6.87 (d, *J* =2.0 Hz, 1H), 3.53 (t, *J* =6.3 Hz, 2H), 2.63 (t, *J* =6.3 Hz, 2H), 1.48 (s, 9H); ^{13}C NMR (126 MHz, CDCl$_3$) δ 171.72, 147.73, 129.25, 117.60, 113.06, 80.83, 39.65, 35.08, 28.10. HRMS (ESI): calcd for C$_{17}$H$_{22}$NO$_2$ [M + H]$^+$: 272.1645, found: 272.1649.

4. Conclusions

In summary, we have developed an effective and environmentally friendly methodology for the synthesis of β-amino acid esters based on aromatic amines catalyzed by lipase TL IM from *Thermomyces lanuginosus* under continuous-flow microreactors. Lipase TL IM from *Thermomyces lanuginosus* was first used to catalyze Michael addition reaction of aromatic amines with acrylates. We studied the effects of various reaction parameters including the reaction medium, reaction temperature, enzyme, substrate molar ratio, residence time/flow rate and substrate structure on the reaction. Through this technique, 17 β-amino acid esters were rapidly synthesized. Compared with traditional methods, the salient features of this method include green reaction solvent (methanol), mild reaction condition (35 °C), short residence time (30 min) and high yield. These features make our methodology an attractive alternative to the current synthesis of β-amino acid esters. Our studies highlight the importance of selecting a reaction system for a specific biotransformation and show that enzymatic reactions can benefit greatly from the continuous-flow microreactor. Our results provide direction for the exploration of new enzyme catalysis processes.

Author Contributions: Research design, R.-J.L., L.-H.D. and X.-P.L.; data curation, M.X., P.-F.C. and M.-J.Y.; writing—original draft, R.-J.L.; supervision, R.-J.L., L.-H.D. and P.-F.C.; writing—review and editing, L.-H.D., R.-J.L., M.X., P.-F.C., M.-J.Y. and X.-P.L. All authors have read and agreed to the published version of the manuscript.

Funding: This research was funded by the Natural Science Foundation of Zhejiang Province grant number [LGN20C200020], the Key Research & Development Projects of Zhejiang Province grant number [2020C03090], the International Cooperation Project 948 grant number [2014-4-29], the National Science and Technology Support Project grant number [2015BAD14B0305], the National Natural Science Foundation of China grant number [2130617], the Science and Technology Research Program of Zhejiang Province grant number [2014C32094] and the APC was funded by the Natural Science Foundation of Zhejiang University of Technology grant number [116004029].

Acknowledgments: The authors would like to thank the Natural Science Foundation of Zhejiang Province and Key Research & Development Projects of Zhejiang Province (LGN20C200020 and 2020C03090), the International Cooperation Project 948 (2014-4-29), the National Science and Technology Support Project (2015BAD14B0305), the National Natural Science Foundation of China (21306172), the Science and Technology Research Program of Zhejiang Province (2014C32094) as well as the Natural Science Foundation of Zhejiang University of Technology (116004029) for financial support.

Conflicts of Interest: The authors declare no conflict of interest.

References

1. Cardillo, G.; Tomasini, C. Asymmetric synthesis of β-amino acids and α-substituted β-amino acids. *Chem. Soc. Rev.* **1996**, *25*, 117. [CrossRef]
2. Wang, W.B.; Roskamp, E.J. Conversion of β-amino esters to β-lactams via tin (II) amides. *J. Am. Chem. Soc.* **1993**, *115*, 9417–9420. [CrossRef]
3. Xiao, H.; Li, P.; Hu, D.; Song, B. Synthesis and anti-TMV activity of novel β-amino acid ester derivatives containing quinazoline and benzothiazole moieties. *Bioorg. Med. Chem. Lett.* **2014**, *24*, 3452–3454. [CrossRef]
4. Villalba, M.L.; Enrique, A.V.; Higgs, J.; Castaño, R.A.; Goicoechea, S.; Taborda, F.D.; Gavernet, L.; Lick, I.D.; Marder, M.; Blanch, L.E.B. Novel sulfamides and sulfamates derived from amino esters: Synthetic studies and anticonvulsant activity. *Eur. J. Pharmacol.* **2016**, *774*, 55–63. [CrossRef] [PubMed]
5. Karypidou, K.; Ribone, S.R.; Quevedo, M.A.; Persoons, L.; Pannecouque, C.; Helsen, C.; Claessens, F.; Dehaen, W. Synthesis, biological evaluation and molecular modeling of a novel series of fused 1,2,3-triazoles as potential anti-coronavirus agents. *Bioorg. Med. Chem. Lett.* **2018**, *28*, 3472–3476. [CrossRef] [PubMed]
6. Ling, F.; Xiao, L.; Fang, L.; Lv, Y.; Zhong, W. Copper catalysis for nicotinate synthesis through β-alkenylation/cyclization of saturated ketones with β -enamino esters. *Adv. Synth. Catal.* **2017**, *360*, 444–448. [CrossRef]
7. Gong, J.-H.; Wang, Y.; Xing, L.; Cui, P.-F.; Qiao, J.-B.; He, Y.-J.; Jiang, H.-L. Biocompatible fluorinated poly (β-amino ester)s for safe and efficient gene therapy. *Int. J. Pharm.* **2018**, *535*, 180–193. [CrossRef]
8. Liu, Y.; Li, Y.; Keskin, D.; Shi, L. Poly (β-amino esters): Synthesis, formulations, and their biomedical applications. *Adv. Heal. Mater.* **2018**, *8*, 1801359. [CrossRef]
9. Liu, Q.; Shi, W.-K.; Ren, S.-Z.; Ni, W.-W.; Li, W.-Y.; Chen, H.-M.; Liu, P.; Yuan, J.; He, X.-S.; Liu, J.-J.; et al. Arylamino containing hydroxamic acids as potent urease inhibitors for the treatment of Helicobacter pylori infection. *Eur. J. Med. Chem.* **2018**, *156*, 126–136. [CrossRef]
10. A Alsharif, M.; Khan, D.; Mukhtar, S.; Alahmdi, M.I.; Naseem, A. KOt bu-mediated aza-Michael addition of aromatic amines or N-phenylurea to 3-nitro-2-phenyl-2H-chromenes and sequential aerobic dehydrogenation. *Eur. J. Org. Chem.* **2018**, *2018*, 3454–3463. [CrossRef]
11. Xu, X.; Zhang, X.; Wang, Z.; Kong, M. HOTf-catalyzed intermolecular hydroamination reactions of alkenes and alkynes with anilines. *RSC Adv.* **2015**, *5*, 40950–40952. [CrossRef]

12. Rani, P.; Srivastava, R.K. Nucleophilic addition of amines, alcohols, and thiophenol with epoxide/olefin using highly efficient zirconium metal organic framework heterogeneous catalyst. *RSC Adv.* **2015**, *5*, 28270–28280. [CrossRef]
13. Payra, S.; Saha, A.; Banerjee, S. On-water magnetic NiFe 2 O 4 nanoparticle-catalyzed Michael additions of active methylene compounds, aromatic/aliphatic amines, alcohols and thiols to conjugated alkenes. *RSC Adv.* **2016**, *6*, 95951–95956. [CrossRef]
14. Fedotova, A.; Crousse, B.; Chataigner, I.; Maddaluno, J.; Rulev, A.Y.; Legros, J. Benefits of a dual chemical and physical activation: Direct aza-Michael addition of anilines promoted by solvent effect under high pressure. *J. Org. Chem.* **2015**, *80*, 10375–10379. [CrossRef] [PubMed]
15. Sun, C.-C.; Xu, K.; Zeng, C.-C. Transition metal- and base-free electrochemical aza-Michael addition of aromatic aza-heterocycles or Ts-protected amines to α,β-unsaturated alkenes mediated by NaI. *ACS Sustain. Chem. Eng.* **2018**, *7*, 2255–2261. [CrossRef]
16. Kim, S.; Kang, S.; Kim, G.; Lee, Y. Copper-catalyzed aza-Michael addition of aromatic amines or aromatic aza-heterocycles to α,β-unsaturated olefins. *J. Org. Chem.* **2016**, *81*, 4048–4057. [CrossRef]
17. Rostamnia, S.; Alamgholiloo, H. Synthesis and catalytic application of mixed valence iron (FeII/FeIII)-based OMS-MIL-100(Fe) as an efficient green catalyst for the aza-Michael reaction. *Catal. Lett.* **2018**, *148*, 2918–2928. [CrossRef]
18. Torrelo, G.; Hanefeld, U.; Hollmann, F. Biocatalysis. *Catal. Lett.* **2014**, *145*, 309–345. [CrossRef]
19. Lee, C.; Sandig, B.; Buchmeiser, M.R.; Haumann, M. Supported ionic liquid phase (SILP) facilitated gas-phase enzyme catalysis – CALB catalyzed transesterification of vinyl propionate. *Catal. Sci. Technol.* **2018**, *8*, 2460–2466. [CrossRef]
20. Bilal, M.; Zhao, Y.; Noreen, S.; Shah, S.Z.H.; Bharagava, R.N.; Iqbal, H.M. Modifying bio-catalytic properties of enzymes for efficient biocatalysis: A review from immobilization strategies viewpoint. *Biocatal. Biotransformation* **2019**, *37*, 159–182. [CrossRef]
21. Steunenberg, P.; Sijm, M.; Zuilhof, H.; Sanders, J.P.M.; Scott, E.L.; Franssen, M.C.R. Lipase-catalyzed aza-Michael reaction on acrylate derivatives. *J. Org. Chem.* **2013**, *78*, 3802–3813. [CrossRef] [PubMed]
22. Strompen, S.; Weis, M.; Gröger, H.; Hilterhaus, L.; Liese, A. Development of a continuously operating process for the enantioselective synthesis of a β-amino acid esterviaa solvent-free chemoenzymatic reaction sequence. *Adv. Synth. Catal.* **2013**, *355*, 2391–2399. [CrossRef]
23. Xu, F.; Wu, Q.; Chen, X.; Lin, X.; Wu, Q. A single lipase-catalysed one-pot protocol combining aminolysis resolution and aza-Michael addition: An easy and efficient way to synthesise β-amino acid esters. *Eur. J. Org. Chem.* **2015**, *2015*, 5393–5401. [CrossRef]
24. Badgujar, K.; Bhanage, B.M. Lipase immobilization on hyroxypropyl methyl cellulose support and its applications for chemo-selective synthesis of β-amino ester compounds. *Process. Biochem.* **2016**, *51*, 1420–1433. [CrossRef]
25. Gu, B.; Hu, Z.-E.; Yang, Z.; Li, J.; Zhou, Z.; Wang, N.; Yu, X. Probing the mechanism of CAL-B-catalyzed aza-Michael addition of aniline compounds with acrylates using mutation and molecular docking simulations. *ChemistrySelect* **2019**, *4*, 3848–3854. [CrossRef]
26. Planchestainer, M.; Contente, M.L.; Cassidy, J.; Molinari, F.; Tamborini, L.; Paradisi, F. Continuous flow biocatalysis: Production and in-line purification of amines by immobilised transaminase from Halomonas elongata. *Green Chem.* **2017**, *19*, 372–375. [CrossRef]
27. Britton, J.; Majumdar, S.; Weiss, G.A. Continuous flow biocatalysis. *Chem. Soc. Rev.* **2018**, *47*, 5891–5918. [CrossRef]
28. Van der Helm, M.P.; Bracco, P.; Busch, H.; Szymańska, K.; Jarzębski, A.B.; Hanefeld, U. Hydroxynitrile lyases covalently immobilized in continuous flow microreactors. *Catal. Sci. Technol.* **2019**, *9*, 1189–1200. [CrossRef]
29. Movsisyan, M.; Delbeke, E.I.P.; Berton, J.K.E.T.; Battilocchio, C.; Ley, S.V.; Stevens, C.V. Taming hazardous chemistry by continuous flow technology. *Chem. Soc. Rev.* **2016**, *45*, 4892–4928. [CrossRef]
30. Plutschack, M.B.; Pieber, B.; Gilmore, K.; Seeberger, P.H. The hitchhiker's guide to flow chemistry. *Chem. Rev.* **2017**, *117*, 11796–11893. [CrossRef]
31. Baumeister, T.; Zikeli, S.; Kitzler, H.; Aigner, P.; Wieczorek, P.P.; Röder, T. Continuous flow synthesis of amine oxides by oxidation of tertiary amines. *React. Chem. Eng.* **2019**, *4*, 1270–1276. [CrossRef]
32. Priego, J.; Ortíz-Nava, C.; Carrillo-Morales, M.; López-Munguía, A.; Escalante, J.; Castillo, E. Solvent engineering: An effective tool to direct chemoselectivity in a lipase-catalyzed Michael addition. *Tetrahedron* **2009**, *65*, 536–539. [CrossRef]

33. De, K.; Legros, J.; Crousse, B.; Bonnet-Delpon, D. Solvent-promoted and -controlled aza-Michael reaction with aromatic amines. *J. Org. Chem.* **2009**, *74*, 6260–6265. [CrossRef] [PubMed]
34. Svedendahl, M.; Hult, K.; Berglund, P. Fast carbon–carbon bond formation by a promiscuous lipase. *J. Am. Chem. Soc.* **2005**, *127*, 17988–17989. [CrossRef] [PubMed]

© 2020 by the authors. Licensee MDPI, Basel, Switzerland. This article is an open access article distributed under the terms and conditions of the Creative Commons Attribution (CC BY) license (http://creativecommons.org/licenses/by/4.0/).

Article

Biocatalysis of Industrial Kraft Pulps: Similarities and Differences between Hardwood and Softwood Pulps in Hydrolysis by Enzyme Complex of *Penicillium verruculosum*

Andrey S. Aksenov [1,*], Irina V. Tyshkunova [2], Daria N. Poshina [2], Anastasia A. Guryanova [1], Dmitry G. Chukhchin [1], Igor G. Sinelnikov [3], Konstantin Y. Terentyev [1,4], Yury A. Skorik [2], Evgeniy V. Novozhilov [1] and Arkady P. Synitsyn [3,5]

1. Northern (Arctic) Federal University, Northern Dvina Embankment 17, 163000 Arkhangelsk, Russia; stasya658@yandex.ru (A.A.G.); dimatsch@mail.ru (D.G.C.); k.terentev@narfu.ru (K.Y.T.); noev50@gmail.com (E.V.N.)
2. Institute of Macromolecular Compounds of the Russian Academy of Sciences, Bolshoy prospect V.O. 31, 199004 St. Petersburg, Russia; tisha19901991@yandex.ru (I.V.T.); poschin@yandex.ru (D.N.P.); yury_skorik@mail.ru (Y.A.S.)
3. Federal State Institution, Federal Research Centre (Fundamentals of Biotechnology) of the Russian Academy of Sciences, Leninsky prospect, 33, build. 2, 119071 Moscow, Russia; sinelnikov.i@list.ru (I.G.S.); apsinitsyn@gmail.com (A.P.S.)
4. Federal Center for Integrated Arctic Research, Russian Academy of Sciences, Northern Dvina Embankment 23, 163000 Arkhangelsk, Russia
5. Chemical Department, Moscow State University, Vorobyevy Gory, 1-11, 119992 Moscow, Russia
* Correspondence: a.s.aksenov@narfu.ru; Tel.: +7-921-2915446

Received: 28 April 2020; Accepted: 11 May 2020; Published: 13 May 2020

Abstract: Kraft pulp enzymatic hydrolysis is a promising method of woody biomass bioconversion. The influence of composition and structure of kraft fibers on their hydrolysis efficiency was evaluated while using four substrates, unbleached hardwood pulp (UHP), unbleached softwood pulp (USP), bleached hardwood pulp (BHP), and bleached softwood pulp (BSP). Hydrolysis was carried out with *Penicillium verruculosum* enzyme complex at a dosage of 10 filter paper units (FPU)/g pulp. The changes in fiber morphology and structure were visualized while using optical and electron microscopy. Fiber cutting and swelling and quick xylan destruction were the main processes at the beginning of hydrolysis. The negative effect of lignin content was more pronounced for USP. Drying decreased the sugar yield of dissolved hydrolysis products for all kraft pulps. Fiber morphology, different xylan and mannan content, and hemicelluloses localization in kraft fibers deeply affected the hydrolyzability of bleached pulps. The introduction of additional xylobiase, mannanase, and cellobiohydrolase activities to enzyme mixture will further improve the hydrolysis of bleached pulps. A high efficiency of never-dried bleached pulp bioconversion was shown. At 10% substrate concentration, hydrolysates with more than 50 g/L sugar concentration were obtained. The bioconversion of never-dried BHP and BSP could be integrated into working kraft pulp mills.

Keywords: kraft pulp; cellulose; xylan; enzymatic hydrolysis; *Penicillium verruculosum*; glucose; xylose

1. Introduction

Lignocelluloses have been the focus of much attention as promising feedstock for the sustainable production of non-food-derived sugars, from the viewpoint of both energy and the environment. Numerous physical, chemical, and biological methods for converting lignocellulose to sugars are under development, but those that rely on enzymes are particularly attractive. Enzymes can potentially serve as industrial catalysts for biomass conversion, providing benefits, such as high specificity, low energy, and reagent consumption, and little environmental pollution [1].

Consequently, the conversion of lignocellulosic biomass into simple carbohydrates, phenolics, aromatics, and other substances remains a major challenge. The resulting conversion of cellulose to glucose can produce a number of useful products, including biofuels, as well as different organic acids. Lignocellulose is a very challenging material for enzymatic attack, due to its complex and compact structure composed of many biopolymers, each with a different chemical composition and physical structure [2].

Bioconversion processes are preferentially carried out through the enzymatic saccharification of pretreated lignocellulosic substrates. The direct conversion of untreated lignocellulose with enzymes has not yet been technologically rendered, mainly due to the well-known recalcitrance of biomass to enzymatic hydrolysis. Pretreatment plays a crucial role in biomass conversion, as it aids in overcoming the chemical and structural difficulties that are associated with lignocelluloses and allows for the cost-effective production of fermentable sugars via enzymatic saccharification [3–5].

The main barrier to enzymatic saccharification of lignocelluloses is lignin. Carbohydrases readily digest the hemicellulose components, whereas lignin can induce nonproductive binding of cellulases by various mechanisms, as well as a reduction of cellulase catalytic activity due to the possible inhibition of small phenolic molecules, thereby considerably hindering the hydrolytic efficiency [6,7]. Strategies to minimize these cellulase–lignin interactions, such as enzyme engineering [6], substrate modification, and additive blocking, have therefore undergone intensive development [7].

One of the most promising pretreatments of wood biomass is alkali-based technology, such as kraft pulping, as this provides highly efficient delignification. Important advantages of kraft pulping as a pretreatment stage of raw wood materials are as follows [8]: first, the inorganic chemicals used for pulping are regenerated; second, dissolved organic matter and residual lignin are burned to produce energy; and third, a well-established wastewater treatment system is used. This is a highly attractive prospect, as it would utilize the equipment and chemical recovery systems that are already well developed in the pulp industry, thereby lowering the capital cost of pretreatment and providing additional benefits for pulp mills [9].

As cellulose is an unbranched crystalline polymer, several cellulases are needed to degrade it efficiently. Cellulases hydrolyze β-1,4-D-glucan bonds, releasing cello-oligosaccharides, cellobiose, or glucose. The complete degradation of cellulose is carried out by an enzymatic complex, which includes endo-β-1,4-glucanases (EC 3.2.1.4), cellobiohydrolases (EC 3.2.1.91 and EC 3.2.1.176), and β-glucosidases (EC 3.2.1.21) [10].

The production of cellulase cocktails has been widely explored; however, there are still some main challenges with enzymes that need to be overcome in order to develop the sustainable production of bioethanol [10]. Trichoderma fungi (T. reesei, T. viride, T, longibrachiatum) have high secretory ability of enzymes with different substrate specificity [11]. Various companies manufacture enzyme preparations that are produced by these fungi. Commercial preparations of Trichoderma fungus strains, like Cellic HTec series (Novozymes, Denmark), have necessary activities for enzymatic hydrolysis of cellulose-containing substrates [12,13]. When comparing to Trichoderma, the Penicillium verruculosum enzyme complex, along with highly active enzymes (endoglucanase, cellobiohydrolase, xylanase), additionally contains significant β-glucosidase activity [14].

Penicillium cellulases are superior in their rate of hydrolysis and the glucose yield from various cellulose-containing substrates at the same dosage for protein concentration [10]. The advantages of Penicillium verruculosum over Trichoderma reesei enzymes are a higher secretory ability (40–50 g/L of extracellular protein), reduced affinity for lignin, and low sensitivity to inhibition by lignin derivatives [6,13,14]. For examples, cellulases from Penicillium verruculosum have shown good performance on different substrates [15–17] and they can successfully compete with the Trichoderma enzymes, which are known for their high cellulolytic activity.

The composition of enzyme complexes includes xylanases that are intended for the destruction of xylan to improve the effect of cellulases. The xylanase activity is quite high in the *Penicillium verruculosum* enzyme complex [12,17]. This ensures the bioconversion of xylan as a biopolymer. Biocatalysis leads to the accumulation of soluble xylan degradation products in hydrolysates in the form of xylose, xylobiose, and xylooligomers [18].

Kraft pulp is technical pulp obtained by kraft pulping of raw wood materials. Kraft mills produce four types of kraft pulp: unbleached hardwood pulp (UHP), unbleached softwood pulp (USP), bleached hardwood pulp (BHP), and bleached softwood pulp (BSP). Woody kraft pulp has been widely studied as a substrate for bioconversion [15,17,19–27]. Some studies have been performed with samples that were obtained after laboratory kraft pulping [23–26]. In the industrial kraft pulping, which is usually carried out as a continuous process, the same residual lignin content in pulp characterized by the kappa number can be obtained by rapid cooking with high chemical consumption and high temperature, or by slow cooking with low chemical consumption and low temperature. The topochemical effect leads to the formation of kraft pulp fiber, the outer part of which has a higher concentration of residual lignin when compared to the average value in the pulp [8]. As a result, lignin in kraft pulp fiber is mostly located in the outer layer of the fiber [28]. From 85% to 90% of the residual lignin in unbleached kraft pulp was isolated as lignin–carbohydrate complexes (LCCs) of three types: xylan–lignin, glucomannan–lignin–xylan, and glucan–lignin [29]. The most stable LCC formed during kraft pulping of pine pulp contained mannan and xylan, which are most recalcitrant and they remain in the pulp with a decreasing kappa number [29]. Kraft pulp fibers are also characterized by an increased concentration of xylan in the outer layers of the cell wall [8]. Non-cellulose components on the surface of these fibers present a physical barrier to cellulases. In addition, the fibers partially retain the primary walls, since the delignification in the kraft process mainly starts from the cell lumen. The cellulose microfibrils of primary walls show high resistance to the action of cellulases [30]. The structure of the surface layers distinguishes kraft pulp fibers from other lignocellulose fibers proposed for the bioconversion of cellulose into glucose.

Kraft pulp remains poorly understood as a substrate for bioconversion, despite conducted research. The influence of the composition and localization of hemicelluloses of hardwood and softwood pulps on enzymatic hydrolysis has not been sufficiently studied. Residual lignin has a negative effect on biocatalysis. For example, a decrease in the lignin content from 18.6% to 4.8% led to a more than double increase in the reducing sugars (RS) yield, from 31.6% to 67.5% [21]. Never-dried USP bioconversion was performed at a substrate concentration of 5% with an enzyme mixture that was prepared with Novozymes Cellic® CTec2 cellulase enzyme and Cellic® HTec2 hemicellulase enzyme at a dosage of 10 FPU/g.

The role of (LCCs with xylan and mannan present in kraft pulp is not well understood [18,29,31]. Researchers used a large dosage of enzyme preparations and reduced the concentration of the substrate to neutralize the influence of lignin. Conventional enzymatic hydrolysis of kraft pulp is typically carried out at a low substrate concentration, as a rule no more than 3% [19,23,24,26].

The research direction of increasing the efficiency of bioconversion due to increasing the dosage of enzyme preparations and reducing the substrate concentration allows for the assessment of the potential impact of catalytic action on kraft pulp. However, such a technique conflicts with the requirements for industrial implementation of enzymatic hydrolysis. A substrate consistency below 5% solid content leads to a sugar concentration below 5% in the hydrolysate, thus a final concentration of ethanol less than 2% (w/w) after fermentation [20].

A significant portion of industrial kraft pulp is produced in bleached form. If up to 90–97% of wood lignin is dissolved in kraft pulping, the lignin is almost completely removed in the process of industrial pulp bleaching. It is obvious that the bioconversion of bleached kraft pulps is faster and more complete under milder conditions in the absence of lignin [17,19]. It was shown that fiber fractions obtained from eucalyptus BHP at a substrate concentration of 2% and an enzyme dosage of

10 FPU/g after 48 h of bioconversion yielded 92.8–93.2% glucose [19]. The role of xylan and mannan is increased in the absence of lignin in bleached pulp.

Industrial dry bleached kraft pulp and various types of paper can be used as substrates for bioconversion into glucose. The drying of cellulose fibers has a significant effect on their structure and properties [32]. Several papers describe changes in cellulose fiber structures related to hornification [15,22,32–36]. Cellulose microfibrils can form additional hydrogen bonds in amorphous areas when water is removed during drying. The formation of irreversible or partially reversible H-bonds is called hornification [33]. A high content of hemicellulose has been shown to reduce the negative effect of hornification by preventing the aggregation of microfibrils in cellulose fibers [34,35].

Drying and the associated hornification decrease the reactivity of kraft pulp due to decreased swelling and cellulase accessibility [15,22]. From this point of view, the enzymatic hydrolysis of never-dried industrial kraft pulp is of particular interest. In our previous study [17], we compared the effectiveness of biocatalysis of industrial hardwood and softwood pulps of cellulases from *Penicillium verruculosum*. The effect of the dosage of the *Penicillium verruculosum* enzyme complex was verified. It was established that a high conversion of never-dried pulp was achieved using a relatively low dosage of 10 FPU/g of pulp. It was shown that the crystallinity degree of BHP and BSP in enzymatic hydrolysis changed through a maximum due to the predominant degradation of the amorphous component at the beginning of the bioconversion process. Further study of the biocatalysis features of hardwood and softwood kraft pulps will provide recommendations for optimizing the composition of enzyme preparations by a set of activities and their ratio. The composition of non-hydrolysable residues and their structure, properties, and possibilities for rational use are not sufficiently studied.

The present study examined the biocatalysis of four substrates (UHP, USP, BHP, and BSP) by the *Penicillium verruculosum* enzyme complex. The aim was to evaluate the impact of the hardwood and softwood kraft pulp composition and structural fiber features on hydrolysis efficiency and to provide recommendations for improving the bioconversion of kraft pulps into hexoses and pentoses.

2. Results and Discussion

2.1. Influence of Lignin on Bioconversion of Never-Dried Kraft Pulp

Most of the studies of kraft pulp bioconversion to glucose have been conducted with unbleached kraft pulps that contain kraft lignin [21,24]. There is a general opinion that the lower the lignin proportion, the higher the bioconversion degree of cellulose into glucose.

In our study, never-dried UHP and USP were taken to assess the effect of kraft lignin on biocatalysis. A 5% substrate concentration was assumed. Other biocatalysis conditions were 24 h duration and dosage of *Penicillium verruculosum* enzyme complex at 10 FPU/g. The lignin content was 2.1% and 3.0% for UHP and USP, respectively. There is a general pattern for unbleached pulps: kraft lignin has a negative effect on the results of bioconversion, even at a low content (Figure 1).

The best results under the selected bioconversion conditions were obtained while using never-dried UHP among the unbleached pulps (Figure 1). The yield of glucose and RS for never-dried UHP sample was 41.9% and 55.3%, respectively. It should be noted that the content of residual lignin in UHP was 1.5 times lower than in USP. The glucose yield in enzymatic hydrolysis of never-dried USP with a higher lignin content was 13% lower and the RS yield was 8% lower when compared to UHP (Figure 1).

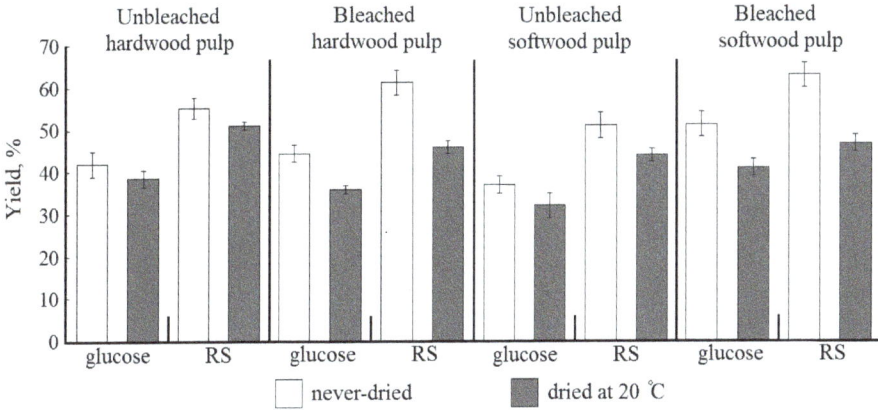

Figure 1. Effect of lignin and drying on bioconversion of kraft pulp.

Even a very small amount of kraft lignin (2–3%) negatively affects the results of enzymatic hydrolysis (Figure 1). The possible reasons for the limited hydrolysis of unbleached kraft pulp include unproductive absorption and inhibition of cellulase action by soluble lignin degradation products. Non-specified sorption of cellulases onto the kraft lignin due to its low content in pulp probably has limited value and it can be compensated by increased dosage of enzyme complex. In UHP hydrolysate, the lignin concentration was 45 mg/L and 53 mg/L at 24 h and 48 h of enzymatic hydrolysis, respectively. In USP hydrolysate, lignin concentration was lower, 21 mg/L and 37 mg/L (Table 1). Probably dissolved lignin does not inhibit the biocatalysis process at such low concentrations. Under the action of the *Penicillium verruculosum* enzyme complex, only 4.3–5.0% UHP lignin and 1.4–2.5% USP lignin passed into the solution. It was obvious that almost all of the kraft lignin remained in the mass during enzymatic hydrolysis. Dissolved lignin represents 4–5% of hardwood pulp lignin and 1–3% of softwood pulp lignin, while most of the lignin was undissolved and remained in non-hydrolyzed residue.

Table 1. Lignin and mannose concentration in hydrolysates of unbleached pulps.

Pulp	Lignin Concentration, mg/L		Mannose Concentration, mg/L
	24 h	48 h	24 h
Hardwood	45 ± 1	53 ± 1	78 ± 2
Softwood	21 ± 1	37 ± 2	72 ± 1

During the kraft cooking to a high kappa number, most of the residual lignin remains bounded to xylan, however when cooking proceeds to a low kappa number, glucomannan-lignin complexes prevail. Lignin–carbohydrate complexes with glucomannan are located mainly on the surface of cellulose microfibrils [37]. Mannan is partially covered by adsorbed xylan inside the cell wall of kraft pulp fibers. This localization reduces the availability of mannan for enzymes [18].

Mannose occurred in hydrolysates, but in significantly smaller amounts than xylan destruction products. After 24 h of UHP enzymatic hydrolysis, the mannose concentration was 78 mg/L (Table 1), and glucomannan conversion into mannose was 47% in hydrolysate. In USP hydrolysate, the concentration was 72 mg/L; therefore, glucomannan conversion into mannose was only 2%. Thus, the main amount of glucomannan remained in USP as part of the LCC. For hardwood pulps with a mannan content of 0.3%, this is less important when compared to softwood pulps with a mannan content of 6.8%. Mannose and mannooligosaccharides were shown to inhibit cellulases and decrease the conversion of substrates [38,39]. However, they exert strong inhibition at higher concentrations (2.5 g/L and higher) than the mannose concentrations that were found in UHP and USP hydrolysates.

It should be noted that the main reason for the limited hydrolysis of unbleached kraft pulp is that the residual lignin in the composition of LCC with mannan and xylan reduces the availability of cellulose for cellulase action. The cellulose content is higher in USP than in UHP. However, the percentage of glucose in hydrolysate after UHP bioconversion is 76.0 ± 1.2% when compared to 73.0 ± 1.4% in USP hydrolysate. Xylanase, part of the *Penicillium verruculosum* enzyme complex, largely destroys the xylan component of LCC, thereby increasing the availability of cellulose for cellulases. This applies equally to UHP and USP. It was shown that the degree of crystallinity of softwood and hardwood kraft pulps increased in the first stage of the process (3–6 h) due to the enhanced destruction of xylan and amorphous cellulose by xylanase and endogluconase, respectively [17]. In our opinion, LCC with glucomannan has a strong negative effect on the enzymatic hydrolysis of cellulose in USP. The glucomannan of the LCC is resistant to the action of *Penicillium verruculosum* enzyme complex, since the biocatalytic mannanase activity in this complex is at a low level. It can be recommended to include mannanase in the enzyme complex to optimize USP bioconversion. For UHP bioconversion, where the glucomannan content is low, it is sufficient to have only xylanase for successful enzymatic hydrolysis.

When bleaching kraft pulp, the main tasks are to remove the lignin and improve the pulp brightness. Carbohydrates, mainly xylan located on the fiber surface, are dissolved in approximately equal measure when LCC is destroyed and lignin is dissolved.

For never-dried hardwood pulp, lignin removal, and LCC destruction during bleaching stimulated slightly improved biocatalysis: glucose yield from BHP increased by 6.2% and RS yield by 11% when compared to UHP.

It can be concluded that the destruction of cellulose and xylan during the conversion of never-dried unbleached and bleached hardwood kraft pulp occurs in approximately equal proportions.

Residual lignin removal and LCC destruction with xylan and mannan in the bleaching leads to a significant increase in the yield of target products during enzymatic hydrolysis of never-dried softwood kraft pulp. Of particular importance is LCC destruction with glucomannan, which is localized near cellulose microfibrils. The yield of glucose from BSP was 51.3% of oven-dry pulp and the RS yield was 62.9%. Glucose yield increased by 38% and RS by 23% when compared to USP (Figure 1). The ratio of glucose and non-glucose sugars in softwood pulp hydrolysate significantly improved; the glucose content was 82%. Never-dried BSP bioconversion gave the best results for all the main targets.

2.2. Effect of Drying on Bioconversion of Unbleached and Bleached Kraft Pulps

Drying and related hornification decrease the reactivity of kraft pulp due to decreased swelling and cellulase accessibility [15,22]. The hornification degree increases with the increasing drying temperature of bleached kraft pulp [40]. In our research, drying, even under mild conditions, significantly reduced the reactivity of unbleached and bleached kraft pulp during enzymatic hydrolysis (Figure 1). The least negative drying effect was shown for UHP, with the yield of glucose and RS decreasing by 8–9%. The decrease in the yield of soluble bioconversion products for USP was 13–14%. It is demonstrated that the lignin content has no significant effect on the hydrolyzability of dried pulps [22]. This can be explained by the fact that lignin in unbleached pulp might play a positive role as an amorphous material, preventing the aggregation of cellulose microfibrils during fiber hornification and reducing hornification caused by drying. Obviously, hemicelluloses in kraft cellulose also have a positive effect on reducing the hornification degree. Thus, the negative effects of drying of unbleached kraft pulps on enzymatic hydrolysis turned out to be relatively small. The most important factor determining the hornification of dry pulps was the supramolecular structure [22,41]. Cellulose conversion to glucose mainly depended on the specific surface area and average pore size. In this study [22], the pore volume of the substrate was determined from fiber saturation point measurements. It is considered that the pore sizes of the fiber wall should be greater than typical dimensions of enzyme molecules, around 10 nm [42]. For dry birch BHP, the average pore size was 19 nm [22].

It should be noted that previously high-glucose yields were obtained from dried UHP and USP [22–25]. In one study [25], it was even concluded that the complete removal of lignin was not necessary in order to obtain a sufficiently high yield of glucose and RS from birch and beech pulps. However, these results were obtained when enzymatic hydrolysis was performed at a low substrate concentration and high enzyme dosage, approximately 67 FPU/g.

It was demonstrated that there was some optimal lignin content in the bioconversion of pine USP [25], since the glucose yield was 11.3% higher at a kappa number up to 47.7 than at a lower or higher kappa number (17.2 and 86.2, respectively). Obviously, this does not agree with the concept that the higher the lignin content, the lower the efficiency of enzymatic hydrolysis. However, in our opinion, the deterioration of enzymatic hydrolysis results at a very low kappa number (17.2) is due to the fact that the lignin content in this case was insufficient for preventing strong compaction of cellulose microfibrils during drying.

Reactivity during drying decreases significantly more for bleached than unbleached kraft pulp (Figure 1). The negative effect of drying in BHP and BSP is shown to be equal, despite the different content of cellulose and hemicellulose. Glucose yield is reduced by 19–20% and RS yield by 25–26%. Such significant decreases in the yield of soluble products of enzymatic hydrolysis occurred when only 2–3% of lignin was removed. Ultrastructural changes that are caused by the destruction of tightly packed regions of the cell wall resulting from delignification play an important role [43]. The ability of cellulose to form highly ordered supramolecular structures supported by many new intermolecular hydrogen bonds is enhanced as a result of lignin removal [22].

Thus, pulp drying leads to the hornification of the fiber and the formation of a dense structure of the cell wall, which reduces the availability of the substrate to enzymes. A comparison of the results for hydrolysis of bleached and unbleached pulp suggests a strong effect of bleaching on improving the hydrolyzability of kraft pulp. Since the best results were obtained while using never-dried bleached kraft pulps, only these pulps were used in further experiments.

2.3. Changes of Kraft Pulp Fibers Length and Width During Enzymatic Hydrolysis

Significant changes in the length and width of kraft pulp fibers occurred at early stages of enzymatic hydrolysis. Figure 2 presents the changes in softwood and hardwood fibers that occurred during the first 5 h of hydrolysis.

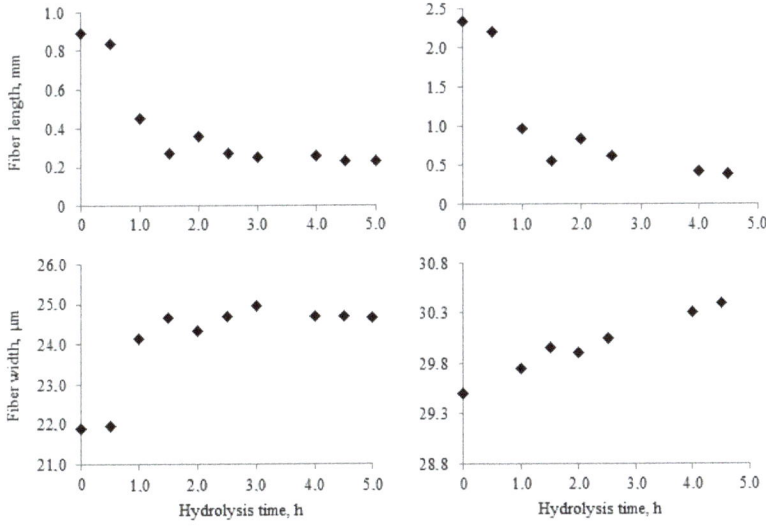

Figure 2. Fiber length and width of bleached hardwood pulp (BHP; **left**) and bleached softwood pulp (BSP; **right**) during enzymatic hydrolysis.

The average length of softwood kraft pulp fibers was almost 2.7 times higher when compared with hardwood fibers. The fiber length decreased during hydrolysis. Enzymatic destruction was the main reason for fiber cutting.

Fiber cross-cutting was the most intensive during the first 1–2 h. Softwood pulp fibers were shortened by almost five times in this period. Hardwood fibers were also shortened, but, to a lesser extent, by three times. The reduction in fiber length improved pulp stirring at the initial stage of hydrolysis. It took place when the average fiber length was about 0.3–0.5 mm. Later the average fiber length continued to decrease, but to a much lesser extent. After 5 h of hydrolysis, the decrease of fiber length was 84% for softwood pulp and 74% for hardwood pulp. Despite this, softwood pulp fiber remained approximately 1.5 times longer than hardwood fiber.

Simultaneously with fiber shortening, the width increased due to swelling. The destruction of fiber surface layers was necessary for the swelling. It occurred under the action of *Penicillium verruculosum* enzyme complex. The action of cellulase and xylanase first destroyed the outer fiber layers that were most accessible to hydrolysis, leading to a partial loosening of their structure.

Softwood and hardwood pulp fibers swelled most intensively in the first hour of hydrolysis. However, the degree and rate of swelling were different for these fibers. Hardwood libriform fibers have thicker cell walls (5.00 µm) when compared with these of the water-conducting softwood tracheids of early (2.01 µm) and late (4.16 µm) wood tissues [44]. A significant difference in fiber cell wall thickness maintained after kraft pulping. The hardwood pulp fiber width increased quickly in the first hour of hydrolysis, but then changed slightly. The fiber width increased by 12% after 5 h of hydrolysis. This effect led to swelling of the largest S_2 layer.

Softwood pulp fiber swelling was gradual for 5 h. The increase in fiber width did not exceed 3% in this period. Differences in the morphology and chemical composition of fiber cell walls could be a factor in such different behavior. Hardwood pulp fiber contains much more adsorbed xylan on the surface. Its rapid destruction in the first stages of hydrolysis could have "open the surface" and led to more intensive swelling. Besides, softwood pulp, unlike hardwood, has a wider fiber cavity, so cell wall swelling could go toward the center of the fiber [45]. It does not cause increased width and the device does not fix it.

Changes in fiber structure during 1–5 h of enzymatic hydrolysis were visualized while using an optical microscope (Figure 3).

Kraft fibers have a shape of long cells with pointed or rounded ends. There are many bends and kinks in the fibers. Dislocations often arise after chemical treatment during pulping and bleaching or after the mechanical stress of pulp transporting and mixing [46]. Cutting mainly occurs in fiber dislocations. Dislocations are defined as local changes in cellulose microfibrils of the S_1 layer, or in the S_1 and S_2 layers of the secondary cell wall. This also includes kinks and bends [47]. This fiber cross-cutting phenomenon has been observed and discussed previously [48,49]. More accessible for enzymes, opened and less ordered structures are formed in fiber dislocations due to such structural features. This provides deeper enzyme penetration into dislocations and better binding to cellulose [47,50]. This was previously confirmed by the observation of endoglucanases, cellobiohydrolases, and cellulose-binding modules binding at dislocations [46].

Only a few initial shapes of long fibers were detected after 1 h of hydrolysis. Increasing the hydrolysis time changed the fibers noticeably. Fiber fragments that had many cracks represented BSP. The cell wall structure was "torn off" at the edges. Softwood pulp fibers were fragmented, but the fragment size was quite large. A significant accumulation of small fiber fragments was noticed after 2–3 h of hydrolysis (Figure 3). The fiber fragments obtained after 3 h of hydrolysis had no significant dislocations. Only small fiber fragments were observed in softwood pulp after 5 h of hydrolysis. It should be noted that most of the fiber fragments at that time point had neither twists nor bends. The prolonged action of the *Penicillium verruculosum* enzyme complex led to the disintegration of the fiber surface layers and increased the amount of fines fraction in the pulp.

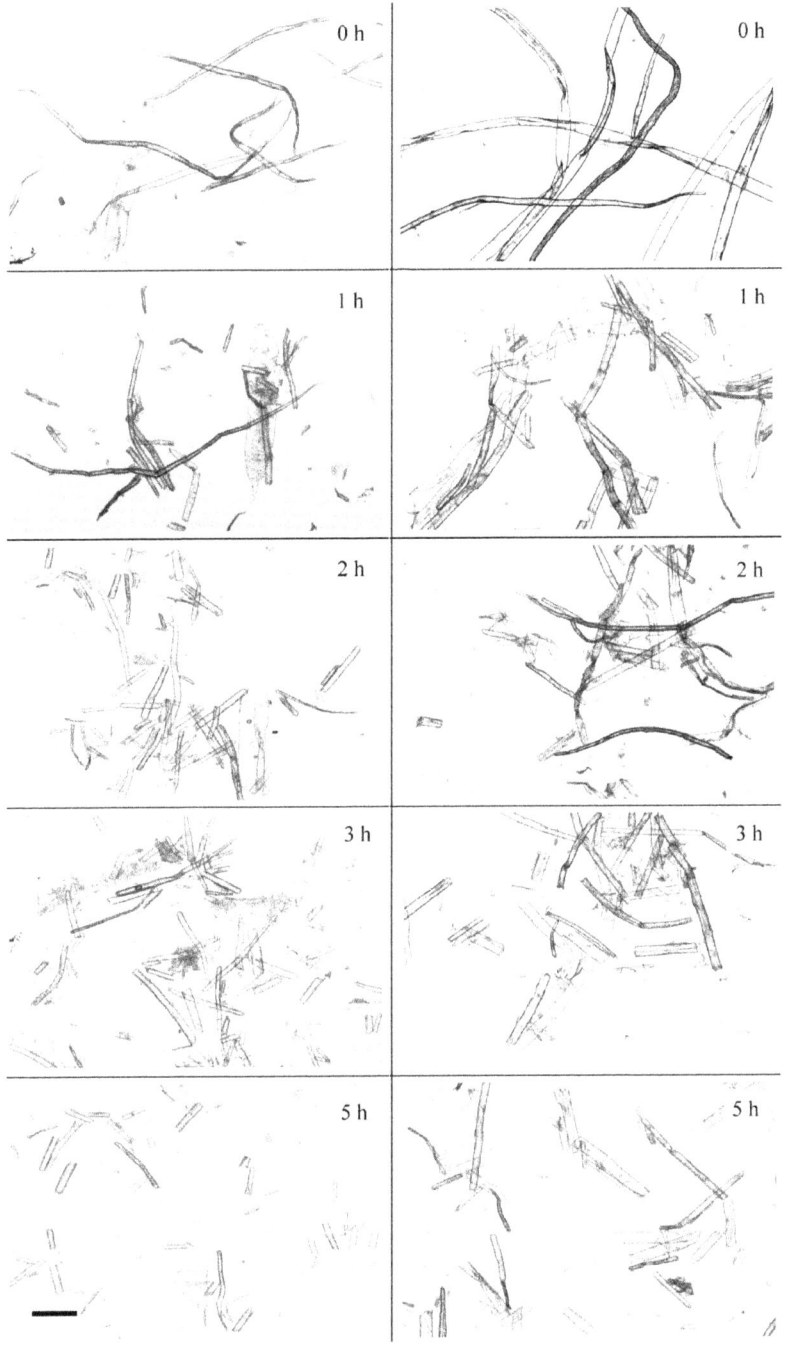

Figure 3. Microscopic images of gradual enzymatic degradation of hardwood (**left**) and softwood (**right**) pulp fibers for 5 h. Bar: 0.1 mm.

Initial hardwood pulp contained thinner and shorter fibers and vessel elements. BHP fibers had bends and twists, like softwood fibers. Significant fragmentation of fibers occurred and multiple breaks were observed after 1 h of enzyme action. Hardwood pulp fibers were fragmented by enzyme action faster and to a greater extent when compared to softwood pulp. Vessels were also partially destroyed (5 h), but with less extension than fibers.

2.4. Behavior of Hardwood Pulp Vessels during Enzymatic Hydrolysis

Vessels in hardwood perform a conducting function [2,51]. Vessels are built from elements that are joined end-on-end, forming long, continuous tubes [51]. Long vessels of hardwood are destroyed during the kraft process in order to separate short fragments, the length of which is comparable to the fibers. However, the width of the vessel fragments is much larger than the width of the fibers.

These fragments in the composition of BHP are rare, about five fragments of vessels per 1000 fibers (Figure 4). The change in the number of vessels in the bioconversion process was studied while using the L&W Fiber Tester. Fragments of vessels were more resistant to selected carbohydrase complex than libriform fibers. At the beginning of the enzymatic hydrolysis process (1–5 h), the ratio of vessel fragments to fibers increased rapidly by about three times (Figure 4). This is because the proportion of fractions of long and medium fibers in enzymatic hydrolysis decreases rapidly, and the amount of fines fraction (less than 0.2 mm) increases. When calculating the ratio of vessels and fibers, the fines fraction of fibers is not taken into account.

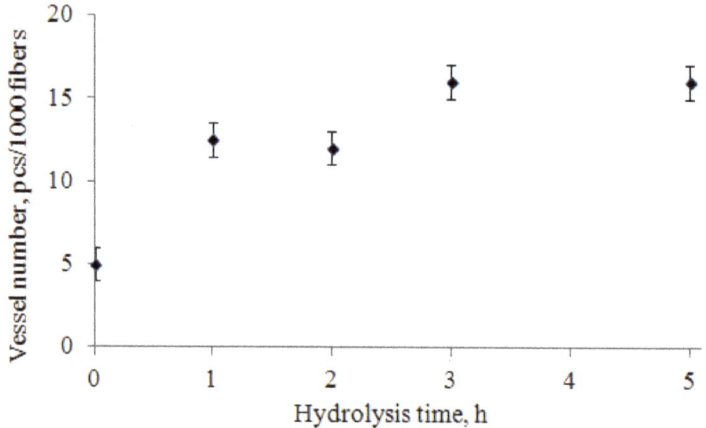

Figure 4. Number of vessels during enzymatic hydrolysis of hardwood pulp.

The cell walls of vessels of birch (Figure 5A) and aspen (Figure 5B) are highly ornamented with pits, the locations of which may differ [51]. The wall structure of vessels is built of three layers of cellulosic fibrils arranged in varying helical angles around the central axis, as in wood fibers. The average thickness of the primary wall of the anatomical elements of hardwood xylem is approximately the same, 0.06–0.07 µm (together with middle lamella) [2,20]. At the same time, the S_2 layer in vessels is the thinnest in comparison with other cell types. For this reason, the average thickness of the vessel's cell walls is also the thinnest, 1.00 µm, whereas, in libriform fibers, it is 5.00 µm [2,44]. While using the UV microscopy method, it was found that the content of lignin in the vessels of white birch (*Betula papyrifera*) was higher (24–28%) than in the libriform fibers (19–22%) [52]. In addition, vessel lignin is closer in composition to lignin in coniferous wood.

After the kraft process and bleaching, the fragments of hardwood pulp vessels looked slightly modified (Figure 5C). During enzymatic hydrolysis, the fiber structure was intensively destroyed, but the vessel fragments retained their walls for 3–8 h of enzymatic hydrolysis (Figure 5D,E). A portion of the vessel wall is observed, even after enzymatic hydrolysis for 48 h (Figure 5F).

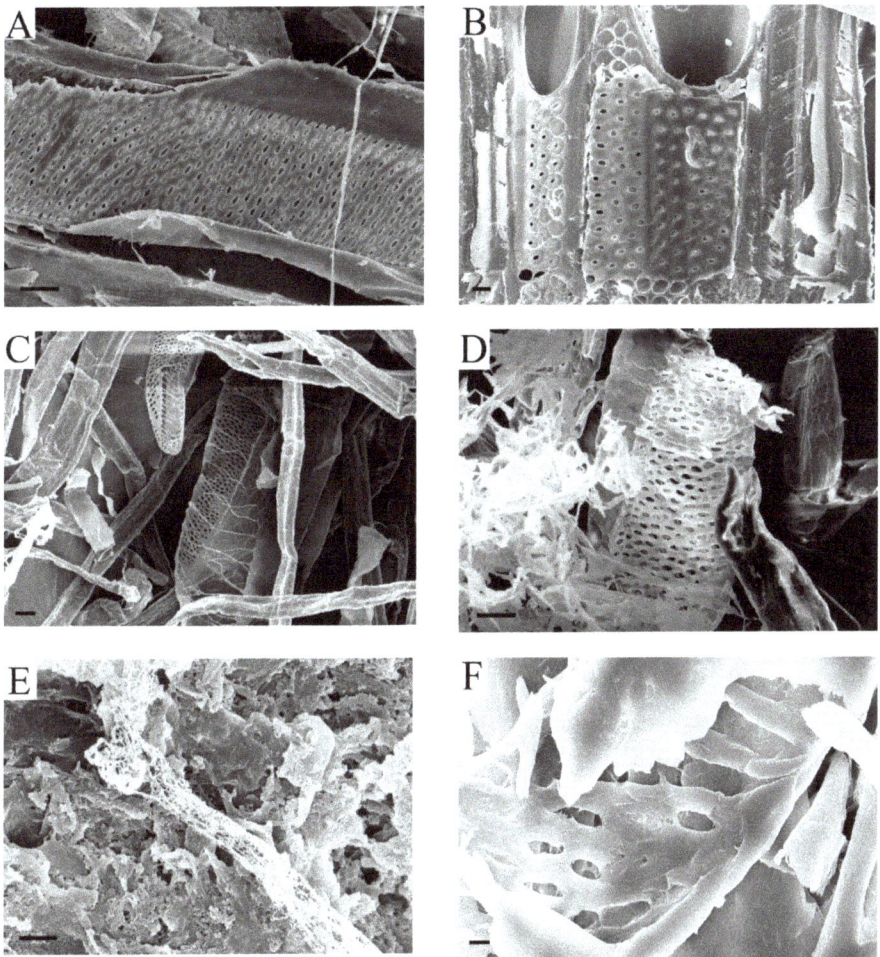

Figure 5. Wood vessels and BHP vessels before and after enzymatic hydrolysis: (**A**) birch wood vessel wall with numerous pits; (**B**) aspen wood vessels with numerous pits of various shapes in the walls; (**C**) vessels fragments of the BHP with numerous pits; (**D**) vessel fragment of BHP after 3 h of enzymatic hydrolysis; (**E**) vessel fragment of the BHP after 8 h of enzymatic hydrolysis; and, (**F**) residue of the vessel wall of BHP after 48 h of enzymatic hydrolysis. Bars: (**A,B,D,E**)—10 µm, (**C,F**)—1 µm.

However, the negative effect of high lignin content in vessels on enzymatic hydrolysis is practically excluded when using bleached pulp as a substrate. The reason for the high resistance of vessel walls to chemical and enzymatic degradation remains unclear. The chemical composition of vessel walls has not been studied enough. Perhaps the primary walls of vessels contain more xyloglucan than the primary walls of fibers. It is established that this hemicellulose polysaccharide can strongly limit the biocatalytic effect of cellulases [53].

2.5. Role of Fiber Pits in Kraft Pulp during Enzymatic Hydrolysis

Pits appear in the cell walls of woody fibers after the formation of the S_1 layer [31]. There are two main types of pits: simple and bordered pits [51]. Simple pits of libriform fibers are slit-shaped and small in size (Figure 6A). Figure 6B shows the bordered pits of tracheids partially destroyed after cryomechanical treatment with liquid nitrogen.

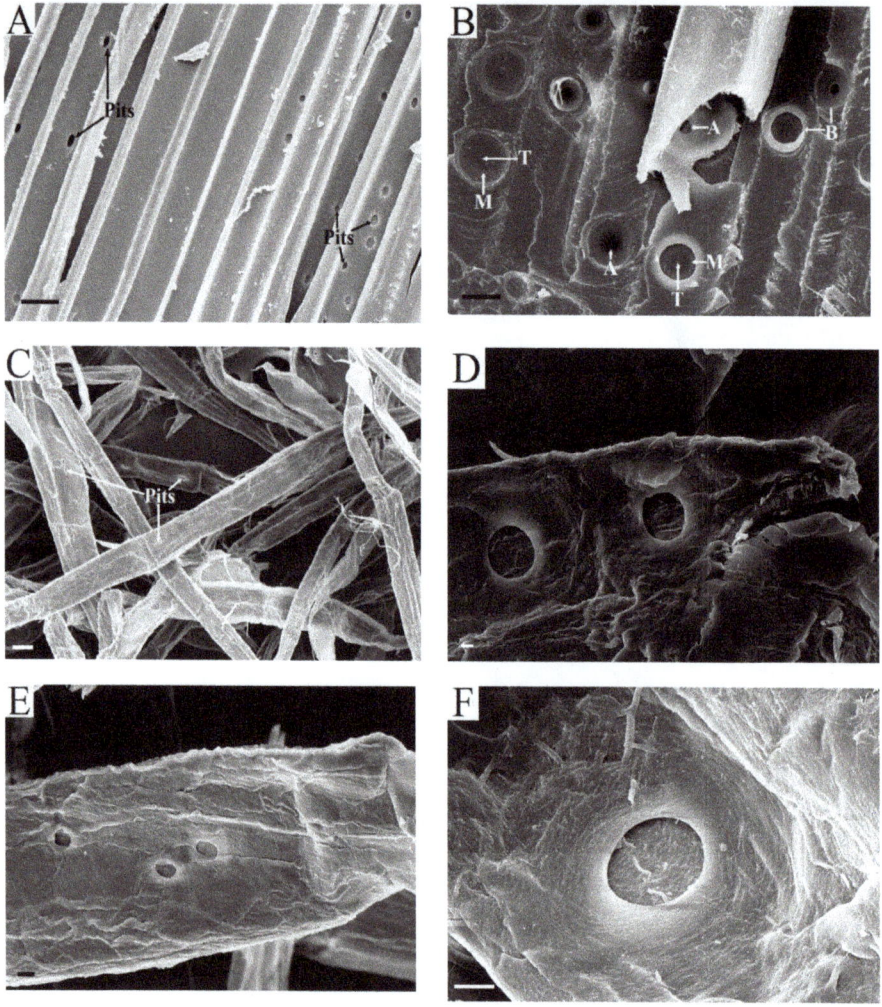

Figure 6. Pits of wood and kraft pulp fibers: (**A**) simple pits in libriform fibers; (**B**) bordered pits in tracheids of conifers; (**C**) pits in hardwood kraft pulp fibers; (**D**) pits in fibers of softwood kraft pulp after 3 h of enzymatic hydrolysis; and, (**E**) pits in fibers of hardwood kraft pulp after 3 h of enzymatic hydrolysis; (**F**) pit in fiber of softwood kraft pulp after 5 h of enzymatic hydrolysis. A—aperture, B—border, BP—bordered pit, M—margo, T—torus. Bars: (**A**–**C**)—10 µm; (**D**)—1 µm; (**E**,**F**)—2 µm.

The separation of the wood matrix into individual fibers during kraft pulping leads to the destruction of pit membranes. As a result, pit apertures are open in the walls of kraft fibers (Figure 6C–F). Through these holes, enzymes can penetrate into the lumen and attack the walls of the fibers from the inside. It is obvious that large pits of softwood kraft fibers are more effective for transporting enzymes than very narrow pits of hardwood pulp fibers. As far as we know, the role of pits of kraft fibers during enzymatic hydrolysis has not been discussed before. It is important to note that enzyme transport through the pits is possible immediately after the beginning of enzymatic hydrolysis and it does not depend on the degree of destruction of the surface layers of fibers or their cutting in the transverse direction. The pit cavities localized in the walls of kraft fibers remained for 3–5 h of enzymatic hydrolysis (Figure 6). In addition, the structure of the cell wall in the presence of pits has obvious defects. For this reason, it cannot be excluded that bends in the fibers mainly occur in the areas of pit localizations.

2.6. Changes of kraft fiber structure during enzymatic hydrolysis

Long pulp fibers were cut into small fragments during enzymatic hydrolysis, as shown in Figure 2. Changes of fiber structure during hydrolysis at the micro level have been investigated in a number of studies [46–49,54–56]. As a result, a two-stage hydrolysis mechanism was proposed [46,56]. The first stage is a quick initial enzyme cross-cutting of the fiber into small particles. The second stage is swelling of fiber parts and their gradual destruction by peeling/erosion of the surface. However, it is believed that these stages are not clearly separated in time and they can take place simultaneously.

Figure 7 shows variants of kraft fiber destruction at an early stage of hydrolysis.

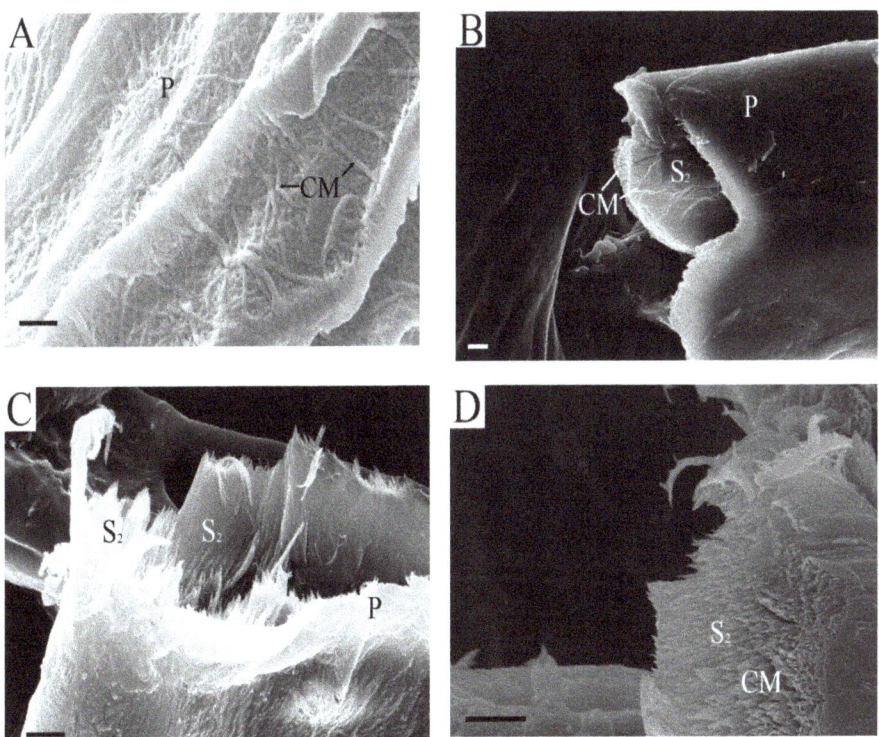

Figure 7. Cont.

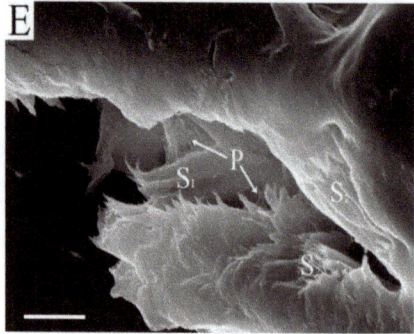 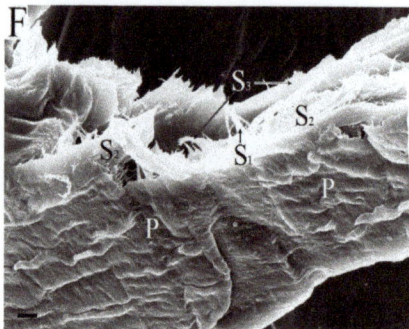

Figure 7. Kraft fibers at different stages of enzymatic degradation: (**A**) fiber of BSP with preserved primary wall; (**B**) cross-cutting of the BSP fiber after 1 h of enzymatic hydrolysis; (**C**) cross-cutting of the BSP fiber after 3 h of enzymatic hydrolysis; (**D**) the cellobiohydrolase catalyzes the hydrolysis of cellulose microfibrils of the S_2 layer; (**E**) primary wall fragments were preserved during enzymatic degradation of S_1 layer and S_2 layer; and, (**F**) enzymatic degradation of the surface of BHP fiber after 5 h of enzymatic hydrolysis. CM—cellulose microfibrils, P—primary wall, S_1—S_1 layer, S_2—S_2 layer. Bars: (**A**)—200 nm; (**B–F**)—1 μm.

The fiber morphology and topochemical processes occurring on its surface determine the enzymatic destruction of the fiber. The enzymes of *Penicillium verruculosum* enzyme complex can immediately penetrate directly into the fiber through large pits in the cell walls (Figure 6) and begin enzymatic destruction of the S_2 layer. It was found that the primary wall, which is resistant to the action of cellulases, is partially preserved after the kraft process and bleaching (Figure 7A). The outer layer of kraft fibers of bleached pulp is open and available for the action of enzymes of the *Penicillium verruculosum* enzyme complex. The destruction of surface-located xylan is faster. The cellulose macromolecules are broken into separate fragments mainly under the action of endo-β-1,4-glucanases. The destruction of surface fiber layers leads to fiber swelling. Fiber cutting occurs simultaneously with the swelling of fiber fragments.

Figure 7B shows the destruction of the fiber at the cross-cutting site at 1 h of hydrolysis. The outer layer of the BSP fiber does not yet have defects on the surface. The smooth end of a cross-cutting fiber fragment is observed. Increasing the number of fiber breaks provides easy access to the S_1 and S_2 layers for enzymes from transverse directions. The enzymatic destruction increases at the site where the fiber breaks (Figure 7C). Fiber fragments are formed with an open surface of the S_1 and S_2 layers in the cut zone.

Cellulose microfibrils of the S_2 layers that were oriented along the fiber axis were hydrolyzed by cellobiohydrolases (Figure 7D). This is the main direction of biocatalytic destruction of the kraft fiber structure. Earlier it was suggested that fiber particle ends are more reactive to enzymatic hydrolysis when compared with the central part [46]. However, primary wall fragments with characteristic randomly oriented cellulose microfibrils were still preserved in the fiber, even when significant degradation of the S_1 and S_2 layers has taken place (Figure 7E). BHP fiber has the least durable surface layer due to its high xylan content, which is rapidly destroyed at the beginning of hydrolysis. Figure 7F shows significant enzymatic degradation of the surface of BHP fiber after 5 h of enzymatic hydrolysis. Thus, after fibers are cut into separate fragments and form new ends of cross-cutting fiber fragments, there is preferential enzymatic destruction of the largest S_2 layer, which is supplemented by hydrolysis of the S_1 layer and partially primary wall.

The SEM images of fibers at different enzymatic hydrolysis stages did not reveal qualitative differences between the hydrolysis mechanisms for the tested substrates (Figure 8). The process followed the same stages for all pulp samples according to the sequence that is described above.

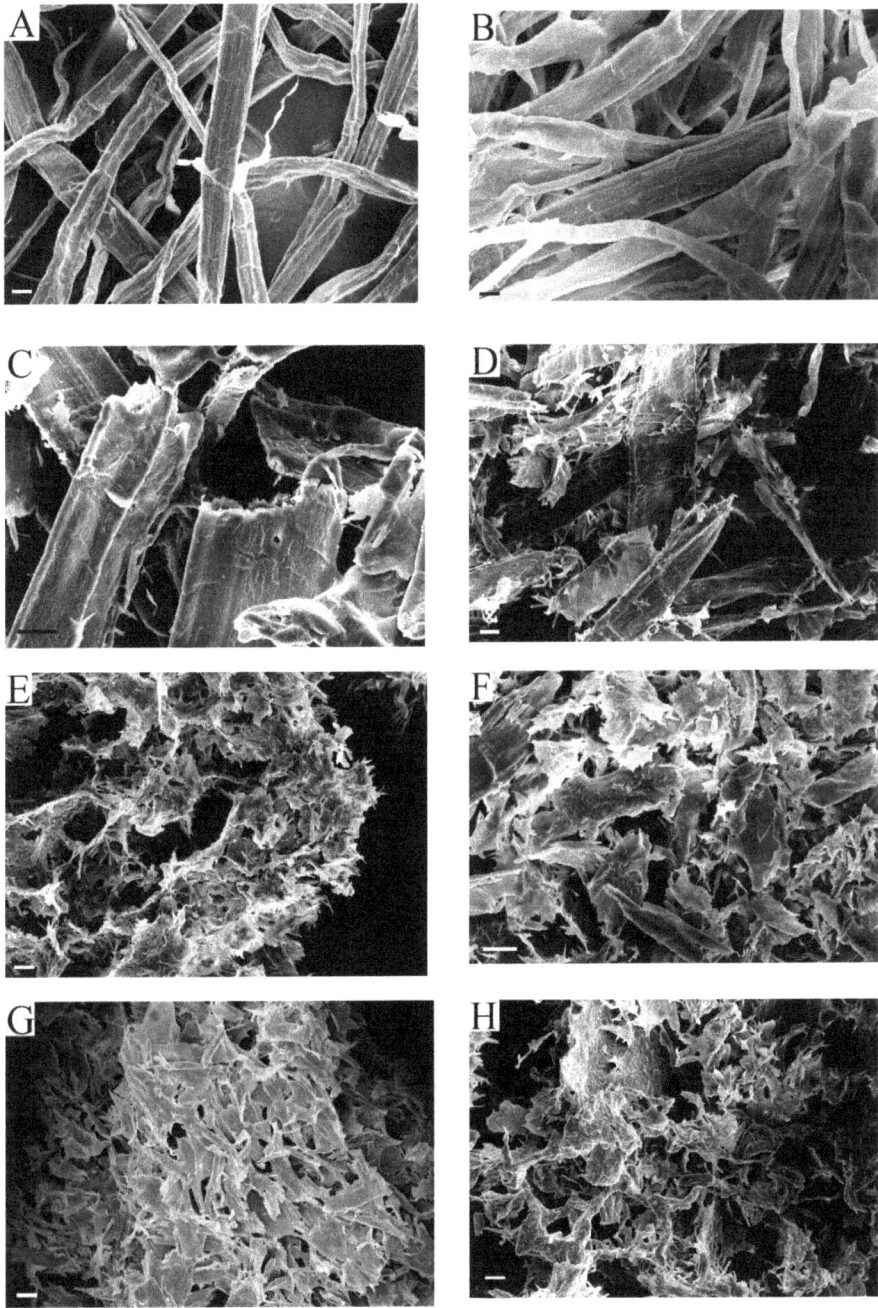

Figure 8. Fibers of BHP and SHP before and after enzymatic hydrolysis: (**A**) fibers of BHP; (**B**) fibers of BSP; (**C**) the BHP fiber after 3 h of enzymatic hydrolysis; (**D**) the BSP fiber after 3 h of enzymatic hydrolysis; (**E**) the BHP fiber after 8 h of enzymatic hydrolysis; (**F**) the BSP fiber after 8 h of enzymatic hydrolysis; (**G**) the BHP fiber after 24 h of enzymatic hydrolysis; and, (**H**) the BSP fiber after 24 h of enzymatic hydrolysis. Bars: (**A–H**)—10 μm.

Figure 8A,B show BHP and BSP fibers before enzymatic hydrolysis. Fiber fragments obtained after 3 h of hydrolysis (Figure 8C,D) had new ends that were formed during cross-cutting. The formation of new ends during fiber cross-cutting gives more new active sites for cellobiohydrolase.

Large fiber fragments were no longer observed after 8 h of enzymatic hydrolysis in hardwood pulp (Figure 8E). The deep degradation of fibers and unwrapped fiber fragment formation were noticed. At the same time, severely shortened and damaged fibers were still present in BSP (Figure 8F). More intensive surface erosion and swelling of BHP fibers were the reason for this difference. This is consistent with the previously noted more intensive increase of BHP fiber width during hydrolysis (Figure 2).

The formation of small fiber fragments with a smooth surface was observed after 24 h of hydrolysis. The particle size in hardwood pulp (Figure 8G) was smaller than in softwood pulp (Figure 8H). It was proposed in order to evaluate the possibility of using non-hydrolysable residues as cellulose powder [57].

It was established that the hydrolysis rate decreases significantly [54] in the last stages of enzyme hydrolysis. It happens despite decreased particle size and increased specific surface of the unhydrolyzed residue. Such a slowdown in hydrolysis rate can happen for various reasons, including the presence of components, for which there are no specific biocatalysts in the *Penicillium verruculosum* enzyme complex, the presence of recalcitrant to hydrolysis components of cell walls, and the accumulation of hydrolysis products that are inhibitors of individual enzymes.

2.7. Characteristics of Residue of Hardwood Bleached Kraft Pulp after Prolonged Enzymatic Hydrolysis

The residue (yield was 16% of BHP) after 72 h of enzymatic hydrolysis was fractionated while using a specially designed separation funnel. Two fractions were isolated that differed in structure. A microscopic study allowed for us to establish the morphological characteristics of each one (Figure 9).

The top fraction was deeply destroyed fragments of pulp fibers (Figure 9A,B). The top fraction contained small flat fragments of a highly degraded secondary cell wall. The crystallinity degree was determined on an XRD-7000S diffractometer (Shimadzu), according to the procedure described by Chukhchin et al. 2016 [58]. A linear relationship between the ratio of crystalline and amorphous parts of various materials was obtained, which confirmed the presence of additivity when evaluating this parameter. The proposed method allows for us to objectively evaluate and compare the crystallinity degree of cellulose of various modifications and degrees of destruction. The crystallinity degree of the components of top fraction was low (28.9%). This is typical for pulp residues after prolonged hydrolysis [17]. They were structured in the form of film-like fragments. The crystallinity degree of the components of the bottom fraction (37.6%) was similar to the crystallinity of BHP (40.7%) [58]. The bottom fraction (Figure 9C,D) contained fragments with a chaotic arrangement of cellulose microfibrils. These were mostly residues of the destroyed primary cell wall. It is established that the primary walls of woody fibers are resistant to enzymatic hydrolysis [44]. This was experimentally confirmed by studying pit formation in the walls of tracheids and fibers of libriform [30]. The action of endo-1,4-β-glucanases delivered by exosomes led to the formation of pit contours in the S_1 layer of the secondary wall. However, the cellulose microfibrils of the primary walls of two neighboring cells were resistant to the action of plant endo-1,4-β-glucanases and preserved as pit membranes. Microfibrils of the primary walls, together with xyloglucan, form a cellulose-xyloglucan network [59,60]. The xyloglucan component of the xyloglucan–cellulose complex was much more accessible to enzymatic hydrolysis by *Streptomyces* endoglucanases than cellulose microfibrils [53]. Evidently, cellulose microfibrils of the primary wall are likely protected from cellulase action by xyloglucan.

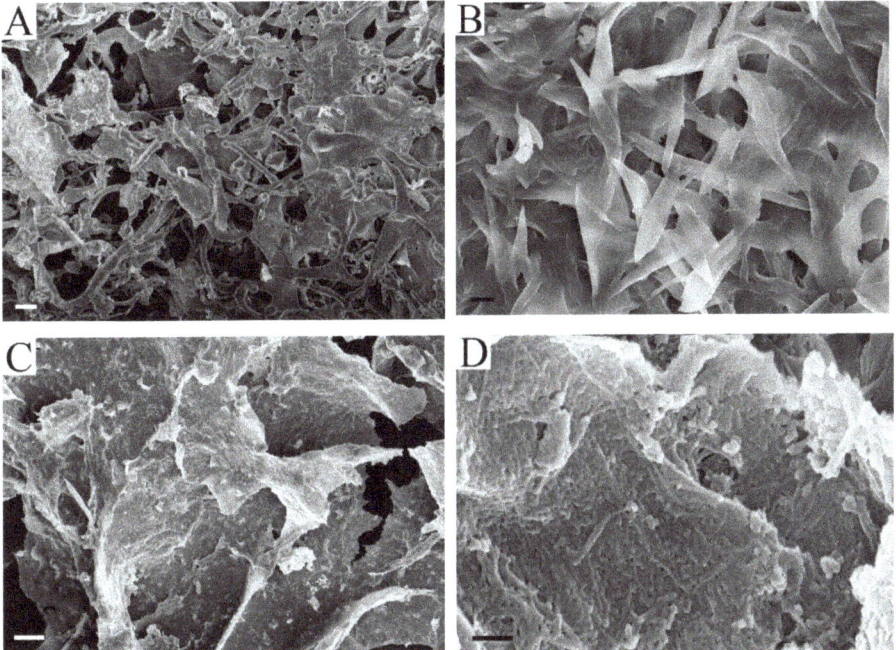

Figure 9. Images of cellulose residue fractions after enzymatic hydrolysis of BHP over 72 h: (**A,B**) the top fraction of cellulose residue in different magnification; (**C,D**) the bottom fraction of cellulose residue in different magnification. Bars: (**A**)—10 µm; (**B**)—1 µm; (**C**)—10 µm; and, (**D**)—200 nm.

2.8. Analysis of Bleached Pulp Hydrolysates

Enzymatic hydrolysis of kraft pulps produced glucose and xylan degradation products, because cellulases and xylanase are the main enzymes of the *Penicillium verruculosum* complex. We studied sugar accumulation during bleached pulp hydrolysis at initial stages of hydrolysis (up to 24 h) at pulp concentrations of 5% and 10%. Glucose is the main sugar formed during the hydrolysis of kraft pulp. The accumulation of glucose in hydrolysates occurred quickly during the first hours and then slowed down later. The glucose concentration became noticeably higher for BSP as compared to BHP after 6 h of hydrolysis (Figure 10). For BSP hydrolysates, it was 29 and 46 g/L, and, for hardwood hydrolysates, it was 21 and 35 g/L after 24 h of hydrolysis at 5 and 10% mass concentration, respectively. A lower glucose yield from BHP when compared to BSP was noted earlier in this study. It was mainly due to lower cellulose content in BHP.

An increase in sugar concentration in hydrolysates can lead to the inhibition of enzymes. It was previously described that cellobiose inhibits cellulases [61]. However, in our experiments, we applied high level of cellobiase activity that prevented the accumulation of cellobiose in kraft pulp hydrolysates.

Xylose, xylobiose, and xylooligosaccharides represented xylan degradation products of hydrolysates. Chromatographic analysis of sugars showed high concentrations of xylan destruction products, xylose and xylobiose. After 24 h of hydrolysis, the amount of xylobiose was higher when compared with xylose for both BHP and BSP (Figure 11). It could be due to a xylanase activity of *Penicillium verruculosum* enzymes complex.

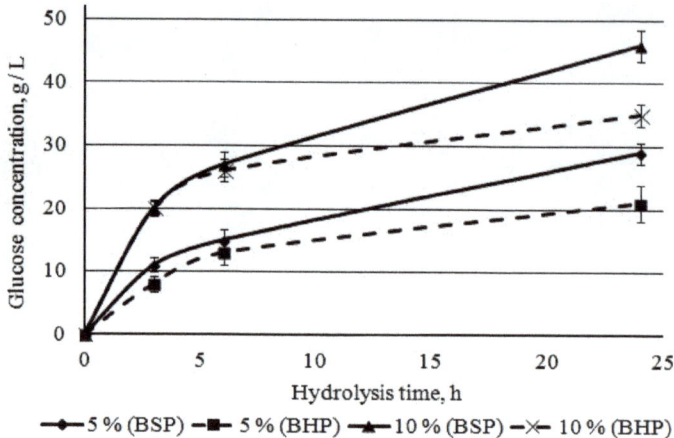

Figure 10. Glucose concentration in pulp hydrolysates during 24 h of hydrolysis.

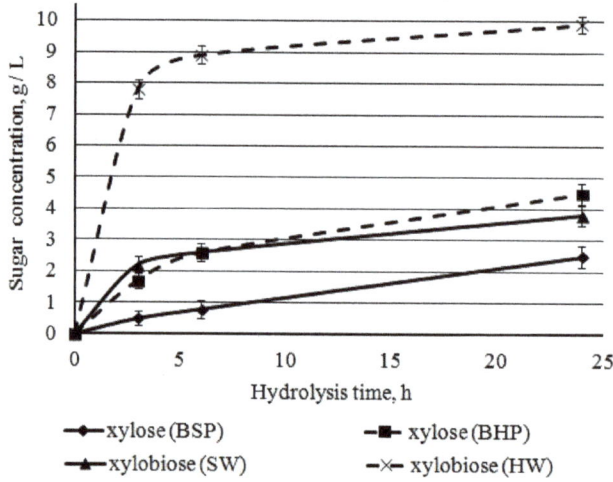

Figure 11. Concentration of xylose and xylobiose in pulp hydrolysates during 24 h of hydrolysis.

The distribution of xylan in kraft fibers promoted fast xylan hydrolysis. This surface xylan is more accessible for *Penicillium verrucolosum* xylanase. It was found that the surface layer of BHP had xylan and cellulose contents of 44.4% and 54.7%, respectively [62]. The mass ratio of xylan/cellulose of our BHP was 0.27 [17], corresponding to the potential mass ratio xylose/glucose in hydrolysates of 0.29. During hydrolysis, the following xylose + xylobiose/glucose ratios were obtained: 0.47 after 3 h, 0.43 after 6 h, and 0.41 after 24 h. This confirms that xylan was rapidly hydrolyzed when compared to cellulose. The surface layer of BSP had xylan and cellulose contents of 12.7% and 81.1%, respectively [62]. The inner layer of BSP had a 20% less xylan than the surface layer. The mass ratio of xylan/cellulose of our BSP was 0.093 [17], corresponding to the potential mass ratio xylose/glucose in hydrolysates of BSP 0.095. However, during hydrolysis the xylose + xylobiose/glucose ratio was 0.13–0.14. Thus, xylan and cellulose are hydrolyzed in equal proportions. It is necessary to include mannanase in the composition of *Penicillium verruculosum* enzyme complex and increase the proportion of cellobiohydrolases to accelerate the enzymatic hydrolysis of BSP.

The xylan conversion that was calculated from xylose and xylobiose concentrations was about 60% and 80% for hardwood and softwood pulp, respectively. The incomplete hydrolysis is related to the presence of resistant xylan in cellulose crystalline parts [63] and xyloglucan in the fiber wall [64].

Xylobiose was intensively accumulated during the first hours of hydrolysis; at the same time, the rate of xylose accumulation remained relatively constant during the first 24 h. A high concentration of xylobiose can inhibit cellulases more strongly than xylose [65]. The inhibition of *Thermoascusaur antiacus* cellobiohydrolase I and *Trichoderma reesei* cellobiohydrolase II with xylobiose and xylotriose has been demonstrated [66–68]; relevant inhibition was observed at oligosaccharide concentration over 2 g/L. In our study, this concentration was already achieved at the first hour of hardwood pulp hydrolysis.

Thus, the lower conversion for BHP as compared to BSP is related, first of all, to inhibition by a high concentration of xylobiose, released in the first hours of hydrolysis and remaining high. The determined relations allow for us to reasonably conclude that enzyme complexes for BHP should be enriched with β-xylosidase (xylobiase) hydrolyzing xylobiose.

The positive effect of high substrate concentration on the enzymatic hydrolysis efficiency (10%) on pulp has been established; sugar concentration of 50 g/L was achieved by acceptable enzyme dosage of 10 FPU/g after 24 h of hydrolysis.

3. Material and Methods

3.1. Substrates

The substrates included industrial samples of never-dried softwood and hardwood kraft pulps produced from spruce and a 1:1 aspen/birch mixture, respectively. In order to evaluate the effect of pulp hornification on enzymatic hydrolysis, it was also performed with air-dried pulp samples. The kraft pulp composition was determined earlier [17]. Hardwood pulps contained 74.5% cellulose for USP and 76.0% for BSP; softwood pulps contained 79.5% cellulose for USP and 81.0% for BSP. Unbleached pulps contained 2.1% (UHP) and 3.0% (USP) of lignin, measured as non-hydrolyzed residue after two-stage acidic treatment. In hardwood pulps, among noncellulosiccompounds, there was more xylan, 21.5% for UHP, and 21.9% for BHP, and less mannan, 0.3% for USP and BSP. The softwood pulps contained 7.3% xylan for USP, 7.5 % for BSP, and 6.8% mannan for USP and 7.0% for BSP.

3.2. Enzymes

The enzymes that were used in this study were cellulolytic enzymes (B1-221-151 and F10) produced by recombinant *Penicillium verruculosum* strains, which mainly contained endoglucanases, cellobiohydrolases, xylanase (B1-221-151), and cellobiase (F10). The enzyme preparation mix contained the complete set of hemicellulases and cellulases, including cellobiohydrolases 1 and 2 (approximately 40% of total protein), β-glucosidase (approximately 25%), endogluconases 1, 2, and 3 (approximately 10%), and xylanase (approximately 2%). Approximately 23% of the enzyme preparation consisted of ballast proteins [12]. Enzyme activity toward different substrates—carboxymethylcellulose (CMC), Avicel, xylan, galactomannan, cellobiose, and Whatman®filter paper No. 1—was determined while using well-established methods [41,69]. Enzymatic activity was determined under the following conditions: pH 5.0 (0.1 M sodium acetate buffer) and 50 °C. The protein content was determined according to Bradford assay [70]. *Penicillium verruculosum* enzyme complex carried high levels of CMC-ase, Avicelase, cellobiase, and xylanase activity [18], but low mannanase activity (<1.0 U/mg). The enzyme dosage was adjusted to give total activity of up to 10 FPU per gram of dry pulp, as in previous research [18].

3.3. Enzymatic Hydrolysis

The enzymatic hydrolysis of pulp at 5% and 10% concentrations was conducted on a laboratory shaker at 50 °C with constant stirring (250 rpm) for 3, 6, 24, 48, and 72 h. pH was maintained at 5.0 with sodium acetate buffer. Non-hydrolyzed residue was separated from the reaction mixture by centrifugation at 4200 rpm for 15 min. The supernatant was collected and analyzed for sugar and lignin content. The non-hydrolyzed residue was carefully washed with distilled water in order to separate soluble sugars. Washing was performed by sequentially repeating cycles of residue mixing and centrifugation at 4200 rpm for 15 min. After each cycle, the supernatant was gently merged and a fresh portion of distilled water was added. Non-hydrolyzed pulp residue was further analyzed for yield calculation, fractionation, and determination of crystallinity degree, fiber size, and morphology.

3.4. Non-Hydrolyzed Residue Yield Calculation

The washed non-hydrolyzed residues were frozen, freeze-dried, and kept at 20 °C to stabilize moisture content. Subsequently, residues were weighed (air-dry weight) and moisture content (%) was determined by drying 1 g samples at 103 ± 2 °C to oven-dry weight. Non-hydrolyzed residue yield (Y_{res}) was calculated by the following formula:

$$Y_{res} = \frac{\text{air dry residue weight (g)} \times (100 - \text{moisture content (\%)})}{\text{initial amount of pulp (g)}} \times 100\% \quad (1)$$

3.5. Non-Hydrolyzed Residue Characterization

3.5.1. Fiber Size and Morphology Analysis

All of the residue samples for fiber size and morphology analysis were frozen and freeze-dried. Afterwards, each one was suspended in 400 mL of distilled water, mixed, and then kept swelling for 15–30 min. Subsequently, 200 mL of the suspension was analyzed on L&W Fiber Tester equipment [71]. The fibers were defined as objects more than 0.2 mm long and less than 0.1 mm wide. Vessel elements were defined as objects wider than 0.1 mm. The device calculated the number of vessels per thousand detected fibers. Two parallel tests were performed for each sample. Changes in the structure of the fibers in water suspension were also investigated using a Carl Zeiss Axio Imager 2 optical microscope (Oberkochen, Germany).

3.5.2. Non-Hydrolyzed Residue Fractionation

A sample of the non-hydrolyzed hardwood pulp residue (72 h) was fractionated while using a specially designed vertical glass separatory funnel 2 L in volume, 1 m in length, and 0.04 m in diameter. The top end of the column was opened and the bottom end was closed by draining taps to provide slow column draining. The wet residue was mixed with a small amount of water, stirred, and then gently added from the top end to the water-filled column. Over time, the residue separated itself by sedimentation into slowly settling (upper) and rapidly settling (lower) fractions. The lower and upper fractions were collected by taking samples through the drain tap at the bottom end of the funnel. The fractions of residue were dewatered by centrifugation, frozen, and freeze-dried, as described previously.

3.5.3. Detecting Crystallinity Degree

The crystallinity degree of freeze-dried fractions was determined on an XRD-7000S diffractometer (Shimadzu, Kyoto, Japan), according to the procedure described by Chukhchin et al. 2016 [58]. The scanning range with respect to angle 2θ was 10°–70°, the scanning rate was 0.5°/min., the step was

0.02°, and the frequency of sample rotation was 30 rpm. Two parallel tests were performed for each sample. IC was calculated while using a special formula:

$$IC = \frac{\int_{2\theta=a}^{b} \left|\frac{dI}{d2\theta}\right| d2\theta}{\int_{2\theta=a}^{b} I d2\theta} \qquad (2)$$

where IC is the crystallinity index (deg^{-1}); a and b are, respectively, the initial and final values of the scanning angle (deg); and, I is the reflection intensity (number of counts).

In order to determine the final value of the crystallinity degree of samples, the result must be compared with the data on an ideal crystal (gold):

$$X_{Au} = \frac{IC}{IC_{Au}} \times 100\% \qquad (3)$$

where IC is the crystallinity index of the sample (deg^{-1}) and IC_{Au} is the crystallinity index of gold (deg^{-1}).

The result must be compared with the data of microcrystalline cellulose from bacterial cellulose to determine the crystallinity degree X_{BC} of the cellulose sample:

$$X_{BC} = \frac{IC_{Cel}}{IC_{BC}} \times 100\% \qquad (4)$$

where IC_{cel} is the crystallinity index of the cellulose sample (deg^{-1}) and IC_{bc} is the crystallinity index of microcrystalline cellulose from bacterial cellulose (deg^{-1}).

3.5.4. Scanning Electron Microscopy Analysis

Scanning electron microscopy (SEM) of initial pulp samples and non-hydrolyzed residues was used to evaluate the destruction of fibers and vessels during enzymatic hydrolysis. Prior to SEM, initial hardwood and softwood pulp fibers and non-hydrolyzed residues were freeze-dried from a water slurry and then sputter-coated with 5 nm Pt/Pd mixture while using a Q150TES spattering system (Quorum, Laughton, East Sussex, UK). The images were obtained using a SIGMA VP instrument (Zeiss, Oberkochen, Germany) operated at 10 kV accelerated voltage.

3.6. Hydrolysate Analysis

3.6.1. Determining Glucose and Reducing Sugars

Reducing sugars (RS) in the hydrolysates were determined with the Somogyi–Nelson assay [72]. Glucose was determined while using the glucose oxidase–peroxidase assay. The concentrations of minor sugars (cellobiose, xylobiose, xylose, and mannose) were determined using an LC-20 Prominence system (Shim-pack ISA-07/S2504 column, Shimadzu, Kyoto, Japan) with post-column derivatization and fluorometric detection.

3.6.2. Sugar Yield Calculations

The yield of glucose (Y_G), xylose (Y_X), mannose (Y_M), and reducing sugars (Y_{RS}) were calculated by the following formulas:

$$Y_G = \frac{\text{glucose in hydrolisate (g)} \times 0.9}{\text{initial amount of glucan in pulp (g)}} \times 100\% \qquad (5)$$

$$Y_X = \frac{\text{xylose in hydrolisate (g)} \times 0.88}{\text{initial amount of xylan in pulp (g)}} \times 100\% \qquad (6)$$

$$Y_M = \frac{\text{mannose in hydrolisate (g)} \times 0.9}{\text{initial amount of mannan in pulp (g)}} \times 100\% \qquad (7)$$

$$Y_{RS} = \frac{\text{RS in hydrolisate (g)} \times 0.9}{\text{initial amount polysaccharides in pulp (g)}} \times 100\% \qquad (8)$$

Glucose was the main sugar in all hydrolysates, so a coefficient of 0.9 for RS yield was chosen.

3.7. Lignin Content

Lignin content in hydrolysates was determined while using UV absorbance at 280 nm wavelength and extinction coefficient of 22.3 L/(g·cm), as used in [73], with a UNICO-2800 spectrophotometer (United Products & Instruments, Dayton, NJ, USA).

4. Conclusions

All of the industrial kraft pulps (UHP, USP, BHP, and BSP) are ready substrates for bioconversion using *Penicillium verruculosum* enzyme complex and give high yields of soluble hydrolysis products. The assumed conditions of enzymatic hydrolysis at a substrate concentration of 5–10% with a *Penicillium verruculosum* enzyme complex dosage of 10 FPU/g of pulp can be acceptable for obtaining hexoses and pentoses in industry.

Even a small amount of kraft lignin (2–3%) negatively affects the results of enzymatic hydrolysis. The residual lignin in LCC with mannan and xylan reducing the availability of cellulose for cellulases is the main reason for the limited hydrolysis of unbleached kraft pulp. Irreversible changes in the structure of fibers during drying reduce the yield of enzymatic hydrolysis products by 8–14% from unbleached pulps and by 19–26% from bleached pulps.

Never-dried BSP showed the best results in glucose and RS yields, while dry USP gave the worst results among the four industrial kraft pulps. This study focused on the bioconversion of never-dried bleached pulps. The availability of fiber cell walls for the enzymes is essential for effective hydrolysis. The enzymes can enter the fiber through large pits in the cell walls. Changes in the morphology and structure of kraft fibers during enzymatic hydrolysis were visualized while using optical and electron microscopy. Fiber cross-cutting occurred at early stages of enzymatic hydrolysis led to the formation of new surfaces for cellulase action. The improved mixing of kraft pulp after rapid shortening of the fibers during enzymatic hydrolysis allows for us to use higher substrate concentrations, 5% or higher. As a result, the concentration of hexoses and pentoses in the hydrolysate increases and the efficiency of their further processing also increases. Vessels are more resistant to enzymatic hydrolysis than hardwood pulp fibers. Treatment by *Penicillium verruculosum* enzyme complex leads to the active destruction of the S_1 and S_2 layers. The primary walls of kraft fibers and vessels are relatively resistant to enzymatic influences.

Features of the composition of hardwood and softwood kraft pulps require the optimization of the composition of *Penicillium verruculosum* enzyme complex. In kraft pulp bioconversion, xylan hydrolysis by xylanase yields degradation products that can inhibit the biocatalytic activity of hydrolases. The composition of this complex for hardwood kraft pulp might be improved by adding xylobiase. It is recommended to include mannanase in the composition of *Penicillium verruculosum* enzyme complex and increase the proportion of cellobiohydrolases to accelerate the enzymatic hydrolysis of softwood kraft pulp.

This study shows the high efficiency of bioconversion of the never-dried bleached kraft pulps. The bioconversion process could be integrated into an existing kraft pulp mill, since it is impractical to transport and store never-dried pulp. The processing of hexoses and pentoses into products with high added value should be organized at the same pulp mill. It is necessary to evaluate the possibility of using non-hydrolysable residues as cellulose powder for rational use in the future.

Author Contributions: Author Contributions: E.V.N., A.S.A. and A.P.S. designed the research study. E.V.N., D.N.P. and K.Y.T. prepared an original draft. Process of enzymatic hydrolysis performed by A.S.A., A.A.G., I.G.S. and D.N.P. SEM study performed by D.G.C. The crystallinity degree measurements by I.V.T. and D.G.C., A.S.A., E.V.N., D.N.P., I.V.T., D.G.C. and Y.A.S. performed the research and analyzed data as well as discussed, reviewed, and commented on the manuscript. All authors have read and agree to the published version of the manuscript.

Funding: The reported study was funded by RFBR, project number 20-04-00457.

Acknowledgments: The research was performed using instrumentation of Core Facility Center 'Arktika' of Northern (Arctic) Federal University.

Conflicts of Interest: The authors declare no conflict of interest.

References

1. Adrio, J.L.; Demain, A.L. Microbial Enzymes: Tools for Biotechnological Processes. *Biomolecules* **2014**, *4*, 117–139. [CrossRef]
2. Ek, M.; Gellerstedt, G.; Henriksson, G. *Wood Chemistry and Biotechnology*; Walter de Gruyter: Berlin, Germany, 2009; Volume 1.
3. Alvira, P.; Tomás-Pejó, E.; Ballesteros, M.; Negro, M. Pretreatment technologies for an efficient bioethanol production process based on enzymatic hydrolysis: A review. *Bioresour. Technol.* **2010**, *101*, 4851–4861. [CrossRef]
4. Xu, H.; Li, B.; Mu, X. A review of alkali-based pretreatment to enhance enzymatic saccharification for lignocellulosic biomass conversion. *Ind. Eng. Chem. Res.* **2016**, *55*, 8691–8705. [CrossRef]
5. Sun, S.; Sun, S.; Cao, X.; Sun, R. The role of pretreatment in improving the enzymatic hydrolysis of lignocellulosic materials. *Bioresour. Technol.* **2016**, *199*, 49–58. [CrossRef] [PubMed]
6. Berlin, A.; Gilkes, N.; Kurabi, A.; Bura, R.; Tu, M.; Kilburn, D.; Saddler, J. Weak Lignin-Binding Enzymes. *Appl. Biochem. Biotechnol.* **2005**, *121*, 163–170. [CrossRef]
7. Liu, H.; Sun, J.; Leu, S.Y.; Chen, S. Toward a fundamental understanding of cellulase-lignin interactions in the whole slurry enzymatic saccharification process. *Biofuels Bioprod. Biorefining* **2016**, *10*, 648–663. [CrossRef]
8. Ek, M.; Gellerstedt, G.; Henriksson, G. *Pulping Chemistry and Technology*; Walter de Gruyter: Berlin, Germany, 2009; Volume 2.
9. Van Heiningen, A. Converting a kraft pulp mill into an integrated forest biorefinery. *Pulp Pap. Can.* **2006**, *107*, 38–43.
10. Contreras, F.; Pramanik, S.; Rozhkova, A.M.; Zorov, I.N.; Korotkova, O.; Sinitsyn, A.P.; Schwaneberg, U.; Davari, M.D. Engineering Robust Cellulases for Tailored Lignocellulosic Degradation Cocktails. *Int. J. Mol. Sci.* **2020**, *21*, 1589. [CrossRef]
11. Skomarovsky, A.; Gusakov, A.; Okunev, O.; Solov'eva, I.; Bubnova, T.; Kondrat'eva, E.; Sinitsyn, A. Studies of hydrolytic activity of enzyme preparations of Penicillium and Trychoderma fungi. *Appl. Biochem. Microbiol.* **2005**, *41*, 182–184. [CrossRef]
12. Chekushina, A.; Dotsenko, G.; Sinitsyn, A. Comparing the efficiency of plant material bioconversion processes using biocatalysts based on Trichoderma and Penicillium verruculosum enzyme preparations. *Catal. Ind.* **2013**, *5*, 98–104. [CrossRef]
13. Proskurina, O.; Korotkova, O.; Rozhkova, A.; Matys, V.Y.; Koshelev, A.; Okunev, O.; Nemashkalov, V.; Sinitsyna, O.; Sinitsyn, A. Application of the "fusion" approach for the production of highly efficient biocatalysts based on recombinant strains of the fungus Penicillium verruculosum for the conversion of cellulose-containing biomass. *Catal. Ind.* **2013**, *5*, 327–334. [CrossRef]
14. Berlin, A.; Balakshin, M.; Gilkes, N.; Kadla, J.; Maximenko, V.; Kubo, S.; Saddler, J. Inhibition of cellulase, xylanase and β-glucosidase activities by softwood lignin preparations. *J. Biotechnol.* **2006**, *125*, 198–209. [CrossRef] [PubMed]
15. Morozova, V.; Semenova, M.; Rozhkova, A.; Kondrat'eva, E.; Okunev, O.; Bekkarevich, A.; Novozhilov, E.; Sinitsyn, A. Influence of the cycle number in processing of cellulose from waste paper on its ability to hydrolysis by cellulases. *Appl. Biochem. Microbiol.* **2010**, *46*, 363–366. [CrossRef]
16. Novozhilov, E.V.; Aksenov, A.S.; Demidov, M.L.; Chukhchin, D.G.; Dotsenko, G.S.; Osipov, D.O.; Sinitsyn, A.P. Application of complex biocatalysts based on recombinant Penicillium verruculosum enzyme preparations in the hydrolysis of semichemical hardwood pulp. *Catal. Ind.* **2014**, *6*, 348–354. [CrossRef]

17. Novozhilov, E.; Sinel'nikov, I.; Aksenov, A.; Chukhchin, D.; Tyshkunova, I.; Rozhkova, A.; Osipov, D.; Zorov, I.; Sinitsyn, A. Biocatalytic conversion of kraft pulp using cellulase complex of Penicillium verruculosum. *Catal. Ind.* **2016**, *8*, 95–100. [CrossRef]
18. Tenkanen, M.; Tamminen, T.; Hortling, B. Investigation of lignin-carbohydrate complexes in kraft pulps by selective enzymatic treatments. *Appl. Microbiol. Biotechnol.* **1999**, *51*, 241–248. [CrossRef]
19. Ramos, L.; Nazhad, M.; Saddler, J. Effect of enzymatic hydrolysis on the morphology and fine structure of pretreated cellulosic residues. *Enzym. Microb. Technol.* **1993**, *15*, 821–831. [CrossRef]
20. Zhang, X.; Qin, W.; Paice, M.G.; Saddler, J.N. High consistency enzymatic hydrolysis of hardwood substrates. *Bioresour. Technol.* **2009**, *100*, 5890–5897. [CrossRef]
21. Wu, S.; Chang, H.; Phillips, R.; Jameel, H. Techno-economic analysis of the optimum softwood lignin content for the production of bioethanol in a repurposed kraft mill. *BioResources* **2014**, *9*, 6817–6830. [CrossRef]
22. Aldaeus, F.; Larsson, K.; Srndovic, J.S.; Kubat, M.; Karlström, K.; Peciulyte, A.; Olsson, L.; Larsson, P.T. The supramolecular structure of cellulose-rich wood pulps can be a determinative factor for enzymatic hydrolysability. *Cellulose* **2015**, *22*, 3991–4002. [CrossRef]
23. Buzała, K.; Przybysz, P.; Rosicka-Kaczmarek, J.; Kalinowska, H. Production of glucose-rich enzymatic hydrolysates from cellulosic pulps. *Cellulose* **2015**, *22*, 663–674. [CrossRef]
24. Buzała, K.; Przybysz, P.; Rosicka-Kaczmarek, J.; Kalinowska, H. Comparison of digestibility of wood pulps produced by the sulfate and TMP methods and woodchips of various botanical origins and sizes. *Cellulose* **2015**, *22*, 2737–2747. [CrossRef]
25. Buzała, K.P.; Przybysz, P.; Kalinowska, H.; Przybysz, K.; Kucner, M.; Dubowik, M. Evaluation of pine kraft cellulosic pulps and fines from papermaking as potential feedstocks for biofuel production. *Cellulose* **2016**, *23*, 649–659. [CrossRef]
26. Przybysz Buzała, K.; Kalinowska, H.; Małachowska, E.; Boruszewski, P.; Krajewski, K.; Przybysz, P. The Effect of Lignin Content in Birch and Beech Kraft Cellulosic Pulps on Simple Sugar Yields from the Enzymatic Hydrolysis of Cellulose. *Energies* **2019**, *12*, 2952. [CrossRef]
27. Branco, R.H.; Amândio, M.S.; Serafim, L.S.; Xavier, A.M. Ethanol Production from Hydrolyzed Kraft Pulp by Mono-and Co-Cultures of Yeasts: The Challenge of C6 and C5 Sugars Consumption. *Energies* **2020**, *13*, 744. [CrossRef]
28. Sjöberg, J.; Kleen, M.; Dahlman, O.; Agnemo, R. Analyses of carbohydrates and lignin in the surface and inner layers of softwood pulp fibers obtained employing various alkaline cooking processes. *Nord. Pulp Pap. Res. J.* **2002**, *17*, 295–301. [CrossRef]
29. Lawoko, M.; Henriksson, G.; Gellerstedt, G. Structural Differences between the Lignin–Carbohydrate Complexes Present in Wood and in Chemical Pulps. *Biomacromolecules* **2005**, *6*, 3467–3473. [CrossRef]
30. Chukhchin, D.G.; Bolotova, K.; Sinelnikov, I.; Churilov, D.; Novozhilov, E. Exosomes in the phloem and xylem of woody plants. *Planta* **2020**, *251*, 12. [CrossRef]
31. Lawoko, M.; Berggren, R.; Berthold, F.; Henriksson, G.; Gellerstedt, G. Changes in the lignin-carbohydrate complex in softwood kraft pulp during kraft and oxygen delignification. *Holzforschung* **2004**, *58*, 603–610. [CrossRef]
32. Wistara, N.; Young, R.A. Properties and treatments of pulps from recycled paper. Part I. Physical and chemical properties of pulps. *Cellulose* **1999**, *6*, 291–324. [CrossRef]
33. Fernandes Diniz, J.M.B.; Gil, M.H.; Castro, J.A.A.M. Hornification—its origin and interpretation in wood pulps. *Wood Sci. Technol.* **2004**, *37*, 489–494. [CrossRef]
34. Oksanen, T.; Buchert, J.; Viikari, L. The role of hemicelluloses in the hornification of bleached kraft pulps. *Holzforsch. Int. J. Biol. Chem. Phys. Technol. Wood* **1997**, *51*, 355–360. [CrossRef]
35. Rebuzzi, F.; Evtuguin, D.V. Effect of Glucuronoxylan on the Hornification of Eucalyptus globulus Bleached Pulps. *Macromol. Symp.* **2005**, *232*, 121–128. [CrossRef]
36. Kontturi, E.; Vuorinen, T. Indirect evidence of supramolecular changes within cellulose microfibrils of chemical pulp fibers upon drying. *Cellulose* **2009**, *16*, 65–74. [CrossRef]
37. Salmén, L. Wood morphology and properties from molecular perspectives. *Ann. For. Sci.* **2015**, *72*, 679–684. [CrossRef]
38. Xiao, Z.; Zhang, X.; Gregg, D.J.; Saddler, J.N. Effects of sugar inhibition on cellulases and β-glucosidase during enzymatic hydrolysis of softwood substrates. *Appl. Biochem. Biotechnol.* **2004**, *115*, 1115–1126. [CrossRef]

39. Kumar, R.; Wyman, C.E. Strong cellulase inhibition by Mannan polysaccharides in cellulose conversion to sugars. *Biotechnol. Bioeng.* **2014**, *111*, 1341–1353. [CrossRef]
40. Laine, J.; Lindström, T.; Bremberg, C.; Glad-Nordmark, G. Studies on topochemical modification of cellulosic fibres. Part 4. Toposelectivity of carboxymethylation and its effects on the swelling of fibres. *Nord. Pulp Pap. Res. J.* **2003**, *18*, 316–324. [CrossRef]
41. Sinitsyn, A.; Gusakov, A.; Vlasenko, E.Y. Effect of structural and physico-chemical features of cellulosic substrates on the efficiency of enzymatic hydrolysis. *Appl. Biochem. Biotechnol.* **1991**, *30*, 43–59. [CrossRef]
42. Liu, Y.-S.; Baker, J.O.; Zeng, Y.; Himmel, M.E.; Haas, T.; Ding, S.-Y. Cellobiohydrolase hydrolyzes crystalline cellulose on hydrophobic faces. *J. Biol. Chem.* **2011**, *286*, 11195–11201. [CrossRef]
43. Agarwal, U.P.; Zhu, J.; Ralph, S.A. Enzymatic hydrolysis of loblolly pine: Effects of cellulose crystallinity and delignification. *Holzforschung* **2013**, *67*, 371–377. [CrossRef]
44. Fengel, D.; Wegener, G. *Wood: Chemistry, Ultrastructure, Reactions*; Walter de Gruyter: Berlin, Germany, 1984.
45. Gharehkhani, S.; Sadeghinezhad, E.; Kazi, S.N.; Yarmand, H.; Badarudin, A.; Safaei, M.R.; Zubir, M.N.M. Basic effects of pulp refining on fiber properties—A review. *Carbohydr. Polym.* **2015**, *115*, 785–803. [CrossRef] [PubMed]
46. Chauve, M.; Barre, L.; Tapin-Lingua, S.; da Silva Perez, D.; Decottignies, D.; Perez, S.; Ferreira, N.L. Evolution and impact of cellulose architecture during enzymatic hydrolysis by fungal cellulases. *Adv. Biosci. Biotechnol.* **2013**, *4*, 1095–1109. [CrossRef]
47. Ander, P.; Hildén, L.; Daniel, G. Cleavage of softwood kraft pulp by HCl and cellulases. *BioResources* **2008**, *3*, 477–490.
48. Clarke, K.; Li, X.; Li, K. The mechanism of fiber cutting during enzymatic hydrolysis of wood biomass. *Biomass Bioenergy* **2011**, *35*, 3943–3950. [CrossRef]
49. Thygesen, L.G.; Hidayat, B.J.; Johansen, K.S.; Felby, C. Role of supramolecular cellulose structures in enzymatic hydrolysis of plant cell walls. *J. Ind. Microbiol. Biotechnol.* **2011**, *38*, 975–983. [CrossRef]
50. Hidayat, B.J.; Felby, C.; Johansen, K.S.; Thygesen, L.G. Cellulose is not just cellulose: A review of dislocations as reactive sites in the enzymatic hydrolysis of cellulose microfibrils. *Cellulose* **2012**, *19*, 1481–1493. [CrossRef]
51. Evert, R.F. *Esau's Plant Anatomy: Meristems, Cells, and Tissues of the Plant Body: Their Structure, Function, and Development*; John Wiley & Sons: Hoboken, NJ, USA, 2006.
52. Fergus, B.; Goring, D. The distribution of lignin in birch wood as determined by ultraviolet microscopy. *Holzforsch. -Int. J. Biol. Chem. Phys. Technol. Wood* **1970**, *24*, 118–124. [CrossRef]
53. Hayashi, T.; Maclachlan, G. Pea xyloglucan and cellulose: I. Macromolecular organization. *Plant Physiol.* **1984**, *75*, 596–604. [CrossRef]
54. Mansfield, S.D.; de Jong, E.; Stephens, R.S.; Saddler, J.N. Physical characterization of enzymatically modified kraft pulp fibers. *J. Biotechnol.* **1997**, *57*, 205–216. [CrossRef]
55. Wang, L.; Zhang, Y.; Gao, P.; Shi, D.; Liu, H.; Gao, H. Changes in the structural properties and rate of hydrolysis of cotton fibers during extended enzymatic hydrolysis. *Biotechnol. Bioeng.* **2006**, *93*, 443–456. [CrossRef] [PubMed]
56. Arantes, V.; Gourlay, K.; Saddler, J.N. The enzymatic hydrolysis of pretreated pulp fibers predominantly involves "peeling/erosion" modes of action. *Biotechnol. Biofuels* **2014**, *7*, 87–97. [CrossRef] [PubMed]
57. Novozhilov, E.; Tyshkunova, I.; Guryanova, A.; Terentyev, K.; Aksenov, A. COMPLEX USAGE OF THE SUBSTRATE IN THE BLEACHED KRAFT PULP ENZYMATIC HYDROLYSIS. *Int. Multidiscip. Sci. Geoconference Sgem* **2019**, *19*, 643–650.
58. Chukhchin, D.; Malkov, A.; Tyshkunova, I.; Mayer, L.; Novozhilov, E. Diffractometric method for determining the degree of crystallinity of materials. *Crystallogr. Rep.* **2016**, *61*, 371–375. [CrossRef]
59. Cosgrove, D.J. Growth of the plant cell wall. *Nat. Rev. Mol. Cell Biol.* **2005**, *6*, 850. [CrossRef] [PubMed]
60. Chebli, Y.; Geitmann, A. Cellular growth in plants requires regulation of cell wall biochemistry. *Curr. Opin. Cell Biol.* **2017**, *44*, 28–35. [CrossRef]
61. Zhao, Y.; Wu, B.; Yan, B.; Gao, P. Mechanism of cellobiose inhibition in cellulose hydrolysis by cellobiohydrolase. *Sci. China Ser. C Life Sci.* **2004**, *47*, 18–24. [CrossRef]
62. Dahlman, O.; Jacobs, A.; Sjöberg, J. Molecular properties of hemicelluloses located in the surface and inner layers of hardwood and softwood pulps. *Cellulose* **2003**, *10*, 325–334. [CrossRef]
63. Vena, P.; García-Aparicio, M.; Brienzo, M.; Görgens, J.; Rypstra, T. Effect of alkaline hemicellulose extraction on kraft pulp fibers from Eucalyptus grandis. *J. Wood Chem. Technol.* **2013**, *33*, 157–173. [CrossRef]

64. Kont, R.; Kurašin, M.; Teugjas, H.; Väljamäe, P. Strong cellulase inhibitors from the hydrothermal pretreatment of wheat straw. *Biotechnol. Biofuels* **2013**, *6*, 135. [CrossRef]
65. Kim, Y.; Ximenes, E.; Mosier, N.S.; Ladisch, M.R. Soluble inhibitors/deactivators of cellulase enzymes from lignocellulosic biomass. *Enzym. Microb. Technol.* **2011**, *48*, 408–415. [CrossRef] [PubMed]
66. Qing, Q.; Yang, B.; Wyman, C.E. Xylooligomers are strong inhibitors of cellulose hydrolysis by enzymes. *Bioresour. Technol.* **2010**, *101*, 9624–9630. [CrossRef] [PubMed]
67. Zhang, J.; Tang, M.; Viikari, L. Xylans inhibit enzymatic hydrolysis of lignocellulosic materials by cellulases. *Bioresour. Technol.* **2012**, *121*, 8–12. [CrossRef] [PubMed]
68. Zhang, J.; Viikari, L. Xylo-oligosaccharides are competitive inhibitors of cellobiohydrolase I from Thermoascus aurantiacus. *Bioresour. Technol.* **2012**, *117*, 286–291. [CrossRef]
69. Ghose, T. Measurement of cellulase activities. *Pure Appl. Chem.* **1987**, *59*, 257–268. [CrossRef]
70. Bradford, M.M. A rapid and sensitive method for the quantitation of microgram quantities of protein utilizing the principle of protein-dye binding. *Anal. Biochem.* **1976**, *72*, 248–254. [CrossRef]
71. Karlsson, H. *Fibre Guide: Fibre Analysis and Process Applications in the Pulp and Paper Industry: A Handbook*; Lorentzen & Wettre: Kista, Sweden, 2006.
72. Nelson, N. A photometric adaptation of the Somogyi method for the determination of glucose. *J. Biol. Chem.* **1944**, *153*, 375–380.
73. Kalliola, A.; Kuitunen, S.; Liitiä, T.; Rovio, S.; Ohra-aho, T.; Vuorinen, T.; Tamminen, T. Lignin oxidation mechanisms under oxygen delignification conditions. Part 1. Results from direct analyses. *Holzforschung* **2011**, *65*, 567–574. [CrossRef]

© 2020 by the authors. Licensee MDPI, Basel, Switzerland. This article is an open access article distributed under the terms and conditions of the Creative Commons Attribution (CC BY) license (http://creativecommons.org/licenses/by/4.0/).

Article

Synthesis of DHA/EPA Ethyl Esters via Lipase-Catalyzed Acidolysis Using Novozym® 435: A Kinetic Study

Chia-Hung Kuo [1,*], Chun-Yung Huang [1], Chien-Liang Lee [2], Wen-Cheng Kuo [3], Shu-Ling Hsieh [1] and Chwen-Jen Shieh [4,*]

1. Department of Seafood Science, National Kaohsiung University of Science and Technology, Kaohsiung 811, Taiwan; cyhuang@nkust.edu.tw (C.-Y.H.); slhsieh@nkust.edu.tw (S.-L.H.)
2. Department of Chemical and Materials Engineering, National Kaohsiung University of Science and Technology, Kaohsiung 807, Taiwan; cl_lee@nkust.edu.tw
3. Department of Mechatronics Engineering, National Kaohsiung University of Science and Technology, Kaohsiung 811, Taiwan; rkuo@nkust.edu.tw
4. Biotechnology Center, National Chung Hsing University, Taichung 402, Taiwan
* Correspondence: kuoch@nkust.edu.tw (C.-H.K.); cjshieh@nchu.edu.tw (C.-J.S.); Tel.: +886-7-361-7141 (ext. 23464) (C.-H.K.); +886-4-2284-0450 (ext. 5121) (C.-J.S.)

Received: 14 April 2020; Accepted: 12 May 2020; Published: 19 May 2020

Abstract: DHA/EPA ethyl ester is mainly used in the treatment of arteriosclerosis and hyperlipidemia. In this study, DHA+EPA ethyl ester was synthesized via lipase-catalyzed acidolysis of ethyl acetate (EA) with DHA+EPA concentrate in n-hexane using Novozym® 435. The DHA+EPA concentrate (in free fatty acid form), contained 54.4% DHA and 16.8% EPA, was used as raw material. A central composite design combined with response surface methodology (RSM) was used to evaluate the relationship between substrate concentrations and initial rate of DHA+EPA ethyl ester production. The results indicated that the reaction followed the ordered mechanism and as such, the ordered mechanism model was used to estimate the maximum reaction rate (Vmax) and kinetic constants. The ordered mechanism model was also combined with the batch reaction equation to simulate and predict the conversion of DHA+EPA ethyl ester in lipase-catalyzed acidolysis. The integral equation showed a good predictive relationship between the simulated and experimental results. 88–94% conversion yields were obtained from 100–400 mM DHA+EPA concentrate at a constant enzyme activity of 200 U, substrate ratio of 1:1 (DHA+EPA: EA), and reaction time of 300 min.

Keywords: lipase; acidolysis; docosahexaenoic acid ethyl ester; eicosapentaenoic acid ethyl ester; ethyl acetate; kinetics

---

1. Introduction

Docosahexaenoic acid (DHA; C22:6) and eicosapentaenoic acid (EPA; C22:6) belong to n-3 polyunsaturated fatty acids (PUFAs) have unique physiological activities and health functions, are considered essential fatty acids. Thanks to increasing health concerns, the fish oil market has expanded and the amount of fish oil used in health products is growing rapidly. In 2014, the consumption of fish oil was about 282,000 tons with a market value of more than 1.69 billion USD [1]. Fish oils are rich in long-chain n-3 PUFAs, especially DHA and EPA [2]. The health benefits of DHA and EPA have been widely investigated, including their impact on cardiovascular disease [3,4], ethanol-induced fatty liver [5], blood pressure [6], asthma [7], and inflammation [8]. With advancements in industrial technology and the widespread use of biochemical technology, the concentration and purity of DHA and EPA are increasing in food products and medicinal grade drugs [8]. Generally, the DHA and EPA content in fish oil is less than 30% [9]. For health products or medicinal applications, fish oil must use physical, chemical, or enzymatic methods to concentrate n-3 PUFAs. Various concentration methods have been reported, including low temperature solvent crystallization [10], urea complexation [11,12], molecular distillation [13], supercritical fluid extraction [14], high-performance liquid chromatography [15], silver nitrate extraction [16], solvent fractionation [17], and enzymatic methods [18–20]. However, concentrated n-3 PUFAs are unstable and difficult to preserve for long periods due to the carboxyl groups enhanced the catalytic effect to form the free radicals by the decomposition of hydroperoxides [21], and therefore must be esterified as triglycerides or ethyl esters to increase storage stability.

Generally, chemical esterification employs a strong acid, alkaline, or metal (Lewis acid) as the chemical catalyst. These catalysts have several problems, such as reaction at a higher temperature (>160 °C), unwanted side reactions that cause darkening or odors, or the product poses environmental threats [22]. An alternative to chemical-based esterification, biocatalysis seems to be an attractive option for producing esters. Lipase (triacylglycerol ester hydrolase, EC 3.1.1.3) hydrolyses ester linkages of triglycerides into glycerol and fatty acids [23,24]. On the other hands, the lipase catalyzes the reverse synthesis reaction to produce esters in the non-aqueous organic system [25,26]. Immobilized lipase is typically used in this reaction since immobilization improves the life and stability of the enzyme, as well as facilitating enzyme recovery, reuse, and continuous operation [27,28]. Immobilization of lipase may alter its observed activity, specificity or selectivity [29]. Lipases exist in two main forms, open and closed forms [30]. The hydrophobic supports fix the structure of lipase at the open conformation [31], which can promote hyperactivation of lipase after immobilization [32–34]. Novozym® 435 is a commercially available immobilized lipase produced by Novozymes. It is based on immobilization via interfacial activation of lipase B from *Candida antarctica* on a macroporous support formed by poly(methyl methacrylate) crosslinked with divinylbenzene, which has been used for synthesis of many esters [35]. Shimada et al. used immobilized lipase from *Candida antarctica* to catalyze DHA and ethanol for the synthesis of DHA ethyl ester; a high yield of 88% was obtained after 24 h [36]. DHA ethyl ester has been synthesized by both lipase-catalyzed transesterification of lipids with ethanol and lipase-catalyzed esterification of fatty acids with ethanol, with conversion yields of 20% and 60% after 24 h, respectively [37]. Bhandari et al. used immobilized lipase from *Candida antarctica* for the esterification of glycerol with n-3 PUFA, and a high degree of esterification (69%) was obtained after 24 h [38]. However, the reaction normally requires a reaction time of around 24 h.

Lipase catalyzed synthesis of ester can be achieved in different ways [39], as depicted in Scheme 1. This process normally takes place between an acid and an alcohol (direct esterification) [40,41], an ester and an alcohol (transesterification or alcoholysis) [42], an ester and an acid (acidolysis) [43], or between two esters (interesterification) [44]. Our previous research determined that ester synthesized by lipase via transesterification (alcoholysis) exhibited higher reaction rates than that obtained via direct esterification [41,42]. In this study, in order to decrease reaction time and increase the conversion yield, we chose ethyl acetate as the ester for the lipase-catalyzed synthesis of DHA+EPA ethyl esters via acidolysis with DHA+EPA concentrate. Lipase-catalyzed acidolysis has been used to incorporate citronellic acid and myristic acid into egg-yolk phosphatidylcholine [45,46], and to incorporate caprylic acid into corn oil for the synthesis of MLM structured lipids [47], but to date, the lipase-catalyzed synthesis of DHA+EPA ethyl esters via acidolysis of EA with DHA+EPA concentrate has not been reported.

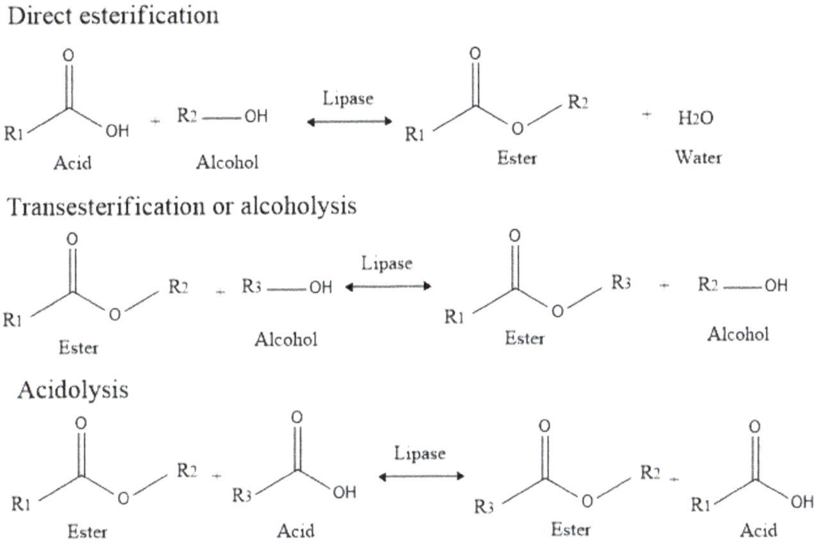

Scheme 1

Direct esterification

R1-COOH (Acid) + R2-OH (Alcohol) ⇌ (Lipase) R1-COO-R2 (Ester) + H2O (Water)

Transesterification or alcoholysis

R1-COO-R2 (Ester) + R3-OH (Alcohol) ⇌ (Lipase) R1-COO-R3 (Ester) + R2-OH (Alcohol)

Acidolysis

R1-COO-R2 (Ester) + R3-COOH (Acid) ⇌ (Lipase) R3-COO-R2 (Ester) + R1-COOH (Acid)

Scheme 1. Lipase-catalyzed synthesis of ester via direct esterification, alcoholysis or acidolysis.

It is useful to know the reaction kinetics and the constants that describe kinetic behavior of the reaction. The ping-pong bi–bi mechanism or an ordered bi–bi mechanism is the most used kinetic models for explaining the reaction of lipase-catalyzed esterification or transesterification [48]. Lipase-catalyzed reactions with free or immobilized enzymes and with substrates (acids and alcohols) of various chain lengths and structures, either in solvent-free systems or in the presence of an organic solvent have been carried out for kinetic studies [49–51]. However, the kinetics and modelling of lipase-catalyzed acidolysis of EA with concentrated n-3 PUFAs have not yet been discussed. The aim of this research was to investigate the kinetics of lipase-catalyzed synthesis of DHA+EPA ethyl esters via acidolysis in order to identify the optimal conditions for ester synthesis. In this study, a statistical experimental design and response surface methodology (RSM) analysis were employed to investigate the kinetic model for determining kinetic constants. Finally, a kinetic model based on the estimated kinetic constants was developed and used for predicting conversion yields of lipase-catalyzed acidolysis in batch reactions.

2. Results and Discussion

2.1. Initial Rate of DHA+EPA Ethyl Ester Production

In order to understand the lipase-catalyzed acidolysis reaction for DHA+EPA ethyl ester synthesis, analysis of the reaction kinetics and the constants is useful to describe the reaction mechanism. In general, the initial rate data were fitted to the proposed kinetic model in order to evaluate the kinetic constants. Therefore, a great number of the initial rate data performed through a wide range of substrate concentrations are required to measurement from systematic experiments. A more efficient technique to solve the problem is response surface methodology (RSM), which can be used for obtaining sufficient data to solve the kinetic parameters under the less number of experimental runs and time. The lipase-catalyzed acidolysis of EA with DHA+EPA concentrate is represented in Scheme 2. With the purpose of finding a kinetic model to illustrate the relationship between the DHA+EPA concentrate and EA in the reaction, the effect of substrate concentrations on the initial rate of DHA+EPA ethyl ester production was investigated. A five-level-two-factor central composite design was employed and combined with RSM in order to better understand the relationship between substrate concentration and initial production rate. The DHA+EPA concentrate prepared by using the acetone fractionation of fatty acid salts method [52], contained 54.4% DHA and 16.8% EPA, was used as raw material for the lipase-catalyzed synthesis of DHA+EPA ethyl ester. The substrate concentrations and their selected levels are listed in Table 1. The SAS statistical software (SAS Institute, Cary, NC, USA) was used to fit the initial rate data in Table 1 to the second-order polynomial equation. The second-order polynomial Equation (1) is as below:

$$Y \text{ (μmol min}^{-1} \text{ U}^{-1}) = -0.011807 + 0.0002X_1 + 0.000265X_2 - 0.000000451X_1X_1 \\ + 0.000000665X_2X_1 - 0.000000757X_2X_2 \quad (1)$$

where Y is the initial rate of DHA+EPA ethyl ester production, and X_1, and X_2 are uncoded values for DHA+EPA concentrate and EA concentration, respectively.

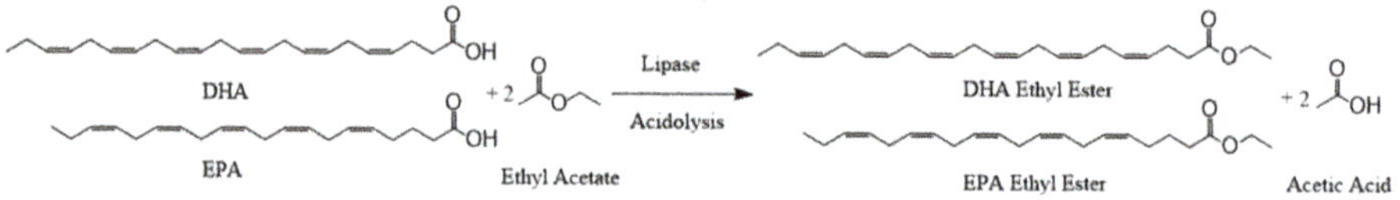

Scheme 2. Lipase-catalyzed synthesis of DHA+EPA ethyl ester via acidolysis of EA with DHA+EPA concentrate.

Table 1. Central composite design and initial experimental rate of DHA+EPA ethyl ester production.

Treatment No. [1]	Factor		Initial Rate of DHA+EPA Ethyl Ester Production (μmol min^{-1} U^{-1} Enzyme)
	X_1 DHA+EPA (mM)	X_2 Ethyl Acetate (mM)	
1	−2 [2] (40)	0 (220)	0.0168 ± 0.0001
2	−1 (130)	−1 (130)	0.0421 ± 0.0032
3	−1 (130)	1 (310)	0.0533 ± 0.0029
4	0 (220)	−2 (40)	0.0291 ± 0.0011
5	0 (220)	0 (220)	0.0627 ± 0.0023
6	0 (220)	0 (220)	0.0639 ± 0.0020
7	0 (220)	2 (400)	0.0494 ± 0.0050
8	1 (310)	−1 (130)	0.0466 ± 0.0041
9	1 (310)	1 (310)	0.0793 ± 0.0054
10	2 (400)	0 (220)	0.0816 ± 0.0027

[1] The treatments were run in a random order; [2] The values −2, −1, 0, 1, and 2 are coded levels.

The determination coefficient of 0.92 in the analysis of variance (ANOVA) data (Table 2) showed that the second-order polynomial model was fitted well to represent the actual relationship between the initial rate of DHA+EPA ethyl ester production and the substrate concentrations. The ANOVA results revealed the fitted model (Equation 1) was significant as the p-value was smaller than 0.05 ($p = 0.0248$). Therefore, this model was adequate for predicting the initial rate within the range of variables employed. The ANOVA results indicated that the quadratic and cross-product terms had less influence ($p > 0.05$) as compared to the linear term had a significant ($p < 0.05$) influence on the initial rate of DHA+EPA ethyl ester production.

The relationship between substrate concentration and initial rate can be better understood by analyzing the contour plot. The effects of the concentrations of both DHA+EPA concentrate and EA on the initial rate of DHA+EPA ethyl ester production is shown in Figure 1. At any given concentration of DHA+EPA, an increase in EA concentration led to an increase in the initial rate. At any given concentration of EA, the increase in concentration of DHA+EPA concentrate led to an increase in the initial rate. However, an inhibitory effect on the initial rate was seen at high concentrations of DHA+EPA concentrate (>300 mM) when EA concentration was low (50 mM). The inhibitory effect on the initial rate might cause by the high concentrations of acid substrate inhibited the lipase activity or due to mass transfer diffusional limitations. Inhibition by high concentrations of acid substrate has been reported for lipase-catalyzed esterification of ethyl oleate [53], propyl caprate [54], isoamyl butyrate [55], and 2-ethylhexyl-2-ethylhexanoate [56].

Table 2. ANOVA for substrate concentration pertaining to the response (initial rate).

Factor [1]	Degree of Freedom	Sum of Squares	Prob > F
Model	5	0.003478	0.0248 *
Linear term	2	0.002728	0.0096 *
Quadratic	2	0.000634	0.1011
Cross-product	1	0.000116	0.2787
			$R^2 = 0.92$

[1] Independent variable X_1: DHA+EPA concentration, X_2: EA concentration; * Significant at p-Value < 0.05

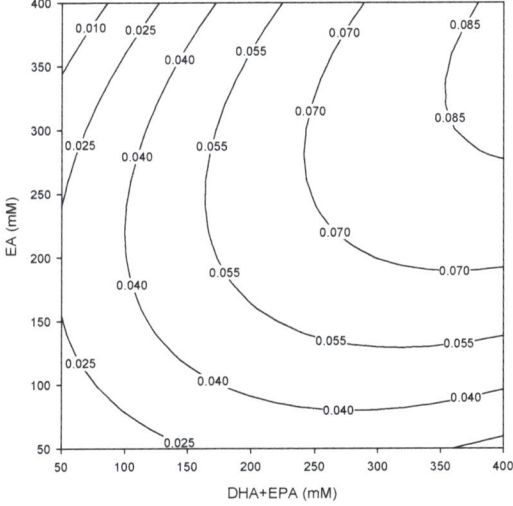

Figure 1. Contour plot showing the initial rate of DHA+EPA ethyl ester production at varying concentrations of DHA+EPA (mM) and EA (mM). Numbers on contours denote initial rate (μmol min^{-1} U^{-1}) under given reaction conditions.

2.2. Kinetic Modeling

Enzyme-catalyzed ester synthesis is usually a two-substrate reaction. The mechanism of this type of reaction is a very broad and extremely complex topic, but the mechanism by which the enzyme reacts with the substrate is crucial to the analysis of enzyme kinetics. Multi-substrate enzyme reactions can mainly be explained by three mechanisms: ping-pong bi-bi mechanism, ordered mechanism, and random-order mechanism. The first mechanism is releasing one or more products before all substrates bind with enzymes. This mechanism is a non-sequential reaction. In the other two mechanisms, the substrates must all be combined with the enzyme before the product releases. The order of substrate binding is random for the random-order mechanism, but substrates must bind in a particular sequence for the ordered mechanism. The ordered and random-order mechanisms are sequential reactions involving the formation of a ternary complex (one involving the enzyme and both substrates). The general rate equation of a two-substrate enzyme-catalyzed reaction, as derived by Alberty [57], is as follows:

$$v = \frac{V_{max}[A][B]}{K_{dA}K_{mB} + K_{mB}[A] + K_{mA}[B] + [A][B]} \qquad (2)$$

where v is the initial reaction rate, Vmax is the maximum initial reaction rate, [A] is the initial concentration of DHA+EPA concentrate, [B] is the initial concentration of EA, and K_{mA} and K_{mB} are the Michaelis constants for DHA+EPA and EA, respectively. K_{dA} is the dissociation constant of the DHA+EPA-lipase complex.

The reaction mechanism can be examined from the Lineweaver-Burk plot, which plots 1/v against 1/[A] at constant [B]. The general rate equation for reactions using a ping-pong bi-bi mechanism is simpler, in that the term of $K_{dA}K_{mB}$ is zero. The slope of the Lineweaver-Burk plot is independent of [B]. A series of Lineweaver-Burk plots obtained at different [B] values would thus be parallel for the ping-pong bi-bi mechanism. In contrast, a series of Lineweaver-Burk plots for ordered and random-order mechanisms at different [B] values showed changes in both the intercept and slope. From the RSM results, reciprocal initial reaction rates (1/v) were plotted against the inverse concentration of DHA+EPA concentrate (1/[A]) at different EA concentrations [B]. The Lineweaver–Burk plot in Figure 2 shows that both the intercept and slope changed at different EA concentrations. Therefore, this reaction followed either the ordered or random-order mechanism.

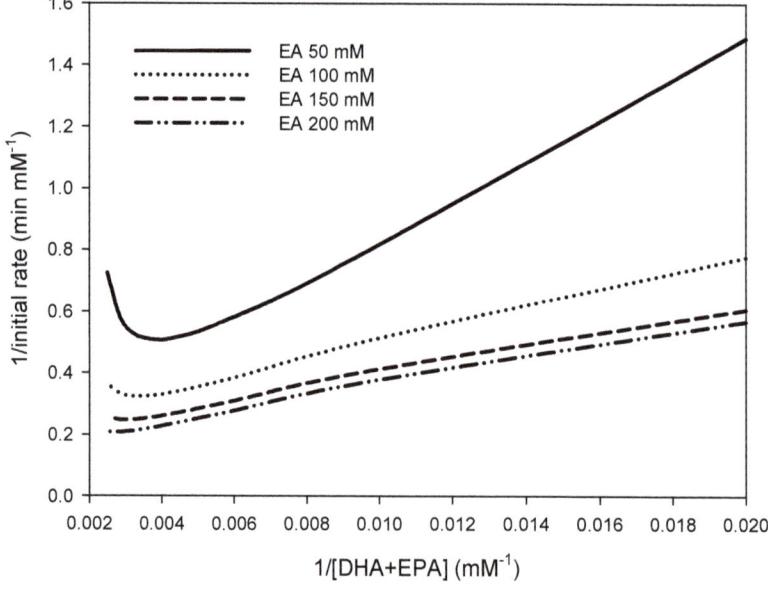

Figure 2. Lineweaver–Burk plot of reciprocal initial reaction rate vs. reciprocal concentrations of DHA+EPA concentrate at various EA concentrations during the synthesis of DHA+EPA ethyl ester via lipase-catalyzed acidolysis.

To determine kinetic constants, the initial rates obtained from RSM results were fitted to Equation (2) by non-linear regression analysis using the Polymath 6.0 program (Polymath Software, Willimantic, CT, USA). The fitness of the kinetic model was confirmed by a satisfactory value of the determination coefficient ($R^2 = 0.9$). The estimated kinetic constants are listed in Table 3. Based on the calculated kinetic constants, K_{mB} was found to be lower than K_{mA}, showing higher EA affinity towards immobilized lipase. These results agree with the ordered mechanism. Initially, EA binds to immobilized lipase (due to higher affinity) and forms an EA-enzyme complex (E-EA). DHA or EPA then combines with E-EA to form a ternary complex (E-EA-DHA or E-EA-EPA) and is further isomerized into a new complex (E-DHA ethyl ester-Acetic acid or E-EPA ethyl ester-Acetic acid). This ternary complex is broken down into acetic acid, while the binary complex subsequently releases DHA ethyl ester or EPA ethyl ester and enzyme. The ordered mechanism has also been proposed for lipase-catalyzed transesterification that uses vinyl acetate as ester in the synthesis of geranyl acetate [58,59], N-(2-hydroxyphenyl) acetamide [60], and (R)-1-(pyridin-4-yl) ethyl acetate [61]. One of the most frequently reactions used to incorporate novel fatty acids into triacylglycerol is lipase-catalyzed acidolysis, several researches have been employed immobilized lipase as catalyst and no water is involved in the acidolysis reaction [62–64]. Kim and Hill Jr have been used both free and immobilized lipases to synthesize structured lipid containing pinolenic acid via lipase-catalyzed acidolysis, the use of immobilized lipase gives the higher degree of incorporation than use of free lipase [65]. This is due to the immobilization may alter the behavior in this reaction [66]. Besides, Verdasco-Martín et. al. have been reported the drying of substrate and immobilized enzyme on the hydrophobic supports are the key parameter to enhance the product yield and decrease the reaction time in the acidolysis reaction [67].

Table 3. Kinetic constants of lipase-catalyzed acidolysis in the synthesis of DHA+EPA ethyl ester.

Parameters	V_{max} (mM min^{-1})	K_{mA} (mM)	K_{mB} (mM)	K_{dA}
Values	14.33	416.97	155.96	49.72

2.3. Modelling of Lipase-Catalyzed Acidolysis Reaction

In this study, the kinetic model of the ordered mechanism was employed to simulate the performance of the batch reactor. The rate equation of the order mechanism described above in Equation (2) was integrated, giving rise to the following Equation (3) [53,68]:

$$kcat\frac{E_T t}{V} = [A]_0 X - K_{mA} \ln(1-X) - K_{mB}\frac{[A]_0}{[B]_0}\ln(1-X) + \frac{K_{dA} K_{mB}}{[B]_0}(\frac{1}{1-X} - 1) \quad (3)$$

where X is the degree of conversion, $kcatE_T$ is the maximum initial reaction rate (V_{max}), E_T is the total amount of enzyme, t is the time, and V is the liquid volume. The time curves were calculated from Equation (3) to predict the conversion of 50 mM DHA+EPA concentrate to DHA+EPA ethyl esters with different concentrations of EA. Figure 3 shows the comparison between the experimental results and the predicted curves, as calculated from Equation (3). A good fit between the experimental and predicted values was obtained, indicating that the proposed kinetic model is valid for this experiment.

2.4. Effect of DHA+EPA Concentration

Figure 4 shows the effects of varying concentrations of DHA+EPA concentrate on the conversion of DHA+EPA ethyl ester were investigated with a constant enzyme activity of 200 U, substrate ratio of 1:1 (DHA+EPA: EA), and reaction time of 300 min. After the reaction, the molar conversions were 93%, 94%, 89%, and 89% for DHA+EPA concentrate of 100 mM, 200 mM, 300 mM, and 400 mM, respectively. Conversion decreased slightly as concentration of DHA+EPA concentrate increased. Conversion at high concentrations of DHA+EPA concentrate (400 mM) was still 89%, indicating the lipase-catalyzed acidolysis of EA with DHA+EPA concentrate was very efficient. The reaction scheme is shown in Scheme 2, and the reaction product is easy to recover. After the reaction, the remaining reactant EA, acetic acid byproduct, and reaction solvent hexane are removed by a vacuum rotary evaporator in order to obtain the DHA+EPA ethyl ester. Lipase-catalyzed reactions for the synthesis of DHA-enriched esters have been widely investigated. Wang et al. reported on the modification of phosphatidylcholine with n-3 PUFA-rich ethyl esters by immobilized MAS1, via lipase-catalyzed transesterification that incorporated 43.55% n-3 PUFA into phosphatidylcholine after 72 h [69]. Zhang et al. used lipase-catalyzed ethanolysis of algal oil to synthesize 2-docosahexaenoylglycerol; product yields of 27–31% were obtained [70]. Bhandari et al. obtained 76.2% esterification after 24 h from the selective esterification of tuna-FFA with butanol using *Rhizopus oryzae* lipase [71]. This study shows the advantages of lipase-catalyzed acidolysis. Not only is DHA+EPA ethyl ester synthesized efficiently, the product can also be easily recovered.

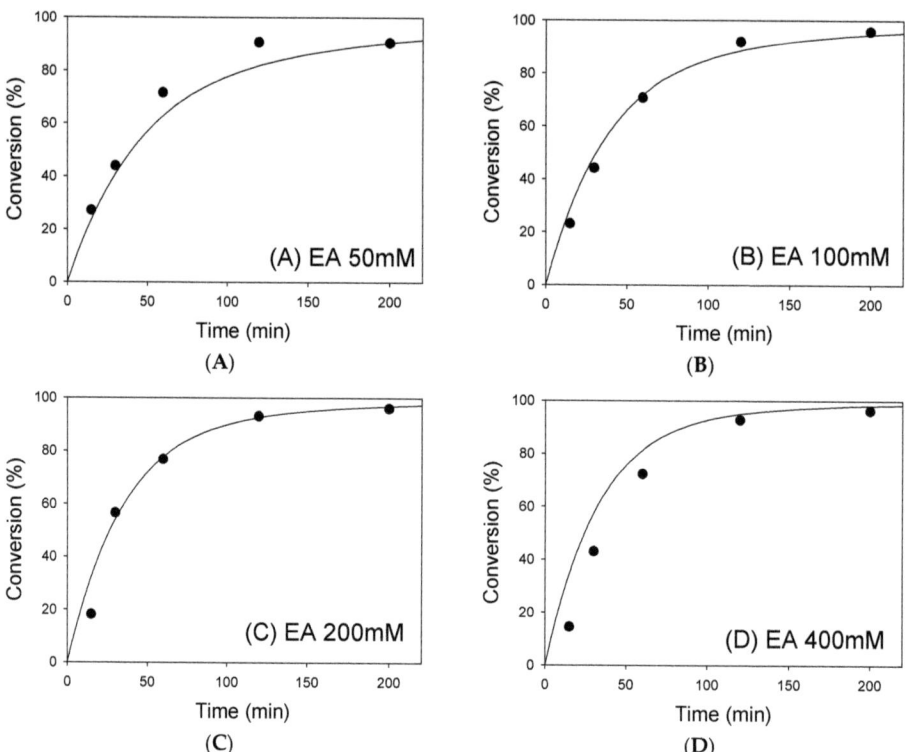

Figure 3. Comparison of experimental results and the simulated conversions obtained from the integrated rate equation (Equation (3)) using 50 mM DHA+EPA concentrate with (**A**) 50, (**B**) 100, (**C**) 200, and (**D**) 400 mM EA. Points are experimental data and lines are predicted from integrated rate equation (Equation (3)).

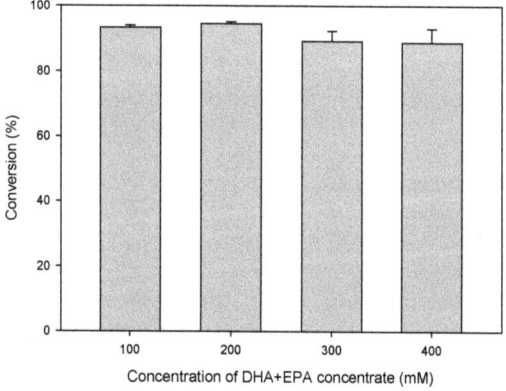

Figure 4. Effect of concentration of DHA+EPA concentrate on the conversion of DHA+EPA ethyl ester at a substrate ratio of 1:1 (DHA+EPA: EA), temperature of 60 °C, and enzyme activity of 200 U.

3. Materials and Methods

3.1. Materials

Cobia livers were collected from Cobiahome Inc. (Pingtung, Taiwan). Novozym® 435 (10,000 U/g; propyl laurate units (PLU)), lipase from *Candida antarctica* B (EC3.1.1.3) immobilized on a macroporous acrylic resin was obtained from Novo Nordisk Bioindustrials Inc. (Copenhagen, Denmark). Ethyl acetate (99.8%) was produced from Merck (Darmstadt, Germany). Fatty acid methyl ester standards (Supelco 37 Component FAME Mix, Catalog No. 47885) and BF_3-methanol reagent (14% BF_3 in CH3OH, *w/v*) were purchased from Sigma-Aldrich (St. Louis, MO, USA). cis-4,7,10,13,16,19-Docosahexaenoic acid used for analysis was purchased from Acros (NJ, USA). cis-5,8,11,14,17-Eicosapentaenoic acid used for analysis was purchased from TCI Co., LTD. (Tokyo, Japan). Unless otherwise specified, all reagents and chemicals used in this study were of analytical grade.

3.2. Cobia Liver Oil Extraction and Preparation of DHA+EPA Concentrate

Cobia liver oil was extracted using the homogenization method, as described elsewhere [72]. The extracted crude oil was saponified to obtain free fatty acids (FFAs), as described by Chen and Ju [17]. The DHA+EPA concentrate was prepared from the FFAs by using the acetone fractionation of fatty acid salts method, as described elsewhere [52]. Briefly, 50 g FFAs was mixed with 695 mL acetone and preheated at 30 °C for 30 min. This was followed by the addition of 8.5 mL 15 N NaOH aqueous solution. The solution was reacted for 90 min with a magnetic stirrer at 550 rpm. A Buchner funnel were used to separate the solid and liquid phases. 400 mL of distilled water was added to the filtrate, and pH was adjusted to 1 by adding 12 N hydrochloric acid for recovering FFAs from the corresponding sodium salts, followed by extraction with 250 mL hexane for 20 min. The hexane layer was collected using a separating funnel, and the hexane was removed by vacuum rotary evaporator at 70 °C to recover the DHA+EPA concentrate. The fatty acid profiles of the DHA+EPA concentrate were analyzed by the GC method, as described elsewhere [72]. The fatty acid profile of DHA+EPA concentrate was 54.4% DHA, 16.8% EPA, 7.0% docosadienoic acid, 14.2% oleic acid, 0.8% linoleic acid, 0.5% linolenic acid, 3.1% palmitoleic acid, 2.7% palmitic acid and 0.5% myristic acid.

3.3. Experimental Design for Determining the Kinetic Constants

Lipase-catalyzed acidolysis of EA with DHA+EPA concentrate was investigated by RSM in order to evaluate the effect of substrate concentration on the reaction rate. A five-level–two-factor central composite design was employed to determine the kinetic constants. Table 1 shows the substrate concentrations (x_i), levels, and experimental design in terms of coded or uncoded. The acidolysis reactions were performed at 60 °C in *n*-hexane for the initial rate measurements. 20 mg immobilized lipase Novozym® 435 was added to 3 mL of the reaction mixture, containing various amounts of DHA+EPA concentrate and EA, for a 20 min incubation period. All reactions were done in duplicate. After the reaction, liquid samples were withdrawn from the reaction mixture to determine the produced DHA+EPA ethyl ester using HPLC. Initial reaction rates are expressed as produced μmol of DHA+EPA ethyl ester per unit of enzyme for 1 min ($\mu mol\ min^{-1}\ U^{-1}$). The polymath 6.0 (Polymath Software, Willimantic, CT, USA) was used to calculate kinetic constants by nonlinear regression using, based on 898 data points obtained from the RSM results.

3.4. Lipase-Catalyzed Synthesis of DHA+EPA Ethyl Ester

Lipase-catalyzed acidolysis was carried out in a capped glass tube (diameter 1.5 cm) with different concentrations of DHA+EPA concentrate and EA, supplemented with *n*-hexane to bring the total volume to 3 mL. 20 mg immobilized lipase Novozym® 435 was then added to initiate the reaction. The reaction mixture was agitated in an orbital shaking bath (150 rpm) at 60 °C. All reactions were done in duplicate. Samples were withdrawn from the reaction mixture at different times and analyzed by HPLC. The amount of DHA+EPA concentrate added was calculated by the following equation:

$$DHA + EPA\ concentrate\ added\ (\mu L) = DHA + EPA\ concentration\ (mM) \times 0.003 \times 312 \div 0.864 \quad (4)$$

where 0.003 is the total reaction volume, 312 is the average molecular weight of DHA+EPA concentrate, and 0.864 is the specific gravity of DHA+EPA concentrate.

3.5. Analytical Methods

The samples were analyzed by Hitachi HPLC system (Hitachi, Tokyo, Japan), consisting of a pump (L-23130), and a UV/VIS detector (L-2420), and Inertsil ODS-3 column (5 µM, 250 mm × 4.6 mm). The mobile phase were deionized water and methanol containing 0.1% acetic acid. The gradient elution was started from 80% to 100% methanol for 10 min, followed by elution at 100% methanol for 20 min. The flow rate and UV detector were set at 1.0 mL min^{-1} and wavelength of 303 nm, respectively. The conversion yield % was calculated from the peak areas of the substrate (DHA+EPA) and product (DHA+EPA ethyl ester).

3.6. Statistical Analysis

Experimental data (Table 1) were analyzed by the response surface regression (RSREG) procedure of SAS (SAS Institute, Cary, NC) to fit the following second-order polynomial equation:

$$Y = \beta_\kappa + \Sigma \beta_{\kappa i} X_i + \Sigma \beta_{\kappa ii} X_i^2 + \Sigma \Sigma \beta_{\kappa ij} X_i X_j \quad (5)$$

where Y is the response, β_{k0}, β_{ki}, β_{kii} and β_{kij} are the constant coefficients, and X_i and X_j are the uncoded independent variables.

4. Conclusions

Lipase-catalyzed acidolysis of EA with DHA+EPA concentrate was first reported for synthesizing DHA+EPA ethyl esters. A 5-level–2-factor central composite design was used to evaluate the initial rate in order to determine the reaction mechanism. Based on the slopes of Lineweaver-Burk plots, the reaction followed the ordered mechanism. The initial rate data calculated from the RSM results was also a good fit to the ordered mechanism kinetic model. The proposed kinetic model integrated with a batch mole balance equation could be used to predict experimental conversions. There was a good correlation between the experimental conversion data and the predicted data from the integrated equation. The high conversion of DHA+EPA ethyl ester (94%) was obtained at the DHA+EPA concentrate of 200 mM, enzyme activity of 200 U, substrate ratio of 1:1 (DHA+EPA: EA), and reaction time of 300 min. Esterification of DHA or EPA concentrate with EA via acidolysis seems to be a more suitable strategy than with ethanol, as the data presented here confirms that lipase-catalyzed acidolysis is a potential reaction pathway for ester synthesis.

Author Contributions: Conceptualization, C.-H.K. and C.-J.S.; software, C.-H.K.; investigation, C.-H.K.; resources, C.-Y.H., C.-L.L., W.-C.K. and S.-L.H.; writing—original draft preparation, C.-H.K.; writing—review and editing, C.-H.K. and C.-J.S. All authors have read and agree to the published version of the manuscript.

Funding: This work was supported by research funding grants provided by the Ministry of Science and Technology of Taiwan (MOST 104-2218-E-022-001-MY2 and MOST 108-2221-E-992-048).

Conflicts of Interest: The authors declare no conflict of interest.

References

1. Finco, A.M.D.O.; Mamani, L.D.G.; Carvalho, J.C.D.; de Melo Pereira, G.V.; Thomaz-Soccol, V.; Soccol, C.R. Technological trends and market perspectives for production of microbial oils rich in omega-3. *Crit. Rev. Biotechnol.* **2017**, *37*, 656–671. [CrossRef]

2. Liu, S.-H.; Chiu, C.-Y.; Wang, L.-P.; Chiang, M.-T. Omega-3 fatty acids-enriched fish oil activates AMPK/PGC-1α signaling and prevents obesity-related skeletal muscle wasting. *Mar. Drugs* **2019**, *17*, 380. [CrossRef]
3. Minihane, A. Impact of genotype on EPA and DHA status and responsiveness to increased intakes. *Nutrients* **2016**, *8*, 123. [CrossRef]
4. Haug, I.J.; Sagmo, L.B.; Zeiss, D.; Olsen, I.C.; Draget, K.I.; Seternes, T. Bioavailability of EPA and DHA delivered by gelled emulsions and soft gel capsules. *Eur. J. Lipid Sci. Technol.* **2011**, *113*, 137–145. [CrossRef]
5. Yamazaki, T.; Li, D.; Ikaga, R. Effective food ingredients for fatty liver: Soy protein β-conglycinin and fish oil. *Int. J. Mol. Sci.* **2018**, *19*, 4107. [CrossRef]
6. Liu, X.; Cui, J.; Li, Z.; Xu, J.; Wang, J.; Xue, C.; Wang, Y. Comparative study of DHA-enriched phospholipids and EPA-enriched phospholipids on metabolic disorders in diet-induced-obese C57BL/6J mice. *Eur. J. Lipid Sci. Technol.* **2014**, *116*, 255–265. [CrossRef]
7. Stoodley, I.; Garg, M.; Scott, H.; Macdonald-Wicks, L.; Berthon, B.; Wood, L. Higher omega-3 index is associated with better asthma control and lower medication dose: A cross-sectional study. *Nutrients* **2020**, *12*, 74. [CrossRef]
8. Jeong, Y.K.; Kim, H. A mini-review on the effect of docosahexaenoic acid (DHA) on cerulein-induced and hypertriglyceridemic acute pancreatitis. *Int. J. Mol. Sci.* **2017**, *18*, 2239. [CrossRef]
9. Mohanty, B.P.; Ganguly, S.; Mahanty, A.; Sankar, T.; Anandan, R.; Chakraborty, K.; Paul, B.; Sarma, D.; Syama Dayal, J.; Venkateshwarlu, G. DHA and EPA content and fatty acid profile of 39 food fishes from India. *Biomed Res. Int.* **2016**, *4027437*. [CrossRef]
10. Mu, H.; Zhang, H.; Li, Y.; Zhang, Y.; Wang, X.; Jin, Q.; Wang, X. Enrichment of DPAn-6 and DHA from *Schizochytrium sp.* oil by low-temperature solvent crystallization. *Ind. Eng. Chem. Res.* **2016**, *55*, 737–746. [CrossRef]
11. Dovale-Rosabal, G.; Rodríguez, A.; Contreras, E.; Ortiz-Viedma, J.; Muñoz, M.; Trigo, M.; Aubourg, S.P.; Espinosa, A. Concentration of EPA and DHA from refined salmon oil by optimizing the urea–fatty acid adduction reaction conditions using response surface methodology. *Molecules* **2019**, *24*, 1642. [CrossRef]
12. Vázquez, L.; Prados, I.M.; Reglero, G.; Torres, C.F. Identification and quantification of ethyl carbamate occurring in urea complexation processes commonly utilized for polyunsaturated fatty acid concentration. *Food Chem.* **2017**, *229*, 28–34. [CrossRef]
13. Solaesa, Á.G.; Sanz, M.T.; Falkeborg, M.; Beltrán, S.; Guo, Z. Production and concentration of monoacylglycerols rich in omega-3 polyunsaturated fatty acids by enzymatic glycerolysis and molecular distillation. *Food Chem.* **2016**, *190*, 960–967. [CrossRef]
14. Montañés, F.; Tallon, S. Supercritical fluid chromatography as a technique to fractionate high-valued compounds from lipids. *Separations* **2018**, *5*, 38. [CrossRef]
15. Jiao, G.; Hui, J.; Burton, I.; Thibault, M.-H.; Pelletier, C.; Boudreau, J.; Tchoukanova, N.; Subramanian, B.; Djaoued, Y.; Ewart, S. Characterization of shrimp oil from *Pandalus borealis* by high performance liquid chromatography and high resolution mass spectrometry. *Mar. Drugs* **2015**, *13*, 3849–3876. [CrossRef]
16. Rincón Cervera, M.Á.; Venegas, E.; Ramos Bueno, R.P.; Suárez Medina, M.D.; Guil Guerrero, J.L. Docosahexaenoic acid purification from fish processing industry by-products. *Eur. J. Lipid Sci. Technol.* **2015**, *117*, 724–729. [CrossRef]
17. Chen, T.-C.; Ju, Y.-H. Enrichment of eicosapentaenoic acid and docosahexaenoic acid in saponified menhaden oil. *J. Am. Oil Chem. Soc.* **2000**, *77*, 425–428. [CrossRef]
18. Castejón, N.; Señoráns, F.J. Strategies for enzymatic synthesis of omega-3 structured triacylglycerols from *Camelina sativa* oil enriched in EPA and DHA. *Eur. J. Lipid Sci. Technol.* **2019**, *121*, 1800412. [CrossRef]
19. Valverde, L.M.; Moreno, P.A.G.; Callejón, M.J.J.; Cerdán, L.E.; Medina, A.R. Concentration of eicosapentaenoic acid (EPA) by selective alcoholysis catalyzed by lipases. *Eur. J. Lipid Sci. Technol.* **2013**, *115*, 990–1004. [CrossRef]
20. Morais Júnior, W.G.; Fernández-Lorente, G.; Guisán, J.M.; Ribeiro, E.J.; De Resende, M.M.; Costa Pessela, B. Production of omega-3 polyunsaturated fatty acids through hydrolysis of fish oil by *Candida rugosa* lipase immobilized and stabilized on different supports. *Biocatal. Biotransform.* **2017**, *35*, 63–73. [CrossRef]
21. Miyashita, K.; Takagi, T. Study on the oxidative rate and prooxidant activity of free fatty acids. *J. Am. Oil Chem. Soc.* **1986**, *63*, 1380–1384. [CrossRef]

22. Hills, G. Industrial use of lipases to produce fatty acid esters. *Eur. J. Lipid Sci. Technol.* **2003**, *105*, 601–607. [CrossRef]
23. Verma, M.L.; Rao, N.M.; Tsuzuki, T.; Barrow, C.J.; Puri, M. Suitability of recombinant lipase immobilised on functionalised magnetic nanoparticles for fish oil hydrolysis. *Catalysts* **2019**, *9*, 420. [CrossRef]
24. Souza, L.T.d.A.; Moreno-Perez, S.; Fernández Lorente, G.; Cipolatti, E.P.; De Oliveira, D.; Resende, R.R.; Pessela, B.C. Immobilization of *moniliella spathulata* r25l270 lipase on ionic, hydrophobic and covalent supports: Functional properties and hydrolysis of sardine oil. *Molecules* **2017**, *22*, 1508. [CrossRef]
25. Huang, S.-M.; Li, H.-J.; Liu, Y.-C.; Kuo, C.-H.; Shieh, C.-J. An efficient approach for lipase-catalyzed synthesis of retinyl laurate nutraceutical by combining ultrasound assistance and artificial neural network optimization. *Molecules* **2017**, *22*, 1972. [CrossRef]
26. Huang, S.-M.; Hung, T.-H.; Liu, Y.-C.; Kuo, C.-H.; Shieh, C.-J. Green synthesis of ultraviolet absorber 2-ethylhexyl salicylate: Experimental design and artificial neural network modeling. *Catalysts* **2017**, *7*, 342. [CrossRef]
27. Mendoza-Ortiz, P.A.; Gama, R.S.; Gómez, O.C.; Luiz, J.H.; Fernandez-Lafuente, R.; Cren, E.C.; Mendes, A.A. Sustainable enzymatic synthesis of a solketal ester—process optimization and evaluation of its antimicrobial activity. *Catalysts* **2020**, *10*, 218. [CrossRef]
28. Yang, H.; Zhang, W. Surfactant imprinting hyperactivated immobilized lipase as efficient biocatalyst for biodiesel production from waste cooking oil. *Catalysts* **2019**, *9*, 914. [CrossRef]
29. Mateo, C.; Palomo, J.M.; Fernandez-Lorente, G.; Guisan, J.M.; Fernandez-Lafuente, R. Improvement of enzyme activity, stability and selectivity via immobilization techniques. *Enzym. Microb. Technol.* **2007**, *40*, 1451–1463. [CrossRef]
30. Derewenda, U.; Brzozowski, A.M.; Lawson, D.M.; Derewenda, Z.S. Catalysis at the interface: The anatomy of a conformational change in a triglyceride lipase. *Biochemistry* **1992**, *31*, 1532–1541. [CrossRef]
31. Manoel, E.A.; dos Santos, J.C.; Freire, D.M.; Rueda, N.; Fernandez-Lafuente, R. Immobilization of lipases on hydrophobic supports involves the open form of the enzyme. *Enzym. Microb. Technol.* **2015**, *71*, 53–57. [CrossRef] [PubMed]
32. Fernandez-Lafuente, R.; Armisén, P.; Sabuquillo, P.; Fernández-Lorente, G.; Guisán, J.M. Immobilization of lipases by selective adsorption on hydrophobic supports. *Chem. Phys. Lipids* **1998**, *93*, 185–197. [CrossRef]
33. Cabrera, Z.; Fernandez-Lorente, G.; Fernandez-Lafuente, R.; Palomo, J.M.; Guisan, J.M. Novozym 435 displays very different selectivity compared to lipase from *Candida antarctica* B adsorbed on other hydrophobic supports. *J. Mol. Catal. B Enzym.* **2009**, *57*, 171–176. [CrossRef]
34. Chen, G.J.; Kuo, C.H.; Chen, C.I.; Yu, C.C.; Shieh, C.J.; Liu, Y.C. Effect of membranes with various hydrophobic/hydrophilic properties on lipase immobilized activity and stability. *J. Biosci. Bioeng.* **2012**, *113*, 166–172. [CrossRef]
35. Ortiz, C.; Ferreira, M.L.; Barbosa, O.; dos Santos, J.C.; Rodrigues, R.C.; Berenguer-Murcia, Á.; Briand, L.E.; Fernandez-Lafuente, R. Novozym 435: The "perfect" lipase immobilized biocatalyst? *Catal. Sci. Technol.* **2019**, *9*, 2380–2420. [CrossRef]
36. Shimada, Y.; Watanabe, Y.; Sugihara, A.; Baba, T.; Ooguri, T.; Moriyama, S.; Terai, T.; Tominaga, Y. Ethyl esterification of docosahexaenoic acid in an organic solvent-free system with immobilized *Candida antarctica* lipase. *J. Biosci. Bioeng.* **2001**, *92*, 19–23. [CrossRef]
37. Poisson, L.; Ergan, F. Docosahexaenoic acid ethyl esters from *Isochrysis galbana*. *J. Biotechnol.* **2001**, *91*, 75–81. [CrossRef]
38. Bhandari, K.; Chaurasia, S.; Dalai, A. Lipase-catalyzed esterification of docosahexaenoic acid-rich fatty acids with glycerol. *Chem. Eng. Commun.* **2015**, *202*, 920–926. [CrossRef]
39. Roby, M.H.H. Synthesis and characterization of phenolic lipids. In *Phenolic Compounds-Natural Sources, Importance and Applications*; IntechOpen: London, UK, 2017; Available online: https://www.intechopen.com/books/phenolic-compounds-natural-sources-importance-and-applications/synthesis-and-characterization-of-phenolic-lipids (accessed on 18 May 2020).
40. Kuo, C.H.; Ju, H.Y.; Chu, S.W.; Chen, J.H.; Chang, C.M.J.; Liu, Y.C.; Shieh, C.J. Optimization of lipase-catalyzed synthesis of cetyl octanoate in supercritical carbon dioxide. *J. Am. Oil Chem. Soc.* **2012**, *89*, 103–110. [CrossRef]
41. Chen, H.C.; Kuo, C.H.; Twu, Y.K.; Chen, J.H.; Chang, C.M.J.; Liu, Y.C.; Shieh, C.J. A continuous ultrasound-assisted packed-bed bioreactor for the lipase-catalyzed synthesis of caffeic acid phenethyl ester. *J. Chem. Technol. Biotechnol.* **2011**, *86*, 1289–1294. [CrossRef]

42. Kuo, C.H.; Chiang, S.H.; Ju, H.Y.; Chen, Y.M.; Liao, M.Y.; Liu, Y.C.; Shieh, C.J. Enzymatic synthesis of rose aromatic ester (2-phenylethyl acetate) by lipase. *J. Sci. Food Agric.* **2012**, *92*, 2141–2147. [CrossRef] [PubMed]
43. Yankah, V.V.; Akoh, C.C. Lipase-catalyzed acidolysis of tristearin with oleic or caprylic acids to produce structured lipids. *J. Am. Oil Chem. Soc.* **2000**, *77*, 495–500. [CrossRef]
44. Mitra, K.; Kim, S.A.; Lee, J.H.; Choi, S.W.; Lee, K.T. Production and characterization of α-linolenic acid enriched structured lipids from lipase-catalyzed interesterification. *Food Sci. Biotechnol.* **2010**, *19*, 57–62. [CrossRef]
45. Rychlicka, M.; Niezgoda, N.; Gliszczyńska, A. Lipase-catalyzed acidolysis of egg-yolk phosphatidylcholine with citronellic acid. New insight into synthesis of isoprenoid-phospholipids. *Molecules* **2018**, *23*, 314. [CrossRef]
46. Chojnacka, A.; Gładkowski, W. Production of structured phosphatidylcholine with high content of myristic acid by lipase-catalyzed acidolysis and interesterification. *Catalysts* **2018**, *8*, 281. [CrossRef]
47. Yue, C.; Ben, H.; Wang, J.; Li, T.; Yu, G. Ultrasonic pretreatment in synthesis of caprylic-rich structured lipids by lipase-catalyzed acidolysis of corn oil in organic system and its physicochemical properties. *Foods* **2019**, *8*, 566. [CrossRef]
48. Arıkaya, A.; Ünlü, A.E.; Takaç, S. Use of deep eutectic solvents in the enzyme catalysed production of ethyl lactate. *Process Biochem.* **2019**, *84*, 53–59. [CrossRef]
49. Cavallaro, V.; Tonetto, G.; Ferreira, M.L. Optimization of the enzymatic synthesis of pentyl oleate with lipase immobilized onto novel structured support. *Fermentation* **2019**, *5*, 48. [CrossRef]
50. Yadav, G.D.; Kamble, M.P. A green process for synthesis of geraniol esters by immobilized lipase from *Candida antarctica* B fraction in non-aqueous reaction media: Optimization and kinetic modeling. *Int. J. Chem. React. Eng.* **2018**, *16*. [CrossRef]
51. Jaiswal, K.S.; Rathod, V.K. Acoustic cavitation promoted lipase catalysed synthesis of isobutyl propionate in solvent free system: Optimization and kinetic studies. *Ultrason. Sonochem.* **2018**, *40*, 727–735. [CrossRef]
52. Kuo, C.H.; Huang, C.Y.; Chen, J.W.; Wang, H.M.D.; Shieh, C.J. Concentration of docosahexaenoic and eicosapentaenoic acid from cobia liver oil by acetone fractionation of fatty acid salts. *Appl. Biochem. Biotechnol.* **2020**. [CrossRef] [PubMed]
53. Oliveira, A.; Rosa, M.; Aires-Barros, M.; Cabral, J. Enzymatic esterification of ethanol and oleic acid—A kinetic study. *J. Mol. Catal. B Enzym.* **2001**, *11*, 999–1005. [CrossRef]
54. Parikh, D.T.; Lanjekar, K.J.; Rathod, V.K. Kinetics and thermodynamics of lipase catalysed synthesis of propyl caprate. *Biotechnol. Lett.* **2019**, *41*, 1163–1175. [CrossRef] [PubMed]
55. Krishna, S.H.; Karanth, N. Lipase-catalyzed synthesis of isoamyl butyrate: A kinetic study. *Biochim. Biophys. Acta-Protein Struct. Mol. Enzym.* **2001**, *1547*, 262–267. [CrossRef]
56. Daneshfar, A.; Ghaziaskar, H.; Shiri, L.; Manafi, M.; Nikorazm, M.; Abassi, S. Synthesis of 2-ethylhexyl-2-ethylhexanoate catalyzed by immobilized lipase in n-hexane: A kinetic study. *Biochem. Eng. J.* **2007**, *37*, 279–284. [CrossRef]
57. Alberty, R.A. The relationship between Michaelis constants, maximum velocities and the equilibrium constant for an enzyme-catalyzed reaction. *J. Am. Chem. Soc.* **1953**, *75*, 1928–1932. [CrossRef]
58. Badgujar, K.C.; Bhanage, B.M. Synthesis of geranyl acetate in non-aqueous media using immobilized *Pseudomonas cepacia* lipase on biodegradable polymer film: Kinetic modelling and chain length effect study. *Process Biochem.* **2014**, *49*, 1304–1313. [CrossRef]
59. Patel, V.; Shah, C.; Deshpande, M.; Madamwar, D. Zinc oxide nanoparticles supported lipase immobilization for biotransformation in organic solvents: A facile synthesis of geranyl acetate, effect of operative variables and kinetic study. *Appl. Biochem. Biotechnol.* **2016**, *178*, 1630–1651. [CrossRef]
60. Magadum, D.B.; Yadav, G.D. Chemoselective acetylation of 2-aminophenol using immobilized lipase: Process optimization, mechanism, and kinetics. *ACS Omega* **2018**, *3*, 18528–18534. [CrossRef]
61. Magadum, D.B.; Yadav, G.D. One-pot synthesis of (R)-1-(pyridin-4-yl) ethyl acetate using tandem catalyst prepared by co-immobilization of palladium and lipase on mesoporous foam: Optimization and kinetic modeling. *Chirality* **2017**, *29*, 811–823. [CrossRef]
62. Wang, J.; Shahidi, F. Acidolysis of p-coumaric acid with omega-3 oils and antioxidant activity of phenolipid products in in vitro and biological model systems. *J. Agric. Food Chem.* **2014**, *62*, 454–461. [CrossRef] [PubMed]

63. Yanık, D.K.; Keskin, H.; Fadıloğlu, S.; Göğüş, F. Acidolysis reaction of terebinth fruit oil with palmitic and caprylic acids to produce low caloric spreadable structured lipid. *J. Am. Oil Chem. Soc.* **2013**, *90*, 999–1009. [CrossRef]
64. Ifeduba, E.A.; Akoh, C.C. Modification of stearidonic acid soybean oil by immobilized *Rhizomucor miehei* lipase to incorporate caprylic acid. *J. Am. Oil Chem. Soc.* **2014**, *91*, 953–965. [CrossRef]
65. Kim, I.H.; Hill, C.G., Jr. Lipase-catalyzed acidolysis of menhaden oil with pinolenic acid. *J. Am. Oil Chem. Soc.* **2006**, *83*, 109–115. [CrossRef]
66. Rodrigues, R.C.; Ortiz, C.; Berenguer-Murcia, Á.; Torres, R.; Fernández-Lafuente, R. Modifying enzyme activity and selectivity by immobilization. *Chem. Soc. Rev.* **2013**, *42*, 6290–6307. [CrossRef]
67. Verdasco-Martín, C.M.; Corchado-Lopo, C.; Fernández-Lafuente, R.; Otero, C. Rapid and high yield production of phospholipids enriched in CLA via acidolysis: The critical role of the enzyme immobilization protocol. *Food Chem.* **2019**, *296*, 123–131. [CrossRef]
68. Kuo, C.-H.; Chen, G.-J.; Chen, C.-I.; Liu, Y.-C.; Shieh, C.-J. Kinetics and optimization of lipase-catalyzed synthesis of rose fragrance 2-phenylethyl acetate through transesterification. *Process Biochem.* **2014**, *49*, 437–444. [CrossRef]
69. Wang, X.; Qin, X.; Li, X.; Zhao, Z.; Yang, B.; Wang, Y. Insight into the modification of phosphatidylcholine with n-3 polyunsaturated fatty acids-rich ethyl esters by immobilized MAS1 lipase. *Molecules* **2019**, *24*, 3528. [CrossRef]
70. Zhang, Y.; Wang, X.; Zou, S.; Xie, D.; Jin, Q.; Wang, X. Synthesis of 2-docosahexaenoylglycerol by enzymatic ethanolysis. *Bioresour. Technol.* **2018**, *251*, 334–340. [CrossRef]
71. Bhandari, K.; Chaurasia, S.; Dalai, A.; Gupta, A. Purification of free DHA by selective esterification of fatty acids from tuna oil catalyzed by *Rhizopus oryzae* lipase. *J. Am. Oil Chem. Soc.* **2013**, *90*, 1637–1644. [CrossRef]
72. Kuo, C.H.; Liao, H.Z.; Wang, Y.H.; Wang, H.M.D.; Shieh, C.J.; Tseng, C.Y. Highly efficient extraction of EPA/DHA-enriched oil from cobia liver using homogenization plus sonication. *Eur. J. Lipid Sci. Technol.* **2017**, *119*, 1600466. [CrossRef]

© 2020 by the authors. Licensee MDPI, Basel, Switzerland. This article is an open access article distributed under the terms and conditions of the Creative Commons Attribution (CC BY) license (http://creativecommons.org/licenses/by/4.0/).

Communication

Enantioselective Epoxidation by Flavoprotein Monooxygenases Supported by Organic Solvents

Daniel Eggerichs [1], Carolin Mügge [1], Julia Mayweg [1], Ulf-Peter Apfel [2,3] and Dirk Tischler [1,*]

[1] Microbial Biotechnology, Faculty of Biology and Biotechnology, Ruhr-Universität Bochum, Universitätsstr. 150, 44780 Bochum, Germany; daniel.eggerichs@rub.de (D.E.); carolin.muegge@rub.de (C.M.); julia.mayweg@gmail.com (J.M.)
[2] Activation of Small Molecules, Faculty of Chemistry and Biochemistry, Ruhr-Universität Bochum, Universitätsstr 150, 44780 Bochum, Germany; ulf.apfel@rub.de
[3] Fraunhofer UMSICHT, Division of Energy, Osterfelder Strasse 3, 46047 Oberhausen, Germany
* Correspondence: dirk.tischler@rub.de; Tel.: +49-234-32-22656

Received: 29 April 2020; Accepted: 14 May 2020; Published: 19 May 2020

Abstract: Styrene and indole monooxygenases (SMO and IMO) are two-component flavoprotein monooxygenases composed of a nicotinamide adenine dinucleotide (NADH)-dependent flavin adenine dinucleotide (FAD)-reductase (StyB or IndB) and a monooxygenase (StyA or IndA). The latter uses reduced FAD to activate oxygen and to oxygenate the substrate while releasing water. We circumvented the need for the reductase by direct FAD reduction in solution using the NAD(P)H-mimic 1-benzyl-1,4-dihydronicotinamide (BNAH) to fuel monooxygenases without NADH requirement. Herein, we report on the hitherto unknown solvent tolerance for the indole monooxygenase from *Gemmobacter nectariphilus* DSM15620 (*Gn*IndA) and the styrene monooxygenase from *Gordonia rubripertincta* CWB2 (*Gr*StyA). These enzymes were shown to convert bulky and rather hydrophobic styrene derivatives in the presence of organic cosolvents. Subsequently, BNAH-driven biotransformation was furthermore optimized with regard to the applied cosolvent and its concentration as well as FAD and BNAH concentration. We herein demonstrate that *Gn*IndA and *Gr*StyA enable selective epoxidations of allylic double bonds (up to 217 mU mg^{-1}) in the presence of organic solvents such as tetrahydrofuran, acetonitrile, or several alcohols. Notably, *Gn*IndA was found to resist methanol concentrations up to 25 vol.%. Furthermore, a diverse substrate preference was determined for both enzymes, making their distinct use very interesting. In general, our results seem representative for many IMOs as was corroborated by in silico mutagenetic studies.

Keywords: styrene monooxygenase; indole monooxygenase; two-component system; chiral biocatalyst; solvent tolerance; biotransformation; epoxidation; NAD(P)H-mimics

---

1. Introduction

Styrene and indole monooxygenases (StyA or IndA) are the initial enzymes in degradation pathways of human-made aromatic pollutants, such as styrenes, within several bacteria [1–3]. Similar to the general detoxification pathways in higher organisms, hydrophobic compounds are typically oxidized in bacteria to increase their water solubility and create a functional moiety, which can be easily metabolized by downstream enzymes [4,5]. For example, styrene monooxygenases perform the oxidation of double bonds with molecular oxygen and an additional reduction equivalent to form an epoxide along with water as side product [6,7].

Due to their role as mainly detoxifying enzymes, SMOs and IMOs accept a large variety of substrates. Although the highest activities were described for conjugated double bonds in direct proximity of an aromatic ring, SMOs were likewise shown to enable the epoxidation of unconjugated double bonds even without the aromatic moiety [8–10]. Furthermore, SMOs and IMOs catalyze sulfoxidations of artificial substrates at increased rates as compared to their supposed natural substrates [11,12].

Apart from their occurrence during biological degradation pathways, chiral epoxides are important building blocks for the industrial production of fine chemicals such as drugs or fungicides (Figure 1) [13,14]. Due to their broad substrate scope, their stability and remarkably high regio- and stereoselectivity, SMOs and IMOs are promising candidates for the industrial application toward chiral biomolecules [13,15–17]. Along this line, several studies highlighted the versatile applications for this enzyme class up to pilot scale yielding >300 g (S)-styrene oxide [18–20].

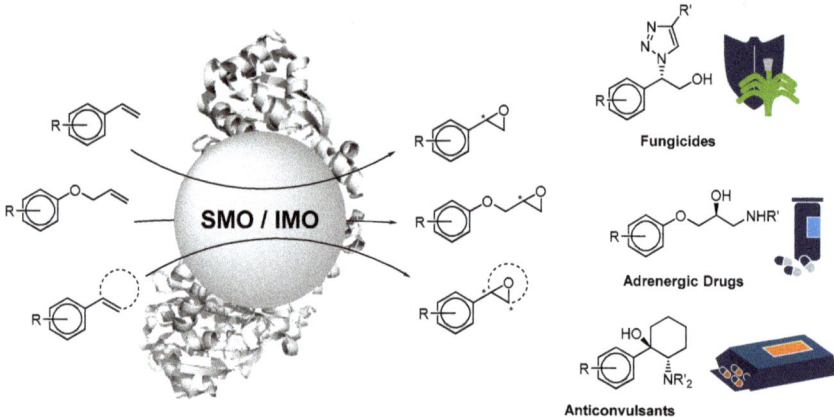

Figure 1. Styrene and indole monooxygenases activate double bonds by chiral epoxidation. Those epoxide intermediates are valuable building blocks in the chemical industry and can be further processed toward fine chemicals.

Nevertheless, there are some significant drawbacks that have to be overcome before SMOs or IMOs can be used on an even larger industrial scale. First, SMOs and IMOs depend on a suitable electron donor to reduce their flavin adenine dinucleotide (FAD) cofactor after each reaction cycle. In nature, nicotinamide adenine dinucleotide (NADH) is consumed stoichiometrically and regenerated in vivo, but it is too expensive to use with isolated enzymes on larger scales.

Among the enzyme family, SMOs and IMOs can be differentiated into mono- and two-component systems [7,21]. The first contain a NADH binding domain which allows them to reduce their FAD cofactor autonomously but limits them exclusively to their natural cosubstrate. In contrast, two-component oxygenases require an additional reductase (StyB or IndB) which supplies the oxygenase with reduced FAD in nature. Furthermore, the self-sufficient mono-component enzymes of these monooxygenases show a lower oxygenase activity than the others [12,22,23].

Notably, in artificial systems, two-component enzymes can also use $FADH_2$ from the solution [24]. Thus, cofactor reduction by chemical methods without the need for NADH as an expensive cosubstrate is possible. Commonly, two-component SMOs and IMOs are cost-efficiently fueled by (a) the reduction of FAD by chemical agents (e.g., sodium thiosulfate or ruthenium complexes [8,25]), (b) the use of NAD(P)H-mimics [26] or (c) the electrochemical reduction of the cofactor [27].

In this study, the NAD(P)H-mimic 1-benzyl-1,4-dihydronicotinamide (BNAH) was preferred as cheap and easy to synthesize electron donor for the FAD cofactor in solution [11].

In addition to the typical need to apply expensive reductants, the limited solubility of many organic substrates in aqueous solutions (e.g., 2.9 mM for styrene) is a key problem and restricts substrate conversion to the lower millimolar range using isolated biocatalysts [28]. For higher amounts, whole cells have to be used in two-phase systems or with cosolvents [29]. In particular, styrene derivatives with hydrophobic side chains have a comparably low water solubility [30]. In addition, only a few enzymes with uncommonly capacious substrate binding sites are known to accept those large styrene derivatives [11]. Nevertheless, chiral styrene oxide derivatives with hydrophobic side chains are important industrial precursors for several drugs (Figure 1). In this study, we maximized the production of these compounds in an overall process optimization by applying various water-miscible solvents to increase the amount of substrate in solution and make it accessible for the enzymes. Our results show remarkably high product formation in reactions catalyzed by the IMO designated as GnIndA from *Gemmobacter nectariphilus* DSM15620 in presence of organic solvents.

To the best of our knowledge, solvent-tolerant SMOs or IMOs have not been reported before. Herein, we describe the systematical reaction optimization toward several water-miscible cosolvents for the probable solvent-tolerant GnIndA and GrStyA from *Gordonia rubripertincta* CWB2. In addition, our work highlights criteria which will help to identify solvent-tolerant enzymes based on their sequence and allows creating more robust mutants.

2. Results

2.1. Reaction Condition Optimization

GrStyA and GnIndA were successfully produced by recombinant expression as described previously [11]. Both enzymes were obtained in pure form after His-tag purification using an N-terminal His$_{10}$-tag and yielded 15 mg L^{-1} for GrStyA and 50 mg L^{-1} for GnIndA (Figure S1). Since both enzymes were described to accept styrene derivatives with large side chains as substrate [11], we aimed to maximize the product amount in a BNAH-driven process.

Because GnIndA and GrStyA belong to the subgroup of two-component systems, they were expected to accept NAD(P)H-mimics and their activity was established with BNAH as reduction equivalent (standard conditions: 20 mM Tris-HCl pH 7.5, 10 mM BNAH, 50 µM FAD, 1 mM dithiothreitol, 5 vol.% glycerol, 20 U/mL catalase and 10 vol.% MeOH for GnIndA or 5 vol.% DMSO for GrStyA). Dithiothreitol was applied to support enzyme stability as reported earlier [22]. For the sake of comparability of in-house performed assays, the protocol was kept unchanged, even though adding dithiothreitol was not mandatory for herein tested enzymes (data not shown). For both enzymes, the optimal BNAH concentration was determined to be 15 mM while for GrStyA a comparable high product amount was also observed at a concentration of 10 mM BNAH. These concentrations are equal to a 200- to 300-fold BNAH excess toward FAD (Figure 2a). For GnIndA, lower product amounts with 10 mM BNAH were observed in contrast to otherwise similar values in comparison with GrStyA.

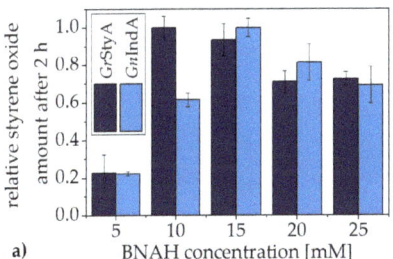
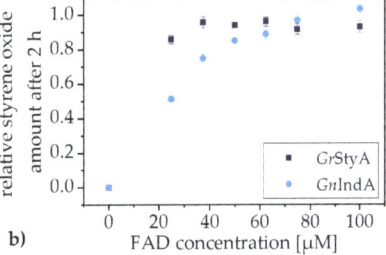

Figure 2. Relative produced styrene oxide amount under standard reaction conditions (20 mM Tris-HCl pH 7.5, 1 mM dithiothreitol, 5 vol.% glycerol, 20 U/mL catalase and 10 vol.% MeOH for GnIndA or 5 vol.% DMSO for GrStyA) in variation of (**a**) BNAH (50 µM FAD added) and (**b**) FAD concentration (10 mM BNAH added) for GrStyA and GnIndA: Both enzymes show their maximal product amount for a supplied BNAH concentration of 10 to 15 mM in presence of 50 µM FAD. The cofactor concentration to reach >90% relative product amount was determined for GrStyA >40 µM and for GnIndA > 60 µM FAD.

In addition, the FAD saturation concentration was found to be higher than for GrStyA (Figure 2b). GrStyA reached >80% of the maximal produced styrene oxide supplied with 25 µM FAD and no further increase of product amount was observed for cofactor concentrations above 40 µM (Figure 2b). In contrast, GnIndA required more than 60 µM FAD to reach >90% of the maximal substrate production and reached the maximal product formation at a FAD concentration of 100 µM.

These findings indicate a lower affinity of GnIndA toward $FADH_2$. The BNAH-driven FAD reduction in solution is enzyme-independent, but the uptake of reduced $FADH_2$ correlates with the affinity of the enzyme toward the cofactor. Since the effect of a lower affinity is only detectable under limiting conditions, different product formation rates can only be observed in low FAD concentration ranges as it was demonstrated for GnIndA (Figure 2).

After the optimization of cofactor and cosubstrate concentrations, we tested the influence of organic solvents on the enzymatic reaction. The cosolvent is required to increase the solubility of the anticipated substrates in this study and to make those styrene derivatives with hydrophobic side chains accessible for the biocatalyst. In addition, BNAH is also hardly soluble in aqueous solutions and likewise requires cosolvents. In total, nine water-miscible solvents with $logP_{O/W}$ values ranging from −1.35 (dimethyl sulfoxide) to +0.46 (tetrahydrofuran) were applied to the enzymatic reaction in concentrations up to 25 vol.% (Figure 3, Table S1). In this concentration range, the dissolved amount of the most hydrophobic substrate 1-phenyl-1-cyclohexene could be increased between a factor of three (methanol) and 100 times (1-propanol). This increase marks the upper limit for effects achieved by the cosolvent regarding the substrate solubility (Figure S2). Since styrene is about 1.6 times better soluble in water by the means of the $logP_{O/W}$ value (Table 1), a less pronounced effect in solubility increase can be expected by solvent addition for the more polar styrene. Nevertheless, the unsubstituted styrene was chosen as standard substrate to quantify the effects of each solvent and its concentration on the overall enzyme reaction. As measure of the effect, the produced amount of styrene oxide after 2 h of reaction time was used.

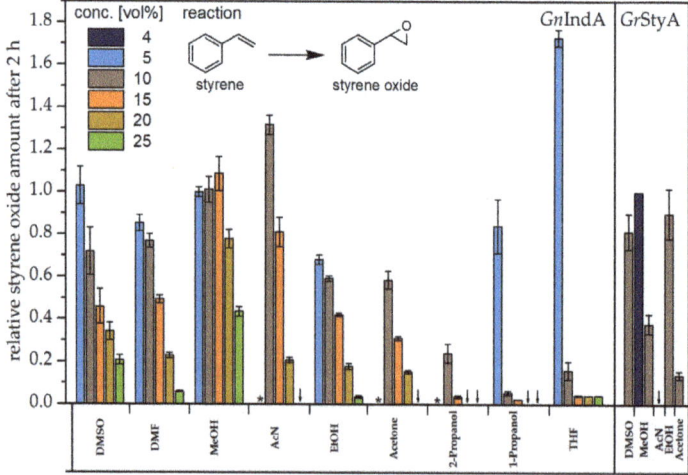

Figure 3. Relative styrene oxide amount formed by GnIndA (left) and GrStyA (right) in presence of nine organic solvents: In general, for GnIndA a remarkably high product formation up to 15 vol.% cosolvent was observed while MeOH and THF were outstanding due to the stable product formation at higher concentrations and the increased product formation in lower concentration ranges, respectively. Except for DMSO and EtOH, a significantly reduced product amount was detected for GrStyA already at 10 vol.% cosolvent. The relative styrene oxide amount is normalized to 5 vol.% MeOH (GnIndA, 0.70 mM styrene oxide) resp. 4 vol.% MeOH (GrStyA, 0.47 mM styrene oxide). * not measured.

Table 1. Conversion of different substrates by flavin-dependent monooxygenases.

Substrate		Water Solubility [30] (logP$_{O/W}$) [28]	Observed Activity [a] [mU mg^{-1}] (Enantiomeric Excess)	
			GrStyA	GnIndA
(styrene structure)	Styrene	2.85 mM (2.95)	50 ± 8 (>98% S)	212 ± 7 (>80% S)
			Product Formation Rate [b] [%]	
(3-methoxystyrene structure)	3-Methoxystyrene	- (2.9) [c]	81 ± 8	60 ± 2
(4-methoxystyrene structure)	4-Methoxystyrene	- (3.1) [c]	<1	n.d.
(allyl phenyl ether structure)	Allyl phenyl ether	- (2.94)	7 ± 1	2 ± 1
(E-β-methylstyrene structure)	E-β-Methylstyrene	1.18 mM (3.35)	88 ± 9	169 ± 3
(Z-β-methylstyrene structure)	Z-β-Methylstyrene	- (3.2) [c]	137 ± 5	65 ± 11
(β,β-dimethylstyrene structure)	β,β-Dimethylstyrene	- (3.8) [c]	122 ± 3	30 ± 1
(1-phenyl-1-cyclohexene structure)	1-Phenyl-1-cyclohexene	- (4.5) [c]	7 ± 1	<1
(1,2-dihydronaphtalene structure)	1,2-Dihydronaphtalene	- (3.2) [c]	100 ± 1	n.d.

[a] initial styrene oxide formation rate under optimal conditions (20 mM Tris-HCl pH 7.5, 1 mM dithiothreitol, 5 vol.% glycerol, 20 U/mL catalase and 10 vol.% MeOH for GnIndA or 5 vol.% DMSO for GrStyA), [b] initial epoxide production rate normalized to formation rate of styrene oxide (100%) under the same conditions, [c] computed values by XLogP3 3.0 (PubChem release 2019.06.18), n.d. = not detected.

This simplification allows no differentiation between the contributions of either the enzyme stability, increased activity, or the substrate accessibility in solution; it rather shows the sum of all effects and is sufficient in order to develop an efficient enzymatic process in the first place.

The obtained styrene oxide amount was normalized to the amount produced in presence of 5 vol.% (GnIndA) or 4 vol.% methanol (GrStyA) (Figure 3). For the enzymatic reaction catalyzed by GnIndA, a remarkably high product formation after addition of various solvents in concentrations up to 15 vol.% was found, while methanol was outstanding with still 43% product production at 25 vol.% organic solvent in solution. In presence of up to 15 vol.% acetonitrile, an increased product amount was observed in comparison to 5 vol.% MeOH and other solvents at the respective concentrations. The highest increase in product formation of +72% relative to the result in 5 vol.% MeOH was found for 5 vol.% THF, which is the solvent with the most positive logP$_{O/W}$ value (+0.46). The detected styrene oxide amount for solvents with positive logP$_{O/W}$ values (THF, 1- and 2-propanol) is comparably high at 5 vol.% in general, but a significant decrease in product formation can be observed for higher concentrations. It has to be highlighted that the enantiomeric excess (ee) was investigated for selected solvents and no variation was observed under any investigated condition (GnIndA: >80% (S)-styrene oxide; Table S2).

In contrast to GnIndA, a decreased product formation in presence of most of the tested solvents was observed for the same reaction catalyzed by GrStyA. Only for DMSO and ethanol a product amount above 80% relative to standard conditions was detected at 10 vol.% cosolvent (Figure 3).

2.2. Substrate Spectrum

After optimization of the reaction conditions, we tested the enzymatic activity of both enzymes on eight substrates which contain aliphatic residues at the vinyl chain or methoxy modifications at the aromatic ring (Table 1). For all substrates, the water solubility is equal to or up to 1.6-times lower than for the unsubstituted styrene by means of the logP$_{O/W}$ value. In addition, all substrates exhibit a larger sterical size than styrene due to their substitution patterns. The substitutions were chosen to cover diverse positions within the molecule in order to compare the influence of variations at these positions on the enzymatic activity. Ether moieties were selected in *para*- and *meta*-position relative to and between the reactive double bond and the phenyl unit to investigate their influence on substrate recognition by the enzyme. The E- and Z-selectivity was investigated using a methyl group at the β-vinyl carbon in E- and Z-β-methylstyrene. In comparison to this small residue, a six-membered ring in 1-phenyl-1-cyclohexene and 1,2-dihydronaphtalene was selected to determine the E-and Z-selectivity as example for larger substituents.

Based on the findings described in the previous section, 10 vol.% methanol was applied as cosolvent for *Gn*IndA in these experiments because it allowed the best combination of produced styrene oxide and substrate concentration in solution (Figure 3 and Figure S2). For *Gr*StyA, a cosolvent concentration of 5 vol.% DMSO was chosen due to the high product formation in presence of DMSO. Furthermore, DMSO dissolved twice the substrate amount than ethanol at 5 vol.% respectively, for which also a stable product formation up to 10 vol.% was detected (Figure S2). The remaining reaction conditions were applied as described in the previous section. For the determination of product formation rates, the respective activity on styrene under optimal conditions was used as 100% value.

Our results show a four-fold higher activity of *Gn*IndA on the standard substrate styrene than observed for *Gr*StyA. In return, for the less active *Gr*StyA, an enantiomeric excess of >98% (S)-styrene oxide was detected while for *Gn*IndA an *ee*-value >80% (S)-styrene oxide was observed. Furthermore, many substrate-dependent differences in the product formation rates were found. For *Gn*IndA with the higher activity for styrene oxidation, the acceptance of larger substrates is overall reduced. For most substrates, a more than 35% reduced activity was observed while no product could be detected for 4-methoxystyrene and 1,2-dihydronaphtalene. The latter is representative for all tested substrates containing a residue in Z-position to the aromatic ring: For all of these substrates, a reduced enzyme activity was observed. In contrast, for E-β-methylstyrene (observed activity ~358 mU mg^{-1}), a 69 ± 3% increased activity relative to styrene (observed activity ~212 mU mg^{-1}) was determined indicating a clear E-selectivity of the enzyme. It is especially noteworthy that *Gr*StyA shows an opposite behavior in the relative activity toward E- and Z-β-methylstyrene, being more active in the conversion of the Z-substrate.

For *Gr*StyA, Z-β-methylstyrene, β,β-dimethylstyrene and 1,2-dihydronaphtalene were found to show the same or higher enzyme activity than toward styrene. Interestingly, the same compounds that contain the previously mentioned residue in Z-position cause a decreased epoxidation activity in *Gn*IndA. In addition, for *Gr*StyA a reduced activity for E-β-methylstyrene was observed which indicates an overall Z-selectivity. Furthermore, *Gr*StyA converts 1,2-dihydronaphtalene at a comparable rate than styrene while 1-phenyl-1-cyclohexene is converted significantly slower (7 ± 1%). Thus, *Gr*StyA was found to be Z-selective.

In general, *Gr*StyA catalyzed the epoxidation of all tested substrates and overall, a lesser reduction in activity was observed for most of the substrates. Nevertheless, the actual activity of *Gn*IndA is still higher than of *Gr*StyA for most tested substrates.

For substrates with ether residues at the aromatic ring, a substitution-specific decrease in activity was observed for both enzymes while the relative product formation was slightly higher for *Gr*StyA. Interestingly, the position of the ether moiety correlates similarly with the activity decrease: A methoxy group in *meta*-position was accepted best by both enzymes while the same substituent in *para*-position caused the lowest conversion rate, if converted at all, for both enzymes. These position-specific correlations in enzymatic activity indicate a similar substrate binding at the aromatic ring while the recognition at the reactive double bond was found to be different due to clear isomer preferences for the enzymes (GrStyA: Z-selective, GrIndA E-selective).

In addition to the substrate spectrum, the conversion of the standard substrate styrene to styrene oxide by both enzymes was investigated in a higher resolved time scale (Figure 4). GnIndA reached a plateau in product amount of 0.70 ± 0.03 mM styrene oxide (35.0% conversion) after 30 min while GrStyA produced a maximal amount of 0.47 ± 0.04 mM styrene oxide (23.5% conversion) after 80 min. Further investigations showed that GrStyA is stable under the reaction conditions and the cosubstrate limits the reaction, although a five-fold excess of BNAH was supplied (Figure S3). The enzymes used most of the supplied BNAH to produce hydrogen peroxide instead of performing the epoxidation of the substrate. It can be estimated that GnIndA has an uncoupling rate of 86% while GrStyA shows 91% uncoupling. This uncoupling effect was also described for other enzymes, is typically influenced by the type of electron donor, and was reported for NAD(P)H-mimics before [31–33]. Nevertheless, this obvious drawback of BNAH can be overcompensated by continuous addition of the cosubstrate (Figure S3) which is justified by the low price and easy provision of the compound.

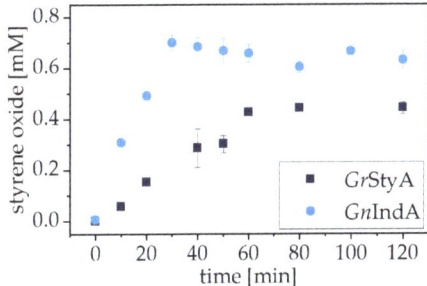

Figure 4. Formation of styrene oxide under optimal conditions (20 mM Tris-HCl pH 7.5, 2 mM styrene, 1 mM dithiothreitol, 5 vol.% glycerol, 20 U/mL catalase and 10 vol.% MeOH for GnIndA or 5 vol.% DMSO for GrStyA): GnIndA produces 0.70 ± 0.03 mM styrene oxide within 30 min (35.0% conversion) while GrStyA produced a maximal amount of 0.47 ± 0.04 mM styrene oxide after 80 min (23.5% conversion).

2.3. In Silico Analysis

Because we found the IMO GnIndA from *Gemmobacter nectariphilus* DSM15620 to be active even in the presence of substantial amounts of organic solvents while the SMO GrStyA from *Gordonia rubripertincta* CWB2 was not, we focused on sequential and structural differences of the enzymes to explain the observed behavior. For this, the structure of the SMO PpStyA from *Pseudomonas putida* S12 (PDB accession: 3IHM) was used as template since it is the only known SMO structure so far [34]. Sharing 60% sequence identity with GrStyA and 29% with GnIndA, PpStyA was used to create valid models using the software Modeler 9.23 without the need for further optimization (Figure S4) [35,36].

The PpStyA structure was subjected to the FireProt and HotSpot Wizard webservers which highlight so-called hot spot amino acids based on energy minimization and evolutionary mutations [37,38]. Compared to the structural model, these amino acids were found to be located mainly at junctions and interaction surfaces of secondary structure elements which are important for the overall protein stability (Figure 5). In addition, FireProt suggested mutations for these locations to increase the thermostability of the structure (Figure 5a, Table S3). The mutations are further differentiated in energy mutants which result from the artificial energy optimization with random mutagenesis of the structure and in evolutionary mutants which highlight natural occurring mutations in determined hot spots.

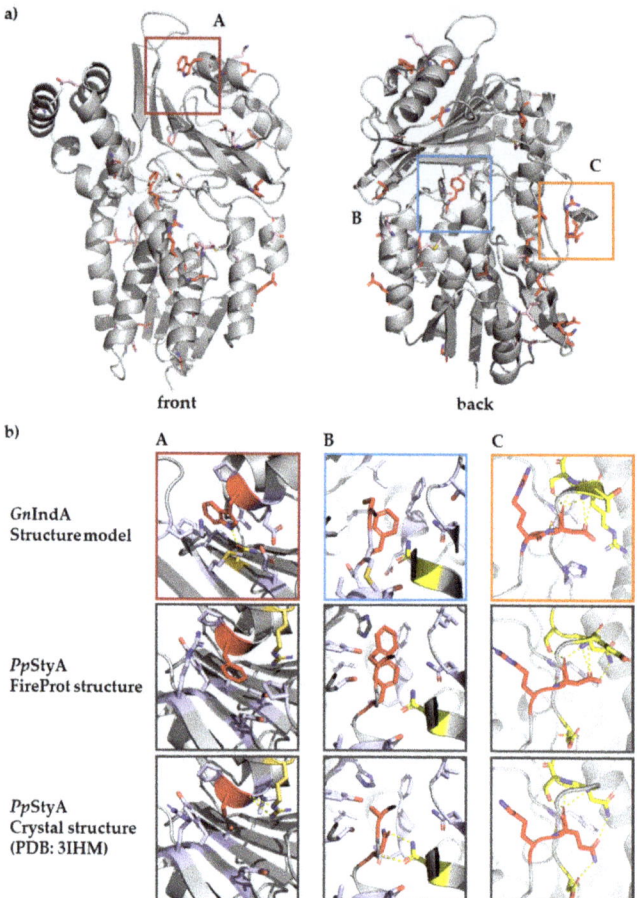

Figure 5. (a) structure model of *Gn*IndA calculated with Modeler 9.23 using the SMO structure *Pp*StyA from *Pseudomonas putida* S12 (PDB: 3IHM) as template. Amino acids targeted by the FireProt webserver on the template structure are highlighted in the *Gn*IndA model: red if mutated, pink if functionally different mutation. (b) Magnification of three structural hot spot regions of the *Gn*IndA structure model (top), as highlighted in (a), in comparison with the altered structure (middle) and the template crystal structure (bottom). Color code: red: altered amino acid, yellow: amino acid with directional interaction, blue: amino acid with non-directional interaction and surrounding amino acids. **A:** S249 does not cause hydrophobic interaction with the β-sheet amino acids. The S249F mutation increases the interaction between both secondary structure elements. In the *Gn*IndA model, W251 causes hydrophobic interactions as well but additionally rigidifies the structure by a hydrogen bond to N192 of the β-sheet below. **B:** In the *Pp*StyA structure, two helices are connected by hydrogen bonds between N51 (red) and N309 while A49 contributes little to the hydrophobic interactions in this central interaction point of three α-helices. By means of FireProt, stronger interactions are achieved by the double mutation A49F/N51F ($\Delta\Delta G = -4.4$ kJ mol^{-1}). The *Gn*IndA model already contains F50 while C48 contributes more to the hydrophobic interactions than A49 in the *Pp*StyA structure. **C:** The mutation Q155D leads to an alternative hydrogen bond network forcing a more rigid α-helical structure while the tightness of the hydration shell is strengthened due to a charge increase by the K154R mutation.

Energy mutation simulations caused in most cases the introduction of large hydrophobic amino acids while the evolutionary mutations are more diverse, albeit expressed in less extreme structural or functional amino acid exchanges (Table S3). Thus, evolutionary mutations strengthened mainly existing interactions by different mechanisms: through an increase of hydrophobic interactions (e.g., S249F or A49F and N51F, Figure 5bA,B), an introduction of electrostatic interactions (e.g., Q155D, Figure 5bC), an increase of rigidity (e.g., N285P, not shown) or through the variation of charges to increase the tightness of the hydration shell (e.g., K154R, Figure 5bC).

As mentioned before, most of the suggested mutations for PpStyA were found at interaction surfaces of secondary structure elements. As an example, S249 is located in the PpStyA structure at the tip of the interaction site of the upper helix with the β-sheet below and does not contribute to any hydrophobic interaction with the surrounding amino acids (Figure 5bA). To increase the interaction between both secondary structure elements and prevent unfolding at this position, the replacement of serine by a phenylalanine (S249F) was suggested by FireProt. In the GnIndA model, tryptophan (W251) was found in this position which allowed not only increased hydrophobic interactions but additionally rigidifies the structure by a hydrogen bond to N192 of the β-sheet. This and further examples (Figure 5bB,C) highlight that mutations suggested by the FireProt and HotSpot Wizard webservers which are found in the native GnIndA sequence fulfill the same purpose with similar contributions to the overall protein stability. Based on these data, we state that the presence of mutations in the PpStyA structure suggested by FireProt which are found in the protein sequence of interest correlate with an increased thermo- and solvent stability of the protein as well.

For quantification of the effects, a multiple sequence alignment of PpStyA with GnIndA together with eight closest related IMOs and GrStyA together with 12 related SMOs revealed that half of the suggested mutations for the PpStyA structure were naturally present in the GnIndA sequence (18 out of 36, Figure S5). Except for one single mutation (Y328F), all evolutionary mutations are present in the GnIndA sequence (17 of 18, Table S3). In the PpStyA structure, theses mutations cause a predicted difference in thermostability of $\Delta\Delta G = -21.5$ kJ mol^{-1}. From the suggested energy mutations, only a single mutation was found in the GnIndA sequence (A251Y) which contributes to structural stability of $\Delta\Delta G = -12.1$ kJ mol^{-1} in the PpStyA structure optimized by FireProt. However, in addition, three correlating residues were found in the GnIndA sequence (C49, V226 and W249) which have the same functionalities than the suggested mutations, so that they probably contribute similarly to improved structural stability.

In contrast, native GrStyA contains only 19% of the mutations suggested by FireProt (7 out of 36). These variations should still result in an increased stability compared to native PpStyA ($\Delta\Delta G = -10.2$ kJ mol^{-1}). Still, GnIndA can be expected to have a higher stability regarding elevated temperatures which will in last consequence correlate with the enhanced tolerance toward organic solvents.

3. Discussion

Two flavin containing monooxygenases, the SMO GrStyA and the IMO GnIndA, were probed for their applicability in the stereoselective epoxidation of styrene and eight differently substituted derivatives. Their relative activity and their tolerance toward the varied substitution patterns differed significantly, equally their tolerance toward organic solvents in reaction scenarios relevant for a potential industrial application of these enzymes. The herein presented data give a highly differential picture of these interesting biocatalysts worthy of more in-depth investigation.

While we observed a difference in oxygenation activity for flavin-based two-component monooxygenases in previous studies [11], an in-depth screening of substrate-, cosubstrate- and cosolvent-dependent reactivity has not been reported so far. The conversion of styrene derivatives with a hydrophobic substitution pattern was confirmed for both enzymes under investigation. While GrStyA shows an overall broader substrate spectrum, a generally higher activity was observed for GnIndA in most cases.

The major difference between both enzymes is their altered product formation rate in the presence of organic solvents. For GnIndA, a constant product amount was detected in presence of most tested solvents up to a concentration of 15 vol.%. Furthermore, in presence of 25 vol.% methanol, 43%

residual styrene oxide production was observed and the addition of 5 vol.% THF resulted in the highest increase in product amount for all solvents and concentrations tested. In contrast, addition of 10 vol.% cosolvent resulted in significantly decreased product formation when GrStyA was used as biocatalyst for all solvents except for DMSO and ethanol. For those, a product amount above 80% relative to unsubstituted styrene was still detected.

To explain the observed differences with respect to supplied solvents, structural calculations were performed. The styrene monooxygenase PpStyA from *Pseudomonas putida* S12 (PDB accession: 3IHM) was used in our in silico approach since it is the only reported SMO structure so far (Figure S5). By means of FireProt one can predict potential mutations to reach a more (thermo)stable enzyme. This tool can be supported by the HotSpot Wizard server, predicting potential sites to improve overall protein stability. Solvent- and thermostability often go hand in hand and thus we employed these tools together to rationalize our findings. The results revealed that GnIndA contains half of the suggested mutations (18 out of 36, FireProt) and shows functionally altered amino acids in highlighted stability hot spots in 86% of all cases (36 out of 42, HotSpot Wizard). By analysis of these amino acids in context of the protein structure, we showed a functional conservation in the model for GnIndA for these highlighted amino acids and demonstrated a structure-sequence dependency. Based on these data, we deduce higher stabilities for proteins that intrinsically contain the in silico suggested mutations in their native sequence. GrStyA naturally contains only 19% of the FireProt (7 out of 36) and 52% HotSpot Wizard sites (22 out of 42). These results indicate a higher thermostability of GnIndA which correlates with the observed increased tolerance toward organic solvents.

The IMO GnIndA produces a remarkably high styrene oxide amount in presence of several water-miscible solvents. In addition, the enzyme contains a unique set of mutations among related IMO which correlate with structural changes suggested by the FireProt webserver in order to generate a more stable protein. Combining both facts, we consider GnIndA to be the first solvent-tolerant IMO.

Interestingly, seven of the 18 crucial amino acids of GnIndA that are altered in comparison to PpStyA are conserved in the other nine aligned IMOs while no such conservation was found for the aligned SMO sequences (Table S3). In general, the chances to find solvent tolerance can be estimated to be higher for IMOs than SMOs. Further investigations on sequence level will show if other enzymes with higher stability than GnIndA exist in the subfamily.

4. Materials and Methods

Chemicals, substrates, and solvents were purchased from commercial supplier (Sigma Aldrich, Darmstadt, Germany; TCI Europe, Zwijndrecht, Belgium; and VWR, Darmstadt, Germany) in the highest purity available. BNAH was synthesized as described previously [39].

Enzyme production and purification. The respective genes for GrStyA (ASR05591) and GnIndA (WP_028028710) were each cloned by NotI and NdeI sites into a pET16bp vector under the control of a T7-promotor (1 mM IPTG for induction) using standard molecular biology tools [11]. The genes were then expressed heterologously in *E. coli* BL21 (DE3), growing by overnight shaking in baffled flasks in lysogeny broth (LB) medium (10 g L^{-1} tryptone, 5 g L^{-1} yeast extract, 10 g L^{-1} NaCl) at 37 °C. The cells were harvested by centrifugation and lysed using ultrasonication. The proteins were purified from cell free crude extract by nickel affinity chromatography and stored at −20 °C in 50 mM Tris-HCl buffer, pH 7.5, containing 50% glycerol and 1 mM dithiothreitol. Purity of the enzymes was verified by SDS-PAGE.

Protein quantification. Protein concentrations were determined by the Bradford method [40] using Bio-Rad protein assay reagent (#5000006). Bovine serum albumin (Sigma Aldrich, Darmstadt, Germany) served as a reference protein.

Biotransformation standard conditions. The enzymatic reaction was performed in a total volume of 200 µL in a 1.5 mL glass vial according to earlier descriptions [12,22,26]. Under standard conditions, the solution contained 20 mM Tris-HCl pH 7.5, 50 µM FAD, 10 mM BNAH, 1 mM dithiothreitol, 5 vol.% glycerol, 20 U/mL catalase from bovine liver and 10 vol.% methanol for GnIndA or 5 vol.% dimethyl sulfoxide for GrStyA. An appropriate amount of enzyme (3.0 µM GrStyA, 2.7 µM GnIndA) was added and the solution was preheated to 30 °C for ten minutes at 750 rpm shaking before the reaction was started by addition of 2 mM substrate. Samples were taken immediately after substrate addition and after 2 h reaction time. If not indicated differently, styrene was used as standard substrate for all comparative studies.

Cofactor, cosubstrate and cosolvent optimization. The enzymatic reaction was performed as described above, but the concentration of the investigated compound was chanced according to the following concentrations. The minimal required amount of dissolved FAD cofactor was identified by applying a concentration range from 0 to 100 µM FAD. For determination of the optimal cosubstrate concentration a range between 5 and 25 mM BNAH was applied in the standard conditions. To determine the optimal cosolvent and its concentration for GnIndA, 5 to 25 vol.% of a variety of different water-miscible solvents were tested in 5 vol.% increments. For GrStyA, 10 vol.% concentrations were tested. All conditions were compared by the amount of produced styrene oxide after 2 h reaction time normalized to the produced amount under the above-described standard conditions.

Initial enzyme activity. The enzymatic reaction for all tested substrates was performed in triplicates under the standard conditions described above. The volume of the reaction increased to 250 µL and 30 µL samples were taken after 5, 10, 20 and 30 min of reaction time and analyzed by HPLC. The initial activity of the respective enzymes toward a certain substrate was determined relative to styrene by comparing the slope of the linear part in product formation to the slope observed for styrene as substrate.

Time-dependent product formation. The enzymatic reaction was performed under the standard conditions described above in a twelve-fold multiplicate. Within the first hour, samples were taken every ten minutes and then every 20 minutes, from three random vials to a maximum of three samples per vial to avoid effects caused by the loss of volume during the experiment.

Synthesis of product standards. For determination of the HPLC retention times for the enzymatic products, the following three reactions were compared: First, the purchased substrate standard was measured. Through addition of excess *meta*-chloroperoxybenzoic acid to the substrate containing measurement vial in solid form, the respective epoxide was synthesized chemically in situ, which resulted in the observation of a second peak corresponding to a more polar substance. The further addition of water containing 0.1 vol.% trifluoroacetic acid led to disappearance of the epoxide peak and to detection of an even more polar compound, which is assigned to the respective diol. The procedure was validated for styrene, for which commercial standards of styrene oxide and 1-phenylethane-1,2-diol were available.

HPLC analysis. The produced epoxide amount was quantified by HPLC measurements (Thermo Scientific, Oberhausen, Germany) using an isocratic method at 60 vol.% acetonitrile in water on a Knauer Eurospher 100-5 C18 Vertex plus column (size: 125 × 4 mm, pore size: 100 Å, particle size: 5 µm; Knauer, Berlin, Germany) as stationary phase. Compounds were detected by UV/vis absorbance at 214 nm. The system was calibrated using commercially available styrene and racemic styrene oxide.

GC-FID analysis. Chiral GC-FID analysis was used to determine the enantiomeric excess of the product. For this, the enzymatically produced epoxidation product was extracted from the reaction solution with one volume ethyl acetate. The organic phase was dried over anhydrous magnesium sulfate and analyzed using a Macherey-Nagel Hydrodex β-6TBDM column (length: 50 m, inner diameter: 0.25 mm; Macherey-Nagel, Düren, Germany) under isothermal conditions at 100 °C by means of GC-FID (Shimadzu, Duisburg, Germany). The calibration was performed using commercially available styrene and (*S*)-styrene oxide.

In silico calculation of hot spot amino acids. The only known structure of the styrene monooxygenase PpStyA from *Pseudomonas putida* S12 (PDB accession number: 3IHM) was used to calculate crucial amino acids for protein stability using the HotSpot Wizard [38] and FireProt [37] webservers. The results were compared by a multiple sequence alignment of 15 SMOs closely related to GrStyA and 10 IMOs related to GnIndA. The alignment was calculated using Mega X (10.1.5) [41]. From the absence or presence of suggested mutations in each sequence in comparison to PpStyA, conclusions about the overall stability were drawn.

Homology modeling. The structural model for GnIndA was constructed using Modeler 9.23 [35]. The structure of PpStyA from *Pseudomonas putida* S12 (PDB accession number: 3IHM) served as template together with the structure files obtained from the FireProt results (see above) as template.

5. Conclusions

To sum up, we successfully optimized the reaction conditions for two flavin containing two-component monooxygenases regarding the conversion of large hydrophobic substrates. The intriguingly different preference for E- respective Z-enantiomeric substrates by GrIndA resp. GrStyA bears a high potential for industrial applications, e.g., when it comes to enantiomer separation in complex mixtures.

Furthermore, GnIndA was found to be remarkably tolerant toward nine tested organic solvents. This can be explained by beneficial mutations in crucial residues matching the predicted increased stability based on in silico calculations. In addition, these suggested mutations can be used to construct the most stable protein sequence for a targeted search for even more stable IMOs, but at the same time give valuable insight into potential stabilization strategies also for SMOs. Respectively, we could simplify and conclude that IMOs are more stable biocatalysts.

Supplementary Materials: The following are available online at http://www.mdpi.com/2073-4344/10/5/568/s1, Figure S1. SDS gel of the protein production, Table S1. Organic solvents used to optimize the enzymatic epoxidation of styrene, Figure S2. Concentration of 1-phenyl-1-cyclohexane in aqueous solution in presence of different organic cosolvents, Table S2. Enantiomeric excess of (S)-styrene oxide produced by GnIndA in presence of selected organic solvents, Figure S3. Continuously fed biotransformation of styrene to styrene oxide by GrStyA, Figure S4. Results of the quality check of the structural GnIndA model, Figure S5. Multiple sequence alignment of GnIndA, GrStyA and PpStyA with 22 related SMOs and IMOs, Table S3. FireProt mutations in comparison with their abundance among IMOs and SMOs.

Author Contributions: Supervision: D.T., conceptualization: D.T., U.-P.A. Investigations: D.E., J.M., Software: D.E., writing—original draft: D.T., D.E. writing—review and editing: C.M., U.-P.A. All authors have read and agreed to the published version of the manuscript.

Funding: D.E. received funding from the Deutsche Bundesstiftung Umwelt (PhD scholarship). The project was supported by the Deutsche Forschungsgemeinschaft (RTG 2341 MiCon). C.M., and D.T. were funded by the Federal Ministry for Innovation, Science and Research of North Rhine–Westphalia (PtJ-TRI/1411ng006—ChemBioCat). U.-P.A. is grateful for financial support from the Deutsche Forschungsgemeinschaft (Emmy Noether grant AP242/2-1) and the Fraunhofer Internal Programs under Grant No. Attract 097-602175.

Acknowledgments: We thank Thomas Heine (TU Bergakademie Freiberg) for fruitful discussions and support during the experimental phase.

Conflicts of Interest: The authors declare no conflict of interest.

References

1. Copley, S.D. Evolution of a metabolic pathway for degradation of a toxic xenobiotic: The patchwork approach. *Trends Biochem. Sci.* **2000**, *25*, 261–265. [CrossRef]
2. Heine, T.; Zimmerling, J.; Ballmann, A.; Kleeberg, S.B.; Rückert, C.; Busche, T.; Winkler, A.; Kalinowski, J.; Poetsch, A.; Scholtissek, A.; et al. On the enigma of glutathione-dependent styrene degradation in *Gordonia rubripertincta* CWB2. *Appl. Environ. Microbiol.* **2018**, *84*, e00154-18. [CrossRef] [PubMed]
3. Tischler, D. *Microbial Styrene Degradation*; Springer International Publishing: Cham, Switzerland, 2015; ISBN 978-3-319-24860-8.
4. Furnes, B.; Schlenk, D. Evaluation of xenobiotic N- and S-oxidation by variant flavin-containing monooxygenase 1 (FMO1) enzymes. *Toxicol. Sci.* **2004**, *78*, 196–203. [CrossRef] [PubMed]
5. Lin, G.-H.; Chen, H.-P.; Shu, H.-Y. Detoxification of indole by an indole-induced flavoprotein oxygenase from *Acinetobacter baumannii*. *PLoS ONE* **2015**, *10*, e0138798. [CrossRef] [PubMed]
6. Montersino, S.; Tischler, D.; Gassner, G.T.; van Berkel, W.J.H. Catalytic and structural features of flavoprotein hydroxylases and epoxidases. *Adv. Synth. Catal.* **2011**, *353*, 2301–2319. [CrossRef]

7. Huijbers, M.M.E.; Montersino, S.; Westphal, A.H.; Tischler, D.; van Berkel, W.J.H. Flavin dependent monooxygenases. *Arch. Biochem. Biophys.* **2014**, *544*, 2–17. [CrossRef] [PubMed]
8. Hollmann, F.; Lin, P.-C.; Witholt, B.; Schmid, A. Stereospecific biocatalytic epoxidation: The first example of direct regeneration of a FAD-dependent monooxygenase for catalysis. *J. Am. Chem. Soc.* **2003**, *125*, 8209–8217. [CrossRef]
9. Lin, H.; Liu, Y.; Wu, Z.-L. Asymmetric epoxidation of styrene derivatives by styrene monooxygenase from *Pseudomonas* sp. LQ26: Effects of α- and β-substituents. *Tetrahedron Asymmetry* **2011**, *22*, 134–137. [CrossRef]
10. Lin, H.; Qiao, J.; Liu, Y.; Wu, Z.-L. Styrene monooxygenase from *Pseudomonas* sp. LQ26 catalyzes the asymmetric epoxidation of both conjugated and unconjugated alkenes. *J. Mol. Catal. B Enzymatic* **2010**, *67*, 236–241. [CrossRef]
11. Heine, T.; Scholtissek, A.; Hofmann, S.; Koch, R.; Tischler, D. Accessing enantiopure epoxides and sulfoxides: Related flavin-dependent monooxygenases provide reversed enantioselectivity. *ChemCatChem* **2020**, *12*, 199–209. [CrossRef]
12. Tischler, D.; Schwabe, R.; Siegel, L.; Joffroy, K.; Kaschabek, S.R.; Scholtissek, A.; Heine, T. VpStyA1/VpStyA2B of *Variovorax paradoxus* EPS: An aryl alkyl sulfoxidase rather than a styrene epoxidizing monooxygenase. *Molecules* **2018**, *23*, 809. [CrossRef] [PubMed]
13. Hwang, S.; Choi, C.Y.; Lee, E.Y. Bio- and chemo-catalytic preparations of chiral epoxides. *J. Ind. Eng. Chem.* **2010**, *16*, 1–6. [CrossRef]
14. Breuer, M.; Ditrich, K.; Habicher, T.; Hauer, B.; Kesseler, M.; Stürmer, R.; Zelinski, T. Industrial methods for the production of optically active intermediates. *Angew. Chem. Int. Ed. Engl.* **2004**, *43*, 788–824. [CrossRef] [PubMed]
15. Lin, H.; Liu, J.-Y.; Wang, H.-B.; Ahmed, A.A.Q.; Wu, Z.-L. Biocatalysis as an alternative for the production of chiral epoxides: A comparative review. *J. Mol. Catal. B Enzymatic* **2011**, *72*, 77–89. [CrossRef]
16. Torres Pazmiño, D.E.; Winkler, M.; Glieder, A.; Fraaije, M.W. Monooxygenases as biocatalysts: Classification, mechanistic aspects and biotechnological applications. *J. Biotechnol.* **2010**, *146*, 9–24. [CrossRef]
17. De Vries, E.J.; Janssen, D.B. Biocatalytic conversion of epoxides. *Curr. Opin. Biotechnol.* **2003**, *14*, 414–420. [CrossRef]
18. Panke, S.; Held, M.; Wubbolts, M.G.; Witholt, B.; Schmid, A. Pilot-scale production of (S)-styrene oxide from styrene by recombinant *Escherichia coli* synthesizing styrene monooxygenase. *Biotechnol. Bioeng.* **2002**, *80*, 33–41. [CrossRef]
19. Gao, P.; Wu, S.; Praveen, P.; Loh, K.-C.; Li, Z. Enhancing productivity for cascade biotransformation of styrene to (S)-vicinal diol with biphasic system in hollow fiber membrane bioreactor. *Appl. Microbiol. Biotechnol.* **2017**, *101*, 1857–1868. [CrossRef]
20. Schmid, A.; Hofstetter, K.; Feiten, H.-J.; Hollmann, F.; Witholt, B. Integrated biocatalytic synthesis on gram scale: The highly enantioselective preparation of chiral oxiranes with styrene monooxygenase. *Adv. Synth. Catal.* **2001**, *2001*, 343.
21. Mascotti, M.L.; Juri Ayub, M.; Furnham, N.; Thornton, J.M.; Laskowski, R.A. Chopping and changing: The evolution of the flavin-dependent monooxygenases. *J. Mol. Biol.* **2016**, *428*, 3131–3146. [CrossRef]
22. Tischler, D.; Eulberg, D.; Lakner, S.; Kaschabek, S.R.; van Berkel, W.J.H.; Schlömann, M. Identification of a novel self-sufficient styrene monooxygenase from *Rhodococcus opacus* 1CP. *J. Bacteriol.* **2009**, *191*, 4996–5009. [CrossRef] [PubMed]
23. Tischler, D.; Gröning, J.A.D.; Kaschabek, S.R.; Schlömann, M. One-component styrene monooxygenases: An evolutionary view on a rare class of flavoproteins. *Appl. Biochem. Biotechnol.* **2012**, *167*, 931–944. [CrossRef] [PubMed]
24. Zhang, W.; Hollmann, F. Nonconventional regeneration of redox enzymes—A practical approach for organic synthesis? *Chem. Commun.* **2018**, *54*, 7281–7289. [CrossRef] [PubMed]
25. Tischler, D.; Schlömann, M.; van Berkel, W.J.H.; Gassner, G.T. FAD C(4a)-hydroxide stabilized in a naturally fused styrene monooxygenase. *FEBS Lett.* **2013**, *587*, 3848–3852. [CrossRef] [PubMed]
26. Paul, C.E.; Tischler, D.; Riedel, A.; Heine, T.; Itoh, N.; Hollmann, F. Nonenzymatic regeneration of styrene monooxygenase for catalysis. *ACS Catal.* **2015**, *5*, 2961–2965. [CrossRef]
27. Hollmann, F.; Hofstetter, K.; Habicher, T.; Hauer, B.; Schmid, A. Direct electrochemical regeneration of monooxygenase subunits for biocatalytic asymmetric epoxidation. *J. Am. Chem. Soc.* **2005**, *127*, 6540–6541. [CrossRef]

28. Hansch, C.; Leo, A.; Hoekman, D. *Hydrophobic, Electronic, and Steric constants*; American Chemical Society: Washington, DC, USA, 1995; ISBN 0841229529.
29. Toda, H.; Imae, R.; Itoh, N. Efficient biocatalysis for the production of enantiopure (*S*)-epoxides using a styrene monooxygenase (SMO) and *Leifsonia* alcohol dehydrogenase (LSADH) system. *Tetrahedron Asymmetry* **2012**, *23*, 1542–1549. [CrossRef]
30. Yalkowsky, H.S.; He, Y.; Jain, P. *Handbook of Aqueous Solubility Data*, 2nd ed.; CRC Press: Boca Raton, FL, USA, 2010; ISBN 978-1-4398-0246-5.
31. Morrison, E.; Kantz, A.; Gassner, G.T.; Sazinsky, M.H. Structure and mechanism of styrene monooxygenase reductase: New insight into the FAD-transfer reaction. *Biochemistry* **2013**, *52*, 6063–6075. [CrossRef]
32. Gassner, G.T. The styrene monooxygenase system. *Meth. Enzymol.* **2019**, *620*, 423–453. [CrossRef]
33. Massey, V. Activation of molecular oxygen by flavins and flavoproteins. *J. Biol. Chem.* **1994**, *269*, 22459–22462.
34. Ukaegbu, U.E.; Kantz, A.; Beaton, M.; Gassner, G.T.; Rosenzweig, A.C. Structure and ligand binding properties of the epoxidase component of styrene monooxygenase. *Biochemistry* **2010**, *49*, 1678–1688. [CrossRef] [PubMed]
35. Webb, B.; Sali, A. Comparative protein structure modeling using MODELLER. *Curr. Protoc. Bioinform.* **2016**, *54*, 5–6. [CrossRef]
36. Eisenberg, D.; Lüthy, R.; Bowie, J.U. VERIFY3D: Assessment of protein models with three-dimensional profiles. *Methods Enzymol.* **1997**, *277*, 396–404. [CrossRef] [PubMed]
37. Musil, M.; Stourac, J.; Bendl, J.; Brezovsky, J.; Prokop, Z.; Zendulka, J.; Martinek, T.; Bednar, D.; Damborsky, J. FireProt: Web server for automated design of thermostable proteins. *Nucleic Acids Res.* **2017**, *45*, W393–W399. [CrossRef] [PubMed]
38. Sumbalova, L.; Stourac, J.; Martinek, T.; Bednar, D.; Damborsky, J. HotSpot Wizard 3.0: Web server for automated design of mutations and smart libraries based on sequence input information. *Nucleic Acids Res.* **2018**, *46*, W356–W362. [CrossRef]
39. Paul, C.E.; Arends, I.W.C.E.; Hollmann, F. Is simpler better? Synthetic nicotinamide cofactor analogues for redox chemistry. *ACS Catal.* **2014**, *4*, 788–797. [CrossRef]
40. Bradford, M.M. A rapid and sensitive method for the quantitation of microgram quantities of protein utilizing the principle of protein-dye binding. *Anal. Biochem.* **1976**, *72*, 248–254. [CrossRef]
41. Kumar, S.; Stecher, G.; Li, M.; Knyaz, C.; Tamura, K. MEGA X: Molecular evolutionary genetics analysis across computing platforms. *Mol. Biol. Evol.* **2018**, *35*, 1547–1549. [CrossRef]

© 2020 by the authors. Licensee MDPI, Basel, Switzerland. This article is an open access article distributed under the terms and conditions of the Creative Commons Attribution (CC BY) license (http://creativecommons.org/licenses/by/4.0/).

Article

Effects of Long-Term Supplementation with Aluminum or Selenium on the Activities of Antioxidant Enzymes in Mouse Brain and Liver

Ilona Sadauskiene [1,*], Arunas Liekis [1], Inga Staneviciene [2], Rima Naginiene [1] and Leonid Ivanov [2]

1. Neuroscience Institute, Lithuanian University of Health Sciences, LT-50161 Kaunas, Lithuania; Arunas.Liekis@lsmuni.lt (A.L.); Rima.Naginiene@lsmuni.lt (R.N.)
2. Department of Biochemistry, Medical Academy, Lithuanian University of Health Sciences, LT-50161 Kaunas, Lithuania; Inga.Staneviciene@lsmuni.lt (I.S.); Leonid.Ivanov@lsmuni.lt (L.I.)
* Correspondence: Ilona.Sadauskiene@lsmuni.lt; Tel.: +370-(37)-302967; Fax: +370-(37)-302959

Received: 8 April 2020; Accepted: 20 May 2020; Published: 23 May 2020

Abstract: The aim of this study was to investigate the effects of aluminum (Al) or selenium (Se) on the "primary" antioxidant defense system enzymes (superoxide dismutase, catalase, and glutathione reductase) in cells of mouse brain and liver after long-term (8-week) exposure to drinking water supplemented with $AlCl_3$ (50 mg or 100 mg Al/L in drinking water) or Na_2SeO_3 (0.2 mg or 0.4 mg Se/L in drinking water). Results have shown that a high dose of Se increased the activities of superoxide dismutase and catalase in mouse brain and liver. Exposure to a low dose of Se resulted in an increase in catalase activity in mouse brain, but did not show any statistically significant changes in superoxide dismutase activity in both organs. Meanwhile, the administration of both doses of Al caused no changes in activities of these enzymes in mouse brain and liver. The greatest sensitivity to the effect of Al or Se was exhibited by glutathione reductase. Exposure to both doses of Al or Se resulted in statistically significant increase in glutathione reductase activity in both brain and liver. It was concluded that 8-week exposure to Se caused a statistically significant increase in superoxide dismutase, catalase and glutathione reductase activities in mouse brain and/or liver, however, these changes were dependent on the used dose. The exposure to both Al doses caused a statistically significant increase only in glutathione reductase activity of both organs.

Keywords: superoxide dismutase (SOD); catalase (CAT); glutathione reductase (GR); aluminum (Al); selenium (Se); mouse; brain; liver

---

1. Introduction

Most chemical elements play a very important role in human life. This is especially true for metals, which may have a major impact on human health [1]. Metals are naturally found in the Earth's crust, whereas humans cause their dissemination into the biosphere. These elements are highly stable, water-soluble, can accumulate in soil, and may enter the human body with food, air, or through the skin. Aluminum (Al) is especially noteworthy, as it plays an exceptional role in modern life due to its wide use in both industrial and household contexts. Al compounds are used in water purification and as antacids, food additives, vaccine adjuvants, and antiperspirants. Al intake does not correlate with Al amount in the body [2]. Al absorption by the gastrointestinal tract varies from 0.01% to 1% of the total Al intake. For a long time, owing to its inertness, Al was regarded as a completely harmless metal, but a growing body of emerging evidence suggests that it may be one of the main factors that cause a number of diseases in humans and animals [3]. Due to its specific chemical properties, Al inhibits more than 200 biologically important functions and causes harmful effects. For instance, chronic exposure to Al has been proven to play a role in the development of neurodegenerative disorders: Parkinson's dementia [4–6] or Alzheimer's disease [7–9]. The mechanisms of Al neurotoxicity are unclear, although research has shown that these mechanisms may mostly be attributed to the ability of Al to: (a) generate reactive oxygen species (ROS) or free radicals when metabolized, and to suppress the activity of antioxidant enzymes and other components of the antioxidant system in various organs, (b) impair signal transduction pathways in the cells, and (c) disturb calcium homeostasis. In addition, Al is one of the causes of hepatotoxicity [10,11] and pathological processes in the testis, kidneys, and lungs [12–14].

However, there is a range of chemical elements that play an important role in the antioxidant processes within cells [15,16]. One such microelement is selenium (Se). The beneficial role of Se in human cells is due to its presence in at least 25 proteins—selenoproteins, part of which are directly involved in redox catalysis. Trace amounts of Se element are essential for cellular functions [17]. Se is an element that reduces the risk of cancer [18], prevents cardiac diseases [19], and protects against the effects of ionizing radiation, heavy metals, and other toxic compounds. There is data indicating that Se strengthens the body's immune system, thus reducing the risk of infection [20]. In large amounts Se and its salts are toxic. Toxic doses of Se are able to negatively affect cellular redox status directly by oxidizing thiols, and indirectly by generating reactive oxygen species (ROS) [21].

Al is the trivalent cation that does not undergo redox changes, but it has a strong prooxidant activity and can potentiate oxidative damages [22]. In addition, high amounts of Se can exert toxic prooxidant properties [21]. The enzymatic and nonenzymatic antioxidant cellular defense systems play a key role in protecting cells from ROS toxicity. Enzymes superoxide dismutase (SOD), catalase (CAT) and glutathione reductase (GR) belong to enzymatic defense system antioxidants. SOD catalyzes the dismutation of two molecules of superoxide anion to molecular oxygen and hydrogen peroxide. Hydrogen peroxide is degraded to water and oxygen by CAT [23]. GR is the essential enzyme for the glutathione redox cycle that maintains the level of most abundant intracellular thiols—glutathione. GR is a nicotinamide adenine dinucleotide phosphate (NADPH)-dependent oxidoreductase, which catalyzes the conversion of oxidized form of glutathione (GSSG) to reduced form (GSH). In its reduced form, glutathione plays key roles in the cellular control of ROS [24].

In our previous experimental studies, we evaluated the effect of acute exposure to Al on oxidative stress and the capacity of the antioxidant system in mouse organs by using the Al intoxication model that involved the injection of $AlCl_3$ solution into the abdominal cavity of the mouse [25–28]. However, the obtained results encouraged us to select a different route of administration of Al: the oral route. This is a natural route of entry of Al into the body, which is also characteristic of humans. In addition, such administration does not cause inflammation at the site of the injection. A number of studies performed with experimental animals have demonstrated changes in the cognitive functions and morphological peculiarities of the CNS following the consumption of water with elevated Al concentrations. Even though the absorption of Al though the gastrointestinal system is very poor, a long-term negative effect of Al cannot be ruled out completely, even if the concentrations of Al that enter the body with potable water are lower. The aim of this study was to evaluate the long-term effect of different doses of Al and Se on the "primary" antioxidant defense system (the enzymes SOD, CAT, and GR) in mouse brain and liver cells after an 8-week oral administration in drinking water supplemented with $AlCl_3$ or Na_2SeO_3. This would complement the results of our previous studies on the effects of Al and Se on oxidative stress in mouse tissues.

2. Results

Data of the activities of SOD, CAT, and GR in the tissues of control and experimental animals are provided in Tables 1–3. Results have shown that Se affects SOD activity in both brain and liver cells (Table 1). The changes in the activity of the enzyme were observed when the animals received the high dose of Se (0.4 mg/L in drinking water); the activity of this enzyme increased in the brain (45%) as well as in the liver (33%), compared to controls. Meanwhile, at 8 weeks since the initiation of the experiment, the administration of the Al solution (50 mg or 100 mg Al/L in drinking water) to the laboratory animals had no effect, i.e., SOD activity in the brain and liver of the experiment animals was the same as in the respective organs of the control group animals.

Table 1. Superoxide dismutase activity in mouse brain and liver.

Group	Superoxide Dismutase Activity (U/mg Protein)	
	Brain	Liver
Control	5.67 ± 1.30	0.87 ± 0.10
Al1	5.53 ± 0.40	0.92 ± 0.24
Al2	5.44 ± 0.86	0.99 ± 0.17
Se1	5.58 ± 1.20	0.96 ± 0.16
Se2	8.23 ± 0.31 *	1.16 ± 0.10 *

The presented data (Mean ± SD) relate to 8–10 experiments. * $p < 0.05$ if compared to the control group. Al1—the mice of the first group; Al2—the mice of the second group; Se1—the mice of the third group; Se2—the mice of the fourth group; C—control group.

Table 2. Catalase activity in mouse brain and liver.

Group	Catalase Activity (U/mg Protein)	
	Brain	Liver
Control	27.37 ± 6.25	963.80 ± 102.04
Al1	24.79 ± 6.88	944.76 ± 34.90
Al2	28.30 ± 9.78	955.62 ± 43.22
Se1	46.48 ± 17.63 *	903.27 ± 103.78
Se2	41.80 ± 1.96 *	1121.29 ± 144.25 *

The presented data (Mean ± SD) relate to 8–10 experiments. * $p < 0.05$ compared to the control group. Al1—the mice of the first group; Al2—the mice of the second group; Se1—the mice of the third group; Se2—the mice of the fourth group; C—control group.

Table 3. Glutathione reductase activity in mouse brain and liver.

Group	Glutathione Reductase Activity (U/mg Protein)	
	Brain	Liver
Control	9.02 ± 0.89	0.88 ± 0.05
Al1	11.05 ± 2.17 *	1.10 ± 0.17 *
Al2	12.03 ± 1.59 *	1.18 ± 0.15 *
Se1	12.28 ± 1.71 *	0.96 ± 0.07 *
Se2	11.20 ± 1.65 *	1.18 ± 0.09 *

The presented data (Mean ± SD) relate to 8–10 experiments. * $p < 0.05$ compared to the control group. Al1—the mice of the first group; Al2—the mice of the second group; Se1—the mice of the third group; Se2—the mice of the fourth group; C—control group.

Similar results (by trend) were obtained when analyzing CAT activity in brain and liver cells of experimental animals. Like in the case of SOD, evident changes in CAT activity were observed in mice that received supplementation with the Se solution for 8 weeks (Table 2): CAT activity was statistically significantly increased in the brain (53%) and liver (16%) of mice that were administered the high dose of Se (0.4 mg/L in drinking water), compared to that observed in the brain and liver of the control mice. However, it is noteworthy that during our experiment, the change in brain CAT activity was also observed in mice that received low doses of Se (0.2 mg/L in drinking water)—in this case, CAT activity increased up to 69%, compared to the control mice. The administration of both doses of Al caused no changes in activity of this enzyme in mouse brain and liver.

The results of the experiment showed that the greatest sensitivity to the effect of the Al and Se was exhibited by GR, and the effect of Al and Se was observed in both the brain and the liver. In addition, an increase in GR activity was registered at all the doses of Al or Se salts administered to the experimental animals (Table 3). Thus, following the administration of Al, GR activity in mouse brain cells increased by 24% and 34% (at the Al dosage of, respectively, 50 mg and 100 mg/L in drinking water), in comparison to control. Se also increased GR activity in the brain by 36% (0.2 mg/L in drinking water) and 24% (0.4 mg/L in drinking water), respectively. The increase in mouse liver GR activity was dose-dependent: 10% and 33% in 0.2 mg/L and 0.4 mg/L drinking water, respectively. Similar changes were observed in the liver GR activity after exposure to Al solutions. The administration of both low and high doses of Al caused a statistically significant increase in this enzyme activity by 22% and 33%, respectively.

3. Discussion

As we have mentioned before, the toxicity of chemical elements (both metals and non-metals) is believed to manifest itself mainly through the formation of ROS or free radicals in the cells of living organisms or through the inhibition of the enzymes that are a part of the antioxidant system. Thus, the cells enter a state that is known as oxidative stress. This stress reflects an imbalance between ROS formation and the ability of the cell's biological system to neutralize these radicals. The antioxidants molecules that constitute the antioxidant defense system act at three different levels: Prevent from radical, scavenge radical and repair radical induced damage [29]. Under usual conditions, a certain pro-oxidant/antioxidant balance is maintained in the cell, thus protecting the cell (its macromolecules) against the destructive effect of the ROS. However, due to various internal and external factors (disease or unfavorable environmental conditions, etc.) this balance may shift in favor of pro-oxidants [30]. Following the depletion of the antioxidant system resources, an uncontrollable increase in ROS in the cell causes oxidative damage to the essential structures of the cell—lipids, nucleic acids, and proteins [31]. Thus, by damaging proteins, nucleic acids and lipids in the cell, ROS plays a key role in the etiology of a number of diseases. In the human organism, oxidative stress is believed to be an important factor in the development of cancer [32], Parkinson's and Alzheimer's diseases, myocardial infarction, heart failure and atherosclerosis [33], the X syndrome [34], autism [35], infection [36], the chronic fatigue syndrome [37], and other diseases.

To ensure protection against oxidative stress, the cells have the so-called antioxidant system. This system is as varied as the radicals that cause oxidative stress. Despite the significant progress in the understanding of the activity of individual enzymes and components of the antioxidant system, the complex composition of the cell's antioxidant system greatly complicates the understanding of the overall functioning of this "defense system". Even though there are several classifications of the antioxidant system in the cells, the classification proposed by Davies [38] in 1988 may be considered one of the most acceptable ones. According to this classification, the antioxidant system is classified into the primary and the secondary ones. The primary antioxidant "defense" system consists of the enzymes such as SOD, CAT, GR and the antioxidant vitamins E, A, C, glutathione, and uric acid. SOD are enzymes that catalyze the dismutation of the superoxide radical, which causes many types of cell damages, into molecular oxygen or hydrogen peroxide. Hydrogen peroxide is also damaging and has degraded by other enzymes such as CAT. GR catalyzes the NADPH-dependent reduction of oxidized glutathione (GSSG) to reduced glutathione (GSH), which plays an important role in the GSH redox cycle that maintains adequate levels of reduced glutathione. The GSH/GSSG ratio determines cell redox status of cells. At rest state healthy cells have a GSH/GSSG ratio greater than 100. When cells are in oxidative stress conditions, the GSH/GSSG ratio drops from 1 to 10 [39]. Glutathione directly scavenges free radicals or is the electron donor for the reduction of peroxides in antioxidant enzymes reactions [40]. Glutathione also is as a thiol buffer maintaining sulfhydryl groups of many proteins in their reduced form.

The secondary antioxidant system includes lipid-cleaving enzymes—phospholipases, proteolytic enzymes—proteases and peptidases, DNA-repairing enzymes—polymerases, glycosylases, endonucleases, exonucleases, and ligases. Concerning the antioxidant "defense" system, of special interest is the enzyme group of this system—superoxide dismutase, catalase, and glutathione peroxidase/glutathione reductase system. They are the elements that form the "first line of defense" against the ROS and free radicals. The increased GR activity improves the cells ability to replenish glutathione, and helps maintain favourable GSH/GSSG ratio during oxidative stress.

As we have mentioned before, the aim of this study was to evaluate the effect of 8-week oral administration of Al and Se salts on the activity of the main enzymes of the antioxidant defense system (SOD, CAT, and GR) in mouse brain and liver. It is noteworthy that our interest in the study on the effect of Al and Se on the elements of the cellular antioxidant system has also been stimulated by the fact that there are a number of studies on this issue, but they use a different animal intoxication scheme [41].

First, we evaluated the activity of the studied enzymes in mouse brain cells. The evaluation showed that after 8-week oral administration with $AlCl_3$ and Na_2SeO_3 solutions, Se ions had the greatest influence on SOD, CAT, and GR activity (by increasing it). During our experiment, the most significant increase in the activity of these enzymes was observed with the administration of the high dose of Se (0.4 mg Se/L in drinking water) by 45%, 53%, and 24%, respectively. Se at low doses (0.2 mg Se/L in drinking water) affected only CAT and GR activity (it increased by, accordingly, 69% and 36%). However, Al (both doses of Al; 50 mg and 100 mg Al/L in drinking water) had virtually no effect on SOD or CAT activity. Only GR proved to be sensitive to the exposure of the experimental mice to both doses of Al. Al ions increased the GR activity by up to 23% and 33%, respectively.

During the second stage of our study, we evaluated the effect of Al and Se on the activity of the antioxidant system enzymes (SOD, CAT, and GR) in hepatocytes. The results of the study showed that the effect of the aforementioned metal salts after 8-week daily oral supplementation caused an increase in SOD, CAT, and GR activity. The effect was analogous to that observed in the brain cells of the same experimental animals. Only a high dose of Se affected SOD and CAT activity (the increase was, accordingly, 33% and 16%), and only GR reacted to the administration of both Al and Se (at all doses). The administration of both low and high doses of Al caused a statistically significant increase in this enzyme activity by 22% and 33%, respectively. Se also increased GR activity in the liver by 10% (low dose) and 33% (high dose), respectively.

It is noteworthy that, compared to the literature data [42–47], our findings partially correlate with those of other studies. For instance, the results of our previous studies showed that exposure to Al tends to reduce the activity of the main enzymes of the antioxidant system [25,26]. Such a trend was observed in both the brain and the liver of laboratory animals [42–45]. Meanwhile, our results showed that the effect of Al on SOD and CAT activity (in brain and liver) was not as significant. GR was an exception here, as its activity statistically significantly increased in both the brain and the liver. Conversely, exposure to Se results in an increase in the activity of all the enzymes of the antioxidant system. In addition, the literature sources have emphasized the antioxidant effect of Se in neutralizing the oxidative stress caused by Al (as well as other factors) [42,43,46,47].

Thus, the results of our study showed that elements (in this case, Al and Se) did affect the components of the antioxidant system in cells of some organs. As expected, the effect depended on the type of the elements. The much weaker effect of Al on some enzymes (SOD and CAT) might have been dependent of its physical and chemical properties; this metal has a relatively difficult entry into the body and very slowly accumulates in organ cells [48]. Al is poorly absorbed in the gastrointestinal tract (0.1%–1.0% of the oral dose) [49]. As a result, symptoms of oxidative stress develop much more slowly. It is known that Al circulates in the blood mainly bound to transferrin (90%) and low molecular mass compounds (e.g., citrate) [50]. Therefore, Al may interfere with Fe homeostasis by displacing it from transferrin; as a result, Fe is released into the bloodstream. Nayak [51] indicated that exposure to Al can impair intestinal Fe absorption. Al ions increase Fe concentration in serum and disrupt normal ferritin level. Al increases uptake of transferrin bound Fe and the transport of non-bound Fe in human glial cells. The higher levels of intracellular Fe can increase oxidative damages [52]. Oxidative damages of cellular lipids, proteins and DNA and disturbed redox status in cells can been characterized by the change in antioxidant enzymes activities and the consumption of sulfhydryl measured by the reduced glutathione (GSH) status [53]. Glutathione reductase is an enzyme which causes a reduction of the oxidized glutathione. Reduced glutathione is a component of the antioxidant system that comprises more than 90% of total intracellular thiols. It is possible that in our experiments using different doses of aluminum (50 mg or 100 mg Al/L in drinking water), Al enters the brain poorly, and so we did not determine the effect of Al ions on the activities of antioxidant enzymes SOD and CAT. It has been found that only 1% of the total body Al accumulates in brain tissue [54]. On the other hand, it is also known that Al increases the uptake of Fe to brain cells. Fe, a redox-active metal, can interact with molecular oxygen to form the superoxide anion, which in turn generates a highly reactive hydroxyl radical [49]. The role of the intracellular antioxidant GSH is important against these radicals. Reduced GSH can be oxidized to glutathione disulfide (GSSG) during oxidative stress. Reduced GSH is regenerated from GSSG during the glutathione redox cycle catalyzed by GR; therefore, we found an increase in the activity of this enzyme in mice brains after long-term (8-week) exposure to drinking water supplemented with AlCl3 (50 mg or 100 mg Al/L in drinking water). We also found similar changes in GR activity in both the liver and the brain.

Se is important for a various biological process in mammalian cells [55]. Its beneficial role is related to low molecular weight Se compounds, as well as to its presence in at least 25 selenoproteins. Some selenoproteins are directly involved in redox catalysis. Se as a cofactor of glutathione peroxidase increases the antioxidant capacity of intracellular systems [56]. The amount of Se in the human brain is low (2.3%) [57], however, the brain uptake of Se has a high preference in the presence of Se dietary deficiency. The decrease of blood Se concentration and normal brain Se concentration was determinate in young rats after Se-deficient 13-week diet [58]. The deficient supply of Se may have a harmful effect on brain cells, can disturb neuronal function and induce cell death [59]. High metabolic activity of the brain causes excessive production of ROS. The nervous tissue is abundant in Fe, a substrate for the Fenton reaction, and is rich in polyunsaturated fatty acids, which are the target for lipid peroxidation. For these reasons, the brain needs highly efficient ROS scavenging that is mainly mediated by Se-required enzymes [60]. Thus, not surprisingly, exposure to this element significantly affects the enzymes of the antioxidant system. However, it is also known, that Se seems to have both beneficial and harmful properties. This element has a two-sided effect depending on its concentration. The toxic doses of Se are able to negatively affect cellular redox status (directly by oxidizing thiols, and indirectly by generating reactive oxygen species) [21]. It was shown that neurotoxicity of inorganic Se compounds is higher in comparison to organic Se compounds [61], with some exceptions, like methylseleninic acid, selenodiglutathione, L-selenocystine. In our experiments, two doses of Se were chosen (0.2 mg or 0.4 mg Se/L in drinking water), but the effect of both doses of Se on the activity of antioxidant enzymes was similar. The data obtained by us are not sufficient to determine whether the effect of Se on the first line of defense is positive, leading to an increase in the activities of CAT, SOD and GR enzymes. Se may act as prooxidant, forcing cells to increase their antioxidant defenses.

Thus, considering the obtained results, we think it is expedient for further study to assess the effect of metals (especially Al) and microelements (Se) on the components of the antioxidant system in organ cells, paying special attention to the distribution of the studied elements in mice organs.

4. Materials and Methods

4.1. Materials

All the reagents and solvents used through experiments were of analytical grade (>99.9%). Nitroblue tetrazolium and tris(hydroxymethyl)aminomethane were purchased from Sigma-Aldrich GmbH (Buchs, Switzerland); aluminum chloride hexahydrate and phenazine methosulfate were obtained from Sigma-Aldrich (Steinheim, Germany). All other chemicals were from Sigma-Aldrich (St. Louis, MO, USA).

4.2. The Object of Research

The experiments were conducted on 4–6-week-old BALB/c laboratory mice (male) weighing 20–25 g. All experimental procedures were performed according to the Law of the Republic of Lithuania Animal Welfare and Protection (License of the State Food and Veterinary Service for working with laboratory animals No. G2-80). The mice were maintained and handled at Lithuanian University of Health Sciences animal house (Kaunas, Lithuania) in agreement with the ARRIVE guidelines.

4.3. The Model of Mouse Intoxication

The mice were given drinking water supplemented with metal salts on a daily basis for 8 weeks. The mice of the first group (Al1) were given drinking water supplemented with $AlCl_3$ (50 mg (1.85 mmol) of Al/L in drinking water). The mice of the second group (Al2) were given drinking water supplemented with $AlCl_3$ (100 mg (3.7 mmol) of Al/L in drinking water). The mice of the third group (Se1) were given drinking water supplemented with Na_2SeO_3 (0.2 mg (2.5 µmol) of Se/L in drinking water). The mice of the fourth group (Se2) were given drinking water supplemented with Na_2SeO_3 (0.4 mg (5.0 µmol) of Se/L in drinking water). The control group mice had free access to drinking water without supplements. Each experimental and control group included 8–10 mice.

After the exposure time, the animals were terminated according to the rules defined by the European Convention for the Protection of Vertebrate Animals Used for Experimental and Other Purposes.

4.4. The Brain and Liver Homogenates Preparation

Following cervical dislocation of the animal the brain and liver were removed, washed, and immediately cooled on ice. The organs were carefully weighed and homogenized in three volumes (relative to organ weight) of buffer (50 mM Tris-HCl, pH 7.6; 250 mM sucrose; 60 mM KCl; 5 mM $MgCl_2$; 10 mM 2-mercaptoethanol). Homogenate was centrifuged at 15,000× g for 15 min (centrifuge „Beckman J2-21", USA), and then postmitochondrial supernatant was used for the measurement of enzymatic activity in organ tissues.

4.5. Protein Concentration Assay

Protein concentration in homogenate samples of the brain and liver was measured by using the Lowry method [62].

4.6. SOD Activity Assay

SOD activity in brain and liver homogenates was evaluated according to the method described in [63]. Activity of enzyme was evaluated spectrofotometrically (spectrophotometer LAMBDA 25, USA) by the inhibition of nitroblue tetrazolium (NBT) recovery rate in the nonenzymatic system with phenazine methosulfate and NADH (nicotinamide adenine dinucleotide/reduced) at 540 nm light wavelength. The SOD activity was expressed as U/mg protein per 1 min, where U was a relative unit of activity defined as the amount of SOD required for the inhibition of NBT reduction by 50% and expressed as a unit of activity in a 1 mg protein sample.

4.7. CAT Activity Assay

CAT activity in brain and liver homogenates was evaluated according to the method described in [63]. CAT activity is measured by hydrogen peroxide reaction with ammonium molybdate which produces a complex that absorbs at 410 nm light wavelength. The results were expressed in U/mg protein. Under these conditions, one unit of catalase (U) decomposes 1 µmol of hydrogen peroxide per 1 min.

4.8. GR Activity Assay

GR activity in the homogenates was evaluated according to the method described in [64]. The GR reaction rate was evaluated according to the reduction in the optical density in the presence of NADPH (nicotinamide adenine dinucleotide phosphate/reduced) oxidation reaction that occurs during glutathione reduction catalyzed by GR. The GR activity was evaluated spectrophotometrically spectrofotometrically (spectrophotometer LAMBDA 25, USA) at 340 nm wavelength. The results were expressed in U/mg protein. The GR activity unit (U) was the amount of the enzyme catalyzing the formation of 1 μmol of the reaction product in 1 min.

4.9. Statistical Analysis

The data were expressed as mean ± SD (standard deviation). Because the mice were grouped into experimental groups according to one factor, statistical significance was assessed using one-way analysis of variance (ANOVA) followed by Tukey's test with the software SPSS Statistics 20.0 (IBM Corporation, NY, USA). The value of $p < 0.05$ was regarded as statistically significant.

5. Conclusions

Se is an element that can increase the capacity of the intracellular antioxidant system; Al might have a lesser effect. Exposure to drinking water supplemented with Na_2SeO_3 (0.2 mg or 0.4 mg Se/L in drinking water) resulted in an increase in antioxidant potential in mouse brain and liver through enhancing the activities of superoxide dismutase, catalase and especially glutathione reductase, which plays a pivotal role in maintaining the redox state of the cell. Exposure to both doses of $AlCl_3$ (50 mg or 100 mg Al/L in drinking water) resulted in a statistically significant increase in brain and liver glutathione reductase activity but caused no changes in the activities of SOD and CAT.

Author Contributions: Data curation, I.S. (Ilona Sadauskiene) and A.L.; Formal analysis, A.L.; Methodology, I.S. (Inga Staneviciene); Project administration, L.I.; Supervision, L.I.; Writing—Original draft, A.L. and R.N.; Writing—review and editing, I.S. (Ilona Sadauskiene) and I.S. (Inga Staneviciene). All authors have read and agreed to the published version of the manuscript.

Funding: This research received no external funding.

Conflicts of Interest: The authors of this study declare that they have no conflict of interest.

Data Availability: The data used to support the findings of this study are available from the corresponding author upon request.

References

1. Jomova, K.; Valk, M. Advances in metal-induced oxidative stress and human disease. *Toxicology* **2011**, *283*, 65–87. [CrossRef]
2. Que, E.L.; Domaille, D.W.; Chang, C.J. Metals in neurobiology: Probing their chemistry and biology with molecular imaging. *Chem. Rev.* **2008**, *108*, 1517–1549. [CrossRef] [PubMed]
3. Schützendübel, A.; Polle, A. Plant responses to abiotic stresses: Heavy metal-induced oxidative stress and protection by mycorrhization. *J. Exp. Bot.* **2002**, *53*, 1351–1365. [CrossRef] [PubMed]
4. Hirsch, E.C.; Brandel, J.P.; Galle, P.; Javoy-Agid, F.; Agid, Y. Iron and Al increase in the substantianigra of patients with Parkinson's disease: An X-ray microanalysis. *J. Neurochem.* **1991**, *56*, 446–451. [CrossRef] [PubMed]
5. Campdelacreu, J. Parkinson disease and Alzheimer disease: Enviromental risk factors. *Neurologia* **2014**, *29*, 541–549. [CrossRef]
6. Chin-Chan, M.; Navarro-Yepes, J.; Quintanilla-Vega, B. Enviromental pollutants as risk factors for neurodegenerative disorders: Alzheimer and Parkinson diseases. *Front. Cell. Neurosci.* **2015**, *9*, 1–22. [CrossRef]
7. Flaten, T.P. Aluminium as a risk factor in Alzheimer's disease, with emphasis on drinking water. *Brain Res. Bull.* **2001**, *55*, 187–196. [CrossRef]

8. Bondy, S.C. Low levels of aluminum can lead to behavioral and morphological changes associated with Alzheimer's disease and age-related neurodegeneration. *Neurotoxicology* **2016**, *52*, 222–229. [CrossRef]
9. Huat, T.J.; Camats-Perna, J.; Newcombe, E.A.; Valmas, N.; Kitazawa, M.; Medeiros, R. Metal toxicity links to Alzheimer's disease and neuroinflammation. *J. Mol. Biol.* **2019**, *431*, 1843–1868. [CrossRef]
10. Geyikoglu, F.; Türkez, H.; Bakir, T.O.; Cicek, M. The genotoxic, hepatotoxic, nephrotoxic, heamatotoxic and histopathological effects in rats after aluminium chronic intoxication. *Toxicol. Ind. Health* **2013**, *29*, 780–791. [CrossRef]
11. Hall, A.R.; Le, H.; Arnold, C.; Brunton, J.; Bertolo, R.; Miller, G.G.; Zello, G.A.; Sergi, C. Aluminum exposure from parenteral nutrition: Early *Bile Canaliculus* changes of the hepatocyte. *Nutrients* **2018**, *10*, 723. [CrossRef] [PubMed]
12. Seidowsky, A.; Dupuis, E.; Drueke, T.; Dard, S.; Massy, Z.A.; Canaud, B. Aluminic intoxication in chronic hemodialysis. A diagnosis rarely evoked nowadays. Clinical case and review of the literature. *Nephrol. Ther.* **2018**, *14*, 35–41. [CrossRef] [PubMed]
13. Ighodaro, O.M.; Omole, J.O.; Ebuehi, O.A.T.; Salawu, F.N. Aluminium-induced liver and testicular damage: Effects of *Piliostigmathonningii* methanolic leaf extract. *Niger. Q. J. Hosp. Med.* **2012**, *22*, 158–162.
14. Bai, K.J.; Chuang, K.J.; Ma, C.M.; Chang, T.Y.; Chuang, H.C. Humanlung adenocarcinoma cells with an EGFR mutation are sensitive to non-autophagic cell death induced by zinc oxide and aluminium-doped zinc oxide nanoparticles. *J. Toxicol. Sci.* **2017**, *42*, 437–444. [CrossRef] [PubMed]
15. Byun, B.J.; Kang, Y.K. Conformational preferences and pK(a) value of selenocysteine residue. *Biopolymers* **2011**, *95*, 345–353. [CrossRef]
16. Valko, M.; Jomova, K.; Rhodes, C.J.; Kuča, K.; Musílek, K. Redox- and non-redox-metal-induced formation of free radicals and their role in human disease. *Arch. Toxicol.* **2016**, *90*, 1–37. [CrossRef]
17. Talbi, W.; Ghazouani, T.; Braconi, D.; Ben Abdallah, R.; Raboudi, F.; Santucci, A.; Fattouch, S. Effects of selenium on oxidative damage and antioxidant enzymes of eukaryotic cells: Wine Saccharomyces cerevisiae. *J. Appl. Microbiol.* **2019**, *126*, 555–566. [CrossRef]
18. Spallholz, J.E. Selenomethionine and Methioninase: Selenium free radical anticancer activity. *Methods Mol. Biol.* **2019**, *1866*, 199–210.
19. Cai, J.; Yang, J.; Liu, Q.; Gong, Y.; Zhang, Y.; Zhang, Z. Selenium deficiency inhibits myocardial development and differentiation by targeting the mir-215-5p/CTCF axis in chicken. *Metallomics* **2019**, *11*, 415–428. [CrossRef]
20. Lauridsen, C. From oxidative stress to inflammation: Redox balance and immune system. *Poult. Sci.* **2019**, *98*, 4240–4246. [CrossRef]
21. Lee, K.H.; Jeong, D. Bimodal actions of selenium essential for antioxidant and toxic pro-oxidant activities: The selenium paradox: A review. *Mol. Med. Rep.* **2012**, *5*, 299–304. [PubMed]
22. Campbell, A.; Prasad, K.N.; Bondy, S.C. Aluminum-induced oxidative events in cell lines: Gloma are more responsible than neuroblostoma. *Free Radic. Biol.* **1999**, *26*, 1166–1671. [CrossRef]
23. Chelikani, P.; Fita, I.; Loewen, P.C. Diversity of structures and properties among catalases. *Cell. Mol. Life Sci.* **2004**, *61*, 192–208. [CrossRef] [PubMed]
24. Couto, N.; Wood, J.; Barber, J. The role of glutathione reductase and related enzymes on cellular redox homoeostasis network. *Free Radic. Biol. Med.* **2016**, *95*, 27–42. [CrossRef]
25. Sadauskiene, I.; Staneviciene, I.; Liekis, A.; Zekonis, G. Catalase activity in mouse brain: The effects of selenium and aluminum ions. *Trace Elem. Electrolytes* **2016**, *33*, 64–69. [CrossRef]
26. Sadauskiene, I.; Viezeliene, D.; Liekis, A.; Zekonis, G.; Liekyte, K. Catalase activity in mouse organs under selenium and/or aluminum ions treatment. *Trace Elem. Electrolytes* **2017**, *34*, 5–9. [CrossRef]
27. Staneviciene, I.; Ivanov, L.; Kursvietiene, L.; Viezeliene, D. Short-term effects of aluminum and selenium on redox status in mice brain and blood. *Trace Elem. Electrolytes* **2017**, *34*, 74–80. [CrossRef]
28. Sadauskiene, I.; Liekis, A.; Staneviciene, I.; Viezeliene, D.; Zekonis, G.; Simakauskiene, V.; Baranauskiene, D.; Naginiene, R. Post-exposure distribution of selenium and aluminum ions and their effects on superoxide dismutase activity in mouse brain. *Mol. Biol. Rep.* **2018**, *45*, 2421–2427. [CrossRef]
29. Ighodaro, O.M.; Akinloye, O.A. First line defence antioxidants-superoxide dismutase (SOD), catalase (CAT) and glutathione peroxidase (GPX): Their fundamental role in the entire antioxidant defence grid. *Alex. J. Med.* **2018**, *54*, 287–293. [CrossRef]
30. Kohen, R.; Nyska, A. Oxidation of biological systems: Oxidative stress phenomena, antioxidants, redox reactions, and methods for their quantification. *Toxicol. Pathol.* **2002**, *30*, 620–650. [CrossRef]

31. Levine, R.L.; Stadtman, E.R. Oxidative modification of proteins during aging. *Exp. Gerontol.* **2001**, *36*, 1495–1502. [CrossRef]
32. Halliwell, B. Oxidative stress and cancer: Have we moved forward? *Biochem. J.* **2007**, *401*, 1–11. [CrossRef] [PubMed]
33. Ramond, A.; Godin-Ribuot, D.; Ribuot, C.; Totoson, P.; Koritchneva, I.; Cachot, S.; Joyeux-Faure, M.; Levy, P. Oxidative stress mediates cardiac infarction aggravation induced by intermittent hypoxia. *Fundam. Clin. Pharmacol.* **2013**, *27*, 252–261. [CrossRef] [PubMed]
34. de Diego-Otero, Y.; Romero-Zerbo, Y.; el Bekay, R.; Decara, J.; Sanchez, L.; Rodriguez-de Fonseca, F.; del Arco-Herrera, I. Alpha-tocopherol protects against oxidative stress in the fragile X knockout mouse: An experimental therapeutic approach for the Fmr1 deficiency. *Neuropsychopharmacology* **2009**, *34*, 1011–1026. [CrossRef]
35. James, S.J.; Cutler, P.; Melnyk, S.; Jernigan, S.; Janak, L.; Gaylor, D.W.; Neubrander, J.A. Metabolic biomarkers of increased oxidative stress and impaired methylation capacity in children with autism. *Am. J. Clin. Nutr.* **2004**, *80*, 1611–1617. [CrossRef]
36. Pohanka, M. Role of oxidative stress in infectious diseases. A review. *Folia Microbiol.* **2013**, *584*, 503–513. [CrossRef]
37. Kennedy, G.; Spence, V.A.; McLaren, M.; Hill, A.; Underwood, C.; Belch, J.J.F. Oxidative stress levels are raised in chronic fatigue syndrome and are associated with clinical symptoms. *Free Radic. Biol. Med.* **2005**, *39*, 584–589. [CrossRef]
38. Davies, K.J. Protein damage and degradation by oxygen radicals. I. general aspects. *J. Biol. Chem.* **1988**, *262*, 9895–9901.
39. Pizzorno, J. Glutathione! *Integr. Med.* **2014**, *13*, 8–12.
40. Marí, M.; Morales, A.; Colell, A.; García-Ruiz, C.; Fernández-Checa, J.C. Mitochondrial glutathione, a key survival antioxidant. *Antioxid. Redox Signal.* **2009**, *11*, 2685–2700. [CrossRef]
41. Abubakar, M.G.; Taylor, A.; Ferns, G.A. The effects of aluminium and selenium supplementation on brain and liver antioxidant status in the rat. *Afr. J. Biotechnol.* **2004**, *3*, 88–93.
42. Krewski, D.; Yokel, R.A.; Nieboer, E.; Borchelt, D.; Cohen, J.; Harry, J.; Kacew, S.; Lindsay, J.; Rondeau, V.; Mahfouz, A.M. Human health risk assessment for aluminium, aluminium oxide, and aluminium hydroxide. *J. Toxicol. Environ. Health B Crit. Rev.* **2007**, *10*, 1–269. [CrossRef] [PubMed]
43. Luo, J.; Li, X.; Li, X.; He, Y.; Zhang, M.; Cao, C.; Wang, K. Selenium-rich yeast protects against aluminum-induced peroxidation of lipide and inflammation in mice liver. *Biometals* **2018**, *31*, 1051–1059. [CrossRef] [PubMed]
44. Shrivastava, S. Combined effect of HEDTA and selenium against aluminum induced oxidative stress in rat brain. *J. Trace Elem. Med. Biol.* **2012**, *26*, 210–214. [CrossRef]
45. El-Maraghy, S.A.; Gad, M.Z.; Fahim, A.T.; Hamdy, M.A. Effect of cadmium and aluminum intake on the antioxidant status and lipid peroxidation in rat tissues. *J. Biochem. Mol. Toxicol.* **2001**, *15*, 207–214. [CrossRef]
46. Swain, C.; Chainy, G.B. Effects of aluminum sulphate and citric acid ingestion on lipid peroxidation and on activities of superoxide dismutase and catalase in cerebral hemisphere and liver of developing young chicks. *Mol. Cell. Biochem.* **1998**, *187*, 163–172. [CrossRef]
47. El-Demerdash, F.M. Antioxidant effect of vitamin E and selenium on lipid peroxidation, enzyme activities and biochemical parameters in rats exposed aluminium. *J. Trace Elem. Med. Biol.* **2004**, *18*, 113–121. [CrossRef]
48. Ozkol, H.; Bulut, G.; Balahoroglu, R.; Tuluce, Y.; Ozkol, H.U. Protective effects of selenium, N-acetylcysteine and vitamin E against acute ethanol intoxication in rats. *Biol. Trace Elem. Res.* **2017**, *175*, 177–185. [CrossRef]
49. Grochowski, C.; Blicharska, E.; Bogucki, J.; Proch, J.; Mierzwińska, A.; Baj, J.; Litak, J.; Podkowiński, A.; Flieger, J.; Maciejewski, R.; et al. Increased aluminum content in certain brain structures is correlated with higher silicon concentration in alcoholic use disorder. *Molecules* **2019**, *24*, 1721. [CrossRef]
50. Yokel, R.A. Brain uptake, retention, and efflux of aluminum and manganese: Review. *Environ. Health Perspect.* **2002**, *110*, 699–704. [CrossRef]
51. Nayak, P. Aluminum: Impacts and disease: A review. *Environ. Res.* **2002**, *89*, 101–115. [CrossRef] [PubMed]
52. Kim, Y.; Olivi, L.; Cheong, J.H.; Maertens, A.; Bressler, J.P. Aluminum stimulates uptake of non-transferrin bound iron and transferrin bound iron in human glial cells. *Toxicol. Appl. Pharmacol.* **2007**, *220*, 349–356. [CrossRef] [PubMed]

53. Jones, D.P. Redox potential of GSH/GSSG couple: Assay and biological significance. *Methods Enzymol.* **2002**, *348*, 93–112. [PubMed]
54. Yokel, R.A.; McNamara, P.J. Aluminium toxicokinetics: An updated minireview. *Pharmacol. Toxicol.* **2001**, *88*, 159–167. [CrossRef]
55. Schomburg, L.; Schweizer, U.; Köhrle, J. Selenium and selenoproteins in mammals: Extraordinary, essential, enigmatic. *Cell. Mol. Life Sci.* **2004**, *61*, 1988–1995. [CrossRef]
56. Rahman, K. Studies on free radicals, antioxidants, and co-factors: A review. *Clin. Interv. Aging* **2007**, *2*, 219–236.
57. Ahmad, A.; Khan, M.M.; Ishrat, T.; Khan, M.B.; Khuwaja, G.; Raza, S.S.; Islam, F.; Shrivastava, P. Synergistic effect of selenium and melatonin on neuroprotection in cerebral ischemia in rats. *Biol. Trace Elem. Res.* **2011**, *139*, 81–96. [CrossRef]
58. Burk, R.F.; Hill, K.E.; Selenoprotein, P. A selenium-rich extracellular glycoprotein. *J. Nutr.* **1994**, *124*, 1891–1897. [CrossRef]
59. Dominiak, A.; Wilkaniec, A.; Wroczyńsk, P.; Adamczyk, A. Selenium in the therapy of neurological diseases. Where is it going? *Curr. Neuropharmacol.* **2016**, *14*, 282–299. [CrossRef]
60. Cardoso, B.R.; Ong, T.P.; Jacob-Filho, W.; Jaluul, O.; Freitas, M.I.; Cozzolino, S.M. Nutritional status of selenium in Alzheimer's disease patients. *Br. J. Nutr.* **2010**, *103*, 803–806. [CrossRef]
61. Song, G.; Zhang, Z.; Wen, L.; Chen, C.; Shi, Q.; Zhang, Y.; Ni, J.; Liu, Q. Selenomethionine ameliorates cognitive decline, reduces tau hyperphosphorylation, and reverses synaptic deficit in the triple transgenic mouse model of Alzheimer's disease. *J. Alzheimers Dis.* **2014**, *41*, 85–99. [CrossRef] [PubMed]
62. Lowry, O.H.; Rosebrough, N.J.; Farr, A.L.; Randall, R.J. Protein measuremen with folinphenol reagent. *J. Biol. Chem.* **1951**, *193*, 265–275. [PubMed]
63. Sadauskiene, I.; Liekis, A.; Bernotiene, R.; Sulinskiene, J.; Kasauskas, A.; Zekonis, G. The effects of Buckwheat leaf and flower extracts on antioxidant status in mice organs. *Oxid. Med. Cell. Longev.* **2018**. [CrossRef] [PubMed]
64. Rachmanova, T.I.; Matasova, L.V.; Semenichina, A.V.; Safonova, A.V.; Makeeva, A.V.; Popova, T.N. *Oxidative Status Evaluation Methods*; Publishing and Printing Center of Voronezh State University: Voronezh, Russia, 2009; p. 62. (In Russian)

© 2020 by the authors. Licensee MDPI, Basel, Switzerland. This article is an open access article distributed under the terms and conditions of the Creative Commons Attribution (CC BY) license (http://creativecommons.org/licenses/by/4.0/).

Article

Development and Optimization of Lipase-Catalyzed Synthesis of Phospholipids Containing 3,4-Dimethoxycinnamic Acid by Response Surface Methodology

Magdalena Rychlicka *, Natalia Niezgoda and Anna Gliszczyńska *

Department of Chemistry, Wrocław University of Environmental and Life Sciences, Norwida 25, 50-375 Wrocław, Poland; natalia.niezgoda@upwr.edu.pl
* Correspondence: rychlicka.magda@wp.pl (M.R.); anna.gliszczynska@wp.pl (A.G.); Tel.: +48-71-320-5183 (M.R.)

Received: 30 April 2020; Accepted: 22 May 2020; Published: 24 May 2020

Abstract: The interesterification reaction of egg-yolk phosphatidylcholine (PC) with ethyl ester of 3,4-dimethoxycinnamic acid (E3,4DMCA) catalyzed by Novozym 435 in hexane as a reaction medium was shown to be an effective method for the synthesis of corresponding structured O-methylated phenophospholipids. The effects of substrate molar ratios, time of the reaction and enzyme load on the process of incorporation of 3,4DMCA into PC were evaluated by using the experimental factorial design of three factors and three levels. The results showed that a substrate molar ratio is a crucial variable for the maximization of the synthesis of 3,4-dimethoxycinnamoylated phospholipids. Under optimized parameters of 1/10 substrate molar ratio PC/E3,4DMCA, enzyme load 30% (w/w), hexane as a medium and incubation time of 3 days, the incorporation of aromatic acid into phospholipid fraction reached 21 mol%. The modified phosphatidylcholine (3,4DMCA-PC) and modified lysophosphatidylcholine (3,4DMCA-LPC) were obtained in isolated yields of 3.5% and 27.5% (w/w), respectively. The developed method of phosphatidylcholine interesterification is the first described in the literature dealing with 3,4DMCA and allows us to obtain new O-methylated phenophospholipids with potential applications as food additives or nutraceuticals with pro-health activity.

Keywords: phosphatidylcholine; 3,4-dimethoxycinnamic acid; enzymatic interesterification; lipase; biocatalysis

1. Introduction

Methoxylated derivatives of cinnamic acid (CA) are constituents of plants which have immense therapeutic potential. They are known in the literature as biomolecules that exhibit antioxidant properties [1,2] and are active as anti-cancer [3,4], anti-diabetic [5–8], hepato- [9,10] and neuroprotective [11–13] agents. It has been proven that the dietary intake of these compounds has a health-promoting effect on the body and effectively supports the prevention of cancer and metabolic or cardiovascular diseases.

Belonging to the group of natural methoxylated derivatives of CA, 3,4-dimethoxycinnamic acid (3,4DMCA) is an ingredient of fruits, spices and herbs [14]. The richest sources of 3,4DMCA are blueberries (725 mg/kg dm), blackberries (501 mg/kg dm) [15] and coffee (*Coffea canephora* var. Robusta) (690 mg/kg dm) [16], where it occurs in two forms, as a free acid and conjugates with quinic acid.

A diet enriched with 3,4DMCA has been highly recommended for patients with cancer of the stomach, prostate and thymus [3]. Watanabe et al., confirmed that injections of 3,4DMCA in saline solution at 0.25 mmol/kg significantly reduce the concentration of polyamines such as putrescine, spermidine and spermine, in the in vivo tests, carried out on a rat model. Its high level is observed in the patient's body under tumorigenesis and after chemotherapy [3]. It has been proven that this natural dimethoxycinnamic acid decreases the serum cholesterol without affecting the concentration of high-density lipoprotein (HDL)-cholesterol exhibiting activity comparable to activity of statins [17]. Based on other study, it was also suggested that 3,4DMCA could be an effective dietary compound in prophylaxis of neurodegenerative diseases, since it was discovered that it is able to act as an antiprion factor [18].

The pharmacokinetics and bioavailability of 3,4DMCA acid have been well evaluated on a group of volunteers whose diet, 24 h before coffee consumption, was low in polyphenolic compounds [19]. Farrell et al. have determined that 3,4DMCA is preferentially absorbed in the free form mainly by passive diffusion in the upper gastrointestinal tract [19], whereas dimethoxycinnamic acid derivatives are de-esterified by enteric microflora esterases. Therefore, a significant amount of free 3,4DMCA is transported to the serosal side increasing its total amount in plasma in comparison to its baseline concentration in coffee solution. Under pharmacokinetics studies, it was also confirmed that 3,4DMCA may undergo reduction with colonic microfloral reductase to dimethoxydihydrocinnamic acid, which small amounts have been quantified in plasma of volunteers [19].

Experiments performed with human hepatic S9 fraction as a model to characterize the metabolic stability and human Caco-2 cell monolayer as a well-accepted model for assessing the cellular permeability of active molecules candidates have proven that methylated polyphenols exhibit higher intestinal permeability as well as higher metabolic stability than their corresponding unmethylated forms [20]. For 3,4DMCA it has been confirmed that the presence of two methoxy groups in its structure determines the activity inhibition of enzymes of first-pass metabolism in the liver. Thereby, this compound reaches the bloodstream in unchanged form in the contrast to the hydroxy derivatives of CA, which are transformed in the human body into sulfonic and glucuronic products that significantly lower their biological activity and oral bioavailability [21]. Farrell et al. reported also that cell membrane permeability of 3,4DMCA is almost 10-fold greater than its conjugates with quinic acid [19].

Although methylated phenols are characterized by even 5- to 8-fold higher rate of absorption than the corresponding unmethylated analogues [20], their therapeutic potential and pro-health activity are still limited by their low content in dietary products, low solubility and rapid metabolism. Therefore, in the last two decades, many attempts have been made to change their physicochemical properties in order to broaden the possibilities of their practical application. The process of lipophilization has been extensively studied as the method for enhancement of polyphenols' bioavailability and therapeutic efficacy after oral administration. In the example of mycophenolic acid triglyceride conjugate, Porter's group confirmed that conjugates of phenols with lipid molecules are targeted to the lymphatic system and are transported through [22,23]. It is an additional advantage in the view of the significant role of the lymphatic system in diseases progression, particularly in therapeutic areas such as autoimmune disorders, cancer or metabolic syndrome [24]. Siliphos®, a phosphatidylcholine complex of silybin has been reported to exhibit 5-fold higher bioavailability after oral administration than free silybin [25]. Recently, we have also shown that the lipophilization of methoxy derivatives of benzoic acid and cinnamic acid based on their direct conjugation with glycerol backbone of phosphatidylcholine leads to greatly increased anticancer activity in the comparison to the free form of these acids [26,27].

Due to the fact that phospholipid complexes are rather unstable in human body and chemical synthesis of phenolipids could be difficult, the promising alternative are the biocatalytic processes of their production. Lipase-catalyzed reactions offer a high selectivity and milder reaction conditions as well as a more natural approach as compared to its chemical counterpart [28]. Many reports on the production of esters and structured triacylglycerols (TAGs) containing a phenolic acid moiety are presented in the literature, where lipases play a key role as biocatalysts [29–31]. However, according to our knowledge enzymatic lipophilization of 3,4DMCA was demonstrated only by two research groups. Synthesis of 3,4-dimethoxycinnamic esters by direct esterification of 3,4DMCA with fatty alcohols using Novozym 435 as a biocatalyst have been described by Guyot et al. Based on the results obtained, the authors indicated that decrease in the esterification yield from 60% to 12% was the result of increasing alcohol carbon chain length from 4 to 8, respectively [32]. In turn, Karboune et al. have shown that it is possible to obtain TAGs structured with 3,4DMCA during the lipase-catalyzed acidolysis of flaxseed oil with this acid. However, the bioconversion efficiency of this process was on a very low level of only 7%, which according to the authors could be the result of spatial hindrance caused by the presence of two methoxy groups in the aromatic ring [33]. Since our previous studies clearly showed that it is possible to obtain in biocatalytic process the phospholipid derivatives of ferulic [34] and anisic [35] acid, we decided to apply this method for lipophilization of 3,4DMCA. For this purpose, we used the egg-yolk phosphatidylcholine (PC) as the lipid substrate which is known to be an effective carrier of substances of hydrophilic nature and is characterized by high compatibility with biological membranes and able to penetrate even the blood–brain barrier [36,37]. In addition, the PC is an easily digestible source of polyunsaturated fatty acid (PUFA) and choline which have a strong health-promoting effect on the body [37,38].

The aim of our research was to develop and optimize a lipase-catalyzed synthesis of structured phospholipids (PLs) containing 3,4-dimethoxycinnamic acid. The activity and selectivity of enzymes and, consequently, the efficiency of enzymatic reaction depends on many parameters such as organic solvent, temperature, substrate molar ratio, reaction time, type and content of biocatalyst. Therefore, it is necessary to evaluate the influence of all of them on the effectiveness of the enzymatic reaction. The traditional method of enzymatic process optimization is based on the study of one factor at a time, which is a time-consuming procedure that incurs higher costs. Nowadays, application of various statistical experimental design to the optimization of enzymatic synthesis is extensively applied. This method is demonstrated to be efficient for understanding the relationship between independent and dependent variables in biocatalytic processes [39,40]. Among many statistical models available, the Box-Behnken design (BBD) is one of those considered to be the most efficient [41]. A significant advantage of BBD is the possibility of construction a second-order polynomial equation and surface plots to predict responses. Therefore, at the beginning of our experiments we performed preliminary tests to determine the initial reaction conditions represented by non-numerical variables (acyl residue donor, biocatalyst and organic solvent), and then we applied Box-Behnken design to optimize numerical variables such as substrate molar ratio, enzyme loading and reaction time.

2. Results

2.1. Effect of 3,4-Dimethoxycinnamonyl Donors on the Transesterification Reaction

The low solubility of phenols and its methoxy derivatives in lipophilic environment may strongly limit the efficiency of reactions catalyzed by lipases. We have already observed this phenomenon during a study on the enzymatic modification of PC with ferulic acid (FA) [34]. Using FA as an acyl donor we obtained much lower incorporation into phosphatidylcholine than under the experiments performed with its ethyl ester (EF). Therefore, we have started our investigation of the lipophilization process from the comparison of the acidolysis and interesterification reactions using 3,4-dimethoxycinnamic acid (3,4DMCA) and its ethyl ester (E3,4DMCA) as the acyl donors. The initial conditions applied for both reaction systems (acidolysis and interesterification) were proposed on the basis of the literature data and were as follows: 1/10 PC/acyl donor molar ratio, 30% enzyme dosage, temperature 50 °C and heptane as the reaction medium. Novozym 435 in a dose 30% (w/w) was chosen due to its high efficiency demonstrated previously in triacylglycerols modification carried out with 3,4DMCA [33]. Optimal temperature reported in the literature for the transesterification reactions catalyzed by Novozym 435 falls in the range of 40–60 °C, however, in our previous study we proved that 50 °C is the most suitable for PLs modifications [35]. Heptane was used as a highly hydrophobic reaction medium, while substrate molar ratio was set as 1/10 PC/acyl donor.

The progress of the reaction was monitored after 1, 2, 3 and 4 days, respectively. To confirm the degree of incorporation of 3,4DMCA into PC at a selected time intervals, samples were taken and purified by solid-phase extraction (SPE) to obtain pure PLs fractions which were subsequently derivatized and analyzed by gas chromatography (GC). At this stage of optimization of reaction parameters, phospholipid products were not fractionated. The incorporation degree of 3,4DMCA has been evaluated in the total phospholipid fraction. The screening experiments have shown that there is a significant difference between the degree of incorporation of 3,4DMCA into PLs during acidolysis and interesterification reactions (Figure 1A). It was observed that the formation of O-methylated phenophospholipids is generally possible only during the reaction with ethyl ester of 3,4-dimethoxycinnamic acid as an acyl donor. Furthermore, under the proposed conditions the interesterification reaction had a linear trend affording the highest degree of incorporation (21 mol%) after 4 days, while in the case of acidolysis it was only 2 mol%. Based on these, E3,4DMCA was selected as a better acyl donor for further studies on the enzymatic synthesis of 3,4-dimethoxycinnamoylated phospholipids.

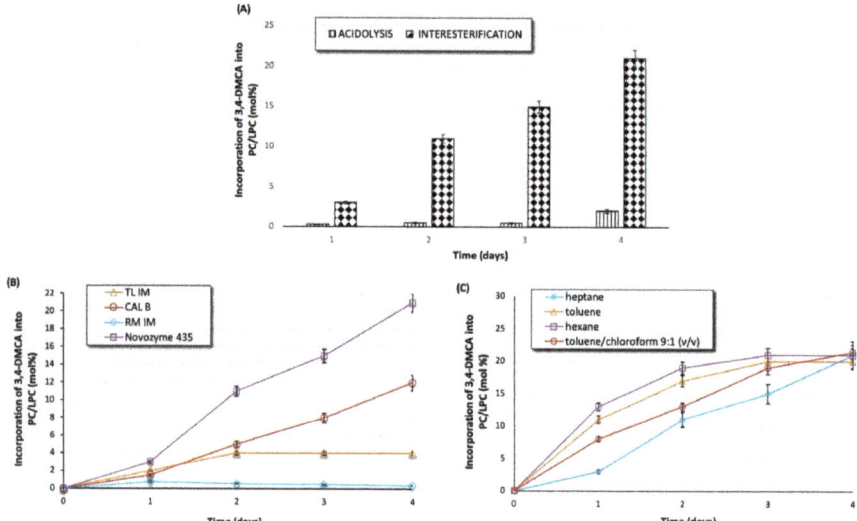

Figure 1. (**A**) Effect of different acyl donors on the incorporation of 3,4DMCA into egg-yolk phosphatidylcholine (PC). Reaction conditions: solvent, heptane; PC/acyl donor molar ratio, 1/10; lipase dosage, Novozym 435 30% (w/w); temperature, 50 °C; (**B**) time course of different lipase-catalyzed transesterification between egg-yolk phosphatidylcholine and 3,4-dimethoxycinnamic acid ethyl ester. Reaction conditions: heptane; PC/E3,4DMCA, 1/10; lipase dosage 30% (w/w); 50 °C; (**C**) effect of different organic solvent on the transesterification reaction of PC with E3,4DMCA. Reaction conditions: PC/E3,4DMCA, 1/10; Novozym 435 30% (w/w); 50 °C.

2.2. Screening of Lipase on the Interesterification Reaction

The reaction of interesterification catalyzed by lipases is quite a new process developed in the early 1980s by Unilever, Novozymes and Fuji, in which lipases are used to exchange of fatty acids residue in lipid molecules [42]. Lipases are the enzymes that catalyzes the complete or partial hydrolysis of ester bounds in TAGs or PLs and under appropriate conditions can promote ester formation as well. Two main classes of lipases used for interesterification can be distinguished: non-specific and 1,3-specific. For modification of egg-yolk phosphatidylcholine with 3,4DMCA, we selected those lipases which are reported in the literature to be highly selective towards the *sn*-1 position of TAGs and PC [29,33,34,43,44], keeping in mind natural therapeutic potential of PC resulting from presence of unsaturated fatty acids in its *sn*-2 position.

In enzymatic processes, lipases can be used in a free powdered form or immobilized on diverse carriers. These second form has many advantages and in positive way influence on their recoverability, reusability and stability, making the interesterification process economically more feasible. In our study we screened four commercially available immobilized lipases: Lipozyme®, Lipozyme TL IM, CALB and Novozym 435 by incubating them in a dosage of 30% (w/w) with 1/10 mole ratio of PC and 3,4DMCA ethyl ester at 50 °C in heptane. Figure 1B shows the effect of different enzymes on 3,4DMCA incorporation into the PLs fraction. Three among the tested enzymes: Novozym 435, CALB and TL IM, were characterized by the ability to incorporate 3,4-DMCA into the PC whereas RM IM was shown not to be active. The incorporation degree of 3,4DMCA attained with various lipases was in order of Novozym 435 > CALB > Lipozyme TL IM. However, it should be noted that only for lipases B from *Candida antarctica* immobilized on different carriers we observed significant activity in this area. In the reaction catalyzed by Novozym 435 the incorporation reached a maximum 21 mol% within 4 days, whereas in CALB-catalyzed interesterification 12 mol% of incorporation was observed. In this investigation, Lipozyme TL IM showed much lower activity giving only 4 mol% incorporation of 3,4DMCA into the PLs fraction. In our previous studies, Novozym 435 was also the most efficient in incorporation of citronellic acid, ferulic acid and anisic acid into phosphatidylcholine in the acidolysis and interesterification processes [34,35,44]. Therefore, this biocatalyst was also used further in this work.

2.3. Effect of Solvent on the Interesterification Reactions

The effectiveness of interesterification reactions can be enhanced by using appropriate solvent as a medium which reduces the reaction viscosity. Then it is also possible to shift equilibrium of lipase-catalyzed reactions towards transesterification products because appropriate organic solvent can limit the water activity in the environment. However, care should be taken to ensure that the solvent is not toxic to the enzyme and dissolves the substrates as well. Based on that, three organic solvents, n-hexane (log P = 3.5), n-heptane (log P = 4) and toluene (log P = 2.5), were selected in the studies. Previously it has been reported that usage of a binary solvent system improves the solubility of ferulic acids in modification of phospholipids [45], therefore we also evaluated the described mixture of organic solvents toluene/chloroform in a volume ratio of 9:1 (log P = 2.5/2.01).

As shown in Figure 1C, in all tested solvents the process of interesterification proceeded effectively giving the degree of incorporation of 3,4-DMCA into the PLs fraction ranging from 19 to 21 mol%. The slowest rate of process was observed in the experiment carried out in n-heptane. In this solvent incorporation of dimethoxycinnamic acid reached 21 mol% after 4 days whereas in n-hexane the same incorporation level was achieved after 3 days of reaction.

2.4. Statistical Analysis of Enzymatic Interesterification of Phosphatidylcholine (PC) with Ethyl Ester of 3,4-Dimethoxycinnamic Acid (E3,4DMCA)

In order to reduce the costs of experiments and, at the same time, better understand the impact of independent variables on the degree of incorporation of 3,4DMCA into phosphatidylcholine and correlations between them after the preliminary selection of biocatalyst and reaction medium, response surface methodology (RSM) and Box-Behnken design were employed. We prepared a 3-level, 3-factor statistical design to evaluate the effects of the selected variables, i.e., substrate molar ratio PC/E3,4DMCA, enzyme loading and reaction time on the synthesis of 3,4-dimethoxycinnamoylated phospholipids (3,4DMCA-PLs). Table 1 shows experimental and predicted values of incorporation expressed in mol%, which were obtained as a result of 15 number of runs with 3 replicates at the central point. The predicted values were obtained from a model fitting technique using the software Statistica 13.3 (StatSoft, Inc. (Tulsa, OK, USA)) and were seen to be sufficiently correlated to the experimental values (Table 1).

Table 1. Experimental and predicted value of incorporation of 3,4-dimethoxycinnamic acid (3,4DMCA) into phosphatidylcholine/lysophosphatidylcholine (PC/LPC) fraction in the Box-Behnken design.

Run	(X_1) Substrate Molar Ratio PC/E3,4DMCA	(X_2) Enzyme Load [%]	(X_3) Reaction Time [days]	Incorporation of 3,4DMCA into PC/LPC [mol%] [a] (Experimental)	Incorporation of 3,4DMCA to PC/LPC [mol%] (Predicted)
1	5	20	3	2 ± 0.3	4
2	15	20	3	15 ± 0.6	14
3	5	40	3	7 ± 0.4	8
4	15	40	3	20 ± 0.8	18
5	5	30	2	5 ± 0.7	5
6	15	30	2	10 ± 0.6	13
7	5	30	4	6 ± 0.9	3
8	15	30	4	13 ± 0.6	13
9	10	20	2	19 ± 0.1	17
10	10	40	2	19 ± 0.1	18
11	10	20	4	12 ± 0.4	13
12	10	40	4	18 ± 0.3	20
13	10	30	3	21 ± 0.5	21
14	10	30	3	21 ± 0.7	21
15	10	30	3	21 ± 0.2	21

[a] Data are presented as mean ± standard deviation (SD) of two independent analyses.

Analysis of variance (ANOVA) was performed for each of the 3 evaluated variables (Table 2). The model of F-value was identified as significant as the *p*-value was less than 0.05. A high value of coefficient of determination value (R^2) 0.93214 (close to 1) indicates that polynomial equation model was highly adequate to represent the actual relationship between the response and variables. Normally, a regression model with the R^2 value above 0.9 is considered as a model having high correlation [46]. Furthermore, the linear distribution visible in Figure 2A, which compares the predicted and experimental values of degree of incorporation, also confirms good fit of the model. The generated model was employed subsequently to study the effect of various parameters and their interactions (X_1 by X_2, X_1 by X_3 and X_2 by X_3) on the incorporation degree. Based on ANOVA results it is visible that from 3 tested variables only substrate molar ratio (Q and L) has a statistically significant effect ($p < 0.05$) on incorporation while the other two parameters: enzyme load (Q, L) and reaction time (Q, L) and the interaction between them are not statistically significant ($p > 0.05$). These data are also visible on the Pareto chart in Figure 2B, which additionally illustrates the direction of impact of individual effects. Most of the independent variables included in this chart have positive value, which means that with an increase in their value, the degree of incorporation also increases. Negative values of a coefficient estimate denote a negative influence of parameters on the incorporation degree.

Table 2. Analysis of variance (ANOVA) for interesterification variables pertaining to the response of percent incorporation of 3,4DMCA to PC/LPC.

Evaluated Factors	Sum of Squares	Degrees of Freedom	Medium Square	F-Value	*p*-Value
(X_1) Substrate molar ratio (L)	180.5000	1	180.5000	24.33708	0.002622
Substrate molar ratio (Q)	342.2500	1	342.2500	46.14607	0.000498
(X_2) Enzyme load (L)	32.0000	1	32.0000	4.31461	0.083060
Enzyme load (Q)	2.2500	1	2.2500	0.30337	0.601667
(X_3) Time of reaction (L)	2.0000	1	2.0000	0.26966	0.622149
Time of reaction (Q)	42.2500	1	42.2500	5.69663	0.054265
X_1 by X_2	0.0000	1	0.0000	0.00000	1.000000
X_1 by X_3	1.0000	1	1.0000	0.13483	0.726077
X_2 by X_3	9.0000	1	9.0000	1.21348	0.312860
Error	44.5000	6	7.4167		
Total error	655.7500	15			
$R^2 = 0.93214$					

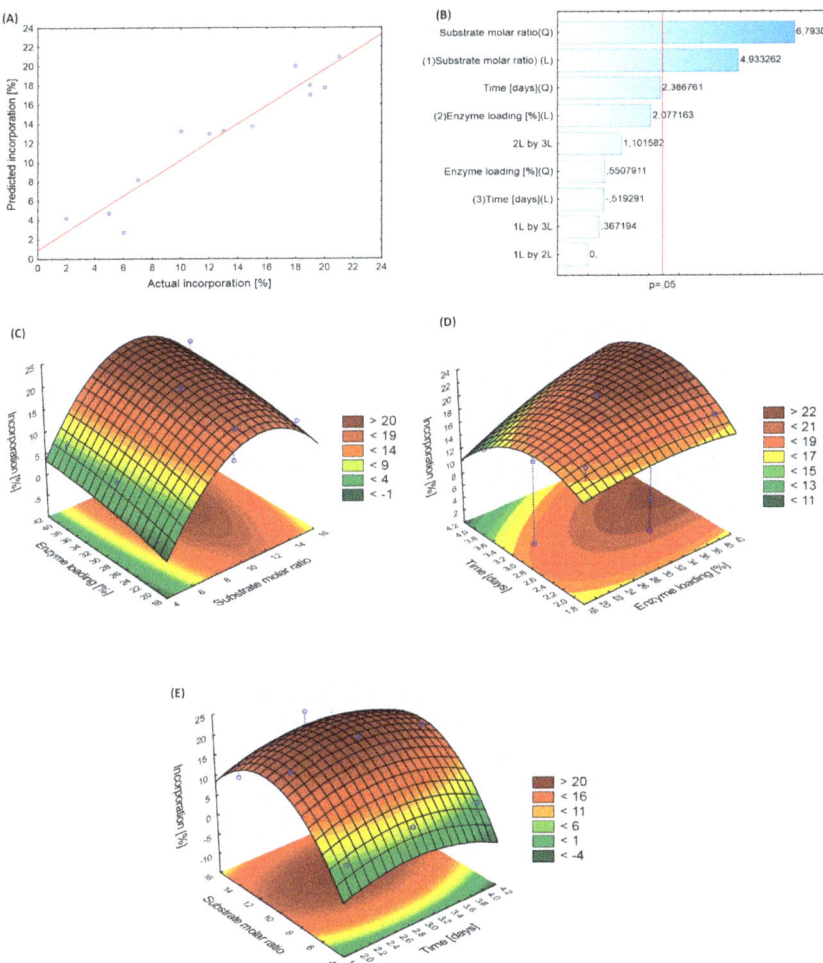

Figure 2. (**A**) Correlation of actual and predicted values of incorporation of 3,4DMCA into phospholipids (PLs) by the response surface model, (**B**) Pareto chart of the analyzed effects for incorporation of 3,4DMCA into PLs, Response surface plot showing effect of (**C**) enzyme loading and substrate molar ratio (**D**) time of reaction and enzyme loading and (**E**) substrate molar ratio and time of reaction on the incorporation of 3,4DMCA into PLs.

To better understand the relationship between reaction parameters and responses, the three-dimensional (3D) response surface plots generated from the model were applied (Figure 2C–E). In Figure 2C the effect of varying of substrate molar ratio and enzyme dosage on incorporation of 3,4DMCA into PLs fraction was studied when time of reaction was kept constant (3 days). The results show that an increase of substrate molar ratio in the range from 1:8 to 1:12 results in increasing level of incorporation from 14 to 21 mol% and the dosage of enzyme does not have significant impact on the reaction. These results are in agreement with the literature data. Yang et al. reported that a high molar ratio of the acyl donor to the lipid shifts the reaction balance towards products and improves the process of incorporation of acyl residues [47]. However, it should be noted that in the case of synthesis of 3,4-dimethoxycinnamoylated phospholipids, a further increase in the PC:E3,4DMCA ratio to 1:16 causes a decrease in the degree of incorporation to almost 10 mol%, which may be the result of the inhibitory effect of too high concentration of substrate on the enzyme activity. Lack of a relationship between higher enzyme dose and reaction productivity have been also observed before by Chen who reported that when the content of the enzyme in the reaction mixture reaches a certain value, further increasing its dose does not cause further increase in the productivity of modified phospholipids [45,48]. Similar observations can be made from Figure 2D, where constant parameter was substrate molar ratio 1:10. It is visible that to short (less than 2 days) or to long (more than 3 days) reaction time causes rapid reduction of the degree of incorporation from the highest value 21 mol% to 10 mol% while the change in the enzyme content in the tested range is of little significance. However, if we consider the relation between varying reaction time and substrate molar ratio (enzyme dosage is constant) depicted in Figure 2E it is visible that in this case reaction time has less impact on the synthesis of structured PC than substrate molar ratio. This observation is in agreement with the conclusions drawn on the basis of ANOVA analysis, which indicates that the substrate molar ratio has the greatest impact on the course of the phospholipid modification process (Figure 2B).

2.5. Identification of the Reaction Products

The optimal parameters for Novozym 435-catalyzed synthesis of 3,4-dimethoxycinnamoylated phospholipids were determined using response surface methodology. The selected optimized parameters were established as PC/E3,4-DMCA molar ratio 1/10, enzyme load 30% (w/w), reaction medium hexane, temperature 50 °C and incubation time 3 days. At these optimal reaction conditions, the interesterification of egg-yolk phosphatidylcholine with ethyl ester of 3,4-dimethoxycinnamic acid was conducted in larger scale. Below we present the reaction scheme of enzymatic interesterification of egg-yolk PC with ethyl ester of 3,4-dimethoxycinnamic acid catalyzed by Novozym 435 (Figure 3).

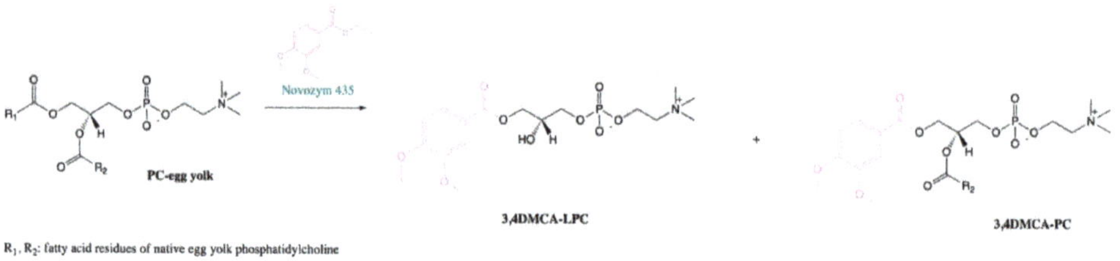

Figure 3. Reaction scheme of lipase-catalyzed interesterification of egg-yolk PC and E3,4DMCA.

The products were identified by thin-layer chromatography (TLC), gas chromatography (GC), high-performance liquid chromatography (HPLC) and nuclear magnetic resonance (NMR) analysis. The products obtained were first qualitatively identified by TLC. Based on the comparison of the R_f values of standards: PC-egg (R_f = 0.45), LPC-egg (R_f = 0.1), 3,4-DMCA (R_f = 0.85) and E3,4-DMCA (R_f = 0.72) with products of the reaction mixtures spotted on the TLC plates we identified two new bands slightly more polar than natural PC and LPC with R_f value 0.35 and 0.07, respectively.

The modified LPC as a more hydrophilic product was analyzed by reversed phase HPLC using a UV/DAD detector under UV light 310 nm in which the aromatic ring attached to the glycerol backbone of LPC produces fluorescent absorption (R_t = 4.991). Formation of the minor more lipophilic product 3,4DMCA-PC was monitored by normal phase HPLC equipped with a UV/CAD detector (at 310 nm) (R_t = 11.433). The HPLC chromatograms of the substrates and products are shown in Figures S5 and S6 of Supplementary Materials.

In the next step, new products were separated from fatty acids and unreacted ester of 3,4-dimethoxycinnamic acid as well as form PC-egg and LPC-egg and fractioned by the column chromatography. The main product 3,4-dimethoxycinnamoylated-LPC (3,4DMCA-LPC) was obtained in high 27.5% of isolated yield whereas second product 3,4-dimethoxycinnamoylated-PC (3,4DMCA-PC) was obtained in trace amount only in 3.5% isolated yield.

The structure of purified product 3,4DMCA-LPC was fully confirmed by NMR spectral data (Supplementary Materials Figure S1). From ^{1}H NMR spectral data of modified LPC, a singlet at δ 2.92 indicates the presence of 9 protons adjacent to nitrogen of the choline moiety. The multiplicity range in δ 6.07–7.36 indicates the aromatic and double bond protons of the aromatic acid. From ^{13}C NMR spectral data of this product, the signal observed at 167.69 ppm indicates the carbonyl group and the signals in the range 111.20–149.21 ppm indicate the aromatic and double bond carbons (Supplementary Materials Figure S2).

The fatty acids composition was also evaluated in the modified phospholipid fraction and product. Table 3 shows the comparison of the fatty acid profile in modified phospholipid fractions and obtained 3,4DMCA-PC. It is visible that incorporation of 3,4DMCA into phospholipid fraction obtained as a result of interesterification reached 21 mol%, whereas GC analysis of the products after their isolation and purification showed, that the content of 3,4DMCA in modified PC (3,4DMCA-PC) was only 8 mol%. It is also noticeable that with the increase of incorporation of phenolic acid, the content of saturated fatty acids (present mainly in the sn-1 position of natural phospholipids) decreases, what confirms the regioselectivity of Novozym 435.

Table 3. Fatty acid composition (% according to gas chromatography (GC)) of phospholipid fractions obtained of interesterification reaction of egg-yolk PC with 3,4-DMCAE.

Fatty and 3,4-DMCA Acids	Native PC [mol%] [a]	Modified Phospholipid Fraction PC/LPC [mol%] [a]	3,4-DMCA-PC [mol%] [a]
C16:0 (PA)	34 ± 0.2	11 ± 0.4	11 ± 0.9
C16:1 (OPA)	1 ± 0.7	1 ± 0.2	2 ± 0.1
C18:0 (SA)	15 ± 0.9	6 ± 0.7	8 ± 0.4
C18:1 (OA)	26 ± 0.7	30 ± 0.7	43 ± 0.3
C18:2 (LA)	20 ± 0.3	25	26 ± 0.2
C20:4 (AA)	4 ± 0.2	6 ± 0.3	2 ± 0.2
3,4-DMCA	-	21	8 ± 0.2

[a] Data are presented as mean ± SD of two independent analysis.

3. Materials and Methods

3.1. Substrates, Chemicals and Enzymes for Enzymatic Reactions

We synthesized 3,4-dimethoxycinnamic acid ethyl ester (E3,4DMCA) in high 86% yield according to the method described previously [49]. Its purity 98% was confirmed by GC whereas spectroscopic data were compared with the literature [50]. Native phosphatidylcholine (PC) was isolated from egg-yolk of Lohman Brown hens and purified as described in an earlier paper [44]. The purity of obtained PC was analyzed by TLC on silica gel-coated plates and further confirmed via the HPLC [51].

Lipases from *Candida antarctica* (Novozym® 435, immobilized, >5000 U/g; CALB, immobilized, >1800 U/g) were purchased from Sigma-Aldrich (St. Louis, MO, USA). Lipase from *Rhizomucor miehei* (Lipozyme® RM IM, immobilized, >30 U/g) was provided by Fluka (Buchs, Switzerland) and lipase from *Thermomyces lanuginosus* (Lipozyme® TL IM, immobilized, 250 U/g) was obtained from the Novozymes A/S (Bagsvaerd, Denmark). A boron trifluoride methanol complex solution (13–15% BF$_3$ × MeOH) and sodium methylate were purchased from Sigma-Aldrich (St. Louis, MO, USA). All organic solvents used in chromatography, silica gel-coated aluminium plates (Kieselgel 6- F254, 0.2 mm) used in thin layer chromatography (TLC) and the silica gel (Kieselgel 60, 230–400 mesh) used in the column chromatography, were purchased from Merck (Darmstadt, Germany).

3.2. Lipase-Catalyzed Reactions of Acidolysis/Interesterification of PC with 3,4DMCA/E3,4DMCA

Experiments have been started with the evaluation of two methods of enzymatic modification of egg-yolk phosphatidylcholine (PC) acidolysis and interesterification with 3,4DMCA and E3,4DMCA, respectively. Lipase-catalyzed reactions of PC (20 mg, 0.026 mmol) with 3,4DMCA/E3,4DMCA (at molar ratio PC/acyl donor, 1/10) were carried out in 2 mL of heptane in 5 mL screw-capped vials on a magnetic stirrer (300 rpm) in N_2 atmosphere at 50 °C using Novozym 435 (30% by weight of substrates) as a biocatalyst. In the next step of the study, the type of lipases (Novozym 435, CALB, RM IM, TL IM) and organic solvent (heptane, hexane, toluene, toluene:chloroform 9:1 (v/v)) were also tested for the interesterification reaction in another set of experiments. In all experiments after selected time intervals (1, 2, 3 and 4 days) the reactions were stopped by enzyme filtration (G4 Shott funnel with Celite layer). Phospholipid fractions were purified from free fatty acid and unreacted ester by the SPE methodology using silica gel columns (Discovery® DSC-Si SPE, 52654-U 500 mg) according to the procedure described by Rychlicka [34]. The purified phospholipid fraction was next analyzed by thin-layer chromatography (TLC) (3.5.1) and then their acid profile was analyzed by gas chromatography (GC) (3.5.2). The incorporation of 3,4DMCA into phospholipid fraction was expressed as mol% and its quantitative analysis was performed on the basis of the peak areas using GC Chemstation Version A.10.02.

3.3. Experimental Factorial Design

After finishing the preliminary tests, a 15-run, 3-factor, 3-level Box-Behnken design was employed to construct polynomial models for the optimization of the process enzymatic interesterification egg-yolk phosphatidylcholine with E3,4DMCA. The independent variables: substrate molar ratio (1:5–1:15), enzyme loading (20–40%), reaction time (2–4 days) were analyzed (Table 1) The STATISTICA 13.3 (StatSoft, Inc.) was used to determine the regression and graphical analysis of the results obtained. Predicted value of dependent variable (% incorporation of 3,4DMCA into phospholipid fraction PC/LPC) was calculated according to the polynomial equation as follows:

$$Y_i = \beta_0 + \beta_{1 \times 1} + \beta_2 X_2 + \beta_3 X_3 + \beta_{12} X_1 X_2 + \beta_{13} X_1 X_3 + \beta_{23} X_2 X_3 + \beta_{11} X_1^2 + \beta_{33} X_3^2$$

where Y_i is predicted response, β_0 is model constant, X_1–X_3 are independent variables and β_1–β_{33} are regression coefficient. All experiments were carried out in two independent analyses and the averages of incorporation were taken as the response.

3.4. Preparative Scale of Lipase-Catalyzed Interesterification of PC with E3,4DMCA

Reaction of egg-yolk phosphatidylcholine (200 mg, 0.26 mmol) with E3,4DMCA (at molar ratio PC/E3,4DMCA, 1/10) was catalyzed by Novozym 435 (30% by weight of substrates) and carried out in 20 mL of hexane on a magnetic stirrer (300 rpm) in N_2 atmosphere at 50 °C. After 3 days interesterification reaction was stopped by enzyme filtration on G4 Shott funnel with Celite layer and organic solvent was evaporated in vacuo. Crude mixture of products was then dissolved in $CHCl_3$ and purified by column chromatography according to the procedure described before [35]. Individual phospholipid fractions PC-egg, LPC-egg, modified phosphatidylcholine (3,4DMCA-PC) and modified lysophosphatidylcholine (3,4DMCA-LPC) were analyzed by TLC plates (3.5.1) GC, HPLC and NMR spectroscopy (^{1}H, ^{13}C, ^{31}P).

3.5. Analytical Methods

3.5.1. Thin-Layer Chromatography (TLC)

Acidolysis or interesterification reaction progress and qualitative analysis of reaction products were controlled by thin-layer chromatography (TLC). For this purpose, samples were spotted on TLC plates and eluted with the mixture of chloroform/methanol/water (65:25:4, $v/v/v$). The products were identify by spraying the TLC plates with the 0.05% primuline solution (acetone: water, 8:2, v/v) and then exposing the plates to UV light (λ = 365 nm).

3.5.2. Gas Chromatography (GC)

Analysis of acid profile of reactions products after SPE or column chromatography purification (PC-egg, LPC-egg, 3,4DMCA-PC, 3,4DMCA-LPC) and standards (PC-egg, 3,4DMCA) were analyzed by gas chromatography after their derivatization to the methyl esters according to the procedure described before [34]. GC analysis were conducted on an Agilent 6890N instrument equipped with DB-WAX column (30 m × 0.32 mm × 0.25 µm) manufactured by Agilent (Santa Clara, CA, USA) using the program and settings previously described [34]. To confirm the fatty acid methyl esters (FAME) profile, their retention times were compared with those of a standard FAME mixture (Supelco 37 FAME Mix) purchased from Sigma Aldrich.

3.5.3. High-Performance Liquid Chromatography (HPLC)

Reactions product 3,4DMCA-PC was also analyzed by HPLC on an DIONEX UltiMate 3000 chromatograph from Thermo Fisher Scientific (Olten, Switzerland) equipped with UV/CAD detector (at 310 nm). A BetaSil DIOL column (Thermo Scientific, 150 × 4.6 mm, 5 µm) was used for analysis. The injection volume, autosampler and column temperature for all analysis were as follows: 15 µL, 20 and 30 °C. Analysis were performed in a gradient mode with a constant flow of 1.5 mL/min. and started with solvent A (1% HCOOH, 0.1% TEA in water) next solvent B (hexane) and solvent C (2-propanol). The elution program was as follows: 3/40/57 (%A/%B/%C ($v/v/v$)), at 5 min = 10/40/50, at 9 min = 10/40/50, at 9.1 min = 3/40/57 and at 19 min = 43/40/57. Total analysis time was 19 min.

To analyze more hydrophilic product 3,4DMCA-LPC RP-HPLC with Ascentis® Express C18 column (150 × 3.0 mm, 5 µm) and UV/DAD detector (310 nm) was used according to the method described before [34]. Analysis was performed in a binary system of solvents: A (water with 3% of acetic acid) and B (acetonitrile) at a flow rate 0.5 mL/min. The gradient mode was as follows: 0 min: 90% A and 10% B; 10–20 min: maintained 30% A and 70% B; 20–21 min changed to 90% A and 10% B. Total analysis time was 31 min. Samples were diluted in a solvent A and the injection volume was 10 µL.

3.5.4. Spectroscopic Spectra (Nuclear Magnetic Resonance, NMR)

NMR analysis was performed to confirm the structure of the products. However, due to the low efficiency of modified PC (3,4DMCA) synthesis NMR experiments was performed only for modified LPC fraction (3,4DMCA-LPC). For this purpose, 3,4DMCA-LPC purified by column chromatography was dissolved in 0.6 mL of $CDCl_3$/MeOH (2:1, v/v) in NMR tube and then analyzed on Bruker Advance II 600 MHz spectrometer (Bruker, Billerica, MA, USA).

1-(3,4-dimethoxy)cinnamoyl-2-hydroxy-sn-glycero-3-phosphocholine (3,4DMCA-LPC)

Colourless greasy solid (26% yield, R_f 0.75); ^{1}H NMR (600 MHz, $CDCl_3$/CD_3OD 2:1 (v/v)), δ: 2.92 (s, 9H, –N(CH_3)$_3$), 3.31 (m, 2H, CH_2-β), 3.60 (2s, 6H, –OCH_3), 3.63–3.67 (m, 2H, CH_2-3′), 3.75 (m, 1H, H-2′), 3.87–3.92 (m, 5H, CH_2-1′,CH_2-α, –OH), 6.07 (d, 1H, J = 15.9 Hz, H-2), 6.58 (m, 1H, H-2″), 6.84 (m, 2H, H-5″, H-6″), 7.36 (d, 1H, J = 15.9 Hz, H-3); ^{13}C NMR (150 MHz, $CDCl_3$/CD_3OD 2:1 (v/v)) δ: 53.84 ((–N(CH_3)$_3$), 55.66 (–OCH_3), 59.24 (C-α), 62.13 (C-1′), 64.98 (C-β), 66.22 (C-3′), 68.18 (C-2′), 111.20 (Ar), 114.83 (C-2), 122.89 (Ar), 125.69 (Ar), 128.70 (Ar), 131.11 (Ar), 145.68 (C-3), 149.21 (Ar), 167.70 (C-1); ^{31}P NMR (243 MHz, $CDCl_3$/CD_3OD 2:1 (v/v)) δ: −2.90.

4. Conclusions

The study showed that the commercial immobilized form of lipase B from *Candida antarctica* (Novozym 435) is a good biocatalyst for the synthesis of 3,4-dimethoxycinnamoylated lysophosphatidylcholine (3,4DMCA-LPC) in the interesterification of egg-yolk phosphatidylcholine with ethyl ester of 3,4DMCA in an organic medium. The optimization of this process was performed by statistical design methods and for this purpose Box-Behnken design and the next surface response methodology (RSM) were successfully applied. The R^2 value of 0.93214 indicated a good fit of the model with experimental findings. The ANOVA implied that the model satisfactorily represented the real relationship of the three main reaction variables and the response. The molar ratio showed a significant effect on the productivity of the reaction. The highest incorporation of 3,4DMCA into phospholipid fraction was achieved at 1/10 PC/E3,4DMCA. We obtained 3,4-dimethoxycinnamoylated lysophosphatidylcholine (3,4DMCA-LPC) in a high isolated yield of 27.5% (w/w), whereas modified phosphatidylcholine (3,4DMCA-PC) was formed during the reaction of interesterification as the minor product. We obtained 3,4-dimethoxycinnamoylated phosphatidylcholine (3,4DMCA-PC) in trace amount in only 3.5% of isolated yield.

Supplementary Materials: The following are available online at http://www.mdpi.com/2073-4344/10/5/588/s1, Figure S1: ^{1}H NMR spectrum of 3,4DMCA-LPC, Figure S2: ^{13}C NMR spectrum of 3,4DMCA-LPC, Figure S3: ^{31}P NMR spectrum of 3,4DMCA-LPC, Figure S4: HPLC chromatogram of 3,4DMCA-PC, Figure S5: HPLC chromatogram of 3,4DMCA-LPC, Figure S6: GC chromatogram of fatty acid composition of PC-egg (standard), Figure S7: GC chromatogram of methyl ester of 3,4DMCA (standard), Figure S8: GC chromatogram of fatty acid composition of modified phospholipid fraction PC/LPC, Figure S9: GC chromatogram of fatty acid composition of isolated product of enzymatic reaction of 3,4DMCA-PC.

Author Contributions: Conceptualization, A.G.; Investigation, M.R.; Methodology, M.R.; A.G.; Visualization M.R.; A.G.; Validation, A.G.; N.N.; Writing—original draft preparation, M.R. and A.G.; Writing—Review and editing, A.G; M.R; N.N.; All authors have read and agreed to the published version of the manuscript.

Funding: Article Processing Charge (APC) was financed under the Leading Research Groups support project from the subsidy increased for the period 2020–2025 in the amount of 2% of the subsidy referred to Art. 387 (3) of the Law of 20 July 2018 on Higher Education and Science, obtained in 2019.

Conflicts of Interest: The authors declare no conflict of interest.

References

1. Natella, F.; Nardini, M.; Di Felice, M.; Scaccini, C. Benzoic and Cinnamic Acid Derivatives as Antioxidants: Structure-Activity Relation. *J. Agric. Food Chem.* **1999**, *47*, 1453–1459. [CrossRef] [PubMed]
2. Sova, M. Antioxidant and Antimicrobial Activities of Cinnamic Acid Derivatives. *Mini Rev. Med. Chem.* **2012**, *12*, 749–767. [CrossRef] [PubMed]
3. Watanabe, S.; Sato, S.; Nagase, S.; Shimosato, K.; Saito, T. Polyamine levels in various tissue of rats treated with 3-hydroxy-4-methoxycinnamic acid 3,4-dimethoxycinnamic acid. *Anticancer Drugs* **1996**, *7*, 866–872. [CrossRef] [PubMed]
4. Gunasekaran, S.; Venkatachalam, K.; Namasivayam, N. Anti-inflammatory and anticancer effects of p-methoxycinnamic acid, an active phenylpropanoid, against 1,2-dimethylhydrazine-induced rat colon carcinogenesis. *Mol. Cell. Biochem.* **2019**, *451*, 117–129. [CrossRef]
5. Adisakwattana, S.; Sookkongwaree, K.; Roengsumran, S.; Petsom, A.; Ngamrojnavanich, N.; Chavasiri, W.; Deesamer, S.; Yibchok-anun, S. Structure-activity relationships of trans-cinnamic acid derivatives on α-glucosidase inhibition. *Bioorganic Med. Chem. Lett.* **2004**, *14*, 2893–2896. [CrossRef]
6. Adisakwattana, S.; Moonsan, P.; Yibchok-Anun, S. Insulin-releasing properties of a series of cinnamic acid derivatives in vitro and in vivo. *J. Agric. Food Chem.* **2008**, *86*, 7838–7844. [CrossRef]
7. Adisakwattana, S. Cinnamic acid and its derivatives: Mechanisms for prevention and management of diabetes and its complications. *Nutrients* **2017**, *9*, 163. [CrossRef]
8. Vinayagam, R.; Jayachandran, M.; Xu, B. Antidiabetic Effects of Simple Phenolic Acids: A Comprehensive Review. *Phytotherapy Res.* **2016**, *30*, 184–199. [CrossRef]
9. Lee, E.J.; Kim, S.R.; Kim, J.; Kim, Y.C. Hepatoprotective phenylpropanoids from Scrophularia buergeriana roots against CCl4-induced toxicity: Action mechanism and structure-activity relationship. *Planta Med.* **2002**, *68*, 407–441. [CrossRef]
10. Fernandez-Martinez, E.; Bobadilla, R.; Morales-Rios, M.; Perez-Alvarez, P. Trans-3-Phenyl-2-Propenoic Acid (Cinnamic Acid) Derivatives: Structure-Activity Relationship as Hepatoprotective Agents. *Med. Chem.* **2007**, *3*, 475–479. [CrossRef]
11. Kim, S.R.; Kim, Y.C. Neuroprotective phenylpropanoid esters of rhamnose isolated from roots of Scrophularia buergeriana. *Phytochemistry* **2000**, *54*, 503–509. [CrossRef]

12. Kim, S.R.; Sung, S.H.; Jang, Y.P.; Markelonis, G.J.; Oh, T.H.; Kim, Y.C. E-p-Methoxycinnamic acid protects cultured neuronal cells against neurotoxicity induced by glutamate. *Br. J. Pharmacol.* **2002**, *135*, 1281–1291. [CrossRef] [PubMed]
13. Kim, S.R.; Kang, S.Y.; Lee, K.Y.; Kim, S.H.; Markelonis, G.J.; Oh, T.H.; Kim, Y.C. Anti-amnestic activity of E-p-methoxycinnamic acid from Scrophularia buergeriana. *Cogn. Brain Res.* **2003**, *17*, 454–461. [CrossRef]
14. Khan, M.G.U.; Nahar, K.; Rahman, M.S.; Hasan, C.M.; Rashid, M.A. Phytochemical and biological investigations of Curcuma longa. *J. Pharm. Sci.* **2009**, *8*, 39–45. [CrossRef]
15. Zadernowski, R.; Naczk, M.; Nesterowicz, J. Phenolic acid profiles in some small berries. *J. Agric. Food Chem.* **2005**, *53*, 2118–2124. [CrossRef]
16. Andrade, P.B.; Leitão, R.; Seabra, R.M.; Oliveira, M.B.; Ferreira, M.A. 3,4-Dimethoxycinnamic acid levels as a tool for differentiation of Coffea canephora var. robusta and Coffea arabica. *Food Chem.* **1998**, *61*, 511–514. [CrossRef]
17. Serna, M.; Wong-Baeza, C.; Santiago-Hernández, J.; Carlos Baeza, I.; Wong, C. Hypocholesterolemic and choleretic effects of three dimethoxycinnamic acids in relation to 2,4,5-trimethoxycinnamic acid in rats fed with a high-cholesterol/cholate diet. *Pharmacol. Rep.* **2015**, *67*, 553–559. [CrossRef]
18. Zanyatkin, I.; Stroylova, Y.; Tishina, S.; Stroylov, V.; Melnikova, A.; Haertle, T.; Muronetz, V. Inhibition of Prion Propagation by 3,4-Dimethoxycinnamic Acid. *Phytother. Res.* **2017**, *31*, 1046–1055. [CrossRef]
19. Farrell, T.L.; Gomez-Juaristi, M.; Poquet, L.; Redeuil, K.; Nagy, K.; Renouf, M.; Williamson, G. Absorption of dimethoxycinnamic acid derivatives in vitro and pharmacokinetic profile in human plasma following coffee consumption. *Mol. Nutr. Food Res.* **2012**, *56*, 1413–1423. [CrossRef]
20. Wen, X.; Walle, T. Methylated Flavonoids Have Greatly Improved Intestinal Absorption and Metabolic Stability. *Drug Metab. Dispos.* **2006**, *34*, 1786–1792. [CrossRef]
21. Walle, U.K.; Walle, T. Bioavailable Flavonoids: Cytochrome P450-Mediated Metabolism of Methoxyflavones. *Drug Metab. Dispos.* **2007**, *35*, 1985–1989. [CrossRef] [PubMed]
22. Han, S.; Hu, L.; Quach, T.; Simpson, J.S.; Trevaskis, N.L.; Porter, C.J.H. Profiling the Role of Deacylation-Reacylation in the Lymphatic Transport of a Triglyceride-Mimetic Prodrug. *Pharm. Res.* **2014**, *32*, 1830–1844. [CrossRef] [PubMed]
23. Han, S.; Hu, L.; Gracia, G.; Quach, T.; Simpson, J.S.; Edwards, G.A.; Trevaskis, N.L.; Porter, C.J.H. Lymphatic Transport and Lymphocyte Targeting of a Triglyceride Mimetic Prodrug Is Enhanced in a Large Animal Model: Studies in Greyhound Dogs. *Mol. Pharm.* **2016**, *13*, 3351–3361. [CrossRef] [PubMed]
24. Trevaskis, N.L.; Kaminskas, L.M.; Porter, C.J.H. From sewer to saviour—Targeting the lymphatic system to promote drug exposure and activity. *Nat. Rev. Drug Discov.* **2015**, *14*, 781–803. [CrossRef] [PubMed]
25. Barzaghi, N.; Crema, F.; Gatti, G.; Pifferi, G.; Perucca, E. Pharmacokinetic studies on IdB 1016, a silybin-phosphatidylcholine complex, in healthy human subjects. *Eur. J. Drug Metab. Pharmacokinet.* **1990**, *15*, 333–338. [CrossRef] [PubMed]
26. Czarnecka, M.; Świtalska, M.; Wietrzyk, J.; Maciejewska, G.; Gliszczyńska, A. Synthesis and biological evaluation of phosphatidylcholines with cinnamic and 3-methoxycinnamic acids with potent antiproliferative activity. *RSC Adv.* **2018**, *8*, 35744–35752. [CrossRef]
27. Czarnecka, M.; Świtalska, M.; Wietrzyk, J.; Maciejewska, G.; Gliszczyńska, A. Synthesis, Characterization, and In Vitro Cancer Cell Growth Inhibition Evaluation of Novel Phosphatidylcholines with Anisic and Veratric Acids. *Molecules* **2018**, *23*, 2022. [CrossRef]
28. Laane, C.; Boeren, S.; Vos, K.; Veeger, C. Rules for optimization of biocatalysis in organic solvents. *Biotechnol. Bioeng.* **1987**, *30*, 81–87. [CrossRef]
29. Sabally, K.; Karboune, S.; St-Louis, R.; Kermasha, S. Lipase-catalyzed transesterification of dihydrocaffeic acid with flaxseed oil for the synthesis of phenolic lipids. *J. Biotechnol.* **2006**, *127*, 167–176. [CrossRef]
30. Sabally, K.; Karboune, S.; St-Louis, R.; Kermasha, S. Lipase-catalyzed synthesis of phenolic lipids from fish liver oil and dihydrocaffeic acid. *Biocatal. Biotransformations* **2007**, *25*, 211–218. [CrossRef]
31. Sun, S.; Shan, L.; Liu, Y.; Jin, Q.; Song, Y.; Wang, X. Solvent-free enzymatic synthesis of feruloylated diacylglycerols ferulated diacylglycerols and kinetic study. *J. Mol. Catal. B Enzym.* **2009**, *57*, 104–108. [CrossRef]
32. Guyot, B.; Bosquette, B.; Pina, M.; Graille, J. Esterification of phenolic acids from green coffe with an immobilized lipase from Candida antarctica in solvent-free medium. *Biotechnol. Lett.* **1997**, *19*, 529–553. [CrossRef]

33. Karboune, S.; St-Louis, R.; Kermasha, S. Enzymatic synthesis of structured phenolic lipids by acidolysis of flaxseed oil with selected phenolic acids. *J. Mol. Catal. B Enzym.* **2008**, *52*, 96–105. [CrossRef]
34. Rychlicka, M.; Maciejewska, G.; Niezgoda, N.; Gliszczyńska, A. Production of feruloylated lysophospholipids via a one-step enzymatic interesterification. *Food Chem.* **2020**, *323*, 126802. [CrossRef]
35. Okulus, M.; Gliszczyńska, A. Enzymatic synthesis of O-methylated phenophospholipids by lipase-catalyzed acidolysis of egg-yolk phosphatidylcholine with anisic and veratric acids. *Catalysts* **2020**, *10*, 538. [CrossRef]
36. Takahashi, K.; Hosokawa, M. Production of tailor-made polyunsaturated phospholipids through bioconversions. *J. Liposome Res.* **2001**, *11*, 343–353. [CrossRef]
37. Hogovest, P. Review—An update on the use of oral phospholipid excipients. *Eur. J. Pharm. Sci.* **2017**, *108*, 1–12. [CrossRef]
38. Suzuki, S.; Yamatoka, H.; Sakai, M.; Kataoka, A.; Furushiro, M.; Kudo, S. Oral administration of soybean lecithin transphosphatidylated phosphatidylserine improves memory impairment in aged rats. *J. Nutr.* **2001**, *131*, 2951–2956. [CrossRef]
39. Gawas, S.D.; Khan, N.; Rathod, V.K. Application of response surface methodology for lipase catalyzed synthesis of 2-ethylhexyl palmitate in solvent free system using ultrasound. *Braz. J. Chem. Eng.* **2019**, *36*, 1007–10017. [CrossRef]
40. Jumbri, K.; Rozy, M.F.A.; Ashari, S.E.; Mohamad, R.; Basri, M.; Masoumi, H.R.F. Optimisation and Characterisation of Lipase Catalysed Synthesis of Kojic Monooleate Ester in a Solvaent-Free System by Response Surface Methodology. *PLoS ONE* **2015**, *10*, e0144664. [CrossRef]
41. Ferreira, S.L.C.; Bruns, R.E.; Ferreira, H.S.; Matos, G.D.; David, J.M.; Brandao, G.C.; Silva, E.G.P.; Portugal, L.A.; Reis, P.S.; Souza, A.S.; et al. Box-Behnken design: An alternative for optimization of analytical methods. *Anal. Chim. Acta* **2007**, *597*, 179–186. [CrossRef] [PubMed]
42. Holm, H.C.; Cowan, W.D. The evolution of enzymatic interesterification in the oils and fats industry. *Eur. J. Lipid Sci. Technol.* **2008**, *110*, 679–691. [CrossRef]
43. Karam, R.; Karboune, S.; St-Louis, R.; Kermasha, S. Lipase-catalyzed acidolysis of fish liver oil with dihydroxyphenylacetic acid in organic solvent media. *Process Biochem.* **2009**, *44*, 1193–1199. [CrossRef]
44. Rychlicka, M.; Niezgoda, N.; Gliszczyńska, A. Lipase-Catalyzed Acidolysis of Egg-Yolk Phosphatidylcholine with Citronellic Acid. New Insight into Synthesis of Isoprenoid-Phospholipids. *Molecules* **2018**, *23*, 314. [CrossRef] [PubMed]
45. Yang, H.; Mu, Y.; Chen, H.; Xiu, Z.; Yang, T. Enzymatic synthesis of feruloylated lysophospholipid in a selected organic solvent medium. *Food Chem.* **2013**, *141*, 3317–3322. [CrossRef] [PubMed]
46. Li, Y.; Lu, J.; Gu, G.; Mao, Z. Characterization of the enzymatic degradation of arabinoxylans in grist containing wheat malt using response surface methodology. *J. Am. Soc. Brew. Chem.* **2005**, *63*, 171–176.
47. Yang, T.; Xub, X.; Hea, C.; Lic, L. Lipase-catalyzed modification of lard to produce human milk fat substitutes. *Food Chem.* **2003**, *80*, 473–481. [CrossRef]
48. Chen, B.; Liu, H.; Guo, Z.; Huang, J.; Wang, M.; Xu, X.; Zheng, L. Lipase-Catalyzed Esterification of Ferulic Acid with Oleyl Alcohol in Ionic Liquid/Isooctane Binary Systems. *J. Agric. Food Chem.* **2011**, *59*, 1256–1263. [CrossRef]
49. Li, N.G.; Shi, Z.H.; Tang, Y.P.; Li, B.Q.; Duan, J.A. Highly efficient esterification of ferulic acid under microwave irradiation. *Molecules* **2009**, *14*, 2118–2126. [CrossRef]
50. Jagdale, A.R.; Reddy, R.S.; Sudalai, A.A. concise enantioselective synthesis of 1-[(S)-3-(dimethylamino)-3,4 dihydro-6,7-dimethoxyquinolin-1(2H)-yl]propan-1-one, (S)-903. *Tetrahedron Asymmetry* **2009**, *20*, 335–339. [CrossRef]
51. Gliszczyńska, A.; Niezgoda, N.; Gładkowski, W.; Czarnecka, M.; Świtalska, M.; Wietrzyk, J. Synthesis and Biological Evaluation of Novel Phosphatidylcholine Analogues Containing Monoterpene Acids as Potent Antiproliferative Agents. *PLoS ONE* **2016**, *11*, e0157278. [CrossRef] [PubMed]

© 2020 by the authors. Licensee MDPI, Basel, Switzerland. This article is an open access article distributed under the terms and conditions of the Creative Commons Attribution (CC BY) license (http://creativecommons.org/licenses/by/4.0/).

Article

Hepatoprotective Effects of *Pleurotus ostreatus* Protein Hydrolysates Yielded by Pepsin Hydrolysis

Liwei Zhang [1], Yuxiao Lu [2], Xiaobin Feng [3], Qinghong Liu [3,*], Yuanhui Li [3], Jiamin Hao [3], Yanqiong Wang [3], Yongqiang Dong [3] and Huimin David Wang [4,5,6,7,*]

1. Department of Microbiology and Immunology, China Agricultural University, Beijing 100193, China; Biozhangliwei@163.com
2. Department of Environment and Chemical Engineering, Tangshan College, Tangshan 063000, China; luyuxiao2006@126.com
3. Department of Vegetables, College of Horticulture, China Agriculture University, Beijing 100193, China; fengxiaobin@126.com (X.F.); lyh188116@163.com (Y.L.); hjm20010221@163.com (J.H.); wyq116113@163.com (Y.W.); yqdong96@163.com (Y.D.)
4. Graduate Institute of Biomedical Engineering, National Chung Hsing University, Taichung 402, Taiwan
5. Graduate Institute of Medicine, College of Medicine, Kaohsiung Medical University, Kaohsiung 807, Taiwan
6. Department of Medical Laboratory Science and Biotechnology, China Medical University, Taichung City 404, Taiwan
7. College of Food and Biological Engineering, Jimei University, Xiamen 361021, China
* Correspondence: qhliu@cau.edu.cn (Q.L.); davidw@dragon.nchu.edu.tw (H.D.W.); Tel.: +86-10-6273-2578 (Q.L.); +886-4-22840733 (ext. 651) (H.D.W.); Fax: +886-4-22852242 (H.D.W.)

Received: 23 April 2020; Accepted: 9 May 2020; Published: 26 May 2020

Abstract: *Pleurotus ostreatus* protein extract (POPE) was prepared by the alkali precipitation method with 0.3% (w/v) NaOH. POPEP-III with a MW of 3000–5000 Da was acquired by pepsin enenzymatic hydrolysis. POPEP-III displayed noteworthy effects of 1,1-diphenyl-2-picrylhydrazyl DPPH and H_2O_2 scavenging activities, Fe^{2+} chelating ability, lipid peroxidation inhibition capacity, and metal reducing power. The administration of POPEP-III in mice significantly prevented prior CCl_4-induced strengthen serum ALT and AST activities, changing from 365.44 ± 36.87 IU/L to 220.23 ± 22.27 IU/L and 352.52 IU/L to 206.75 ± 17.26 IU/L, respectively ($p < 0.001$), and suppressed hepatic malondialdehyde (MDA) formation from 15.28 ± 3.47 nmol/mg prot to 10.04 ± 2.06 nmol/mg prot ($p < 0.001$). Mice treated with POPEP-III demonstrated augmented activities of superoxide dismutase (SOD) in the liver, from 187.49 ± 19.81 U/mg prot to 233.35 ± 34.23 U/mg prot, and of glutathione peroxidase (GSH-Px), from 84.01 ± 14.54 U/mg prot to 115.9 ± 16.57 U/mg prot ($p < 0.05$). POPEP-III also prevented CCl_4-induced oxidative liver histological alteration. The results suggest that POPEP-III can protect the liver from CCl_4-induced oxidative damage.

Keywords: *Pleurotus ostreatus*; enenzymatic hydrolysis; peptide; antioxidant; hepatoprotective activity

1. Introduction

As one of the most important organs, the liver plays crucial roles in protein, lipid, saccharide, and drug metabolism in the human body [1,2]. When the liver is exposed to excessive dietary fats, sugars and proteins, alcohols, or drugs, the heavy burden applied on the liver creates difficulty and might lead to liver diseases. In liver metabolism, some of these substances are transformed into free radical-mediated ROS by cytochrome P450 [3]. Liver diseases are common, inducing death at a high ratio of approximately 2% of all death [4]. Liver diseases, such as subclinical icteric hepatitis, hepatic fibrosis, cirrhosis and hepatocellular carcinoma, are related to free radical-mediated ROS. ROS can be generated in the liver mitochondria [5] and participate in the process of liver disease induction and development [6–8].

Free radical-mediated reactive oxygen species (ROS) have attracted appreciable attention based on their role in liver diseases and current methods for scavenging of ROS. CCl_4, a classical hepatotoxin, is reductively de-halogenated by cytochrome P450 to trichloromethyl free radicals ($CCl_3OO\cdot$ or $\cdot CCl_3$) and depletion of antioxidant enzymes in hepatic parenchyma cells and initiates the process of lipid peroxidation [8]. To protect the liver from damaging oxidation, numerous antioxidants have been explored. Aqueous, alcoholic, or supercritical CO_2 extracts with antioxidant activity from selected medicinal herbs demonstrated hepatoprotective effects, although the substances in these extracted mixtures are structurally unclear [9–12]. One antioxidant chemical, silymarin from the *Silybum marianum* seeds, is commonly applied for hepatoprotection [13] and has side effects that include gastrointestinal complaints. Currently, mushrooms are increasingly under examination for their hepatoprotective values. Among the mushrooms explored for hepatoprotective substances, polysaccharides were ranked as the predominant component, and certain mixtures extracted from mushrooms by alcohol, water or other solvents have become topics of research [14–16]. Notably little research has focused on the hepatoprotective peptides from mushrooms [17], especially the hydrolysates from mushroom proteins.

Pleurotus ostreatus is widely cultivated around the world and is reported to contain bioactive compounds that can protect health due to their antioxidant properties. Among these compounds, *P. ostreatus* polysaccharides have attracted increasing attention [18,19]. Selected compounds extracted by ethanol from *P. ostreatus* showed high antioxidant activities, but the constituent was unclear [20,21]. An extraction with water from *P. ostreatus* solid culture presented immune-nutritional recovery [22]. Polysaccharides from *P. ostreatus* are of interest due to their multiple bioactivities, including antitumor, anti-inflammatory and cytotoxic effects [23]. Proteins in *P. ostreatus* have many bioactivities, including anti-tumor effects and use as an adjuvant for HBV DNA vaccine [24,25]. However, most of the previous works were relative to the nutritional evaluation for mushroom proteins, and few current studies focus on the bioactive natural component of low molecular weight peptides with multi-action bioactivity hydrolyzed by enzymes from the *P. ostreatus* proteins or other edible mushrooms. As the output of *P. ostreatus* is tremendous, it is very important to develop the procedure of health-keeping to extend the industry chain, and improve people's health with the natural products respectively. In our study, the antioxidant activity of *P. ostreatus* protein hydrolysates obtained with five proteases was investigated in vitro using biochemical assays. The highest antioxidant activity fractionated hydrolysates were refined through ultrafiltration. Moreover, the fraction with the highest antioxidant activity was investigated via an in vivo model of CCl_4 intoxication. Physical characteristics, such as the degree of hydrolysis and amino acid composition of the hydrolysates fractions, were explored.

2. Results

2.1. Degree of Hydrolysis

Different efficiency was acquired in hydrolyzing POPE for the peptide bonds, as the five enzymes recognize different bonds and function differently. Among pepsin, trypsin, dispase, papain, and bromelin, pepsin had the highest efficiency in POPE hydrolysis (Table 1). The statistics indicated a mild DH of 16.01 ± 0.12% after POPE incubation with pepsin. Hydrolysis proceeded for 1 h at 37 °C and with pH 2.0. DH is defined as the percentage of peptide bonds cleaved. It was well established that DH can affect the functionality of hydrolysates, including antioxidant activity or reducing power. Generally, these effects become more evident when DH is modified more extensively.

Table 1. Degree of Hydrolysis for POPE hydrolyzed by Pep, Pepsin-treated; Try, trypsin-treated; Dis, dispase-treated; Pap, papain-treated and Bro, bromelin-treated. The fractions (mw < 10 kDa) were designed as peptide, and the DH was calculated as DH = (weight of fraction (5–10 kDa) + weight of fraction(3–5 kDa) + weight of fraction (<3 kDa))/weight of POPE. The values are presented as the mean ± SD.

	POPE (mg)	>10 kDa (mg)	5–10 kDa (mg)	3–5 kDa (mg)	<3 kDa (mg)	DH (%)
Pep	500	319.96 ± 3.15	45.76 ± 1.15	23.15 ± 0.86	11.14 ± 0.19	16.01 ± 0.12
Try	500	431.77 ± 5.31	34.79 ± 2.19	19.68 ± 1.17	10.32 ± 1.39	12.96 ± 0.33
Dis	500	429.63 ± 7.31	44.39 ± 0.98	17.93 ± 2.87	9.68 ± 2.63	14.40 ± 0.19
Pap	500	434.75 ± 4.39	39.37 ± 2.33	20.19 ± 1.75	10.17 ± 3.18	13.95 ± 0.22
Bro	500	430.07 ± 8.63	40.33 ± 1.77	19.47 ± 0.96	8.65 ± 1.19	13.67 ± 0.96

2.2. Amino Acid Composition

The amino acid compositions of the hydrolysates suggested that they are rich in Leu, Phe, Glu, Met, Ala, Val, and Asp (Table 2). In addition, the THAA content in POPEP-III was higher than the other three fractions. For protein hydrolysate and peptides, the higher content in hydrophobicity the higher solubility in lipid they have. Moreover, high solubility enhances their antioxidant activity [26]. These highest antioxidant activities were presumably because of the highest content of THAA.

Table 2. Amino acid compositions of POPE and ultrafiltrated POPEP fractions.

Amino Acid (g/100 g)		POPE	POPEP-I (>10 kDa)	POPEP-II (5–10 kDa)	POPEP-III (3–5 kDa)	POPEP-IV (<3 kDa)
Hydrophobic	Phe	8.02	7.89	5.39	14.27	7.78
	Leu	8.56	8.82	5.84	7.11	11.6
	Ala	5.67	5.83	4.39	5.13	6.13
	Met	13.8	11.04	11.95	11.52	13.02
	Val	6.52	5.57	5.08	6.05	5.99
	Ile	5.45	4.84	4.17	4.94	4.84
	Gly	3.96	3.62	4.75	4.48	3.08
	Pro	3.53	2.57	3.51	4.02	3.19
Hydrophilic	Thr	3.09	4.40	2.89	3.53	3.48
	Ser	3.64	3.84	3.94	3.23	4.27
	Cys	2.35	2.20	4.35	1.88	1.21
	Tyr	6.74	8.20	4.91	4.66	4.6
	Lys	3.1	2.75	3.19	2.27	1.94
	Arg	4.28	4.28	3.81	3.35	4.6
	His	1.39	1.45	1.39	1.09	1
	Asp	6.74	7.22	6.71	7.21	6.18
	Glu	9.2	8.91	13.37	11.32	8.31

2.3. Preparation of Protein Hydrolysatess and Antioxidant Activity

Five different proteases, i.e., pepsin, trypsin, dispase, papain and bromelin, originating from different sources were used in the enzymatic hydrolysis of POPE. The antioxidant activities of the hydrolysate hydrolyzed by each protease are shown in Figure 1 and Table 3. Among the hydrolysates, POPEP derived from POPE hydrolyzed by pepsin showed the strongest antioxidant activity. This suggested that when disrupted by proteases, the active amino acids of protein might be exposed, and the antioxidant activity increases [27].

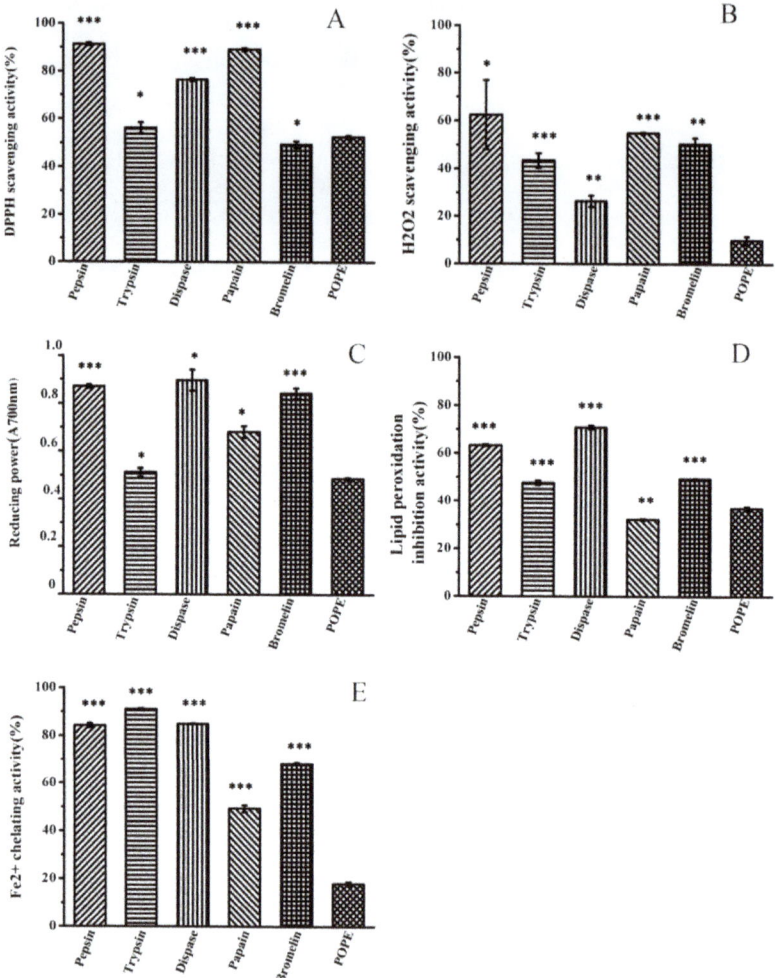

Figure 1. Comparison of in vitro antioxidant effects of *P. ostreatus* protein hydrolysate: Pep, Pepsin-treated; Try, trypsin-treated; Dis, dispase-treated; Pap, papain-treated and Bro, bromelin-treated. (**A**) DPPH scavenging activity. (**B**) Hydrogen peroxide scavenging activity. (**C**) Reducing power. (**D**) Inhibitory effect on lipid peroxidation. (**E**) Fe^{2+} ion chelating activity. The values are presented as the mean values ± SD (n = 3). (***) $p < 0.001$, (**) $p < 0.01$, and (*) $p < 0.05$ compared with the POPE group.

POPEP was further separated using three kinds of UF membranes (10,000, 5000, and 3000 Da MWCO) according to molecular size, and four kinds of fractions, POPEP-I (>10,000 Da), POPEP-II (5000–10,000 Da), POPEP-III (3000–5000 Da) and POPEP-IV (<3000 Da) were obtained. POPEP-III had the highest antioxidant activity among the four fractions (Figure 2, Table 4). The scavenging activity of POPEP-III for DPPH and H_2O_2 was the highest among the samples. The reducing power and Fe^{2+} chelating activity of POPEP-III exhibited the highest antioxidant activity among these samples. However, all the samples had no significant differences in inhibitory effects on lipid peroxidation. POPEP-III has certain similarities with peptides with alcalase-derived hydrolysate from alfalfa leaf protein [28]. Considering that POPEP-III was found to possess the highest antioxidant activity, this fraction was analyzed for molecular weight distribution. The chromatographic data suggested that POPEP-III was composed of molecular weights of 4076 Da and 3287 Da two major fractions.

Table 3. Comparison of in vitro antioxidant effects of *P. ostreatus* protein hydrolysate: Pep, Pepsin-treated; Try, trypsin-treated; Dis, dispase-treated; Pap, papain-treated and Bro, bromelin-treated. The values are presented as the mean ± SD.

	DPPH Scavenging Activity (%)	H_2O_2 Scavenging Activity (%)	Reducing Power (A700nm)	Lipid Peroxidantion Inhibitionactivity (%)	Ferrous Ion Chelating Activity (%)
Pep	91.32 ± 0.60 ***	62.52 ± 14.42 *	0.87 ± 0.008 ***	63.34 ± 0.34 ***	84.28 ± 0.92 ***
Try	56.27 ± 2.30 *	43.47 ± 3.09 ***	0.51 ± 0.02 *	47.56 ± 0.88 ***	91.06 ± 0.36 ***
Dis	76.51 ± 0.68 ***	26.37 ± 2.28 ***	0.90 ± 0.04 *	70.89 ± 0.78 ***	84.98 ± 0.19 ***
Pap	89.19 ± 0.66 ***	54.93 ± 0.42 ***	0.68 ± 0.02 *	32.17 ± 0.37 **	49.37 ± 1.44 ***
Bro	49.46 ± 1.39 *	50.34 ± 2.67 **	0.84 ± 0.02 ***	49.26 ± 0.10 ***	68.22 ± 0.42 ***
POPE	52.64 ± 0.62	10.01 ± 1.84	0.49 ± 0.004	36.89 ± 0.66	18.05 ± 0.81

(***) $p < 0.001$, (**) $p < 0.01$, and (*) $p < 0.05$ compared with the POPE group.

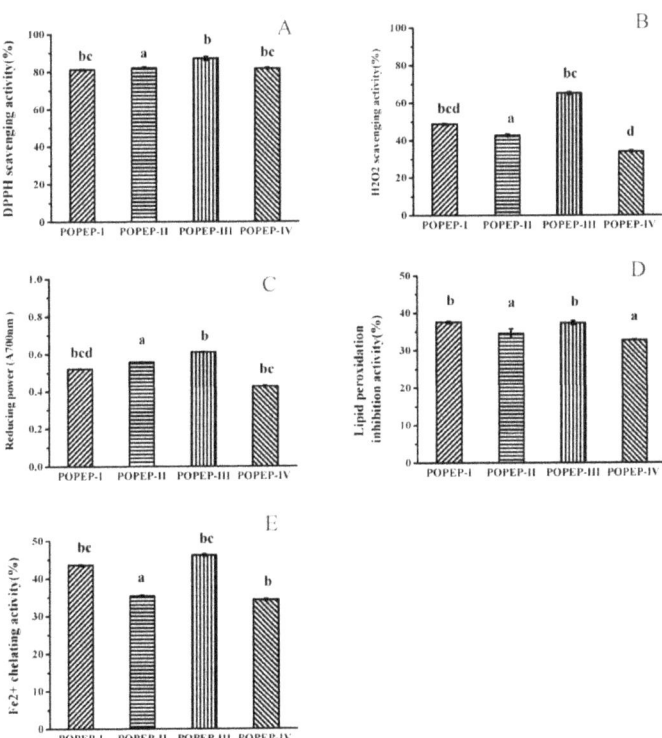

Figure 2. Comparison of in vitro antioxidant effects of different POPEP fractions: (**A**) DPPH scavenging activity. (**B**) Hydrogen peroxide scavenging activity (**C**) Reducing power. (**D**) Inhibitory effect on lipid peroxidation. (**E**) Fe^{2+} chelating activity. The values are presented as the mean values ± SD (n = 3), and the same letters above the bars indicate statistically insignificant differences between the groups ($p < 0.05$).

Table 4. Comparison of in vitro antioxidant effects of different POPEP fractions. The values are presented as the mean ± SD (n = 3). The same letters after the mean indicate statistically insignificant differences between the groups ($p < 0.05$).

	DPPH Scavenging Activity (%)	H2O2 Scavenging Activity (%)	Reducing Power (A700nm)	Lipid Peroxidantion Inhibitionactivity (%)	Ferrous Ion Chelating Activity (%)
POPEP-I	81.26 ± 0.40 [b,c]	48.79 ± 0.34 [b,c,d]	0.52 ± 0.004 [b,c,d]	37.53 ± 0.24 [b]	43.56 ± 0.31 [b,c]
POPEP-II	82.1 ± 1.00 [a]	42.87 ± 0.60 [a]	0.56 ± 0.04 [a]	34.56 ± 1.21 [a]	35.28 ± 0.35 [b,c]
POPEP-III	87.15 ± 0.55 [b]	65.22 ± 0.78 [b,c]	0.61 ± 0.004 [b]	37.38 ± 0.50 [b]	46.19 ± 0.39 [b,c]
POPEP-IV	81.73 ± 0.37 [b,c]	34.18 ± 0.50 [d]	0.43 ± 0.08 [b,c]	32.77 ± 0.09 [a]	34.25 ± 0.35 [b]

[a-c] values followed by different letters in the table are significantly different ($p < 0.05$), where c is the lowest value. Data are presented as mean ± standard deviation (n = 7).

2.4. Effects of POPEP-III on Liver Weight, HI and Body Weight in Mice

The changes in liver weight, HI and body weight, in different experimental groups of mice are shown in Table 5. The liver weight ($p < 0.05$) and HI ($p < 0.05$) increased significantly compared with the normal control after CCl_4 treatment. However, it could be mitigated by pretreatment with POPEP-III, similar to the results of pretreatment with bifendate. Additionally, from the data, we note that the HI was partially dose-dependent on POPEP-III. A decrease in body weight was observed in the CCl_4-treated groups ($p < 0.05$), but to a certain extent, POPEP-III could alleviate the weight decrease.

Table 5. Effects of POPEP-III on body weight, liver weight, and hepatosomatic index (HI) in CCl_4-treated mice.

Treatments	Dose (mg/kg)	Body Weight (g)	Liver Weight (g)	HI (%)
Normal	-	34.32 ± 2.98 [a]	1.21 ± 0.14 [b]	35.1 ± 2.27 [c]
CCl_4 only	-	33.35 ± 2.15 [a]	1.43 ± 0.27 [a]	42.49 ± 5.29 [a]
CCl_4 + Bifendate	200	34.32 ± 2.35 [a]	1.24 ± 0.1 [b]	36.1 ± 1.86 [c]
CCl_4 + POPEP-III	100	34.02 ± 2.28 [a]	1.26 ± 0.13 [a]	36.36 ± 4.45 [c]
CCl_4 + POPEP-III	200	34.02 ± 2.28 [a]	1.26 ± 0.13 [a]	36.97 ± 2.54 [c]
CCl_4 + POPE	200	32.83 ± 2.41 [a]	1.26 ± 0.16 [a]	38.21 ± 2.61 [b]

[a-c] values followed by different letters in the column are significantly different ($p < 0.05$), where c is the lowest value. Data are presented as mean ± standard deviation (n = 7).

2.5. Effects of POPEP-III on AST, ALT, MDA, GSH-Px, and SOD Activities

Some acquainted enzymes in serum, such as ALT and AST, were used as biochemical criteria for early acute hepatic damage. As displayed in Figure 3 and Table 6, CCl_4 treatment significantly increased the activities of ALT and AST in serum. The activity of ALT is 24.68 ± 5.64 IU/L and that of AST is 44.62 ± 5.25 IU/L in group I (the normal group). However, the activities of ALT and AST are 365.44 ± 36.87 IU/L and 352.52 ± 13.44 IU/L respectively, in group II treated with CCl_4 only ($p < 0.001$). The pretreatment of POPEP-III reduced the activities of ALT and AST in a dose-dependent manner as caused by CCl_4 ($p < 0.05$ vs CCl_4-intoxicated group II). At a dose of 200 mg/kg.bw POPEP-III reduced the ALT and AST activities to 220.23 ± 22.27 IU/L and 206.75 ± 17.26 IU/L, respectively, showing less efficiency than bifendate (200 mg/kg.bw), which decreased ALT and AST activities to 114.18 ± 18.36 IU/L and 144.57 ± 21.81 IU/L, respectively.

CCl_4 affected the antioxidant parameters of the liver tissue of mice. As shown in Figure 4A and Table 7, MDA, the product of membrane lipid peroxidation, its levels increased from 5.28 ± 1.55 nmol/mg prot in group I to 15.28 ± 3.47 nmol/mg prot ($p < 0.001$) in group II. When treated with POPEP-III (100 mg/kg, 200 mg/kg), levels of MDA decreased to 12.07 ± 2.16 nmol/mg prot in group IV and to 10.04 ± 2.06 nmol/mg prot in group V, but all are higher than the level of group III treated with bifendate. More interestingly, the level of group VI treated with POPE was the same as that of group II and did not demonstrate positive effects. These results indicate that hydrolysatess are more effective in membrane protection than unhydrolysated POPE but less effective than bifendate.

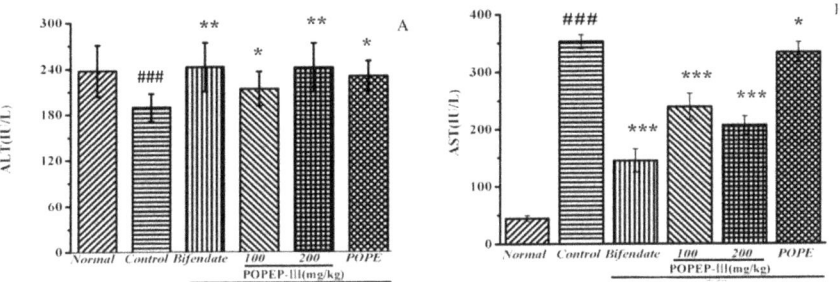

Figure 3. Effects of POPEP-III on the activities of serum ALT (**A**) and AST (**B**) in mice. Values are expressed as the mean values ± SD (n = 7): (###) $p < 0.001$ compared with the normal group; (***) $p < 0.001$, (**) $p < 0.01$, and (*) $p < 0.05$ compared with the CCl$_4$-treated control group.

Table 6. Effects of POPEP-III on the activities of serum ALT and AST in mice.

	ALT (IU/L)	AST (IU/L)
Normal	24.68 ± 5.64	44.62 ± 5.25
Control	365.44 ± 36.87 ###	352.52 ± 13.44 ###
Bifendate	114.18 ± 18.36 **	144.57 ± 21.81 ***
POPEP-III 100	312.53 ± 24.04 *	238.83 ± 25.17 ***
POPEP-III 200	220.23 ± 22.27 **	206.75 ± 17.26 ***
POPE	341.93 ± 27.56 *	333.40 ± 19.04 *

Values are presented as the mean ± SD: (###) $p < 0.001$ compared with the normal group; (***) $p < 0.001$, (**) $p < 0.01$, and (*) $p < 0.05$ compared with the CCl$_4$-treated control group.

SOD and GSH-Px protect liver cells from oxidative damage by scavenging superoxide anions. In this work (Figure 4B,C and Table 7), an apparent decrease was observed in the SOD level from 237.65 ± 36.06 U/mg prot ($p < 0.01$) (Group I) to 187.49 ± 19.81 U/mg prot (Group II) and in GSH-Px level from 152.11 ± 12.58 U/mg prot ($p < 0.01$) (Group I) to 84.01 ± 14.54 U/mg prot (Group II). When treated with POPEP-III at a dosage of 100 mg/kg, the contant of SOD and GSH-Px in Group IV are similar to the level of Group III treated with bifendate, but the contant of SOD and GSH-Px in Group V are lower compared with Group III. Similar to the effect of MDA adjustment, POPE did not demonstrate the positive effect.

As expected, a significant decrease of the MDA and increase of the GSH-Px and SOD enzyme activities occurred with pretreatment of POPEP-III. MDA content, SOD and GSH-Px activity reached 10.04 ± 2.06 nmol/mg prot, 233.35 ± 34.23 U/mg prot and 115.9 ± 16.57 U/mg prot ($p < 0.05$), respectively, and 100 mg/kg.bw of POPEP-III had a relatively slighter influence.

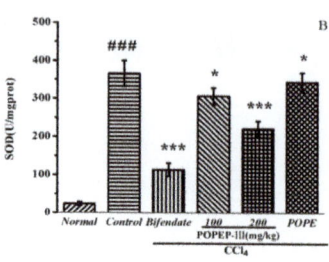
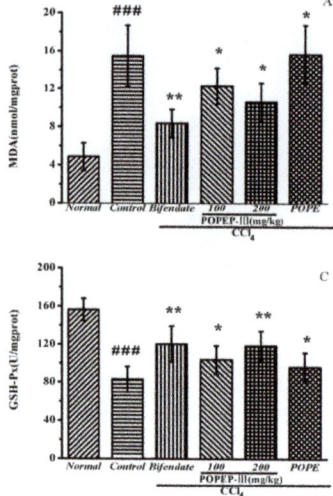

Figure 4. Effects of POPEP-III on hepatic MDA (**A**), SOD (**B**) and GSH-Px (**C**) in mice. Values are expressed as the mean values ± SD (n = 7): (###) $p < 0.001$ compared with normal group; (***) $p < 0.001$, (**) $p < 0.01$, and (*) $p < 0.05$ compared with the CCl$_4$-treated control group.

Table 7. Effects of POPEP-III on hepatic MDA, SOD and GSH-Px in mice.

	MDA (nmol/mg prot)	SOD (U/mg prot)	GSH-Px (U/mg prot)
Normal	5.28 ± 1.55	237.65 ± 36.06	152.11 ± 12.58
Control	15.28 ± 3.47 ###	187.49 ± 19.81 ###	84.01 ± 14.54 ###
Bifendate	8.9 ± 1.58 **	240.11 ± 34.22 ***	118.12 ± 20.44 **
POPEP-III 100	12.07 ± 2.16 *	205.65 ± 24.31 *	100.86 ± 16.37 *
POPEP-III 200	10.04 ± 2.06 *	233.35 ± 34.23 ***	115.9 ± 16.57 **
POPEP	15.01 ±3.32 *	213.45 ± 21.45 *	86.61 ± 16.35 *

Values are presented as the mean ± SD: (###) $p < 0.001$ compared with normal group; (***) $p < 0.001$, (**) $p < 0.01$, and (*) $p < 0.05$ compared with the CCl$_4$-treated control group.

2.6. Histopathology Examination

In this seaction, histopathological observation of the liver was performed to supply further support for the biochemical analysis. Compared with the hepatic cellular architecture of mice tissue from the normal control group (Figure 5A), administration of CCl$_4$ to mice resulted in extensive liver damage characterized by formation of large vacuoles, infiltration of inflammatory cells, severe cellular degeneration, necrosis and nuclear fragmentation (Figure 5B). However, bifendate pretreatment effectively protected the liver against CCl$_4$-induced damage (Figure 5C), and pathological observation showed only hepatocyte swelling around portal areas. In addition, POPEP-III pretreatment showed slighter hepatic injury and ameliorated liver damage by diminution of large vacuole formation and attenuation of cellular degeneration (Figure 5D,E). For the undigested protein POPE group (Figure 5F), POPE pretreatment did not markedly improve CCl$_4$-induced acute liver injury.

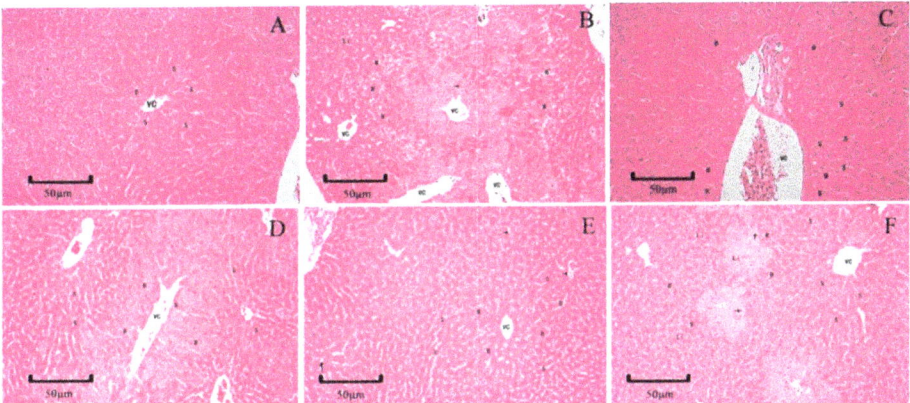

Figure 5. Effects of POPEP-III on CCl$_4$-induced liver histopathological changes in mice (×20 H&E staining): (**A**) Normal group. (**B**) CCl$_4$ only group (240 μL/kg.bw). (**C**) Positive group (200 mg/kg.bw). (**D**) POPEP-III + CCl$_4$ (100 mg/kg.bw). (**E**) POPEP-III + CCl$_4$ (200 mg/kg.bw). (**F**) POPE + CCl$_4$ (200 mg/kg.bw).

3. Materials and Methods

3.1. Chemicals and Reagents

The diagnostic kits for activity assays of ALT, AST, MDA, SOD, and GSH-Px were purchased from Nanjing Jiancheng Bioengineering Institute (Nanjing, China). Linoleic acid, DPPH, BSA, and ferrozine were purchased from Sigma-Aldrich (Steinheim, Germany). The UF membrane reactor system for fractionation of hydrolysate was sourced from Sartoyius AG (Goettingen, Germany). For the proteases used in enzymatic hydrolysis, pepsin, trypsin, and dispase were acquired from Sigma-Aldrich (Steinheim, Germany), while papain and bromelin were purchased from Solarbio (Beijing, China). All of the other chemicals used in this study were of analytical grade.

3.2. Preparation of Crude P. ostreatus Protein

The procedures used for isolating POPE are those from Souza [29] with some modifications. The fresh fruiting bodies of *P. ostreatus* were purchased from a local farm in Beijing. The preparation of *P. ostreatus* protein extract (POPE) involved a multistep procedure that entailed soaking treatment with deionized water, centrifugation, and extraction with NaOH solution. Fresh *P. ostreatus* fruiting bodies were homogenized in deionized water (1:1, w/v) whith a Waring blender. The homogenate was centrifuged (10,000× g, 4 °C, 30 min) after overnight extraction at 4 °C. The supernatant was added with 20% (w/v) NaOH solution until the concentration of NaOH reached 0.3% (w/v) at 25 °C. The dispersion was centrifuged at 10,000× g for 20 min at 4 °C after a 5-h stand at room temperature. The obtained precipitate was dialyzed in deionized water for 72 h to remove free NaOH and was subsequently lyophilized. This product was POPE.

3.3. Enzymatic Hydrolysis of POPE

The lyophilized POPE was dissolved in distilled water to reach a 0.05 g/mL concentration. The pH of the POPE solution was adjusted to the optimal condition for the enzymes used (Table 8). Each enzyme (pepsin, trypsin, dispase, papain, and bromelin) was added proportionally to the POPE solution according to the enzyme/substrate ratio based on the original form of specific enzyme. At the end of the incubation, the solution was heated at 100 °C for 15 min to deactivate the enzyme. Hydrolysate was subsequently obtained by centrifugation (10,000× g, 4 °C, and 10 min). hydrolysate was lyophilized for 96 h at −70 °C and 0.04 Mbar. All hydrolysates powders were stored at −20 °C for further analysis.

Table 8. Optimal reaction conditions of five proteases.

Protease	Activity (U/g)	Optimal Conditions			
		Temp. (°C)	pH	Time (h)	E/S (w/w)
Pepsin (Porcine stomach mucosa)	30,000	37	2.0	3	0.5%
Trypsin (Porcine pancreas)	25,000	45	8.0	5	0.5%
Dispase (Bacillus subtilis)	>60,000	50	7.0	5	0.5%
Papain (Carica papaya)	800,000	45	6.2	5	0.5%
Bromelin (Pineapple)	600,000	45	6.0	5	0.5%

3.4. Ultrafiltration

POPEP was sequentially filtrated using three different MWCO ranges UF membranes of 10,000, 5000, and 3000 Da, respectively. The fractions were designed as follows: POPEP-I with MW distribution >10,000 Da, POPEP-II with MW distribution of 5000–10,000 Da, POPEP-III with MW distribution of 3000–5000 Da and POPEP-IV with MW distribution <3000 Da. All four fractions of POPEP were lyophilized. After that, the four fractions were stored at −20 °C for further analysis.

3.5. Degree of Hydrolysis (DH)

Protein sample solubility was evaluated according to the method of Saeed [30]. Aqueous POPEP solution (1% w/v) was prepared at pH 7 (adjusted with either 2 M HCl or NaOH). Solutions were stirred at room temperature for 30 min and centrifuged at 1200× g for 10 min at 4 °C. The protein concentration of the supernatant was determined according to the Lowry method [31]. Solubility was expressed as the percent of total protein of the original sample.

3.6. Analysis of Amino Acid Composition and Molecular Weight Distribution of POPEs

The lyophilized fractions POPE, POPEP-I, POPEP-II, POPEP-III and POPEP-IV were digested in HCl (6.0 M) solution at 110 °C for 24 h under nitrogen atmosphere. An automated amino acid analyzer (L-8800; Hitachi High Technologies Corp., Tokyo, Japan) equipped with a visible analytical detector was used before post-column derivatization with ninhydrin. Each sample (1 µL) was injected into an analytical column (sulfonic cation-exchange resin, 4.6 mm i.d., 60 mm length, 3 µm particle size; Tokyo, Japan) at 135 °C with detection at 570 nm. An amino acid standard solution containing 16 amino acids was used as the internal standard.

The POPEP fraction which has the highest antioxidant activity was further analyzed in terms of molecular weight distribution by gel filtration chromatography. The peptides were loaded onto a Sephadex G-25 column (1.0 cm × 10 cm, GE-Healthcare, Boston, MA, USA) and eluted with deionized water at a flow rate of 0.2 mL/min. The fractions were collected and monitored at 220 nm. A molecular weight calibration curve was obtained from the following standards: vitamin B12 (1355 Da), L-glutathione (oxidized) (612 Da), L-tyrosine (181 Da), and BSA (6700 Da).

3.7. Analysis of Antioxidant Activities In Vitro

3.7.1. Free Radical Scavenging Activity

The capacity of all protein hydrolysate fractions tracking free radicals was determined using DPPH [32,33]. The scavenging activity of hydrolysate samples is showed as the percentage disappearance of DPPH relative to a control mixture with deionized water instead of the sample. The mathematical calculations followed Equation (1):

$$\text{Scavenging of DPPH radical (\%)} = ((A_{control} - A_{sample})/A_{control}) \times 100\% \qquad (1)$$

where $A_{control}$ is the absorbance of the system using deionized water instead of the sample, and A_{sample} is the absorbance of the system in the presence of the tested sample.

3.7.2. Hydrogen Peroxide Scavenging Activity

The ability to scavenge hydrogen peroxide (H_2O_2) was measured following Ganie [34] with some medications. The detection system included 200 μL of phosphate buffer (0.1 M, pH 7.4), 40 μL of H_2O_2 solution (0.3%), and 100 μL of sample solution (5 mg/mL), and the absorbance was measured at 230 nm after 10 min. The scavenging of H_2O_2 was calculated as similar to Equation (1).

3.7.3. Reducing Power

The reducing power of the peptide samples was measured following our previous work [35]. An amount of 1 mL of 5 mg/mL sample solution was mixed with 1 mL of 0.2 M phosphate buffer saline (pH 6.6) and 1 mL of 0.1% $K_3Fe(CN)_6$ solution and incubated at 50 °C for 20 min. Then, 1 mL of 10% trichloroacetic acid solution was subsequently added. The mixture was centrifuged at 3000× g for 10 min. Next, 1 mL of the upper layer was mixed with 1 mL of distilled water and 0.2 mL of 0.3% $FeCl_3$ solution, and the absorbance was measured spectrophotometrically at 700 nm.

3.7.4. Inhibitory Effect on Lipid Peroxidation

The inhibitory activity on lipid peroxidation was determined using a slightly modified method of conjugated diene [36]. First, 25 μL of sample (5 mg/mL) was incubated with 500 μL of 10 mM linoleic acid emulsion in 0.2 M sodium phosphate buffer (pH 6.6) in the dark at 37 °C for 24 h. Second, 1.5 mL of 60% methanol was added. Finally, the mixture absorbance was measured at 234 nm. The inhibitory activity on lipid peroxidation was measured as similar to Equation (1).

3.7.5. Ferrous Ion Chelating Activity

The chelating activity on Fe^{2+} was detected following the procedure described by our earlier study [37] with some modifications. First, 200 μL of samples (5 mg/mL) was mixed with 550 μL water and 10 μL ferrous chloride (2 mM). The reaction was initiated by adding 40 μL of ferrozine (5 mM). The reactant was shaken to evenly mix and held at 25 °C for 10 min. Finally, the mixture absorbance was measured at 562 nm against a blank. The Fe^{2+} chelating activity was expressed as similar to Equation (1).

3.8. Animal Experiments

Male Kunming mice (22–25 g) were obtained from Beijing HFK Bioscience Company. The animals were housed for 7 days to acclimatize them to the environment and temperature (20–25 °C) with a 12 h light/dark cycle. The animals were fed with a standard diet and given water ad libitum.

The in vivo hepatoprotective activity of the POPEP-III was examined against CCl_4-induced hepatotoxicity in Kunming mice. Forty-two mice were randomly allocated into six groups (I-VI), with seven mice in each group as follows:

Group normal: normal, normal saline (200 mg/kg, i.g.);
Group control: negative control, CCl_4 (12 mL/kg, i.g.);
Group bifendate: positive control, bifendate (200 mg/kg + CCl_4, i.g.);
Group POPEP-III 100: POPEP-III (100 mg/kg + CCl_4, i.g.);
Group POPEP-III 200: POPEP-III (200 mg/kg + CCl_4, i.g.);
Group POPE: POPE (200 mg/kg + CCl_4, i.g.).

All groups accepted allocated intragastric administration treatment for 10 successive days. In the normal group and control group, animals received formal saline at a dosage of (12 mL/kg, i.g.) once daily, and the mice in bifendate group were treated with bifendate (200 mg/kg in distilled water, i.g.) once daily. In the three remaining experimental groups, the mice were pretreated with POPEP-III (100 and 200 mg/kg, i.g.) and POPE (200 mg/kg, i.g.) once daily. On the 11th day, all mice except those in group I received an intraperitoneal gastric perfusion of a CCl_4/olive oil mixture (2%, 12 mL/kg, i.g.), and the normal control group received an equal amount of olive oil alone [38].

All mice were treated with anesthetic via intraperitoneal injection of pentobarbital sodium (30 mg/kg). Approximately 1 mL of blood was drawn from the mouse abdominal artery. The blood was allowed to coagulate for 30 min at room temperature. The serum was isolated by centrifugation at 978× g for 10 min at 4 °C and used in biochemical index detection. The removed livers were washed repeatedly with saline until clean. A small portion of the liver was conserved in 10% (v/v) formaldehyde for further histopathological analysis, and the remaining liver was homogenized with pre-cooled saline (1:9, w/v) for biochemical index detection.

3.9. Biochemical Assays

The activities of the serum ALT and AST enzymes were assayed according to the guidelines of the diagnostic kits, and the results are showed in IU/L. The liver homogenate in saline solution (1:9, w/v) was centrifuged at 8000× g for 10 min at 4 °C, and the supernatant was collecte. The supernatant was stored at −20 °C. The activities of SOD, GSH-Px, and MDA content and hepatic protein concentrations in the liver were measured according to the diagnostic kits instructions. The activities of SOD, GSH-Px, and MDA content were normalized with reference to protein, and the results are showed in units per milligram of protein, units per milligram of protein, and nanomoles per milligram of protein. The HI was calculated as the Equation (2):

$$HI = (Liver\ weight/Body\ weight) \times 100\% \qquad (2)$$

3.10. Histopathological Study on the Liver

The fresh liver was removed from each mouse after dissection, and liver tissue was fixed whith 10% formalin solution. Then the fixed liver tissues were dehydrated in alcohol and embedded in paraffin using the standard microtechnique. Sections (1.5 cm × 1.5 cm × 5 μm) stained with hematoxylin and eosin (H&E) were observed under light microscopy (×20 magnification) for histopathological studies [39].

3.11. Statistical Analysis

Experimental results are expressed as the mean ± SD (standard deviation). The p-values < 0.05 were treated as statistically significant (SPSS, version 20.0).

4. Conclusions

In this study, we obtained a peptide named POPEP-III from *P. ostreatus* by pepsin hydrolysis. POPEP-III exhibit better antioxidant activity in vivo and in vitro. It can significant increase the activity of SOD and GSH-Px, and decrease the content of MAD in serum in mice prior to CCl_4. Serum ALT and AST are index of liver, many liver diseases could lead the serum ALT and AST level increase. In our study, POPEP-III can significance decrease the activity of AlT and AST induced by CCl_4. The results demonstrate that the POPEP-III could be a liver protect reagent to protect the liver from oxidative damage.

In summary, we hydrolyzed proteins from *P. ostreatus* fruiting bodies using pepsin and tested the hepatoprotective effects of POPE and POPEP hydrolysate for potential application in health support. The experiments showed that as an antioxidant ingredient, POPEP-III plays a role by halting free radical chain reactions. Additionally, POPEP-III promoted the activities of SOD and GSH-Px, decreased the level of MDA, and protected the liver from the attacks of free radical-mediated ROS.

Author Contributions: Q.L. and H.D.W. conceived and designed the experiments; L.Z., Y.L. (Yuxiao Lu), Q.L. and X.F. performed the experiments and analyzed the data; Y.L. (Yuanhui Li), Y.W., Y.D. and J.H. contributed the reagents, materials, and analysis tools; and L.Z., X.F., Q.L. and H.D.W. wrote the manuscript. All authors have read and agreed to the published version of the manuscript.

Funding: This research was supported by the National Key R&D Program (Project No. 2018YFD0400200) from the Ministry of Science and Technology of China. The authors are also grateful for the projects of China Agriculture Research System (Project No. CARS-20-08B) from the Ministry of Agriculture and Rural Affairs of China.

Conflicts of Interest: The authors declare no conflicts of interest.

Abbreviations

DPPH	1,1-diphenyl-2-picrylhydrazyl
ALT	alanine aminotransferase
AST	aspartate aminotransferase
MDA	malondialdehyde
SOD	superoxide dismutase
GSH-Px	glutathione peroxidase
ROS	reactive oxygen species
HI	hepatosomatic index
THAA	total hydrophobic amino acid
BSA	bovine serum albumin

References

1. Gao, B.; Jeong, W.; Tian, Z.G. Liver: An organ with predominant innate immunity. *Hepatology* **2008**, *47*, 729–736. [CrossRef] [PubMed]
2. Taub, R.A. Liver regeneration: From myth to mechanism. *Nat. Rev. Mol. Cell Biol.* **2004**, *5*, 836–847. [CrossRef] [PubMed]
3. Stoyanovsky, D.A.; Cederbaum, A.I. Thiol oxidation and cytochrome P450-dependent metabolism of CCl_4 triggers Ca^{2+} release from liver microsome. *Biochemistry* **1996**, *35*, 15839–15845. [CrossRef] [PubMed]
4. Byass, P. The global burden of liver disease: A challenge for methods and for public health. *BMC Med.* **2014**, *12*, 159. [CrossRef] [PubMed]
5. Ronchi, J.A.; Vercesi, A.E.; Castilho, R.F. Reactive oxygen species and permeability transition pore in rat liver and kidney mitoplasts. *J. Bioenerg. Biomembr.* **2011**, *43*, 709–715. [CrossRef]
6. Fernandez-Checa, J.C.; Kaplowitz, N. Hepatic mitochondrial glutathione: Transport and role in disease and toxicity. *Toxicol. Appl. Pharm.* **2005**, *204*, 263–273. [CrossRef]
7. Paik, Y.H.; Kim, J.; Aoyama, T.; Minicis, S.D.; Bataller, R.; Brenner, D.A. Role of NADPH oxidases in liver fibrosis. *Antioxid. Redox Sign.* **2014**, *20*, 2854–2872. [CrossRef]
8. Choi, J.; James Ou, J.H. Mechanisms of Liver Injury. III. Oxidative stress in the pathogenesis of hepatitis C virus. *Am. J. Physiol.-Gastrointest. Liver Physiol.* **2006**, *290*, 847–851. [CrossRef]
9. Akanitapichat, P.; Phraibung, K.; Nuchklang, K.; Prompitakkul, S. Antioxidant and hepatoprotective activities of five eggplant varieties. *Food Chem. Toxicol.* **2010**, *48*, 3017–3021. [CrossRef]
10. Panda, V.; Ashar, H.; Srinath, S. Antioxidant and hepatoprotective effect of *Garcinia indica* fruit rind in ethanol induced hepatic damage in rodents. *Interdiscip. Toxicol.* **2012**, *5*, 207–213. [CrossRef]
11. Zarezade, V.; Moludi, J.; Mostafazadeh, M.; Mohammadi, M.; Veisi, A. Antioxidant and hepatoprotective effects of *Artemisia dracunculus* against CCl_4-induced hepatotoxicity in rats. *Avicenna J. Phytomed.* **2018**, *8*, 51–62. [CrossRef] [PubMed]
12. Bak, M.J.; Jun, M.; Jeong, W.S. Antioxidant and hepatoprotective effects of the red ginseng essential oil in H_2O_2-treated HepG2 cells and CCl_4-treated mice. *Int. J. Mol. Sci.* **2012**, *13*, 2314–2330. [CrossRef] [PubMed]
13. Polyak, S.J.; Morishima, C.; Lohmann, V.; Pal, S.; Lee, D.Y.W.; Liu, Y.Z.; Graf, T.N.; Oberlies, N.H. Identification of hepatoprotective flavonolignans from silymarin. *Proc. Natl. Acad. Sci. USA* **2010**, *107*, 5995–5999. [CrossRef] [PubMed]
14. Liu, Q.; Zhu, M.J.; Geng, X.R.; Wang, H.X.; Ng, T.B. Characterization of polysaccharides with antioxidant and hepatoprotective activities from the edible mushroom *Oudemansiella radicata*. *Molecules* **2017**, *22*, 234. [CrossRef]
15. Liu, Q.; Tian, G.T.; Yan, H.; Geng, X.R.; Cao, Q.P.; Wang, H.X.; Ng, T.B. Characterization of polysaccharides with antioxidant and hepatoprotective activities from the wild edible mushroom *Russula vinosa* Lindblad. *J. Agric. Food Chem.* **2014**, *62*, 8858–8866. [CrossRef]

16. Zhao, H.J.; Zhang, J.J.; Liu, X.C.; Yang, Q.H.; Dong, Y.H.; Jia, L. The antioxidant activities of alkaliextractable polysaccharides from *Coprinus comatus* on alcohol induced liver injury in mice. *Sci. Rep.* **2018**, *8*, 11695. [CrossRef]
17. Sun, J.; He, H.; Xie, B.J. Novel antioxidant peptides from fermented mushroom *Ganoderma lucidum*. *J. Agric. Food Chem.* **2004**, *52*, 6646–6652. [CrossRef]
18. Zhu, B.; Li, Y.Z.; Hu, T.; Zhang, Y. The hepatoprotective effect of polysaccharides from *Pleurotus ostreatus* on carbon tetrachloride-induced acute liver injury rats. *Int. J. Biol. Macromol.* **2019**, *131*, 1–9. [CrossRef]
19. Xia, F.G.; Fan, J.H.; Zhu, M.; Tong, H.B. Antioxidant effects of a water-soluble proteoglycan isolated from the fruiting bodies of *Pleurotus ostreatus*. *J. Taiwan Inst. Chem. E.* **2011**, *42*, 402–407. [CrossRef]
20. Thanaswkaran, J.; Muniyan, S.; Thomas, P.A.; Geraldine, P.L. *Pleurotus ostreatus*, an oyster mushroom, decreases the oxidative stress induced by carbon tetrachloride in rat kidneys, heart and brain. *Chem.-Biol. Interactions* **2008**, *176*, 108–120. [CrossRef]
21. Thanaswkaran, J.; Ramesh, E.; Geraldine, P. Antioxidant activity of the oyster mushroom, *Pleurotus ostreatus*, on CCl_4-induced liver injury in rats. *Food Chem. Toxicol.* **2006**, *44*, 1989–1996. [CrossRef]
22. Llauradó, G.; Morris, H.J.; Lebeque, Y.; Venet, G.; Fong, O.; Marcos, J.; Fontaine, R.; Cos, P.; Bermúdez, R.C. Oral administration of an aqueous extract from the oyster mushroom *Pleurotus ostreatus* enhances the immunonutritional recovery of malnourished mice. *Biomed. Pharmacother.* **2016**, *83*, 1456–1463. [CrossRef] [PubMed]
23. Vieira Gomes, D.C.; de Alencar, M.V.O.B.; Dos Reis, A.C.; de Lima, R.M.T.; de Oliveira Santos, J.V.; da Mata, A.M.O.F.; Soares Dias, A.C.; da Costa, J.S.J.; de Medeiros, M.D.G.F.; Paz, M.F.C.J.; et al. Antioxidant, anti-inflammatory and cytotoxic/antitumoral bioactives from the phylum Basidiomycota and their possible mechanisms of action. *Biomed. Pharmacother.* **2019**, *112*, 108643. [CrossRef] [PubMed]
24. Gao, W.J.; Sun, Y.H.; Chen, S.W.; Zhang, J.Y.; Kang, J.J.; Wang, Y.Q.; Wang, H.X.; Xia, G.L.; Liu, Q.H.; Kang, Y.M. Mushroom lectin enhanced immunogenicity of HBV DNA vaccine in C57BL/6 and HBsAg-transgenic mice. *Vaccine* **2013**, *31*, 2273–2280. [CrossRef] [PubMed]
25. Wang, H.X.; Gao, J.Q.; Ng, T.B. A new lectin with highly potent antihepatoma and antisarcoma activities from the oyster mushroom *Pleurotus ostreatus*. *Biochem. Bioph. Res. Co.* **2000**, *275*, 810–816. [CrossRef] [PubMed]
26. Rajapakse, N.; Mendis, E.; Jung, W.K.; Je, J.Y.; Kim, S.K. Purification of a radical scavenging peptide from fermented mussel sauce and its antioxidant properties. *Food Res. Int.* **2005**, *38*, 175–182. [CrossRef]
27. Sun, Q.; Shen, H.X.; Luo, Y.K. Antioxidant activity of hydrolysates and peptide fractions derived from porcine hemoglobin. *J. Food Sci. Tech.* **2011**, *48*, 53–60. [CrossRef]
28. Xie, Z.J.; Huang, J.R.; Xu, X.M.; Jin, Z.Y. Antioxidant activity of peptides isolated from alfalfa leaf protein hydrolysate. *Food Chem.* **2008**, *111*, 370–376. [CrossRef]
29. Souza, D.; Sbardelotto, A.F.; Ziegler, D.R.; Marczak, L.D.; Tessaro, I.C. Characterization of rice starch and protein obtained by a fast alkaline extraction method. *Food Chem.* **2016**, *191*, 36–44. [CrossRef]
30. Saeed, M.; Cheryan, M. Sunflower protein concentrates and isolates low in polyphenols and phytate. *J. Food Sci.* **1988**, *53*, 1127–1131. [CrossRef]
31. Lowry, O.H.; Rosebrough, N.J.; Farr, A.L.; Randall, R.J. Protein measurement with the Folin phenol reagent. *J. Biol. Chem.* **1951**, *193*, 265–275. [PubMed]
32. Cheng, I.C.; Liao, J.X.; Ciou, J.Y.; Huang, L.T.; Chen, Y.W.; Hou, C.Y. Characterization of Protein Hydrolysates from Eel (*Anguilla marmorata*) and Their Application in Herbal Eel Extracts. *Catalysts* **2020**, *10*, 205. [CrossRef]
33. Wen, T.; Yan, D.D.; Meng, J.; Liu, J.; Xu, H.Y. The enzyme-like property and photocatalytic effect on α, α-diphenyl-β-picrylhydrazyl (DPPH) of CuPt nanocomposite. *Catalysts* **2019**, *9*, 813. [CrossRef]
34. Pushparaj, F.S.; Urooj, A. Antioxidant activity in two pearl millet (*Pennisetum typhoideum*) cultivars as influenced by processing. *Antioxidants* **2014**, *3*, 55–66. [CrossRef] [PubMed]
35. Li, J.; Huang, S.Y.; Deng, Q.Y.; Li, G.L.; Su, G.C.; Liu, J.W.; Wang, H.M.D. Extraction and characterization of phenolic compounds with antioxidant and antimicrobial activities from pickled radish. *Food Chem Toxicol.* **2020**, *136*, 111050–1111055. [CrossRef] [PubMed]
36. Tseng, Y.H.; Yang, J.H.; Mau, J.L. Antioxidant properties of polysaccharides from *Ganoderma tsugae*. *Food Chem.* **2008**, *107*, 732–738. [CrossRef]
37. Chou, H.Y.; Wang, H.M.D.; Kuo, C.H.; Lu, P.H.; Wang, L.; Kang, W.Y.; Sun, C.L. Antioxidant graphene oxide nanoribbon as a novel whiting agent inhibits microphthalmia-associated transcription factor related melanogenesis mechanism. *ACS Omega* **2020**, in press. [CrossRef]

38. Lu, X.S.; Zhao, Y.; Sun, Y.F.; Yang, S.; Yang, X.X. Characterisation of polysaccharides from green tea of *Huangshan Maofeng* with antioxidant and hepatoprotective effects. *Food Chem.* **2013**, *141*, 3415–3423. [CrossRef]
39. Al-Megrin, W.A.; Alkhuriji, A.F.; Yousef, A.O.S.; Metwally, D.M.; Habotta, O.A.; Kassab, R.B.; Abdel Moneim, A.E.; El-Khadragy, M.F. Antagonistic efficacy of luteolin against lead acetate exposure-associated with hepatotoxicity is mediated via antioxidant, anti-inflammatory, and anti-apoptotic activities. *Antioxidants* **2020**, *9*, 10. [CrossRef]

© 2020 by the authors. Licensee MDPI, Basel, Switzerland. This article is an open access article distributed under the terms and conditions of the Creative Commons Attribution (CC BY) license (http://creativecommons.org/licenses/by/4.0/).

Article

Whole-Cells of *Yarrowia lipolytica* Applied in "One Pot" Indolizine Biosynthesis

Andreea Veronica Botezatu (Dediu) [1], Georgiana Horincar [2], Ioana Otilia Ghinea [1], Bianca Furdui [1,*], Gabriela-Elena Bahrim [2], Vasilica Barbu [2], Fanica Balanescu [1], Lidia Favier [3,*] and Rodica-Mihaela Dinica [1,*]

[1] Department of Chemistry, Physics and Environment, "Dunărea de Jos" University of Galati, 111 Domnească Street, 800201 Galati, Romania; veronica.dediu@ugal.ro (A.V.B.); ioana.ghinea@ugal.ro (I.O.G.); fanica.balanescu@ugal.ro (F.B.)
[2] Department of Food Science, Food Engineering and Applied Biotechnology, "Dunărea de Jos" University of Galati, 111 Domnească Street, 800201 Galati, Romania; georgiana.parfene@ugal.ro (G.H.); gabriela.bahrim@ugal.ro (G.-E.B.); vasilica.barbu@ugal.ro (V.B.)
[3] Ecole Nationale Supérieure de Chimie de Rennes, CNRS, UMR 6226, 11 Allée de Beaulieu CS 50837, CEDEX 7, 35708 Rennes, France
* Correspondence: bianca.furdui@ugal.ro (B.F.); lidia.favier@ensc-rennes.fr (L.F.); rodica.dinica@ugal.ro (R.-M.D.)

Received: 9 May 2020; Accepted: 3 June 2020; Published: 5 June 2020

Abstract: A series of yeast strains was tested in order to evaluate their catalytic potential in biocatalysis of one-pot indolizine's synthesis. Yeast cultivation was performed in a submerged system at 28 °C for 72 h at 180 rpm. An assessment of the reagents' toxicity on yeast viability and metabolic functionality concluded that the growth potential of three *Yarrowia lipolytica* strains were least affected by the reactants compared to the other yeast strains. Further, crude fermentation products (biomass and cell-free supernatant)—obtained by submerged cultivation of these yeasts—were used in multistep cascade reactions for the production of fluorescent indolizine compounds with important biologic activities. A whole–cell catalyzed multicomponent reaction of activated alkynes, α-bromo-carbonyl reagents and 4,4′-bipyridine, at room temperature in buffer solution led to the efficient synthesis of bis-indolizines **4a**, **4b** and **4c**, in good-to-excellent yields (47%–77%). The metabolites of the selected *Y. lipolytica* strains can be considered effective biocatalysts in cycloaddition reactions and the high purity and bioconversion yields of the synthesized indolizines indicates a great potential of this type of "green" catalysts. Seeds of *Triticum estivum* L. were used to investigate the impact of the final products on the germination and seedling growth. The most sensitive physiological parameters suggest that indolizines, at the concentrations tested, have non-toxic effect on germination and seedling growth of wheat, fact also confirmed by confocal laser scanning microscopy images.

Keywords: *Yarrowia lipolytica*; whole–cell biocatalysis; indolizine; cycloaddition

1. Introduction

The application of whole microbial-cell biocatalysis methods is steadily increasing as an environment-friendly alternative to dangerous solvents and catalysts [1–4]. The use of enzymes in synthetic transformations has many advantages, such as mild and eco-friendly conditions, high regio- and stereoselectivities [5–9], but it also has drawbacks in the preparation of enzymes that require the cultivation of microorganisms, purification and crystallization of enzymes [2,10]. Therefore, the use of whole-cell systems for conducting chemical reactions offers some great advantages such as decreased reaction time, considerable decrease of costs, fewer steps, elimination of toxic solvents, mild conditions, high purity and yields of the obtained compounds and ecofriendly transformations [10–13].

The *Yarrowia lipolytica* yeast species is widespread in nature. It considered a non-pathogenic microorganism, the U.S. Food and Drug Administration has given its metabolites the GRAS status ("generally recognized as safe") [14] and they are isolated from dairy products, dry fermented sausages or mediums rich in fats or hydrocarbons [15]. The potential applications of this yeast are related to its specific ability to use hydrophobic substrates effectively and the high biosynthesis capacity of numerous enzymes, such as lipases/esterases (LIP genes), cytochromes P450 (ALK genes) and peroxisomal acyl-CoA oxidases (POX genes), its ability to produce surfactants [16,17] and was successfully used in biocatalysis [14,15,18,19].

The indolizine core is present in many biologically active compounds and can be considered an important scaffold for the preparation of new pharmaceuticals [20,21]. These class of compounds holds valuable biologic activities [21], such as antimicrobial [22,23], antioxidant [24,25], anticancer [26–28], enzyme inhibitors [29,30], anti-inflammatory [21] and anti-HIV-1 [31]. Consequently, different pathways were described in the literature for their synthesis [32–34]. Therefore, the development of novel and efficacious methods for the synthesis of these valuable compounds are of significant concern.

This work describes the ability of yeast crude products (whole cells and cultural liquid) to biocatalyze the synthesis of the indolizine ring, in a cascade transformation which involves the "one pot" reaction of three components, a pyridine compound, a phenacyl bromide and an activated alkyne, without the isolation and purification of the intermediate, at room temperature in buffer solution. Literature data pointed out the necessity to assess various culture conditions to detect sustainable, selective and efficient biocatalysts, therefore herein we describe the screening of nine microorganism strains. The selection was based on their physiological functionality in the presence of reagents by toxicity evaluation and possible functionality as biocatalyst for in situ chemical conditions, in reactions involving cycloaddition processes in aqueous media. The high level of lipase activity exhibited by *Yarrowia lipolytica* during submerged cultivation on nutrient broth supplemented with the reagents suggests that the three reactants do not exhibit a negative effect on the metabolites' biosynthesis of this yeast. In this context, these microbial enzymes were used in biocatalysis aims due to their broad specificity, high stability under different reaction conditions, high regio- and enantioselectivity. The lipase catalytic promiscuity is well known [35], playing a role not only in hydrolysis reactions, but also in other type of chemical synthesis such as condensation [36–38] and chemoenzymatic synthesis of polymeric materials [39]. This methodology is an environmentally-friendly regioselective manner to obtain the indolizine core structure as we previously accomplished by using commercial lipases [5]. The one-pot reaction using the crude multienzyme biocatalyst is described in Figure 1. The final products impact on wheat (*Triticum estivum* L.) germination and seedling growth is also analyzed, in order to assess the harmful effects to the environment of the biosynthesized indolizine compounds if they will be used as scaffold for pharmaceutical industry. Toxicity studies were performed on plant seed germination as early studies, taking into account the potential hazard of releasing these synthetic chemicals into the environment during production, transport, use or most importantly during disposal. For this reason, it is very important to predict this type of adverse effects directly on the environment or indirectly on humans, through crops or food, before the substance is released into the pharmaceutical industry.

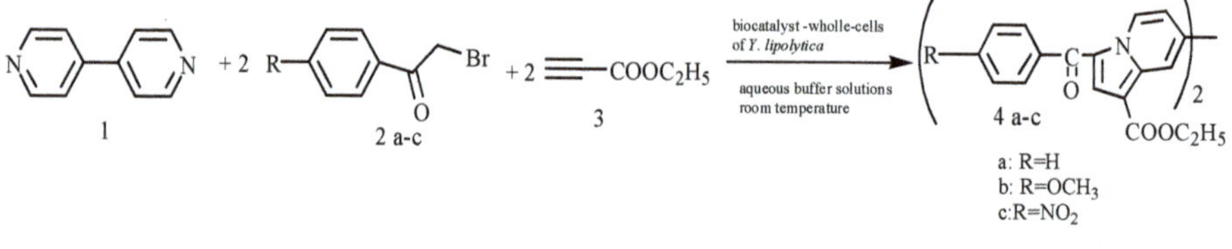

Figure 1. Indolizine synthesis in one-pot bioformation reactions using *Yarrowia lipolytica* crude biocatalysts.

2. Results and Discussion

2.1. Evaluation of the Toxicity of Reactants on the Yeasts' Physiological Activity

An evaluation of the chemicals' influence on the yeast cells metabolism was performed in order to select the most resistant strains. The selected strains were subsequently used as biocatalysts in indolizines synthesis. In a medium supplemented with 0.1%, 0.2% and 0.3% of each stock solution of all three reactants, 4,4'-bipyridine, 2-bromoacetophenones and ethyl propiolate, (compounds **1, 2, 3** from Figure 1), it was observed that *Yarrowia lipolytica* tested strains survived in a high rate, compared to other yeasts, developing the largest colonies, with diameters up to 18 mm (Figure 2).

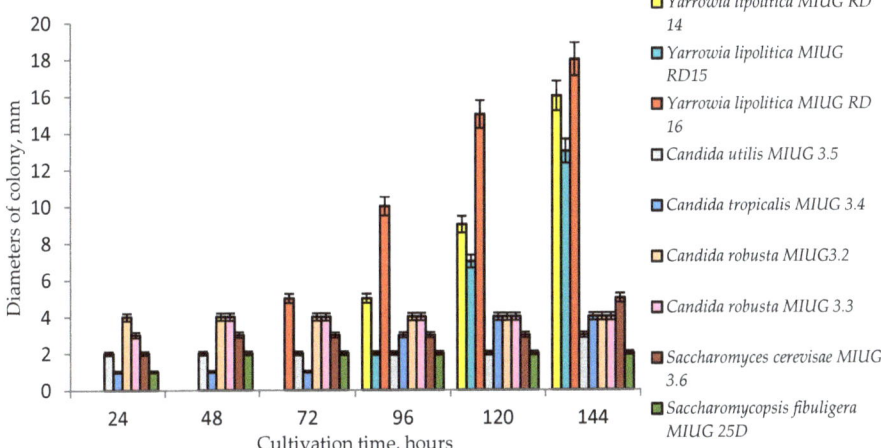

Figure 2. Impact of the 0.1% reactants on the growth of different strains of yeast.

By increasing the concentrations of all the three chemicals, in the culture medium, three strains of *Yarrowia lipolytica* MIUG RD 14, MIUG RD 15 and MIUG RD 16 proved to be the most resistant (Figure 3a,b) developing the largest colonies, while other strains such as *Candida robusta* MIUG 3.3, *Saccharomycopsis figuligera* MIUG 25D or *Saccharomyces cerevisiae* MIUG 3.6 have grown very slowly or have not developed any visible colonies. Consequently, because of their high survival rates only the *Yarrowia lipolytica* strains were selected to be used in indolizine synthesis.

2.2. Growth and Bioconversion Potential of Yarrowia lipolytica in Aerobe Submerged Cultivation Conditions in the Presence of Chemicals

Some reports concerning the lipolytic activity of *Yarrowia lipolytica* are present in literature. Parfene et al. (2011) [40] studied in situ lipase production during solid state cultivation of selected strains on media supplemented with vegetable fat. Destain et al. (1997) [41] selected a *Yarrowia lipolytica* mutant obtained by chemical mutagenesis which produced a lipase with a 35-fold increase in activity, compared to the wild type of enzyme. Pignede et al. (2000) [42] described the LIP2 gene responsible for extracellular lipases produced by *Yarrowia lipolytica* W29, which Fickers et al. (2005) [43] amplified in a *Yarrowia lipolytica* mutant and obtained high lipase activities of 26.450 U mL^{-1} in batch operation and 158.246 U mL^{-1} in fed-batch.

Our previous studies on selected lipase-producing *Yarrowia lipolytica* strains, using vegetable fats (palm, coconut and shea) as substrate, showed that these strains are active producers of extracellular lipases. Strain *Yarrowia lipolytica* MIUG RD 14 was more active than the strains *Yarrowia lipolytica* MIUG RD15 and *Yarrowia lipolytica* MIUG RD16 [40]. The lipase activity potential of these strains was demonstrated by in situ cultivation of whole biomass and also by extracellular enzyme activity evaluation of cell-free crude supernatants.

Dynamics of culture multiplication and extracellular lipase production of *Yarrowia lipolytica* selected strains are shown in Figure 4. The high yield of extracellular lipase production was detected after 72 h of submerged cultivation for all three strains. Thus, the extracellular lipase from *Yarrowia lipolytica* MIUG RD 15 has shown an activity of 375 U mL^{-1}. This biosynthesis yield was similar with the one previously obtained (390 U mL^{-1}) with this strain under normal cultivation conditions [40]. The high level of lipase activity exhibited by the strain *Yarrowia lipolytica* MIUG RD 14 during submerged cultivation (180 rpm at 28 °C) on nutrient broth supplemented with the reactants suggests that the three compounds do not have a toxic effect on the metabolites biosynthesis of this yeast strains.

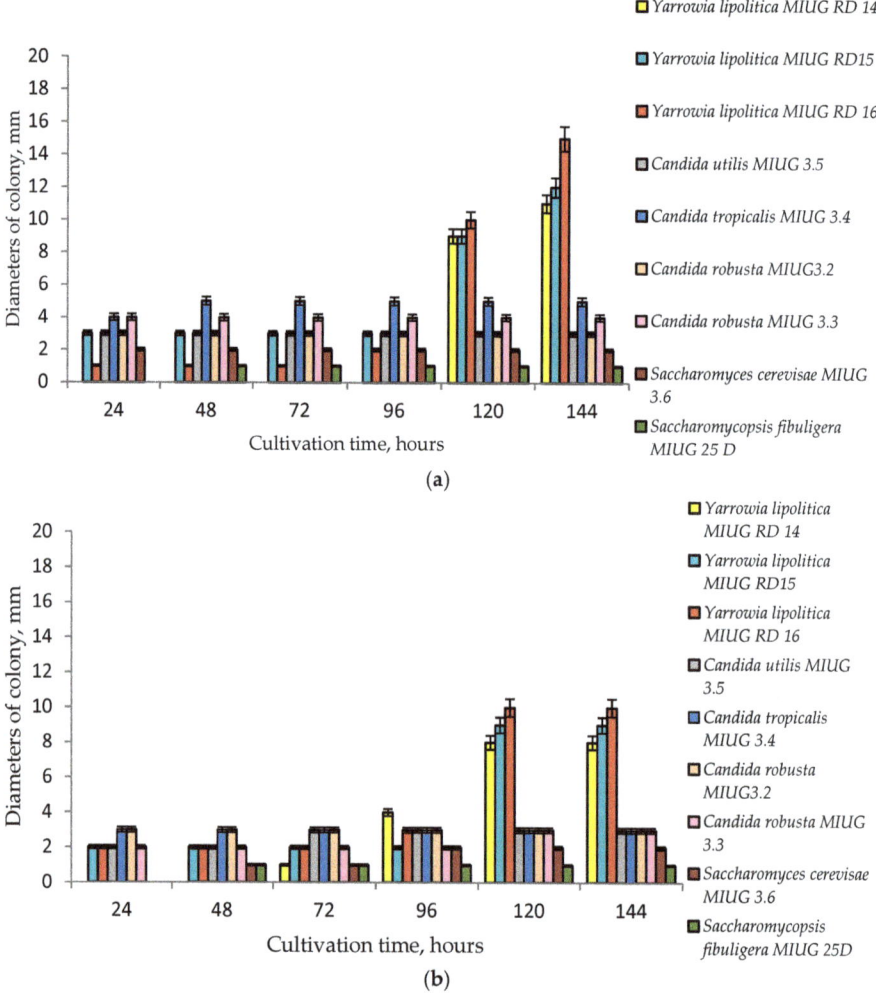

Figure 3. (a) Impact of 0.2% reactants on the growth of different strains of yeast; (b) impact of 0.3% reactants on the growth of different strains of yeast.

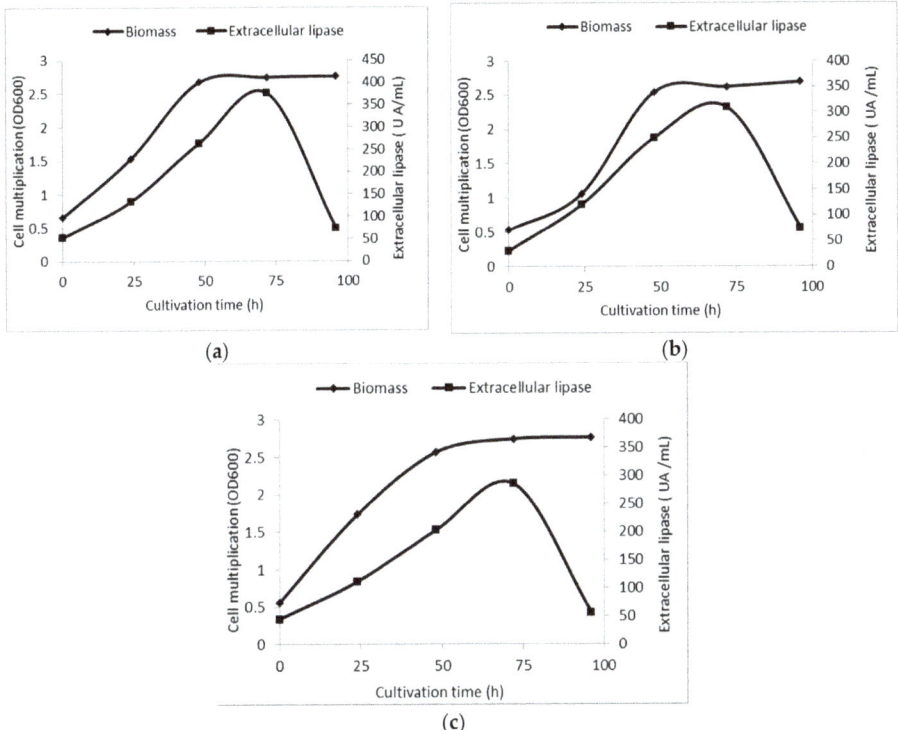

Figure 4. Kinetics of biomass accumulation and extracellular lipase production by *Yarrowia lipolytica* strains, coded MIUG RD 14 (**a**), MIUG RD 15; (**b**) MIUG RD 16; (**c**) during submerged cultivation (180 rpm at 28 °C) on nutrient broth supplemented with chemicals.

2.3. Indolizine Synthesis by Bioconversion

The formation of indolizines compounds in some of the samples was qualitatively assessed through the appearance of yellow-brown fluorescent color in the reaction media, observed under UV light irradiation. In the reactions with no color change, the desired reaction products were not formed.

A major drawback to the application of whole cell biomass in biotransformation reactions is the permeability barrier exhibited by the cell membrane to enzymes or reaction substrates. The presence of water is necessary for lipase interfacial activation involving lid opening at oil–water interface [44]. Nevertheless, *Yarrowia lipolytica* is a nonconventional yeast with the capacity of transforming both polar and nonpolar substrates [45].

As represented in Figures 5 and 6 all the crude fermentation products from the three selected *Yarrowia lipolytica* strains used as biocatalysts lead to the bioformation of the desired reaction product, with a maximum yield achieved after 144 h of reaction.

The best bioconversion yield was observed in the presence of the cell-free supernatant obtained after biomass separation by centrifugation from 144 h of submerged cultivation of *Yarrowia lipolytica* MIUG RD 14, recommending this strain for further research. This crude biocatalyst led to a bioconversion percentage of almost 80% for indolizine **4a**, after 4 days of reaction time in an aqueous medium (Figure 5). These results are correlated with metabolites production potential of these strains and confirm the possible implication of them in bioconversion.

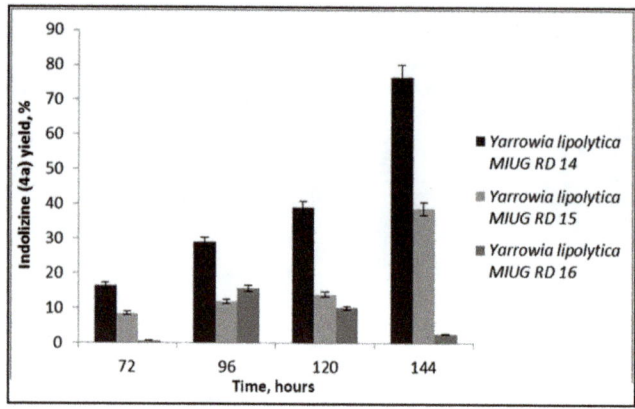

Figure 5. Yield of indolizine **4a**, obtained using *Yarrowia lipolytica* cell-free supernatant as biocatalyst.

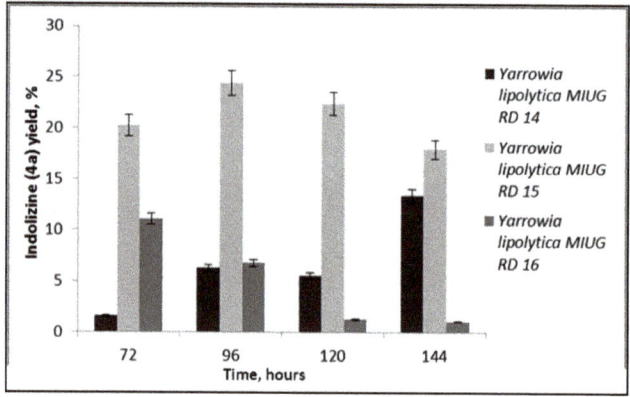

Figure 6. Yield of indolizine **4a**, obtained using *Yarrowia lipolytica* biomass as biocatalysts.

When biomass was used as biocatalyst, it was observed that *Yarrowia lipolytica* MIUG RD 15 strain had higher biotransformation yield, suggesting the capacity of viable cells to release the enzyme in stationary conditions, during the bioconversion process (Figure 6). The bioconversion yields are lower than in the case when the cell-free supernatant was used, according to the extracellular enzyme production yield and also due to the partial biosorbtion of the bioderivates on the cell surface, demonstrated by fluorescence analysis, which cannot be fully extracted with the method employed. The reaction products, the bis-indolizines, after 144 h were obtained in moderate to good yields of 77% (**4a**), 47% (**4b**) and 48% (**4c**) in biocatalysis. The one-pot bioconversion products and their purity were further confirmed by HPLC, FTIR, NMR and elemental analysis.

The structures of all compounds were confirmed by UV-Vis absorption spectra, FTIR, NMR, MS and elemental analysis. The synthesized bis-indolizines have the expected structure and purity, the results of the performed analyses being in concordance with those of the same bis-indolizines formerly synthesized through classical chemical synthesis and characterized [46]. The HPLC chromatograms of the products are shown in Figures S1–S3 of the Supplementary Material. Characteristics of the bioconversion products, from the Supplementary Materials (Figures S4–S9) are as follows:

1,1′-Diethyldicarboxylate-3,3′-benzoyl-7,7′-bis(indolizine) (**4a**), yellow powder, mp 273–275 °C; IR (ATR, cm^{-1}): 1071 (s, C-N); 1167, 1216 (s, C-O-C); 1341 (s, CO$_{ester}$); 1427, 1459, 1479, 1512, 1524 (s, C=N, C=C$_{arom}$); 1645 (s, C=O); 1700 (s, C=O$_{ester}$); 2979 (l, CH$_{alif}$); 3139 (l, CH$_{arom.}$); ^{1}H-NMR

(400 MHz, CDCl$_3$, TMS) δ/ppm: 1.43 (t, J = 7.2, 6H); 4.4 (q, J = 7.2, 4H); 7.42–7.73 [m, 8H: 4H, 2H, 2H (7.52, J = 7.5, J = 1.2)]; 7.78–7.88 [m, 6H: 4H, 2H (7.82)]; 8.77(d, J = 1.2, 2H); 9.98 (d, J = 7.5, 2H); ^{13}C-NMR (400 MHz, CDCl$_3$, TMS δ/ppm): 14.59 (CH$_3$), 60.37 (CH$_2$), 107.55 (C), 113.53 (CH), 116.96 (C), 122.90 (C), 128.52 (CH), 129.05 (CH), 129.33 (CH), 129.53 (CH), 131.74 (CH), 136.61 (C), 139.70 (C), 139.78 (C), 163.97 (C=O$_{ester}$), 185.67 (C=O); MS (APCI+): m/z: 585 [M + H$^+$]; anal. C 73.72; H 5.14; N 4.76%, calcd. for C$_{36}$H$_{28}$N$_2$O$_6$, C 73.96, H 4.83, N 4.79%.

1,1′-Diethyldicarboxylate-3,3′-bis(p-methoxy-benzoyl)-7,7′-bis(indolizine) (**4b**), yellow powder, mp 290–291 °C; IR (ATR, cm^{-1}): 1077 (s, C-N); 1249, 1220, 1167 (s, C-O-C); 1339 (s, CO$_{ester}$); 1463, 1508, 1597 (s, C=N, C=C$_{arom}$); 1665 (s, C=O); 1700 (s, C=O$_{ester}$); 2977 (l, CH$_{alif}$); 3140 (l, CH$_{arom.}$); ^{1}H-NMR (400 MHz, CDCl$_3$, TMS) δ/ppm: 1.43 (t, J = 7.2, 6H); 3.93 (s, 6H); 4.43 (q, J = 7.2, 4H); 7.03 (d, J = 8.8, 4H); 7.33 (d, J = 8.8, 4H); 7.76 (s, 2H); 7.79 (d, J = 7.5, 2H); 8.80 (d, J = 1.2, 2H); 9.98 (d, J = 7.5, 2H); ^{13}C-NMR (400 MHz, CDCl$_3$, TMS δ/ppm): 14.73 (CH$_3$), 56.03 (OCH$_3$), 61.09 (CH$_2$), 107 (C), 113.98 (CH), 114 (CH), 117.72 (CH), 122 (CH), 122.2 (CH), 129.04 (CH), 131.81 (CH), 132.42 (C), 136.57 (CH), 138.57 (C), 163.71 (C), 165.28 (C=O$_{ester}$), 186.06 (C=O); MS (APCI+): m/z: 645 [M + H$^+$]; anal. C 70.92; H 4.88; N 4.28%, calcd. for C$_{38}$H$_{32}$N$_2$O$_8$, C 70.80, H 5.00, N 4.35%.

1,1′-Diethyldicarboxylate-3,3′-bis(p-nitro-benzoyl)-7,7′-bis(indolizine) (**4c**), dark yellow powder, mp 271–272 °C; IR (ATR, cm^{-1}): 1077 (s, C-N);1169, 1208 (s, C-O-C); 1338, 1521 (s, NO$_2$); 1428, 1464 (s, C=N, C=C$_{arom}$); 1656 (s, C=O);1700 (s, C=O$_{ester}$); 2978 (l, CH$_{alif}$); 3359 (l, CH$_{arom.}$); ^{1}H-NMR (400 MHz, CDCl$_3$, TMS) δ/ppm: 1.43 (t, J = 7.2, 6H); 4.4 (q, J = 7.2, 4H); 7.42 (s, 2H); 7.48 (d, J = 7.5, 2H); 7.8 (d, J = 8.8, 4H); 8.03 (d, J = 8.8, 4H); 8.99 (d, J = 1.2, 2H); 10.13 (d, J = 7.5, 2H); ^{13}C-NMR (400 MHz, CDCl$_3$, TMS δ/ppm): 14.29 (CH$_3$), 60.53 (CH$_2$); 108.16 (C); 113.07 (CH); 118.28 (CH); 122.74 (C); 123.52 (CH); 127.46 (CH); 129.14 (CH); 129.86 (CH); 132.60 (C); 138.14 (C);143.06 (C); 150.16 (C), 164.52 (C=O$_{ester}$), 184.15 (C=O); MS (APCI+): m/z: 675 [M + H$^+$]; anal. C 64.15; H 4.11; N 8.26%, calcd. for C$_{36}$H$_{26}$N$_4$O$_{10}$, C 64.09, H 3.88, N 8.31%.

2.4. Evaluation of the Toxicity of Indolizines on the Germination of Triticum estivum L.

Seed germination has gained much attention in recent years due to their simplicity, time effectiveness and low cost, as an environmental exposure model to measure the toxicity of organic compounds at different concentrations [47–49]. Herein, we evaluated the potential of synthetic indolizines to cause particular forms of toxicity in case of their use as pharmaceuticals by plants seeds germination test. The evaluation of germination most sensitive indicators was performed according to our previous method [50] with slight modifications. The test was made in accordance with the Organization for Economic Cooperation and Development (OECD) Guideline for testing of chemicals [51]. The seed germination (SG) values, very similar to those of the control samples, but also relative radicle growth (RRG%) and relative seed germination (RSG%) values greater than 100% for the indolizines treated samples, according to the literature data, means that the presence of the assessed compounds did not exert a harmful influence on seed germination [48,52]. The germination index (GI) values greater than those of the control samples (Table 1), reveals that the tested indolizines have no phytotoxicity [53]. The application of the indolizine **4a** at both concentrations tested, had some positive influence by increasing the average number of roots in wheat germination process (3.7 ± 0.92; 3.6 ± 0.78), in comparison to the control (3.1 ± 0.42). In conclusion, the germination physiological indexes showed that the tested compounds at different concentrations, have no ecotoxicological effects on wheat seedlings (Table 1), compared to control.

The confocal microscopic analysis of the cross-cut sections through the roots and stems of the sprouts resulting from the germination of wheat caryopses, under control conditions (ultrapure water) (Figure 7) or in the presence of indolizines **4a** (Figure 8) or **4c** (Figure 9) with different concentrations, was performed in order to observe possibles morpho-structural changes of plant tissues. The microscopic examinations showed that there were no major differences in the tissue appearance of the sections from the germination of wheat exposed to indolizine compounds compared to control. No lysed or turgid cells were observed in the microscopic analysis. The cells in the cortical parenchyma of roots and stems had intact cell walls (CW), sometimes resulting with a slight tendency to thicken, probably due to the attachment or the adsorption on the surface of wheat seeds of the compounds to which they were exposed during germination. The underlying adsorption mechanism still needs to be analyzed in more detail in further work. For example, in Figure 8a, in the section through the root, it can be seen the annular collenchyma, with uniformly thickened cell walls (3.25–6.28 µm). This collenchyma cells were annexed to the vascular cambium for increasing structural support and integrity. The parenchymal cells of the control sample reached dimensions of 47.24 µm (Figure 7b)—and in the case of samples germinated in the presence of indolizines, the cells had slightly smaller dimensions, on average between 15–25 µm (Figures 8a and 9a,b) and they were more compact, with smaller intercellular spaces. The normal cellular components were visible in both experimental cases: nucleus (N), proplastids/chloroplasts (in orange or red in Figures 8b and 9b) as well as epidermal formations such as stomata (S) type *Zea mays*, hairs (Figure 8b) or tracheids (T) with spiral thickenings of the secondary wall and with diameters of 12–15 µm (Figures 7a and 9b). The **4a** and **4c** indolizines at the tested concentrations (10^{-3} M, 10^{-4} M) did not have a cytotoxic effect on plant organisms, their adsorption did not affect the structure of plant tissues or the physiological processed such as cell division, growth and cell differentiation. Therefore, the results from the microscopic analysis are in agreement with the most sensitive physiological parameters evaluated in Table 1. Such in vivo assays in plant cell models are particularly useful to validate various subsequent applications of the compounds in the pharmaceutical industry [47,51].

Table 1. Evaluation of the impact on the seed germination of *Triticum estivum* L. exposed to different concentrations of indolizine compounds. The results represent the mean values ± standard deviation of four replicates.

Physiological Parameters	Concentration of the Synthetic Indolizines				Control
	4a 10^{-3} M	4a 10^{-4} M	4c 10^{-3} M	4c 10^{-4} M	
SG%	80.33 ± 4.16	86.67 ± 4.73	81.00 ± 2.01	81.33 ± 6.11	80.00 ± 6.32
RRG%	110.26 ± 3.11	108.7 ± 2.75	101.7 ± 2.98	102.91 ± 2.48	100.00 ± 0.00
RSG%	104.68 ± 4.2	113.19 ± 5.3	105.48 ± 3.8	108.33 ± 5.1	100.00 ± 0.00
GI%	88.57 ± 3.6	94.21 ± 3.7	82.38 ± 2.49	83.69 ± 4.2	80.00 ± 6.2
VI%	85.77 ± 0.87	79.13 ± 0.48	89.06 ± 0.73	93.74 ± 0.96	85.80 ± 0.52

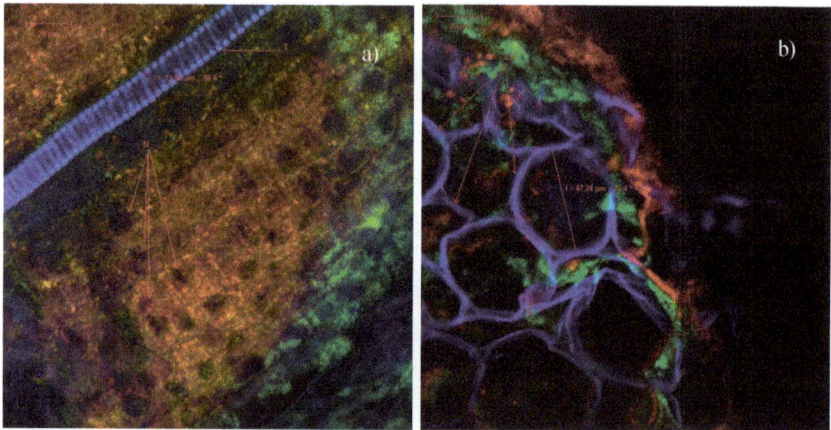

Figure 7. Confocal laser scanning microscopy (CLSM) images of the *Triticum estivum* L. sprout sections-control sample (germination in ultrapure water): (**a**) root, (**b**) shoot.

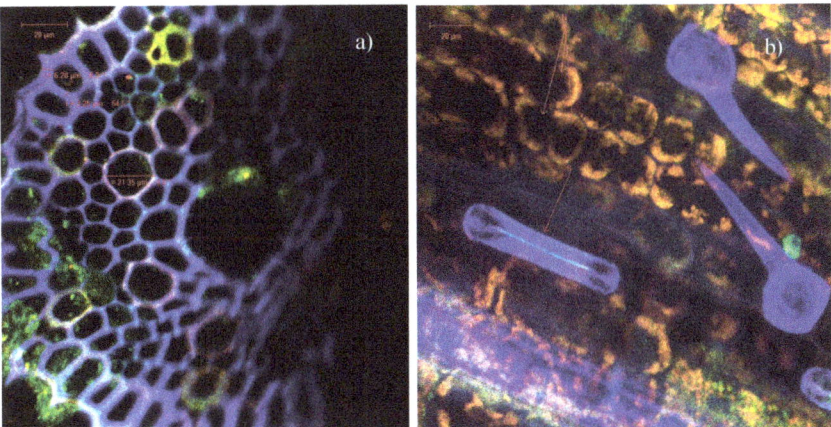

Figure 8. CLSM images of the wheat sprout sections of the indolizine **4a** treated samples at two different concentrations: (**a**) 10^{-3} M, (**b**) 10^{-4} M.

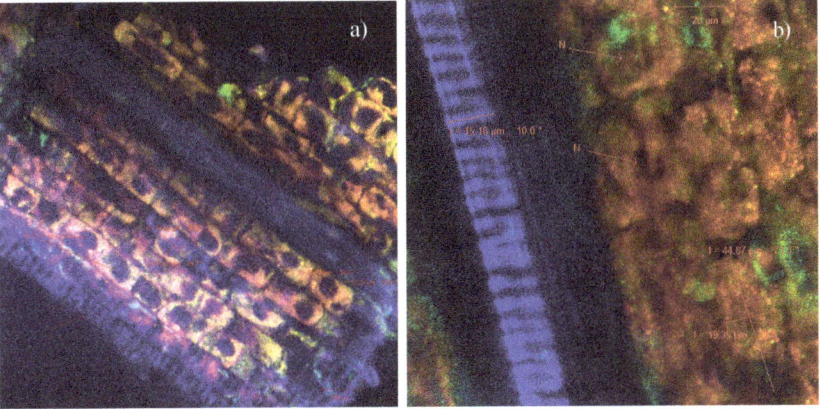

Figure 9. CLSM images of the *Triticum estivum* L. sprout sections of the indolizine **4c** treated samples at two different concentrations: (**a**) 10^{-3} M, (**b**) 10^{-4} M.

3. Materials and Methods

3.1. Microorganisms and Chemicals

All the nine yeast strains, *Candida robusta* MIUG 3.2, *Candida robusta* MIUG 3.3, *Candida tropicalis* MIUG 3.4, *Candida utilis* MIUG 3.5, *Sacharomyces cerevisae* MIUG 3.6, *Yarrowia lipolytica* MIUG RD 14, *Yarrowia lipolytica* MIUG RD 15, *Yarrowia lipolytica* MIUG RD 16 and *Saccharomycopsis fibuligera* MIUG 25D, used in this study, were provided from the Microorganims Collection (acronym MIUG) of the Research and Education Platform (Bioaliment) of „Dunărea de Jos" University of Galati, Romania. The pure stock cultures were maintained by cultivation on yeast extract peptone dextrose (YPD) agar medium and preserved at 4 °C.

3.2. Cultivation Conditions

The basic medium for yeasts cultivation was YPD broth containing (g L^{-1}): peptone (20), yeast extract (10), glucose (10), pH = 5.5. YPD with 20 g L^{-1} agar was also used for pure culture preservation and multiplication. All the reagents were purchased from Merck KGaA (Darmstadt, Germany).

The inoculum was prepared from fresh biomass, 72 h of incubation. The sterile YPD liquid medium was inoculated (10^6 CFU mL^{-1}) and the culture was placed on an orbital shaker (Companion Comecta S.A., Spain) at 180 rpm and 28 °C for 72 h.

3.3. Evaluation of the Toxicity of the Reactants on the Microorganisms' Physiological Functionality

In the first screening step, the toxicity of the reactants which were used for indolizine synthesis (three starting chemicals, 4,4'-bipyridine, 2-bromoacetophenones and ethyl propiolate) upon yeast cells metabolic functionality was tested by stationary cultivation on YPD agar medium supplemented with reagent solutions in different concentrations (0.1%, 0.2%, 0.3%). The reagents ratio in the reaction media was 1:3:3, so this proportion was kept in the culture medium thus obtaining 0.5, 1.0 and 1.5 mmol L^{-1} of 4,4'-bipyridine and 1.5-, 3.0- and 4.5-mmol L^{-1} of 2-bromoacetophenones and ethyl propiolate, respectively, in sterile Petri dishes. The yeast cells from pure active culture were seeded by pinching on the surface of the solide medium, using a sterile microbiologic loop. The inoculated plates were incubated at 28 °C for 144 h. The toxicity of the tested chemicals was evaluated by measuring the diameter of the formed colonies, expressed in mm.

3.4. Yeast Multiplication by Submerged Cultivation in the Presence of Chemicals

The *Yarrowia lipolytica* yeast strains coded MIUG RD 14, MIUG RD 15 and MIUG RD 16 selected as the most resistant to chemicals toxicity were then cultivated in submerged condition in YPD liquid medium supplemented with 0.1% (v/v) of 4,4'-bipyridine (10^{-2} mol·L^{-1}), 2-bromoacetophenones (3×10^{-2} mol·L^{-1}) and ethyl propiolate (3×10^{-2} mol·L^{-1}). After homogenization, the medium was inoculated with 5% of inoculum (10^6 CFU mL^{-1}). Control samples were obtained by yeast cultivation in medium without chemicals. The samples were incubated in an orbital shaker (Companion Comecta S.A., Spain) at 180 rpm and 28 °C, for 100 h. Cells multiplication dynamic was evaluated by measurement of the optical density at 600 nm, at every 24 h, during of 100 h of submerged cultivation. The rate of yeast multiplication inhibition was calculated by comparing the growth potential in YPD liquid medium with and without chemicals.

After yeast cultivation, the biomass was separated by centrifugation (Hettich, Tuttlingen, Germany) for 20 min at 6000 rpm. Cell-free supernatants were assayed for crude extracellular lipase activity.

3.5. Lipase Assay

Lipase activity was determined using the titrimetical method of MaJumder et al. (2009) [36], modified. Briefly, 1 mL of culture supernatant was added to the assay substrate, containing 10 mL of 10% (v/v) homogenized palm oil in 10% (w/v) acacia gum, 2.0 mL of 0.6% CaCl$_2$ solution and 5 mL of 250-mmol L^{-1} phosphate buffer (pH 7.0). All the reagents were purchased from Merck KGaA (Darmstadt, Germany). The biocatalyst and substrate mixture was incubated on orbital shaker with 800 rpm at 30 °C for 20 min. Twenty-five milliliters of ethanol were added to the reaction mixture. The liberated fatty acids were titrated with 0.1-mmol L^{-1} NaOH using phenolphthalein as an indicator. One lipase unit is defined as the amount of enzyme which releases one micro mole fatty acid per minute under specified assay conditions and expressed as units per mL (U mL^{-1}).

3.6. Preparation of Crude Biocatalysts for Cycloaddition Reactions

50 mL of YPD liquid medium was inoculated with 5% (v/v) of *Y. lipolytica* inoculum (10^6 CFU mL^{-1}). Cultivation took place in submerged conditions on an orbital shaker (Companion Comecta S.A., Spain) at 180 rpm and 28 °C, during 144 h. The whole culture was centrifuged for 20 min at 6000 rpm and 4 °C. After centrifugation, the cell-free supernatant was collected in a sterile tube, filtered (Whatman No.1, 0.45 μm) and was used as such as crude enzyme extract. The biomass obtained was washed twice with sterile saline (0.9% NaCl) and used as a whole cell biocatalyst after resuspension in 5 mL 250-mmol L^{-1} phosphate buffer, pH 7.0.

3.7. Indolizine Bioformation Procedure

Two types of biocatalysts—biomass and cell-free supernatant—obtained after 72, 96, 120 and 144 h of submerged cultivation in aerobe conditions, of the three selected *Yarrowia lipolytica* strains were tested for use in reactions of cycloaddition to obtain fluorescent indolizines. To an Eppendorf containing 1 mL of 250-mmol L^{-1} phosphate buffer solutions (pH 7.0) and crude enzymes (0.5 mL culture supernatant or biomass 10%, v/v) was added the pyridinium heterocycle compounds, (4,4'-bipyridine) **1** (15.6 mg, 0.1 mmol), halide derivatives **2** (60 mg, 0.3 mmol) and ethyl propiolate, **3** (30.6 µL, 0.3 mmol). The reaction mixture was stirred at room temperature on an orbital shaker (Biosan, Riga, Latvia) for 8 days.

The evolution of each reaction was followed by TLC and HPLC-MS (Thermo Scientific, Waltham, MA, USA) analysis. The composition in relative percentages of indolizines was computed through the normalization method from the HPLC peak areas, and the bioconversion yield was used as a performance indicator.

3.8. Separation and Purification of the Indolizines Bioconversion Products

All the chemicals and the solvents were of analytical grade and commercially available from Merck (Darmstadt, Germany). The biocatalytic reactions were performed in a thermoshaker Biosan TS 100 (Biosan, Riga, Latvia). The biotransformations were analyzed by a Thermo Scientific system, high performance liquid chromatography coupled with a diode array detector. The electrospray ionization (ESI) mass spectra were measured on Thermo Scientific HPLC/MSQ Plus. HPLC analysis was performed with a C18 HPLC column (50 mm × 0.2 mm, film thickness 0.32 µm) using acetonitrile, flow rate 0.3 mL min^{-1} with UV detector at wavelength λ = 254 nm. ^{1}H NMR and ^{13}C NMR spectra were recorded with a Bruker 400 Ultrashield (400 MHz) spectrometer (Daltonics, Hamburg, Germany) operating at room temperature. Abbreviations for data quoted are m, multiplet; s, singlet; dd, doublet of doublets; d, doublet; t, triplet; q, quartet. $CDCl_3$ was used as solvent. For the NMR data analysis, MestReNova software (version 10.0, Thermo Scientific, Waltham, USA) was used. Melting points were recorded with a Büchi Melting Point B-540 (Essen, Germany). Elemental analyses (C, H, N) were performed with a Fisons Instruments 1108 (Thermo Scientific, Waltham, MA, USA) CHNS-O elemental analyzer. The infrared spectra were collected using a Nicolet iS50 FT-IR spectrometer (Thermo Scientific, Waltham, MA, USA) equipped with a built-in ATR accessory, DTGS detector and KBr beamsplitter. A total of 32 scans were co-added over the range of 4000–400 cm^{-1} with a resolution of 4 cm^{-1}. After each spectrum, the ATR plate was cleaned with ethanol solution. The FT-IR spectrometer was situated in a room that was air conditioned with controlled temperature (21 °C). For the FT-IR data analysis, Omnic software (version 9.6, Thermo Scientific, Waltham, USA) was used.

The extraction of the bioconversion products was carried out with chloroform (3 × 10 mL). The combined extract was washed with distilled water (3 × 10 mL), in order to remove the main intermediate, a bisquaternary pyridinium salt. The organic phase was dried over anhydrous sodium sulphate and filtered using Whatman No.1 filter paper. The solvent was removed under reduced pressure and the crude product was precipitated with methanol to give pure reaction products the bis-indolizine in moderate to good yields for *Y. lipolytica* MIUG RD14 strain biocatalysis. The concentrations of obtained products were analyzed by HPLC-MS using as standards the bis-indolizines obtained by classical method [54].

3.9. Evaluation of the Toxicity of Indolizines on the Germination of Triticum azestivum L.

In this study, healthy and mature wheat seeds were pretreated with sodium hypochlorite (1%) two times for 10 min in order to sterilize them, after which they were rinsed five times with ultrapure water. Germination boxes with 24 germination compartments were used, and in each germination compartment were inserted two layers of filter papers and 25 seeds in the presence and in the absence of two concentrations of biologic interest of indolizines compounds (10^{-3}-M and 10^{-4} M). Control seeds were treated with distilled water, and four replicates were analyzed for each condition. The seeds were incubated in a dark chamber with relative humity and temperature control (70% ± 3% and 23 ± 2 °C), for 5 days and the appropriate water per day was added (Figure 10). Seeds were regarded to have germinated when the radicle was over 2.0 mm in length. The parameters evaluated were the seed germination (SG), the seed germination index (GI), the relative radicle growth (RRG), the relative seed germination (RSG) and the vigor index (VI%) [48–50] as follows:

$$SG = (\text{Number of germinated seeds after 5 days/Number of total seeds}) \times 100\%$$
$$RSG = (\text{Number of germinated seeds(sample)/Number of germinated seeds (control)}) \times 100\%$$
$$RRG = (\text{Total radicle length of germinated seeds (sample)/Total radicle length of germinated seeds (control)}) \times 100\%$$
$$GI = (\% \ SG) \times (\% \ RRG)/100\%$$
$$VI = (\% \ SG) * \text{mean of radicle length/mean of shoot length}$$

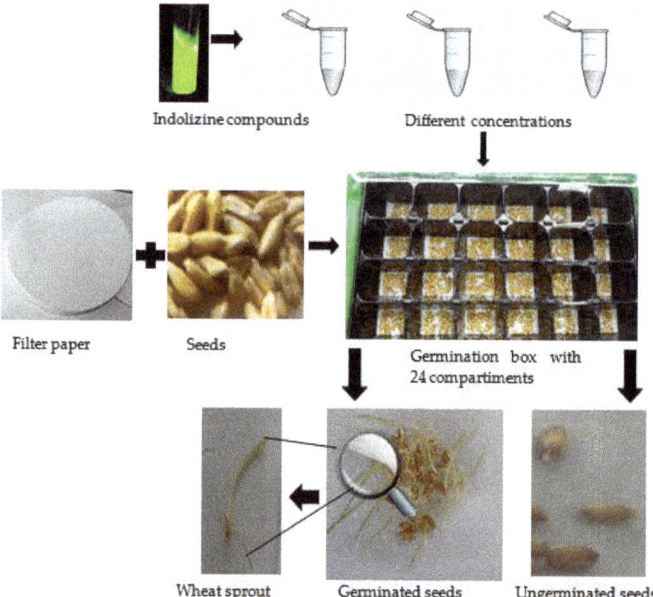

Figure 10. Phytotoxic analysis of the germination of wheat seeds exposed to indolizines compounds at different concentrations.

Confocal laser scanning microscopy (CLSM) using a Zeiss system (LSM 710) (Carl Zeiss Microscopy, GmbH, Jena, Germany) equipped with a diode laser (405 nm), Ar laser (458, 488, 514 nm), DPSS laser (diode pumped solid state e 561 nm) and He–Ne laser (633 nm) was used to analyze the germination of the wheat seeds. The microscopic glass slides with sections through shoot and root of samples were observed using a Zeiss Axio Observer Z1 inverted microscope equipped with a 40×apochromatic objective (numerical aperture 1.4) and the FS49, FS38 and FS15 filters. The parameters used for image acquisition were line scan mode, mean method, average number 8, speed 6, 12-bit-depth. The images were analyzed with the ZEN 2012 SP1 software (Black Edition, Carl Zeiss Microscopy, GmbH, Jena, Germany).

3.10. Statistical Analysis

The data presented here represents the mean ± standard deviation of three or four replicates. Data were subjected to statistical analysis in Microsoft Excel program. Data related to the indolizine biosynthesis was subjected to analysis of variance (one-way ANOVA) in Duncan multiple range test using SPSS (version 10, IBM, Armonk, NY, USA) statistical software. The differences with $p < 0.05$ were considered significant.

4. Conclusions

In summary, we managed to synthesize indolizine compounds with important biologic properties in good yields through a new method by using whole cells biocatalysis. The high purity and yields of the synthesized bis-indolizines denotes a remarkable potential of selected *Y. lipolytica* strains to biosynthesis a multi-enzymatic catalytic system, consisting of oxidases, lipases and cytochromes P450, responsible for the bioconversion. These effective biocataltic properties were exhibited in cycloaddition reactions. This multicomponent reaction that takes place under environmentally friendly conditions has provided a novel and interesting route to construct indolizine nucleus—important scaffolds for the preparation of pharmaceuticals. Furthermore, this "one-pot" reaction highlights relevant features such as mild conditions, selectivity and high efficiency as compared to classical methods that take place in steps with the formation of intermediate compounds, where toxic, expensive and hard-to-recover catalysts are usually used. These indolizine compounds do not exhibit toxic action on plants, evaluated by the effect on seed germination of wheat. The effect of indolizine compounds on mammalian cells and various microorganisms and underlying molecular mechanisms will be explored in our following studies. The present work highlights the role of yeasts species in novel products discovery and drug development.

Supplementary Materials: The following are available online at http://www.mdpi.com/2073-4344/10/6/629/s1. Figure S1: HPLC chromatogram of bis-indolizine **4a**, Figure S2. HPLC chromatogram of bis-indolizine **4b**, Figure S3. HPLC chromatogram of bis-indolizine **4c**, Figure S4. FTIR spectrum of compound **4a**, Figure S5. FTIR spectrum of compound **4b**, Figure S6. FTIR spectrum of compound **4c**, Figure S7. ^{1}H NMR spectrum of bis-indolizine **4a**, Figure S8. ^{1}H NMR spectrum of bis-indolizine **4b**, Figure S9. ^{1}H NMR spectrum of bis-indolizine **4c**.

Author Contributions: Conceptualization, A.V.B., B.F.; and R.-M.D.; methodology, B.F., R.-M.D., I.O.G., G.H., F.B.; software, G.H.; formal analysis I.O.G., V.B., G.H., L.F., A.V.B.; resources, R.-M.D., G.-E.B.; writing—original draft preparation, A.V.B., I.O.G., G.H., L.F., R.-M.D.; writing—review and editing, All; supervision, R.-M.D., B.F., L.F., G.B.; project administration, R.-M.D., B.F.; funding acquisition, R.-M.D., G.-E.B. Investigation, A.V.B., V.B., F.B.; All authors have read and agreed to the published version of the manuscript.

Funding: This work was supported by a grant of the Romanian National Authority for Scientific Research, CNCS-UEFISCDI, project number PN-II-ID-PCE-2011-3-0226.

Acknowledgments: Authors are indebted to the University of "Dunarea de Jos" of Galati, Romania. M.C. thanks the project "Excellence, performance and competitiveness in the Research, Development and Innovation activities at "Dunarea de Jos" University of Galati", acronym "EXPERT", financed by the Romanian Ministry of Research and Innovation in the framework of program 1—Development of the national research and development system, Sub-program 1.2—Institutional Performance—Projects for financing excellence in Research, Development and Innovation, Contract no. 14PFE/17.10.2018. The authors are grateful for the technical support offered by Center MoRAS developed through Grant POSCCE ID 1815, cod SMIS 48,745 (www.moras.ugal.ro).

Conflicts of Interest: The authors declare no conflicts of interest.

References

1. Zhao, Q.; Ansorge-Schumacher, M.B.; Haag, R.; Wu, C. Living whole-cell catalysis in compartmentalized emulsion. *Bioresour. Technol.* **2020**, *295*, 122221. [CrossRef]
2. Feng, G.; Wu, H.; Li, X.; Lai, F.; Zhao, G.; Yang, M.; Liu, H. A new, efficient and highly-regioselective approach to synthesis of 6-O-propionyl-d-glucose by using whole-cell biocatalysts. *Biochem. Eng. J.* **2015**, *95*, 56–61. [CrossRef]
3. Anteneh, Y.S.; Franco, C.M.M. Whole cell actinobacteria as biocatalysts. *Front. Microbiol.* **2019**, *10*, 1–15. [CrossRef] [PubMed]

4. Garzón-Posse, F.; Becerra-Figueroa, L.; Hernández-Arias, J.; Gamba-Sánchez, D. Whole cells as biocatalysts in organic transformations. *Molecules* **2018**, *23*, 1265. [CrossRef]
5. Dinica, R.M.; Furdui, B.; Ghinea, I.O.; Bahrim, G.; Bonte, S.; Demeunynck, M. Novel one-pot green synthesis of indolizines biocatalysed by Candida antarctica lipases. *Mar. Drugs* **2013**, *11*, 431–439. [CrossRef]
6. Matsuda, T.; Yamanaka, R.; Nakamura, K. Recent progress in biocatalysis for asymmetric oxidation and reduction. *Tetrahedron Asymmetry* **2009**, *20*, 513–557. [CrossRef]
7. Kudalkar, G.P.; Tiwari, V.K.; Lee, J.D.; Berkowitz, D.B. A Hammett Study of Clostridium acetobutylicum Alcohol Dehydrogenase (CaADH): An enzyme with remarkable substrate promiscuity and utility for organic synthesis. *Synlett* **2020**, *31*, 237–247. [CrossRef]
8. Antonopoulou, I.; Dilokpimol, A.; Iancu, L.; Mäkelä, M.R.; Varriale, S.; Cerullo, G.; Hüttner, S.; Uthoff, S.; Jütten, P.; Piechot, A.; et al. The synthetic potential of fungal feruloyl esterases: A correlation with current classification systems and predicted structural properties. *Catalysts* **2018**, *8*, 242. [CrossRef]
9. Yin, D.H.; Liu, W.; Wang, Z.X.; Huang, X.; Zhang, J.; Huang, D.C. Enzyme-catalyzed direct three-component aza-Diels–Alder reaction using lipase from Candida sp. 99–125. *Chin. Chem. Lett.* **2017**, *28*, 153–158. [CrossRef]
10. Carballeira, J.D.; Quezada, M.A.; Hoyos, P.; Simeó, Y.; Hernaiz, M.J.; Alcantara, A.R.; Sinisterra, J.V. Microbial cells as catalysts for stereoselective red-ox reactions. *Biotechnol. Adv.* **2009**, *27*, 686–714. [CrossRef]
11. Ratnayake, N.D.; Theisen, C.; Walter, T.; Walker, K.D. Whole-cell biocatalytic production of variously substituted β-aryl- and β-heteroaryl-β-amino acids. *J. Biotechnol.* **2016**, *217*, 12–21. [CrossRef] [PubMed]
12. Birolli, W.G.; Ferreira, I.M.; Alvarenga, N.; De Santos, D.A.; De Matos, I.L.; Comasseto, J.V.; Porto, A.L.M. Biocatalysis and biotransformation in Brazil: An overview. *Biotechnol. Adv.* **2015**, *33*, 481–510. [CrossRef]
13. Wachtmeister, J.; Rother, D. Recent advances in whole cell biocatalysis techniques bridging from investigative to industrial scale. *Curr. Opin. Biotechnol.* **2016**, *42*, 169–177. [CrossRef] [PubMed]
14. Rywińska, A.; Juszczyk, P.; Wojtatowicz, M.; Robak, M.; Lazar, Z.; Tomaszewska, L.; Rymowicz, W. Glycerol as a promising substrate for Yarrowia lipolytica biotechnological applications. *Biomass Bioenergy* **2013**, *48*, 148–166. [CrossRef]
15. Zieniuk, B.; Fabiszewska, A. Yarrowia lipolytica: A beneficial yeast in biotechnology as a rare opportunistic fungal pathogen: A minireview. *World J. Microbiol. Biotechnol.* **2019**, *35*, 1–8. [CrossRef]
16. Fickers, P.; Benetti, P.H.; Waché, Y.; Marty, A.; Mauersberger, S.; Smit, M.S.; Nicaud, J.M. Hydrophobic substrate utilisation by the yeast Yarrowia lipolytica, and its potential applications. *FEMS Yeast Res.* **2005**, *5*, 527–543. [CrossRef]
17. Gonçalves, F.A.G.; Colen, G.; Takahashi, J.A. Yarrowia lipolytica and its multiple applications in the biotechnological industry. *Sci. World J.* **2014**, *2014*, 1–14. [CrossRef]
18. Casas-Godoy, L.; Marty, A.; Sandoval, G.; Ferreira-Dias, S. Optimization of medium chain length fatty acid incorporation into olive oil catalyzed by immobilized Lip2 from Yarrowia lipolytica. *Biochem. Eng. J.* **2013**, *77*, 20–27. [CrossRef]
19. Facin, B.R.; Melchiors, M.S.; Valério, A.; Oliveira, J.V.; De Oliveira, D. Driving immobilized lipases as biocatalysts: 10 years state of the art and future prospects. *Ind. Eng. Chem. Res.* **2019**, *58*, 5358–5378. [CrossRef]
20. Arvin-Berod, M.; Desroches-Castan, A.; Bonte, S.; Brugière, S.; Couté, Y.; Guyon, L.; Feige, J.J.; Baussanne, I.; Demeunynck, M. Indolizine-based scaffolds as efficient and versatile tools: Application to the synthesis of biotin-tagged antiangiogenic drugs. *ACS Omega* **2017**, *2*, 9221–9230. [CrossRef]
21. Sharma, V.; Kumar, V. Indolizine: A biologically active moiety. *Med. Chem. Res.* **2014**, *23*, 3593–3606. [CrossRef]
22. Narayanaswamy, K.; Id, V.; Chandrashekharappa, S.; Pillay, M.; Abdallah, H.H.; Mahomoodally, F.M.; Bhandary, S.; Chopra, D.; Attimarad, M.; Aldhubiab, B.E.; et al. Computational, crystallographic studies, cytotoxicity and anti-tubercular activity of substituted 7-methoxy-indolizine analogues. *PLoS ONE* **2019**, *14*, e0217270. [CrossRef]
23. Venugopala, K.N.; Tratrat, C.; Chandrashekharappa, S. Anti-tubercular potency and computationally assessed drug-likeness and toxicology of diversely substituted indolizines. *Indian J. Pharm. Educ.* **2019**, *53*. [CrossRef]

24. Uppar, V.; Chandrashekharappa, S.; Basarikatti, A.I.; Banuprakash, G.; Mahendra, K.; Chougala, M.; Mudnakudu-nagaraju, K.K.; Ningegowda, R.; Padmashali, B. Synthesis, antibacterial, and antioxidant studies of 7-amino-3-(4-fluorobenzoyl)indolizine-1-carboxylate derivatives. *J. Appl. Pharm. Sci.* **2020**, *10*, 77–85.
25. Narajji, C.; Karvekar, M.D.; Das, A.K. Synthesis and antioxidant activity of 3,3'-diselanediylbis (N,N-disubstituted indolizine-1-carboxamide) and derivatives. *S. Afr. J. Chem.* **2008**, *61*, 53–55.
26. Moon, S.H.; Jung, Y.; Kim, S.H.; Kim, I. Synthesis, characterization and biological evaluation of anti-cancer indolizine derivatives via inhibiting β-catenin activity and activating p53. *Bioorganic Med. Chem. Lett.* **2016**, *26*, 110–113. [CrossRef]
27. Belal, A.; Gouda, A.M.; Ahmed, A.S.; Abdel Gawad, N.M. Synthesis of novel indolizine, diazepinoindolizine and Pyrimidoindolizine derivatives as potent and selective anticancer agents. *Res. Chem. Intermed.* **2015**, *41*, 9687–9701. [CrossRef]
28. Ghinet, A.; Abuhaie, C.M.; Gautret, P.; Rigo, B.; Dubois, J.; Farce, A.; Belei, D.; Bîcu, E. Studies on indolizines. Evaluation of their biological properties as microtubule-interacting agents and as melanoma targeting compounds. *Eur. J. Med. Chem.* **2015**, *89*, 115–127. [CrossRef]
29. Weide, T.; Arve, L.; Prinz, H.; Waldmann, H.; Kessler, H. 3-Substituted indolizine-1-carbonitrile derivatives as phosphatase inhibitors. *Bioorganic Med. Chem. Lett.* **2006**, *16*, 59–63. [CrossRef]
30. Sharma, V.; Kumar, V. Pharmacophore mapping studies on indolizine derivatives as 15-LOX inhibitors. *Bull. Fac. Pharm. Cairo Univ.* **2015**, *53*, 63–68. [CrossRef]
31. Huang, W.; Zuo, T.; Luo, X.; Jin, H.; Liu, Z.; Yang, Z.; Yu, X.; Zhang, L.; Zhang, L. Indolizine derivatives as HIV-1 VIF-elonginc interaction inhibitors. *Chem. Biol. Drug Des.* **2013**, *81*, 730–741. [CrossRef]
32. De Souza, C.R.; Gonçalves, A.C.; Amaral, M.F.Z.J.; Dos Santos, A.A.; Clososki, G.C. Recent synthetic developments and reactivity of aromatic indolizines. *Targets Heterocycl. Syst.* **2016**, *20*, 365–392.
33. Sadowski, B.; Klajn, J.; Gryko, D.T. Recent advances in the synthesis of indolizines and their π-expanded analogues. *Org. Biomol. Chem.* **2016**, *14*, 7804–7828. [CrossRef]
34. Ghinea, I.O.; Dinica, R.M. Breakthroughs in indole and indolizine chemistry-new synthetic pathways, new applications. In *Scope of Selective Heterocycles from Organic and Pharmaceutical Perspective*; Chapter 5; Ravi, V., Ed.; IntechOpen: London, UK, 2016; pp. 115–142. [CrossRef]
35. Kapoor, M.; Gupta, M.N. Lipase promiscuity and its biochemical applications. *Process. Biochem.* **2012**, *47*, 555–569. [CrossRef]
36. Majumder, A.B.; Ramesh, N.G.; Gupta, M.N. A lipase catalyzed condensation reaction with a tricyclic diketone: Yet another example of biocatalytic promiscuity. *Tetrahedron Lett.* **2009**, *50*, 5190–5193. [CrossRef]
37. Reetz, M.T.; Mondière, R.; Carballeira, J.D. Enzyme promiscuity: First protein-catalyzed Morita-Baylis-Hillman reaction. *Tetrahedron Lett.* **2007**, *48*, 1679–1681. [CrossRef]
38. Wang, Z.; Wang, C.Y.; Wang, H.R.; Zhang, H.; Su, Y.L.; Ji, T.F.; Wang, L. Lipase-catalyzed Knoevenagel condensation between α,β unsaturated aldehydes and active methylene compounds. *Chin. Chem. Lett.* **2014**, *25*, 802–804. [CrossRef]
39. Yang, Y.; Zhang, J.; Wu, D.; Xing, Z.; Zhou, Y.; Shi, W.; Li, Q. Chemoenzymatic synthesis of polymeric materials using lipases as catalysts: A review. *Biotechnol. Adv.* **2014**, *32*, 642–651. [CrossRef]
40. Parfene, G.; Horincar, V.B.; Bahrim, G.; Vannini, L.; Gottardi, D.; Maria Elisabetta, G. Lipolytic activity of lipases from different strains of Yarrowia lipolytica in hydrolysed vegetable fats at low temperature and water activity. *Rom. Biotechnol. Lett.* **2011**, *16*, 46–52.
41. Destain, J.; Roblain, D.; Thonart, P. Improvement of lipase production from Yarrowia lipolytica. *Biotechnol. Lett.* **1997**, *19*, 105–108. [CrossRef]
42. Pignede, G.; Wang, H.J.; Fudalej, F.; Seman, M.; Gaillardin, C.; Nicaud, J.M. Autocloning and amplification of LIP2 in Yarrowia lipolytica. *Appl. Environ. Microbiol.* **2000**, *66*, 3283–3289. [CrossRef] [PubMed]
43. Fickers, P.; Fudalej, F.; Nicaud, J.M.; Destain, J.; Thonart, P. Selection of new over-producing derivatives for the improvement of extracellular lipase production by the non-conventional yeast Yarrowia lipolytica. *J. Biotechnol.* **2005**, *115*, 379–386. [CrossRef] [PubMed]
44. Białecka-Florjańczyk, E.; Krzyczkowska, J.; Stolarzewicz, I.; Kapturowska, A. Synthesis of 2-phenylethyl acetate in the presence of Yarrowia lipolytica KKP 379 biomass. *J. Mol. Catal. B Enzym.* **2012**, *74*, 241–245. [CrossRef]

45. Napora, K.; Wrodnigg, T.M.; Kosmus, P.; Thonhofer, M.; Robins, K.; Winkler, M. Yarrowia lipolytica dehydrogenase/reductase: An enzyme tolerant for lipophilic compounds and carbohydrate substrates. *Bioorganic Med. Chem. Lett.* **2013**, *23*, 3393–3395. [CrossRef] [PubMed]
46. Druta, I.; Dinica, R.M.; Bacu, E.; Humelnicu, I. Synthesis of 7, 7′-bisindolizines by the reaction of 4,4′-bipyridinium-ylides with activated alkynes. *Tetrahedron* **1998**, *54*, 10811–10818. [CrossRef]
47. Mazumder, P.; Khwairakpam, M.; Kalamdhad, A.S. Bio-inherent attributes of water hyacinth procured from contaminated water body–effect of its compost on seed germination and radicle growth. *J. Environ. Manag.* **2020**, *257*, 109990. [CrossRef]
48. Li, R.; He, J.; Xie, H.; Wang, W.; Bose, S.K.; Sun, Y.; Hu, J.; Yin, H. Effects of chitosan nanoparticles on seed germination and seedling growth of wheat (*Triticum aestivum* L.). *Int. J. Biol. Macromol.* **2019**, *126*, 91–100. [CrossRef]
49. Cudalbeanu, M.; Furdui, B.; Cârâc, G.; Barbu, V.; Iancu, A.V.; Campello, M.P.C.; Leitão, J.H.; Sousa, S.A.; Dinica, R.M. Antifungal, Antitumoral and Antioxidant Potential of the Danube Delta Nymphaea alba Extracts. *Antibiotion* **2019**, *9*, 7. [CrossRef]
50. Tabacaru, A.; Botezatu-Dediu, A.-V.; Horincar, G.; Furdui, B.; Dinica, R.M. Green Accelerated Synthesis, Antimicrobial Activity. *Molecules* **2019**, *24*, 2424. [CrossRef] [PubMed]
51. OECD. OECD test guideline 208: Terrestrial plant test—Seedling emergence and seedling growth test. *Guidel. Test Chem. Terr. Plant Test Seedl. Emerg. Seedl. Growth Test* **2006**, *227*, 1–21.
52. Luo, Y.; Liang, J.; Zeng, G.; Chen, M.; Mo, D.; Li, G.; Zhang, D. Seed germination test for toxicity evaluation of compost: Its roles, problems and prospects. *Waste Manag.* **2018**, *71*, 109–114. [CrossRef] [PubMed]
53. Tiquia, S.M.; Tam, N.F.Y.; Hodgkiss, I.J. Effects of composting on phytotoxicity of spent pig-manure sawdust litter. *Environ. Pollut.* **1996**, *93*, 249–256. [CrossRef]
54. Barth, G.; Gaillardin, C. Yarrowia lipolytica. In *Nonconventional Yeasts in Biotechnology: A Handbook*; Springer: Berlin/Heidelberg, Germany, 1996; pp. 313–388, ISBN 978-3-642-79856-6.

© 2020 by the authors. Licensee MDPI, Basel, Switzerland. This article is an open access article distributed under the terms and conditions of the Creative Commons Attribution (CC BY) license (http://creativecommons.org/licenses/by/4.0/).

Review

Kramers' Theory and the Dependence of Enzyme Dynamics on Trehalose-Mediated Viscosity

José G. Sampedro [1,*], Miguel A. Rivera-Moran [1] and Salvador Uribe-Carvajal [2]

1. Instituto de Física, Universidad Autónoma de San Luis Potosí, Manuel Nava 6, Zona Universitaria, San Luis Potosí C.P. 78290, Mexico; miguel.rivera.moran@gmail.com
2. Instituto de Fisiología Celular, Universidad Nacional Autónoma de México, Ciudad de México C.P. 04510, Mexico; suribe@ifc.unam.mx
* Correspondence: sampedro@dec1.ifisica.uaslp.mx; Tel.: +52-(444)-8262-3200 (ext. 5715)

Received: 29 April 2020; Accepted: 29 May 2020; Published: 11 June 2020

Abstract: The disaccharide trehalose is accumulated in the cytoplasm of some organisms in response to harsh environmental conditions. Trehalose biosynthesis and accumulation are important for the survival of such organisms by protecting the structure and function of proteins and membranes. Trehalose affects the dynamics of proteins and water molecules in the bulk and the protein hydration shell. Enzyme catalysis and other processes dependent on protein dynamics are affected by the viscosity generated by trehalose, as described by the Kramers' theory of rate reactions. Enzyme/protein stabilization by trehalose against thermal inactivation/unfolding is also explained by the viscosity mediated hindering of the thermally generated structural dynamics, as described by Kramers' theory. The analysis of the relationship of viscosity–protein dynamics, and its effects on enzyme/protein function and other processes (thermal inactivation and unfolding/folding), is the focus of the present work regarding the disaccharide trehalose as the viscosity generating solute. Finally, trehalose is widely used (alone or in combination with other compounds) in the stabilization of enzymes in the laboratory and in biotechnological applications; hence, considering the effect of viscosity on catalysis and stability of enzymes may help to improve the results of trehalose in its diverse uses/applications.

Keywords: trehalose; viscosity; enzymes; protein dynamics; Kramers' theory; protein stabilization; enzyme inhibition

---

1. Trehalose in Biology

The disaccharide trehalose has been used as food or therapeutic agent by humans since ancient times [1]. Trehalose is naturally found in the sweet cocoon (the capsule named trehala-manna) that protects the larvae of weevils, insects in the genus *Larinus* [1–4]. Pure trehalose crystals were isolated first from the ergot of rye by Wiggers (1832), then the carbohydrate was identified as the main component of trehala-manna (from which it takes the name trehalose) by Berthelot (1858) and identified as a non-reducing disaccharide formed by two glucose units [5].

Trehalose (α-D-glucopyranosyl-1, 1-α-D-glucopyranoside) was found at high concentrations in diverse organisms [6] and under stress conditions [4], including yeast and fungi [7,8], plants [9], and others [6]. Trehalose, in addition to forming part of cellular structures [6], has been claimed to behave as a reserve carbohydrate as it is usually found in spores of yeast, fungi, and bacteria [4,6–8,10], and also as a biostructure stabilizer, namely of proteins and membranes during harsh environmental conditions [4], and even in the pathogenesis of virulent bacteria (like actinomycetes and mycobacteria) and fungi [11,12]. Currently, the metabolic pathways for trehalose biosynthesis and hydrolysis, and their regulation are well known [8–10,13,14], and in some organisms, trehalose metabolic pathways seem to have a major role in the regulation of other metabolic processes [7,8].

Among the diverse stress conditions where trehalose is synthesized and accumulated, dehydration (or the phenomenon of anhydrobiosis) has been that which has attracted the most interest (because of its potential industrial applications) [5,15–17], namely in food and pharmaceutics. Van Leeuwenhoek (1702) first described the phenomenon in bdelloid rotifers [18,19]. Since then, other organisms have been recognized to endure the absence of water, such as nematodes, brine shrimp cysts, yeast, and some plants [4,6,18].

Experiments have demonstrated the effectiveness of trehalose to preserve the structure of proteins and membranes in the dry state [16,20,21]. Nonetheless, the role of intrinsically disordered proteins (stress proteins) protecting organisms that synthesize little to no trehalose seems to be important [22,23]. In nature, the dehydration process in anhydrobiotic organisms seems to be slow, i.e., water is removed in minute amounts leading to cell desiccation [16]. At the same time, trehalose begins to be synthesized, reaching high concentrations in the cytoplasm, while a small amount is exported (trehalose is required at both surfaces of the plasma membrane) [24]. Although cell survival does correlate with the amount of trehalose present in the cell [16,19], other biochemical responses contributing to anhydrobiosis (or acting synergically with trehalose) seem to take place simultaneously to trehalose accumulation [16], e.g., the synthesis of heat shock proteins [25].

Physically, under optimum growth conditions (nutrients, temperature, etc.) the cell cytoplasm is a crowded space containing a great diversity of molecules, hence inherently highly viscous [26–29]. Then, during dehydration, the accumulation of trehalose increases viscosity even further as water-loss proceeds [30–34]. Hence, viscosity increases up to finally reach the glassy state [34,35]. Here, although the interaction of trehalose with proteins has been subjected to some controversy [36], the effect of trehalose on the maintenance of the integrity of membranes (via the interaction with phospholipid headgroups) is relatively well understood [16,37]. Nonetheless, what is certain is that viscosity (the glassy state) generated by trehalose is high. Here proteins and other bio-structures are "arrested", which in turn preserves both the structure and function of proteins [34,38–40].

In microorganisms, trehalose is also synthesized in the stationary phase of growth. In yeast cultures when nutrients are exhausted, cells enter a quiescent state known as the stationary phase [41], where the metabolic rate is slowed and cell division stops (arrested cell cycle) [42,43]. Notably, during entry to the stationary phase, trehalose is accumulated at high concentrations (up to 20% in dry weight), as well as glycogen (up to 8% in dry weight) [42]. The cells at the stationary phase thus become denser, heavier, and cytoplasm viscosity increases [31,32,42]. In this regard, trehalose seems to have a role in the exit of yeast from the stationary phase besides [42]; however, no specific role of trehalose in cell metabolism under the quiescent state has been described. Instead, cells in the stationary phase display high thermotolerance, possibly due to trehalose accumulation [4,21].

2. Trehalose, Water, and Proteins

Trehalose/water interaction is the basis for protein stabilization by trehalose [36,44–46]. Water is the medium where most biochemical processes take place and the hydration water has a fundamental role in protein structure and function [47–49]. In this regard, water molecules seem to be more uniformly distributed on the surface of globular proteins when compared to those on intrinsically disordered proteins (IDPs) [48,50]; despite that, the arrangement of tetrahedral water molecules is more disordered on the surface of globular proteins than on IDPs [50]. The above is probably due to the highly irregular surface of globular proteins. Interestingly, the dynamics of hydration water on proteins are different from bulk water molecules [47,48,51–53]. In solution, trehalose molecules display a relatively rigid conformation and slow translational motion (self-diffusion) [46,54]. Besides, trehalose molecules interact through trehalose–trehalose hydrogen bonds [44,55,56]; while trehalose–water hydrogen bonds are weaker [56]. This seems to facilitate vitrification [46], i.e., by allowing water molecules to escape from the system. Additionally, trehalose appears to be physically (preferentially) excluded from protein surfaces [57,58], thus promoting preferential hydration [45,59–61]. This is while maintaining the disordered arrangement of hydration water molecules [50]. Nonetheless, the dynamics of hydration water appears to be slowed down by trehalose and consequently, dynamics of the amino acid residues in proteins decrease [45,48,61–67], e.g., in C-phycocyanin (CP), trehalose slows down the dynamics of the hydration water by an order of magnitude when compared to the dynamics of bulk water [68]. As a result, protein processes where structural dynamics are a major component become affected by the presence of trehalose in the medium, namely folding, unfolding, thermal and cold inactivation, oligomerization, aggregation, catalysis, and even the equilibrium between catalytic states (Figure 1). The thermodynamic effect of trehalose would be to increase the energetic barrier between different structural states (Figure 1).

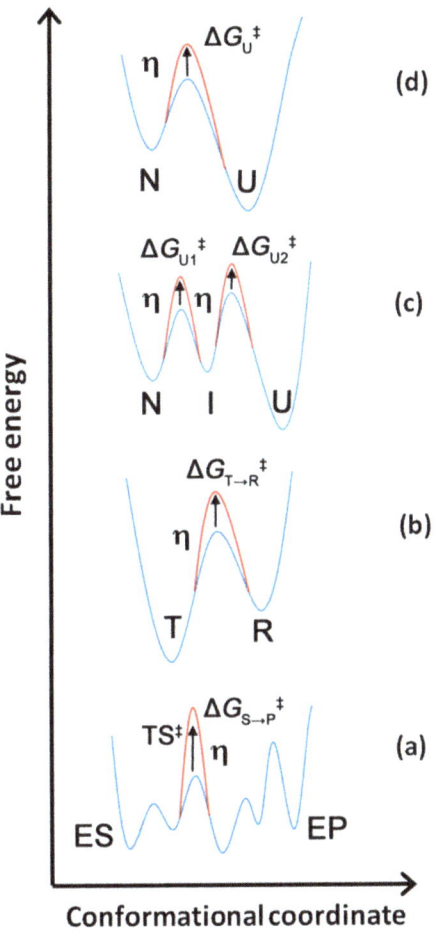

Figure 1. Energy landscape of diffusion-dependent protein processes. High medium viscosity (η) increases the free energy height (red line) of the interconversion between conformational states in proteins (blue line). (**a**) Structural conformational states in enzyme catalysis, (**b**) conformational equilibrium between the T and R states in a cooperative enzyme regarding the model of Monod, Wyman, and Changeux [69,70], (**c**) three-state protein unfolding (N ↔ I ↔ U); formation of the intermediary (I) may involve the dissociation of monomers from the oligomer, (**d**) two-state protein unfolding (N ↔ U).

3. Protein Dynamics and Catalysis

Proteins, like most polymers, display inherent movements inside their structure [71,72]. Some movements participate in the function, e.g., catalysis [71,73], while others are part of the protein vibrational state specific to the medium temperature [74–76]. Indeed, the function of proteins is defined by the dynamics of the three-dimensional structure [77–81]. The inter-conversion between different structural states determines the function of the protein [82]. For example, the motions of the enzyme structure are a major component in catalysis [73,78,83–86]. There exists ample evidence on the role of enzyme structural dynamics in catalysis [73,87–89]; currently, experimental and computational techniques have demonstrated enzymes are highly dynamic entities [90]. Importantly, the rate of the structural dynamics in enzymes whether internal, specific (like loops and domain), as a structural network, or global (collective) has been determined as the limit for the turnover rate (k_{cat}) [71,91]. In this regard, protein crystallography has provided the structural details of enzymes at the atomic level [74,92,93]. Unfortunately, such a structure is just a snapshot of the protein structure that yields limited information about function [77,94]. Nonetheless, efforts have resulted in the successful crystallization of enzymes in complex with ligands which demonstrates the existence of structural dynamics related to catalysis [74,95]. In some fortunate cases, the crystallographic structures of enzymes obtained under different conditions allow the depiction of the catalytic cycle [71,73,96–98]; however, it is worth noting that these states usually are a (small) fraction of the infinite conformations the protein structure may sample [75,82]. Experimental methods to study protein dynamics include NMR relaxation, hydrogen-deuterium exchange, fluorescence, UV-Vis spectroscopy, Raman spectroscopy, infrared spectroscopy, and molecular dynamics simulations (MDS) [73,77,99]; nonetheless, the timescale of measurement usually is variable, e.g., MDS ranges in the ns scale and less, while hydrogen-deuterium exchange ranges between ms and s scale [77]. Importantly, most biological processes (such as enzyme catalysis) lay on the μs to ms time scale [71,73,77]. Here protein dynamics have been evaluated by different methods, e.g., in rotary proteins like FoF1-ATPase [100,101]. MDS has emerged as a very helpful technique to study protein dynamics and function [72,88,102]. The position of any given atom in the polypeptide chain may be known with more precision at any given moment [72], although, a high-resolution three-dimensional (3D) structure is a requisite [103]. Still, when the 3D-structure is not available, modeling is an option [104]. However, the in vitro experimental validation of the in silico results is desirable [49,80,105].

In enzymes, it is known that under isobaric conditions the structural dynamics are modulated by variations in temperature (thermal energy) and solvent composition (cosolvents, ions, polymers, etc.) [38,49,76,106–108], i.e., in Michaelian enzymes, some conformational states may be favored as the kinetics of state interconversion are affected [77,109,110]. The diverse conformational states in a "resting" protein/enzyme (in the absence of ligands) in solution result from side-chain fluctuations, movements of loops and secondary structures, structural domains, and collective global arrangements [82]; thus, before catalysis, enzymes probe multiple conformations [77,109,111]. In this sense, even the whole catalytic cycle may occur in some enzymes in the absence of ligands [73,112]. Notably, in the kinetics of cooperative enzymes, the Monod, Wyman, and Changeux model is used on the assumption that enzymes fluctuate between two states with high ("relaxed", R) and low substrate affinity ("tight", T) [69,113], while binding of ligands, allosteric effectors, and even phosphorylation of some amino acid residues, changes the equilibrium towards the high- or low-affinity state, accordingly [114].

4. Dependence of Enzyme Catalysis on Medium Viscosity (Kramers' Theory)

The viscosity of the crowded cell cytoplasm is higher than pure water [26,27,115–117]. Viscosity varies by cell type, e.g., in *Saccharomyces cerevisiae* cytoplasm it has been calculated to be ~10 cP [115]. Certainly, the viscosity of the cytoplasm is considerably higher in organisms at the quiescent state and during dehydration, osmotic stress, and at heat-shock as they accumulate viscosity generating solutes such as trehalose, glycerol, and others [115]. In the physics of fluids, viscosity is defined as the friction between molecules or the resistance to flow (molecular diffusion) in a liquid (or gaseous) system [115].

In this environment, the three-dimensional (3D) structure of enzymes fluctuates between different conformations [71,118,119]. Protein structural fluctuations are coupled to the movement of water molecules at the hydration shell, and these in turn, to the dynamics of the molecules present in the bulk solution [49,120]. Recently, studies on the translational diffusion of hydration waters concluded that protein function may be modulated by sugar molecules [48,64]. Hence, the inhibition or hindering of protein motions, whether by a decrease in temperature or by changing the composition of the suspending medium, leads to decreased catalysis [85,121–124].

Specifically, trehalose promotes the hydration of the protein surface (preferential hydration) [60,61,125], but it also diminishes the dynamics of hydration water [61,64]. As a result, the motion of amino acid residues is slowed down, and the rate constant of enzyme structural fluctuations becomes dependent on the viscosity of the medium [47]. Certainly, the effect of medium viscosity on enzyme catalysis has already been considered [118], when the dependence of catalysis on viscosity was rationalized to the inhibition of enzyme fluctuations according to Kramers' theory [106,118,120,123,126–128], as described by Equation (1):

$$k = \frac{A}{\eta} e^{-\Delta/k_B T} \qquad (1)$$

where Δ is the height of the potential barrier, k is the rate constant of catalysis (k_{cat}), A is a function of structural parameters characterizing the potential energy profile, k_B is the Boltzmann constant, and η is the viscosity of the medium. Equation (2) is an actualized version of Equation (1) when using the enzyme kinetic parameter V_{max} ($V_{max} = k_{cat} \cdot [E_t]$).

$$V_{max} = \frac{A}{\eta} e^{-\Delta U/RT} \qquad (2)$$

where $-\Delta U$ is the height of the potential barrier, and R is the gas constant. Notably, Equation (2) has been useful to analyze the inhibition of enzyme catalysis by the presence of trehalose [31,32,129,130]; thus indicating that trehalose does inhibit the rate of catalysis by hindering the structural fluctuations of enzymes during the catalytic cycle [31,32,108,130]. V_{max} becomes inversely proportional to the viscosity ($V_{max} \propto 1/\eta$) of the medium, and a straight line is usually observed when plotting V_{max} versus η^{-1} [31,32,108,130]; i.e., as viscosity increases, enzyme activity decreases [32]. Importantly, the viscosity generated by trehalose in the bulk solution would affect mainly the largest structural motions involved in catalysis (Figure 1) [31,129,130]. For example, in the plasma membrane H^+-ATPase from yeast, the motion of the N-domain (a globular structure of ~16.5 kDa) toward the P-domain is required to transfer the γ-phosphate of ATP to the Asp residue [131]. Notably, the activity of the H^+-ATPase decreases linearly as the viscosity generated by trehalose increases (Figure 2a) [31]. This behavior is in agreement with Equation (2), as described by Kramers' theory [31]. Similarly, the H^+-ATPase from plants is inhibited linearly by the viscosity generated by the presence of sucrose (the natural disaccharide in plants) [132]. Additionally, the kinetics of pyruvate reduction by lactate dehydrogenase (LDH) from rabbit muscle is dependent on medium viscosity, as LDH catalysis depends on the movement of a loop on the active site [130,133]. The LDH V_{max} is dependent on medium viscosity as expected from Kramers' theory (Figure 2b) [130]. Notably, the viscosity generated by trehalose increases the activation energy (E_a) in both enzymes, H^+-ATPase, and LDH [31,130] (Figure 3), as suggested in Figure 1a. Therefore, enzyme inhibition by trehalose is expected when a relatively large structural motion is involved in the catalytic mechanism.

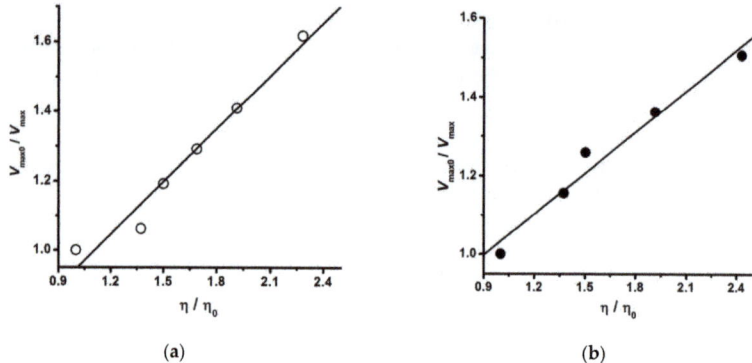

Figure 2. Dependence of enzyme catalysis (30 °C) on medium viscosity as described by Kramers' theory. (a) Viscosity dependence of V_{max} of plasma membrane H^+-ATPase from *Kluyveromyces lactis*. Adapted with permission from Sampedro et al. [31]. Copyright © 2020, American Society for Microbiology; (b) viscosity dependence of V_{max} of lactate dehydrogenase (LDH) from rabbit muscle. Adapted with permission from Hernández-Meza and Sampedro [130]. Copyright © 2020, American Chemical Society.

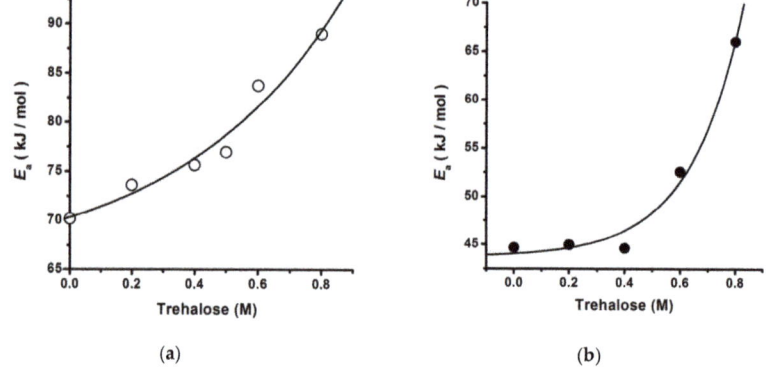

Figure 3. Plot of the activation energy (E_a) of enzyme catalysis versus trehalose concentration. (a) Plasma membrane H^+-ATPase [31]. Reprinted with permission from Sampedro et al. [31]. Copyright © 2020, American Society for Microbiology; (b) LDH [130]. Reprinted with permission from Hernández-Meza and Sampedro [130]. Copyright © 2020, American Chemical Society. The activation energies were calculated as described by the Arrhenius equation ($k_{cat} = A \cdot e^{-E_a/R \cdot T}$) using the enzymes V_{max} values ($V_{max} = k_{cat} \cdot [E_t]$).

Enzymes do display catalysis dependence on viscosity as described by Kramers' theory even when using viscogens others than trehalose; e.g., a decrease in the rate constants (k_{cat}) is observed in the CO_2 hydration and HCO^{3-} dehydration by carbonic anhydrase (CA) [134], the ATPase hydrolysis in chloroplast coupling factor (CF1), myosin and meromyosin [135,136], the deacylation step of subtilisin BPN' catalyzed ester hydrolysis [137], the thioesterase activity of fatty acid synthetase (FAS) [138], the carboxypeptidase-A (CPA)-catalyzed benzoylglycyl-L-phenyl lactate hydrolysis [139], the cysteine protease activity of human ribosomal protein S4 (RPS4) [140], the DNA polymerizing activity of polymerase β (POLB) [141], the sugar cleaving enzyme oligo-1,6-glucosidase 1 (MalL) [142], the indole-3-glycerol phosphate synthase (IGPS) from *Sulfolobus solfataricus* [105], and in the electron transfer reactions of sulfite oxidase (SO), respiratory and photosynthetic complexes [143–145]. In extreme situations of viscosity, i.e., enzymes embedded in a glass-state environment where viscosity is infinitely high, the absence of protein motions leads halts catalysis [120]. The modulation of protein/enzyme dynamics by viscosity has been also recently demonstrated by molecular dynamics simulation (MDS) [146] using the small protein factor Xa [147,148], and ssIGPS [105].

Viscosity generating solutes also affects the enzyme Michaelis-Menten constant ($K_m = \frac{k_{-1}+k_2}{k_1}$) or the binding ($k_1$) and dissociation ($k_{-1}$) of the substrate from the binding site, when in the enzyme a structural diffusion is involved in either of these processes, e.g., glycerol increases the K_m in phosphorylase B [149], sucrose increases the K_m of polymerizing myosin [135], and trehalose decreases the K_m of glucose oxidase (GOx) and glucoamylase from *Aspergillus niger* [150,151]. Therefore, the catalytic efficiency (k_{cat}/K_m) of enzymes varies as substrate affinity is decreased/increased by viscogens [149–151] by hampering structural motions in either ligand binding or dissociation. Interestingly, enzyme activation by trehalose has been reported in thermophilic enzymes like neutral glucoamylase from the mold *Thermomucor indicae-seudaticae* [152]. Similarly, in pyruvate kinase (PK) from *Geobacillus stearothermophilus* (a cooperative enzyme), trehalose increases the velocity of the enzyme reaction (Figure 4a) by increasing substrate affinity (K_R and K_T) in both "tight" and "relaxed" (T and R) states, and hindering the population of catalytically unproductive conformational states; nonetheless, trehalose does affect the equilibrium between the R and T states as described in Figure 1b, without changing the enzyme cooperativity (Hill number, n) (Figure 4b). This phenomena (enzyme activation) may be explained either by the effect of viscosity on the enzyme structural fluctuations (i.e., in the transition between R and T states) [70], by favoring the population of a given oligomeric state [153], by decreasing the protein volume [154] thus leading to the formation of the binding site [155], by preventing the sample of catalytically unproductive enzyme conformations [85], or by a combination of all of them. In this regard, the compressing effect of protein structure by viscosity generating solutes has already been observed with glycerol [155,156], sucrose [155,157], and trehalose [67,155], e.g., the volume of lysozyme decreases ~2% in the presence of trehalose [67].

Kinetics of protein processes others than catalysis are also affected by viscosity [158] as described by Kramers' theory, some of these are the escape of O_2 from respiratory proteins [159] and hemerythrin [160], the heme pocked relaxation in hemoglobin after CO photolysis [161], the dissociation of CO from horse ferrocytochrome c (FCC) [162], and the CO binding to horse myoglobin [163].

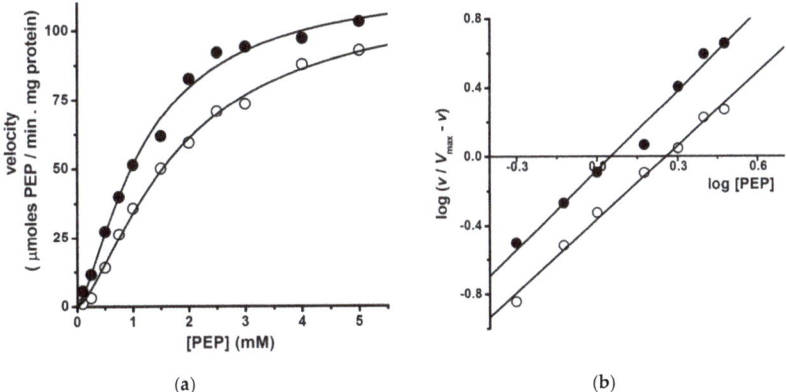

(a) (b)

Figure 4. Activation of pyruvate kinase (PK) from *G. stearothermophilus* by trehalose [164]. (a) Cooperative kinetics of PK in the absence (○) and presence (●) of trehalose. Considering the model of Monod, Wyman, and Changeux [69] in enzyme kinetics analysis: V_{max} = 112 and 118 µmoles PEP·(min·mg protein)$^{-1}$ in the absence and presence of trehalose, respectively. Affinity constants: K_T = 328 and 11, and K_R = 40 × 10^{-4} and 5 × 10^{-4} mM phosphoenolpyruvate (PEP) in the absence and presence of trehalose, respectively. The Hill number (n) was ≈1.5, both in the absence and presence of trehalose. The ratio $[T_0]/[R_0]$ (L) increases ~13 times in the presence of trehalose. (b) Hill plot of PK cooperative kinetics of data in a (symbols as in a). The slope (n) value in both straight lines is ≈1.5. Adapted with permission from Rivera-Moran [164]. Copyright © 2020, M.A. Rivera Moran.

5. The Application of Kramers' Theory on Heat-Mediated Enzyme Inactivation, Protein Folding, and Unfolding

Heat-shock is another stress condition where trehalose may accumulate in the cytoplasm of some organisms [4,6,8,14,24,129,165], e.g., in yeast, trehalose accumulation (~0.5 M) has been related to the ability of the cells to survive at relatively high temperatures [21,165]. The number of conformational states a protein may sample is related to the surrounding temperature [75]. When raising the temperature, the structural fluctuations increase critically up to unfold/denature the three-dimensional (3D) structure of the protein as described in Figure 1c,d; in the denatured state, the number of possible conformations becomes immense [23,166,167]. The first event to occur in enzymes when undergoing unfolding is the loss of catalytic activity [168–171]. It has been proposed that the high viscosity generated by trehalose stabilizes proteins by inducing preferential hydration of the protein surface [60,61], preserving the hydration shell and decreasing its dynamics [172], increasing surface tension on the trehalose-water phase [59], and hampering protein motions that lead to unfolding [61,129,173]. Hence, thermal stabilization mediated by trehalose has been observed in a diversity of enzymes, such as bovine intestine alkaline phosphatase (BIALP) [174], human brain-type creatine kinase (hBBCK) [175], ornithine carbamoyltransferase (OCTase) [176], polyphenol oxidase (PPO) from yacon roots (*Smallanthus sonchifolius*) [177], rabbit muscle phosphofructokinase (PFK) [153], citrate synthase (CS) [25], cutinase from *Fusarium solani pisi* [178], larval *Manduca sexta* fat body glycogen phosphorylase B (GPb) [179], *Renilla* luciferase (Rluc) from *Renilla reniformis* [180], yeast cytosolic pyrophosphatase (PPi) [181], and others. In this regard, the MDS of thermal inactivation and unfolding of Rluc in the presence of trehalose showed that Rluc is stabilized through the inhibition of the structural fluctuations generated by high temperature [180], while in thermal inactivation of lysozyme and glucoamylase (GA) from *A. niger* and α-amylase (AA) from *Bacillus sp*, it was found that trehalose prevents the loss of protein hydrophobic interactions and helical structure, thus avoiding protein aggregation [150,182,183].

The kinetics of enzyme thermal inactivation in the presence of trehalose has been determined experimentally for some enzymes where usually first-order kinetics is observed [178,184,185]. Kinetic analysis showed that trehalose decreases the inactivation rate constant (k_i) of xanthine oxidase (XO) from *Arthrobacter* M3 [184], PK from rabbit muscle [186], GOx from *A. niger* [151], and mushroom tyrosinase (MT) from *Agaricus bisporus* [185]. Therefore, trehalose increases the thermal stability of the enzymes by decreasing the inactivation rate constant (k_i) [129]. Interestingly, thermal inactivation of the plasma membrane H$^+$-ATPase from the yeast *Kluyveromyces lactis* is irreversible, showing biphasic kinetics with the presence of an active intermediary [129,173], i.e., a three-state mechanism (N → I → U) (Figure 1c) with respective inactivation rate constants for each phase, k_{i1} and k_{i2}. Notably, physiological concentrations of trehalose stabilize the H$^+$-ATPase by decreasing k_{i1} and k_{i2} [173,187]. The high viscosity generated by trehalose has been proposed as the mechanism of thermal stabilization of H$^+$-ATPase [129,173]. Detailed analysis of the H$^+$-ATPase thermal inactivation kinetics shows that the inactivation rate constants (k_{i1} and k_{i2}) are inversely proportional to medium viscosity ($k_i \propto 1/\eta$) as described by Kramers' theory and Equation (3) [129]:

$$k_i = \frac{A}{\eta} e^{-\Delta U/RT} \tag{3}$$

where k_i is the first-order inactivation rate constant, η the viscosity of the medium, A has the same meaning as in Equation (1) but regarding the enzyme inactivation process, ΔU is the potential energy barrier for inactivation, R is the gas constant, and T is the absolute temperature [129]. The plot of k_i versus η^{-1} usually shows a straight line when a diffusive component on the thermal inactivation mechanism of the enzyme exists [129,173,188], e.g., the dissociation of monomers from a large oligomer [188]. Analysis of the decrease of the inactivation rate constant by trehalose shows that trehalose increases the activation energy (E_a) for thermal inactivation (Figure 1) of AA [183], H$^+$-ATPase [173], and GOx [151].

Particularly, in GOx where catalysis is dependent on the presence of the coenzyme FAD in the active site [151], the thermal inactivation occurs first through the formation of a molten globule and after that FAD is released (Figure 5a) [189], resulting in the irreversible loss of GOx activity [151,189]. Interestingly, trehalose decreases the rate constant of the thermally generated large structural changes hence preventing the formation of the molten globule [151]. Due to the hampering of the FAD release, the inactivation rate constant (k_i) of GOx decreases [151] (Figure 5b), the $T_{1/2}$ for FAD release is increased to higher temperatures (Figure 5b). In this unfolding mechanism, trehalose seems to act by compacting the protein structure of GOx, thus preventing the release of FAD, and as a consequence, GOx retains the catalytic activity [151]. A similar stabilizing pattern by trehalose was observed in OCTase against thermal inactivation [176].

The dependence of protein unfolding on the viscosity (Figure 1d) generated by trehalose presence has been observed in cytochrome c (Cytc) [140], phosphoglycerate kinase (PGK), β-lactoglobulin (βLG) [190], cutinase, and the small protein barstar (Bs) [169]. Like in thermal inactivation, trehalose decreases the rate constant of thermal unfolding (k_U) by promoting the compactness of the unfolded state [191] and inhibiting large structural fluctuations [180]. Regarding protein folding, the high viscosity generated by trehalose affects the folding rate by inhibiting the structural dynamics as described by Kramers' theory [192–197].

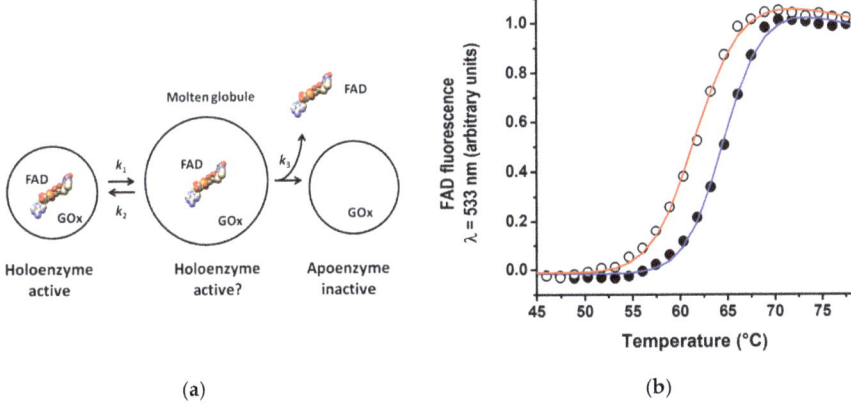

Figure 5. Trehalose prevents the release of FAD during thermal inactivation/unfolding of glucose oxidase (GOx) from *Aspergillus niger* [198]. (**a**) Thermal inactivation mechanism of GOx [151,189]; (**b**) Effect of trehalose on the release of FAD from GOx during thermal inactivation; Trehalose (○) 0.0, and (●) 0.6 M. The data were fitted to the Pace equation [199] by non-linear regression; $T_{1/2}$ = 61.3 and 64.3 °C in the absence and presence of 0.6 M trehalose, respectively. Adapted with permission from Paz-Alfaro [198]. Copyright © 2020, K.J. Paz-Alfaro.

Finally, the effect of the high viscosity generated by trehalose on macromolecular folding is being currently extended to other important biopolymers such as DNA [200,201] and RNA, e.g., in human telomere sequence [202] and RNA tertiary motif of the GAAA tetraloop receptor (TLR) [203], human telomerase hairpin (hTR HP) and H-type pseudoknot from the beet western yellow virus (BWYV) [204].

In cells, trehalose synthesis and accumulation helps to cope with harsh stress conditions. Trehalose affects enzyme activity (catalysis) and stability by hindering protein mobility through the high viscosity generated. Viscosity-mediated effects of trehalose (and other solutes) on enzymes are explained by Kramers' theory, i.e., the rate constant (k) of a given diffusive process in the enzyme structure is inversely proportional to the media viscosity. Currently, trehalose is widely used in enzyme applications (in the laboratory and biotechnology), whether alone or in combination with other compounds; therefore, it would be safe to consider the effect of viscosity on enzyme structural dynamics to improve the results of the use of trehalose.

Author Contributions: Conceptualization, J.G.S.; Methodology, J.G.S.; Validation, J.G.S. and M.A.R.-M.; Formal Analysis, J.G.S., M.A.R.-M., and S.U.-C.; Investigation, J.G.S. and M.A.R.-M.; Resources, J.G.S. and S.U.-C.; Data Curation, J.G.S.; Writing—Original Draft Preparation, J.G.S.; Writing—Review & Editing, J.G.S., M.A.R.-M., and S.U.-C.; Visualization, J.G.S.; Supervision, J.G.S.; Project administration, J.G.S.; Funding Acquisition, J.G.S. and S.U.-C. All authors have read and agreed to the published version of the manuscript.

Funding: This research was partially funded by FAI-UASLP México grant number C19-FAI-05-89.89 to J.G.S. and by a grant to S.U.-C.: DGAPA/PAPIIT Project IN203018.

Acknowledgments: M.A.R.-M. is a recipient of a Ph.D. fellowship of CONACyT México.

Conflicts of Interest: The authors declare no conflict of interest.

Abbreviations

Molecular dynamics simulation: MDS; intrinsically disordered proteins: IDPs; C-phycocyanin: CP; lactate dehydrogenase: LDH; carbonic anhydrase: CA; chloroplast coupling factor CF1; fatty acid synthetase: FAS; carboxypeptidase-A: CPA; ribosomal protein S4: RPS4; polymerase β, POLB; oligo-1,6-glucosidase 1: MalL; indole-3-glycerol phosphate synthetase: IGPS; sulfite oxidase: SO; glucose oxidase: GOx; pyruvate kinase: PK; ferrocytochrome *c*: FCC; bovine intestine alkaline phosphatase: BIALP; human brain-type creatine kinase: hBBCK; ornithine carbamoyltransferase: OCTase; polyphenol oxidase: PPO; phosphofructokinase: PFK; citrate synthase: CS; glycogen phosphorylase B: GPb; *Renilla* luciferase: Rluc; pyrophosphatase: PPi; glucoamylase: GA; and α-amylase: AA; xanthine oxidase: XO; mushroom tyrosinase: MT; cytochrome *c*: Cytc; phosphoglycerate kinase: PGK; β-lactoglobulin: βLG; barstar: Bs; tetraloop receptor: TLR; human telomerase hairpin: hTR HP; beet western yellow virus: BWYV.

References

1. Bodenheimer, F.S. Chapter, V. Asia. In *Insects as Human Food*; Bodenheimer, F.S., Ed.; Springer: Dordrecht, The Netherlands, 1951; pp. 208–280. ISBN 978-94-017-5767-6.
2. Leibowitz, J. A new source of trehalose. *Nature* **1943**, *152*, 414. [CrossRef]
3. Gültekin, L.; Shahreyary-Nejad, S. A new trehala-constructing Larinus Dejean (Coleoptera: Curculionidae) from Iran. *Zool. Middle East* **2015**, *61*, 246–251. [CrossRef]
4. Singer, M.A.; Lindquist, S. Thermotolerance in Saccharomyces cerevisiae: The Yin and Yang of trehalose. *Trends Biotechnol.* **1998**, *16*, 460–468. [CrossRef]
5. Richards, A.; Krakowka, S.; Dexter, L.; Schmid, H.; Wolterbeek, A.P.; Waalkens-Berendsen, D.; Shigoyuki, A.; Kurimoto, M. Trehalose: A review of properties, history of use and human tolerance, and results of multiple safety studies. *Food Chem. Toxicol.* **2002**, *40*, 871–898. [CrossRef]
6. Elbein, A.D. The metabolism of α,α-trehalose. *Adv. Carbohydr. Chem. Biochem.* **1974**, *30*, 227–256. [CrossRef]
7. Gancedo, C.; Flores, C.L. The importance of a functional trehalose biosynthetic pathway for the life of yeasts and fungi. *FEMS Yeast Res.* **2004**, *4*, 351–359. [CrossRef]
8. Voit, E.O. Biochemical and genomic regulation of the trehalose cycle in yeast: Review of observations and canonical model analysis. *J. Theor. Biol.* **2003**, *223*, 55–78. [CrossRef]
9. Lunn, J.E.; Delorge, I.; Figueroa, C.M.; Van Dijck, P.; Stitt, M. Trehalose metabolism in plants. *Plant J.* **2014**, *79*, 544–567. [CrossRef]
10. Elbein, A.D.; Pan, Y.T.; Pastuszak, I.; Carroll, D. New insights on trehalose: A multifunctional molecule. *Glycobiology* **2003**, *13*, 17R–27R. [CrossRef]
11. Thammahong, A.; Puttikamonkul, S.; Perfect, J.R.; Brennan, R.G.; Cramer, R.A. Central role of the trehalose biosynthesis pathway in the pathogenesis of human fungal infections: Opportunities and challenges for therapeutic development. *Microbiol. Mol. Biol. Rev.* **2017**, *81*, e00053-16. [CrossRef]
12. Kalscheuer, R.; Koliwer-Brandl, H. Genetics of mycobacterial trehalose metabolism. *Microbiol. Spectr.* **2014**, *2*. [CrossRef] [PubMed]
13. Smallbone, K.; Malys, N.; Messiha, H.L.; Wishart, J.A.; Simeonidis, E. Building a kinetic model of trehalose biosynthesis in Saccharomyces cerevisiae. In *Methods in Enzymology*; Academic Press Inc.: San Diego, CA, USA, 2011; Volume 500, pp. 355–370. ISBN 978-0-12-385118-5.
14. Eleutherio, E.; Panek, A.; De Mesquita, J.F.; Trevisol, E.; Magalhães, R. Revisiting yeast trehalose metabolism. *Curr. Genet.* **2015**, *61*, 263–274. [CrossRef] [PubMed]
15. Schiraldi, C.; Di Lernia, I.; De Rosa, M. Trehalose production: Exploiting novel approaches. *Trends Biotechnol.* **2002**, *20*, 420–425. [CrossRef]

16. Crowe, J.H. Anhydrobiosis: An unsolved problem with applications in human welfare. *Subcell. Biochem.* **2015**, *71*, 263–280. [CrossRef]
17. McClements, D.J. Modulation of globular protein functionality by weakly interacting cosolvents. *Crit. Rev. Food Sci. Nutr.* **2002**, *42*, 417–471. [CrossRef]
18. Tunnacliffe, A.; Lapinski, J. Resurrecting Van Leeuwenhoek's rotifers: A reappraisal of the role of disaccharides in anhydrobiosis. *Philos. Trans. R. Soc. Lond. Ser. B Biol. Sci.* **2003**, *358*, 1755–1771. [CrossRef]
19. Hengherr, S.; Heyer, A.G.; Köhler, H.R.; Schill, R.O. Trehalose and anhydrobiosis in tardigrades—Evidence for divergence in responses to dehydration. *FEBS J.* **2008**, *275*, 281–288. [CrossRef]
20. Sampedro, J.G.; Guerra, G.; Pardo, J.P.; Uribe, S. Trehalose-mediated protection of the plasma membrane H+-ATPase from Kluyveromyces lactis during freeze-drying and rehydration. *Cryobiology* **1998**, *37*, 131–138. [CrossRef]
21. De Virgilio, C.; Hottiger, T.; Dominguez, J.; Boller, T.; Wiemken, A. The role of trehalose synthesis for the acquisition of thermotolerance in yeast. I. Genetic evidence that trehalose is a thermoprotectant. *Eur. J. Biochem.* **1994**, *219*, 179–186. [CrossRef]
22. Boothby, T.C.; Tapia, H.; Brozena, A.H.; Piszkiewicz, S.; Smith, A.E.; Giovannini, I.; Rebecchi, L.; Pielak, G.J.; Koshland, D.; Goldstein, B. Tardigrades use intrinsically disordered proteins to survive desiccation. *Mol. Cell* **2017**, *65*, 975–984.e5. [CrossRef]
23. Boothby, T.C.; Pielak, G.J. Intrinsically disordered proteins and desiccation tolerance: Elucidating functional and mechanistic underpinnings of anhydrobiosis. *BioEssays* **2017**, *39*, 1700119. [CrossRef]
24. Magalhães, R.S.S.; Popova, B.; Braus, G.H.; Outeiro, T.F.; Eleutherio, E.C.A. The trehalose protective mechanism during thermal stress in Saccharomyces cerevisiae: The roles of Ath1 and Agt1. *FEMS Yeast Res.* **2018**, *18*. [CrossRef]
25. Viner, R.I.; Clegg, J.S. Influence of trehalose on the molecular chaperone activity of p26, a small heat shock/α-crystallin protein. *Cell Stress Chaperones* **2001**, *6*, 126–135. [CrossRef]
26. Hu, J.; Jafari, S.; Han, Y.; Grodzinsky, A.J.; Cai, S.; Guo, M. Size- and speed-dependent mechanical behavior in living mammalian cytoplasm. *Proc. Natl. Acad. Sci. USA* **2017**, *114*, 9529–9534. [CrossRef]
27. Luby-Phelps, K. Cytoarchitecture and physical properties of cytoplasm: Volume, viscosity, diffusion, intracellular surface area. In *International Review of Cytology*; Academic Press Inc.: San Diego, CA, USA, 1999; Volume 192, pp. 189–221. ISBN 978-0-12-364596-8.
28. Feig, M.; Yu, I.; Wang, P.H.; Nawrocki, G.; Sugita, Y. Crowding in cellular environments at an atomistic level from computer simulations. *J. Phys. Chem. B* **2017**, *121*, 8009–8025. [CrossRef]
29. Welch, G.R.; Somogyi, B.; Matkó, J.; Papp, S. Effect of viscosity on enzyme-ligand dissociation II. Role of the microenvironment. *J. Theor. Biol.* **1983**, *100*, 211–238. [CrossRef]
30. Rampp, M.; Buttersack, C.; Lüdemann, H.-D. c, T-Dependence of the viscosity and the self-diffusion coefficients in some aqueous carbohydrate solutions. *Carbohydr. Res.* **2000**, *328*, 561–572. [CrossRef]
31. Sampedro, J.G.; Muñoz-Clares, R.A.; Uribe, S. Trehalose-mediated inhibition of the plasma membrane H+-ATPase from Kluyveromyces lactis: Dependence on viscosity and temperature. *J. Bacteriol.* **2002**, *184*, 4384–4391. [CrossRef]
32. Uribe, S.; Sampedro, J.G. Measuring solution viscosity and its effect on enzyme activity. *Biol. Proced. Online* **2003**, *5*, 108–115. [CrossRef]
33. Magazù, S.; Maisano, G.; Migliardo, P.; Middendorf, H.D.; Villari, V. Hydration and transport properties of aqueous solutions of α-α-trehalose. *J. Chem. Phys.* **1998**, *109*, 1170–1174. [CrossRef]
34. Wyatt, T.T.; Golovina, E.A.; van Leeuwen, R.; Hallsworth, J.E.; Wösten, H.A.B.; Dijksterhuis, J. A decrease in bulk water and mannitol and accumulation of trehalose and trehalose-based oligosaccharides define a two-stage maturation process towards extreme stress resistance in ascospores of N eosartorya fischeri (A spergillus fischeri). *Environ. Microbiol.* **2015**, *17*, 383–394. [CrossRef]
35. Cicerone, M.T.; Soles, C.L. Fast dynamics and stabilization of proteins: Binary glasses of trehalose and glycerol. *Biophys. J.* **2004**, *86*, 3836–3845. [CrossRef]
36. Jain, N.K.; Roy, I. Effect of trehalose on protein structure. *Protein Sci.* **2009**, *18*, 24–36. [CrossRef]
37. Crowe, J.H. Trehalose as a "chemical chaperone": Fact and fantasy. In *Advances in Experimental Medicine and Biology*; Springer: New York, NY, USA, 2007; Volume 594, pp. 143–158. ISBN 9780387399744.
38. Lubchenko, V.; Wolynes, P.G.; Frauenfelder, H. Mosaic energy landscapes of liquids and the control of protein conformational dynamics by glass-forming solvents. *J. Phys. Chem. B* **2005**, *109*, 7488–7499. [CrossRef]

39. Malferrari, M.; Francia, F.; Venturoli, G. Retardation of protein dynamics by trehalose in dehydrated systems of photosynthetic reaction centers. Insights from electron transfer and thermal denaturation kinetics. *J. Phys. Chem. B* **2015**, *119*, 13600–13618. [CrossRef]
40. Malferrari, M.; Savitsky, A.; Lubitz, W.; Möbius, K.; Venturoli, G. Protein immobilization capabilities of sucrose and trehalose glasses: The effect of protein/sugar concentration unraveled by high-field EPR. *J. Phys. Chem. Lett.* **2016**, *7*, 4871–4877. [CrossRef]
41. Zhang, J.; Martinez-Gomez, K.; Heinzle, E.; Wahl, S.A. Metabolic switches from quiescence to growth in synchronized Saccharomyces cerevisiae. *Metabolomics* **2019**, *15*, 121. [CrossRef]
42. Valcourt, J.R.; Lemons, J.M.S.; Haley, E.M.; Kojima, M.; Demuren, O.O.; Coller, H.A. Staying alive. *Cell Cycle* **2012**, *11*, 1680–1696. [CrossRef]
43. Gray, J.V.; Petsko, G.A.; Johnston, G.C.; Ringe, D.; Singer, R.A.; Werner-Washburne, M. "Sleeping Beauty": Quiescence in Saccharomyces cerevisiae. *Microbiol. Mol. Biol. Rev.* **2004**, *68*, 187–206. [CrossRef]
44. Winther, L.R.; Qvist, J.; Halle, B. Hydration and mobility of trehalose in aqueous solution. *J. Phys. Chem. B* **2012**, *116*, 9196–9207. [CrossRef]
45. Shiraga, K.; Adachi, A.; Nakamura, M.; Tajima, T.; Ajito, K.; Ogawa, Y. Characterization of the hydrogen-bond network of water around sucrose and trehalose: Microwave and terahertz spectroscopic study. *J. Chem. Phys.* **2017**, *146*, 105102. [CrossRef]
46. Liu, J.; Chen, C.; Li, W. Protective mechanisms of α,α-trehalose revealed by molecular dynamics simulations. *Mol. Simul.* **2018**, *44*, 100–109. [CrossRef]
47. Bellissent-Funel, M.-C.; Hassanali, A.; Havenith, M.; Henchman, R.; Pohl, P.; Sterpone, F.; van der Spoel, D.; Xu, Y.; Garcia, A.E. Water determines the structure and dynamics of proteins. *Chem. Rev.* **2016**, *116*, 7673–7697. [CrossRef]
48. Schirò, G.; Fichou, Y.; Gallat, F.-X.; Wood, K.; Gabel, F.; Moulin, M.; Härtlein, M.; Heyden, M.; Colletier, J.-P.; Orecchini, A.; et al. Translational diffusion of hydration water correlates with functional motions in folded and intrinsically disordered proteins. *Nat. Commun.* **2015**, *6*, 6490. [CrossRef]
49. Nakagawa, H.; Kataoka, M. Rigidity of protein structure revealed by incoherent neutron scattering. *Biochim. Biophys. Acta Gen. Subj.* **2020**, *1864*, 129536. [CrossRef]
50. Aggarwal, L.; Biswas, P. Hydration water distribution around intrinsically disordered proteins. *J. Phys. Chem. B* **2018**, *122*, 4206–4218. [CrossRef]
51. Nucci, N.V.; Pometun, M.S.; Wand, A.J. Site-resolved measurement of water-protein interactions by solution NMR. *Nat. Struct. Mol. Biol.* **2011**, *18*, 245–250. [CrossRef]
52. Zhang, L.; Wang, L.; Kao, Y.-T.; Qiu, W.; Yang, Y.; Okobiah, O.; Zhong, D. Mapping hydration dynamics around a protein surface. *Proc. Natl. Acad. Sci. USA* **2007**, *104*, 18461–18466. [CrossRef]
53. Houston, P.; Macro, N.; Kang, M.; Chen, L.; Yang, J.; Wang, L.; Wu, Z.; Zhong, D. Ultrafast dynamics of water-protein coupled motions around the surface of eye crystallin. *J. Am. Chem. Soc.* **2020**, *142*, 3997–4007. [CrossRef]
54. Lins, R.D.; Pereira, C.S.; Hünenberger, P.H. Trehalose-protein interaction in aqueous solution. *Proteins Struct. Funct. Bioinform.* **2004**, *55*, 177–186. [CrossRef]
55. Paul, S.S.; Paul, S.S. The influence of trehalose on hydrophobic interactions of small nonpolar solute: A molecular dynamics simulation study. *J. Chem. Phys.* **2013**, *139*, 1–9. [CrossRef]
56. Soper, A.K.; Ricci, M.A.; Bruni, F.; Rhys, N.H.; McLain, S.E. Trehalose in water revisited. *J. Phys. Chem. B* **2018**, *122*, 7365–7374. [CrossRef]
57. Xie, G.; Timasheff, S.N. The thermodynamic mechanism of protein stabilization by trehalose. *Biophys. Chem.* **1997**, *64*, 25–43. [CrossRef]
58. Shimizu, S.; Matubayasi, N. Preferential solvation: Dividing surface vs. excess numbers. *J. Phys. Chem. B* **2014**, *118*, 3922–3930. [CrossRef]
59. Lin, T.-Y.; Timasheff, S.N. On the role of surface tension in the stabilization of globular proteins. *Protein Sci.* **2008**, *5*, 372–381. [CrossRef]
60. Timasheff, S.N. Protein hydration, thermodynamic binding, and preferential hydration. *Biochemistry* **2002**, *41*, 13473–13482. [CrossRef]
61. Olsson, C.; Genheden, S.; García Sakai, V.; Swenson, J. Mechanism of trehalose-induced protein stabilization from neutron scattering and modeling. *J. Phys. Chem. B* **2019**, *123*, 3679–3687. [CrossRef]

62. Fedorov, M.V.; Goodman, J.M.; Nerukh, D.; Schumm, S. Self-assembly of trehalose molecules on a lysozyme surface: The broken glass hypothesis. *Phys. Chem. Chem. Phys.* **2011**, *13*, 2294–2299. [CrossRef]
63. Magno, A.; Gallo, P. Understanding the mechanisms of bioprotection: A comparative study of aqueous solutions of trehalose and maltose upon supercooling. *J. Phys. Chem. Lett.* **2011**, *2*, 977–982. [CrossRef]
64. Corradini, D.; Strekalova, E.G.; Eugene Stanley, H.; Gallo, P.; Stanley, H.E.; Gallo, P. Microscopic mechanism of protein cryopreservation in an aqueous solution with trehalose. *Sci. Rep.* **2013**, *3*, 1218. [CrossRef]
65. Giuffrida, S.; Cottone, G.; Bellavia, G.; Cordone, L. Proteins in amorphous saccharide matrices: Structural and dynamical insights on bioprotection. *Eur. Phys. J. E* **2013**, *36*, 79. [CrossRef] [PubMed]
66. Fogarty, A.C.; Laage, D. Water dynamics in protein hydration shells: The molecular origins of the dynamical perturbation. *J. Phys. Chem. B* **2014**, *118*, 7715–7729. [CrossRef] [PubMed]
67. GhattyVenkataKrishna, P.K.; Carri, G.A. The effect of complex solvents on the structure and dynamics of protein solutions: The case of lysozyme in trehalose/water mixtures. *Eur. Phys. J. E* **2013**, *36*, 14. [CrossRef] [PubMed]
68. Gabel, F.; Bellissent-Funel, M.C. C-phycocyanin hydration water dynamics in the presence of trehalose: An incoherent elastic neutron scattering study at different energy resolutions. *Biophys. J.* **2007**, *92*, 4054–4063. [CrossRef]
69. Changeux, J.-P. Allostery and the Monod-Wyman-Changeux model after 50 years. *Annu. Rev. Biophys.* **2012**, *41*, 103–133. [CrossRef]
70. Waldauer, S.A.; Stucki-Buchli, B.; Frey, L.; Hamm, P. Effect of viscogens on the kinetic response of a photoperturbed allosteric protein. *J. Chem. Phys.* **2014**, *141*, 22D514. [CrossRef]
71. Eisenmesser, E.Z. Enzyme dynamics during catalysis. *Science* **2002**, *295*, 1520–1523. [CrossRef]
72. Alpert, B.; Rivet, E. Protein Dynamics. In *Encyclopedia of Analytical Chemistry*; Meyers, R.A., Ed.; John Wiley & Sons, Ltd.: Chichester, UK, 2011; pp. 1–48. ISBN 9780471976707.
73. Eisenmesser, E.Z.; Millet, O.; Labeikovsky, W.; Korzhnev, D.M.; Wolf-Watz, M.; Bosco, D.A.; Skalicky, J.J.; Kay, L.E.; Kern, D. Intrinsic dynamics of an enzyme underlies catalysis. *Nature* **2005**, *438*, 117–121. [CrossRef]
74. Frauenfelder, H.; Petsko, G.A.; Tsernoglou, D. Temperature-dependent X-ray diffraction as a probe of protein structural dynamics. *Nature* **1979**, *280*, 558–563. [CrossRef]
75. Frauenfelder, H.; Sligar, S.; Wolynes, P. The energy landscapes and motions of proteins. *Science* **1991**, *254*, 1598–1603. [CrossRef]
76. Frauenfelder, H.; Fenimore, P.W.; Young, R.D. Protein dynamics and function: Insights from the energy landscape and solvent slaving. *IUBMB Life* **2007**, *59*, 506–512. [CrossRef]
77. Henzler-Wildman, K.; Kern, D. Dynamic personalities of proteins. *Nature* **2007**, *450*, 964–972. [CrossRef] [PubMed]
78. Agarwal, P.K.; Doucet, N.; Chennubhotla, C.; Ramanathan, A.; Narayanan, C. Conformational sub-states and populations in enzyme catalysis. In *Methods in Enzymology*; Academic Press Inc.: San Diego, CA, USA, 2016; Volume 578, pp. 273–297. ISBN 978-0-12-811107-9.
79. Callender, R.; Dyer, R.B. The dynamical nature of enzymatic catalysis. *Acc. Chem. Res.* **2015**, *48*, 407–413. [CrossRef]
80. Orellana, L. Large-scale conformational changes and protein function: Breaking the in silico barrier. *Front. Mol. Biosci.* **2019**, *6*, 117. [CrossRef] [PubMed]
81. Hammes-Schiffer, S.; Benkovic, S.J. Relating protein motion to catalysis. *Annu. Rev. Biochem.* **2006**, *75*, 519–541. [CrossRef] [PubMed]
82. James, L.C.; Tawfik, D.S. Conformational diversity and protein evolution—A 60-year-old hypothesis revisited. *Trends Biochem. Sci.* **2003**, *28*, 361–368. [CrossRef]
83. Agarwal, P.K.; Geist, A.; Gorin, A. Protein dynamics and enzymatic catalysis: Investigating the peptidyl-prolyl cis-trans isomerization activity of cyclophilin A. *Biochemistry* **2004**, *43*, 10605–10618. [CrossRef] [PubMed]
84. Duff, M.R.; Borreguero, J.M.; Cuneo, M.J.; Ramanathan, A.; He, J.; Kamath, G.; Chennubhotla, S.C.; Meilleur, F.; Howell, E.E.; Herwig, K.W.; et al. Modulating enzyme activity by altering protein dynamics with solvent. *Biochemistry* **2018**, *57*, 4263–4275. [CrossRef]
85. Doshi, U.; McGowan, L.C.; Ladani, S.T.; Hamelberg, D. Resolving the complex role of enzyme conformational dynamics in catalytic function. *Proc. Natl. Acad. Sci. USA* **2012**, *109*, 5699–5704. [CrossRef]
86. Kamerlin, S.C.L.; Warshel, A. At the dawn of the 21st century: Is dynamics the missing link for understanding enzyme catalysis. *Proteins Struct. Funct. Bioinform.* **2010**, *78*, 1339–1375. [CrossRef]

87. Kohen, A. Role of dynamics in enzyme catalysis: Substantial versus semantic controversies. *Acc. Chem. Res.* **2015**, *48*, 466–473. [CrossRef] [PubMed]
88. Karplus, M.; Kuriyan, J. Molecular dynamics and protein function. *Proc. Natl. Acad. Sci. USA* **2005**, *102*, 6679–6685. [CrossRef] [PubMed]
89. MacKerell, A.D.; Rigler, R.; Nilsson, L.; Hahn, U.; Saenger, W. Protein dynamics. A time-resolved fluorescence, energetic and molecular dynamics study of ribonuclease T1. *Biophys. Chem.* **1987**, *26*, 247–261. [CrossRef]
90. Nevin Gerek, Z.; Kumar, S.; Banu Ozkan, S. Structural dynamics flexibility informs function and evolution at a proteome scale. *Evol. Appl.* **2013**, *6*, 423–433. [CrossRef] [PubMed]
91. Tokuriki, N.; Tawfik, D.S. Protein dynamism and evolvability. *Science* **2009**, *324*, 203–207. [CrossRef] [PubMed]
92. Meisburger, S.P.; Case, D.A.; Ando, N. Diffuse X-ray scattering from correlated motions in a protein crystal. *Nat. Commun.* **2020**, *11*, 1271. [CrossRef]
93. Grimes, J.M.; Hall, D.R.; Ashton, A.W.; Evans, G.; Owen, R.L.; Wagner, A.; McAuley, K.E.; Von Delft, F.; Orville, A.M.; Sorensen, T.; et al. Where is crystallography going? *Acta Crystallogr. Sect. D Struct. Biol.* **2018**, *74*, 152–166. [CrossRef]
94. Kang, Y.; Gao, X.; Zhou, X.E.; He, Y.; Melcher, K.; Xu, H.E. A structural snapshot of the rhodopsin-arrestin complex. *FEBS J.* **2016**, *283*, 816–821. [CrossRef]
95. Shahlaei, M.; Madadkar-Sobhani, A.; Mahnam, K.; Fassihi, A.; Saghaie, L.; Mansourian, M. Homology modeling of human CCR5 and analysis of its binding properties through molecular docking and molecular dynamics simulation. *Biochim. Biophys. Acta Biomembr.* **2011**, *1808*, 802–817. [CrossRef]
96. Schlichting, I.; Berendzen, J.; Chu, K.; Stock, A.M.; Maves, S.A.; Benson, D.E.; Sweet, R.M.; Ringe, D.; Petsko, G.A.; Sligar, S.G. The catalytic pathway of cytochrome P450cam at atomic resolution. *Science* **2000**, *287*, 1615–1622. [CrossRef]
97. Kanai, R.; Ogawa, H.; Vilsen, B.; Cornelius, F.; Toyoshima, C. Crystal structure of a Na+-bound Na+,K+-ATPase preceding the E1P state. *Nature* **2013**, *502*, 201–206. [CrossRef] [PubMed]
98. Inesi, G.; Lewis, D.; Ma, H.; Prasad, A.; Toyoshima, C. Concerted conformational effects of Ca2+ and ATP are required for activation of sequential reactions in the Ca^{2+} ATPase (SERCA) catalytic cycle. *Biochemistry* **2006**, *45*, 13769–13778. [CrossRef] [PubMed]
99. Barbato, G.; Ikura, M.; Kay, L.E.; Pastor, R.W.; Bax, A. Backbone dynamics of calmodulin studied by nitrogen-15 relaxation using inverse detected two-dimensional NMR spectroscopy: The central helix is flexible. *Biochemistry* **1992**, *31*, 5269–5278. [CrossRef]
100. Diez, M.; Zimmermann, B.; Börsch, M.; König, M.; Schweinberger, E.; Steigmiller, S.; Reuter, R.; Felekyan, S.; Kudryavtsev, V.; Seidel, C.A.M.; et al. Proton-powered subunit rotation in single membrane-bound F0F1-ATP synthase. *Nat. Struct. Mol. Biol.* **2004**, *11*, 135–141. [CrossRef] [PubMed]
101. Su, T.; Cui, Y.; Zhang, X.; Liu, X.; Yue, J.; Liu, N.; Jiang, P. Constructing a novel nanodevice powered by δ-free FoF1-ATPase. *Biochem. Biophys. Res. Commun.* **2006**, *350*, 1013–1018. [CrossRef] [PubMed]
102. Karplus, M.; McCammon, J.A. Molecular dynamics simulations of biomolecules. *Nat. Struct. Biol.* **2002**, *9*, 646–652. [CrossRef]
103. Lindahl, E.R. Molecular dynamics simulations. In *Molecular Modeling of Proteins (Methods in Molecular Biology)*; Kukol, A., Ed.; Humana Press: Totowa, NJ, USA, 2008; Volume 443, pp. 3–23. ISBN 978-1-58829-864-5.
104. Náray-Szabó, G. *Protein Modelling*; Náray-Szabó, G., Ed.; Springer: Heidelberg, Germany, 2014; ISBN 978-3-319-09975-0.
105. Schlee, S.; Klein, T.; Schumacher, M.; Nazet, J.; Merkl, R.; Steinhoff, H.-J.; Sterner, R. Relationship of catalysis and active site loop dynamics in the (β α) 8 -barrel enzyme indole-3-glycerol phosphate synthase. *Biochemistry* **2018**, *57*, 3265–3277. [CrossRef]
106. Beece, D.; Eisenstein, L.; Frauenfelder, H.; Good, D.; Marden, M.C.; Reinisch, L.; Reynolds, A.H.; Sorensen, L.B.; Yue, K.T.; Marden, M.C.; et al. Solvent viscosity and protein dynamics. *Biochemistry* **1980**, *19*, 5147–5157. [CrossRef]
107. Caliskan, G.; Mechtani, D.; Roh, J.H.; Kisliuk, A.; Sokolov, A.P.; Azzam, S.; Cicerone, M.T.; Lin-Gibson, S.; Peral, I. Protein and solvent dynamics: How strongly are they coupled? *J. Chem. Phys.* **2004**, *121*, 1978–1983. [CrossRef]
108. Caliskan, G.; Kisliuk, A.; Tsai, A.M.; Soles, C.L.; Sokolov, A.P. Protein dynamics in viscous solvents. *J. Chem. Phys.* **2003**, *118*, 4230–4236. [CrossRef]

109. Pan, X.; Schwartz, S.D. Conformational heterogeneity in the michaelis complex of lactate dehydrogenase: An analysis of vibrational spectroscopy using Markov and hidden Markov models. *J. Phys. Chem. B* **2016**, *120*, 6612–6620. [CrossRef] [PubMed]
110. Min, W.; Xie, X.S.; Bagchi, B. Two-dimensional reaction free energy surfaces of catalytic reaction: Effects of protein conformational dynamics on enzyme catalysis. *J. Phys. Chem. B* **2008**, *112*, 454–466. [CrossRef] [PubMed]
111. Świderek, K.; Tuñón, I.; Martí, S.; Moliner, V. Protein conformational landscapes and catalysis. Influence of active site conformations in the reaction catalyzed by L-lactate dehydrogenase. *ACS Catal.* **2015**, *5*, 1172–1185. [CrossRef] [PubMed]
112. Ma, B.; Nussinov, R. Enzyme dynamics point to stepwise conformational selection in catalysis. *Curr. Opin. Chem. Biol.* **2010**, *14*, 652–659. [CrossRef] [PubMed]
113. Thirumalai, D.; Hyeon, C.; Zhuravlev, P.I.; Lorimer, G.H. Symmetry, rigidity, and allosteric signaling: From monomeric proteins to molecular machines. *Chem. Rev.* **2019**, *119*, 6788–6821. [CrossRef]
114. Campitelli, P.; Modi, T.; Kumar, S.; Ozkan, S.B. The role of conformational dynamics and allostery in modulating protein evolution. *Annu. Rev. Biophys.* **2020**, *49*, 267–288. [CrossRef]
115. Puchkov, E.O. Intracellular viscosity: Methods of measurement and role in metabolism. *Biochem. Suppl. Ser. A Membr. Cell Biol.* **2013**, *7*, 270–279. [CrossRef]
116. Luby-Phelps, K. The physical chemistry of cytoplasm and its influence on cell function: An update. *Mol. Biol. Cell* **2013**, *24*, 2593–2596. [CrossRef]
117. Chung, S.Y.; Lerner, E.; Jin, Y.; Kim, S.; Alhadid, Y.; Grimaud, L.W.; Zhang, I.X.; Knobler, C.M.; Gelbart, W.M.; Weiss, S. The effect of macromolecular crowding on single-round transcription by Escherichia coli RNA polymerase. *Nucleic Acids Res.* **2019**, *47*, 1440–1450. [CrossRef]
118. Rickey Welch, G.; Somogyi, B.; Damjanovich, S. The role of protein fluctuations in enzyme action: A review. *Prog. Biophys. Mol. Biol.* **1982**, *39*, 109–146. [CrossRef]
119. Pal, N.; Wu, M.; Lu, H.P. Probing conformational dynamics of an enzymatic active site by an in situ single fluorogenic probe under piconewton force manipulation. *Proc. Natl. Acad. Sci. USA* **2016**, *113*, 15006–15011. [CrossRef] [PubMed]
120. Frauenfelder, H.; Chen, G.; Berendzen, J.; Fenimore, P.W.; Jansson, H.; McMahon, B.H.; Stroe, I.R.; Swenson, J.; Young, R.D. A unified model of protein dynamics. *Proc. Natl. Acad. Sci. USA* **2009**, *106*, 5129–5134. [CrossRef] [PubMed]
121. Schlitter, J. Viscosity dependence of intramolecular activated processes. *Chem. Phys.* **1988**, *120*, 187–197. [CrossRef]
122. Doster, W. Viscosity scaling and protein dynamics. *Biophys. Chem.* **1983**, *17*, 97–103. [CrossRef]
123. Gavish, B.; Werber, M.M.; Gavish, B. Viscosity-dependent structural fluctuations in enzyme catalysis. *Biochemistry* **1979**, *18*, 1269–1275. [CrossRef]
124. Siddiqui, K.S.; Bokhari, S.A.; Afzal, A.J.; Singh, S. A novel thermodynamic relationship based on Kramers theory for studying enzyme kinetics under high viscosity. *IUBMB Life* **2004**, *56*, 403–407. [CrossRef]
125. Ajito, S.; Hirai, M.; Iwase, H.; Shimizu, N.; Igarashi, N.; Ohta, N. Protective action of trehalose and glucose on protein hydration shell clarified by using X-ray and neutron scattering. *Phys. B Condens. Matter* **2018**, *551*, 249–255. [CrossRef]
126. Kramers, H.A. Brownian motion in a field of force and the diffusion model of chemical reactions. *Physica* **1940**, *7*, 284–304. [CrossRef]
127. Gavish, B. Position-dependent viscosity effects on rate coefficients. *Phys. Rev. Lett.* **1980**, *44*, 1160–1163. [CrossRef]
128. Fanghänel, J. Enzymatic catalysis of the peptidyl bond rotation: Are transition state formation and enzyme dynamics directly linked? *Angew. Chem. Int. Ed.* **2003**, *42*, 490–492. [CrossRef]
129. Sampedro, J.G.; Uribe, S. Trehalose-enzyme interactions result in structure stabilization and activity inhibition. The role of viscosity. *Mol. Cell. Biochem.* **2004**, *256*, 319–327. [CrossRef] [PubMed]
130. Hernández-Meza, J.M.; Sampedro, J.G. Trehalose mediated inhibition of lactate dehydrogenase from rabbit muscle. The application of Kramers' theory in enzyme catalysis. *J. Phys. Chem. B* **2018**, *122*, 4309–4317. [CrossRef] [PubMed]

131. Hilge, M.; Siegal, G.; Vuister, G.W.; Güntert, P.; Gloor, S.M.; Abrahams, J.P. ATP-induced conformational changes of the nucleotide-binding domain of Na,K-ATPase. *Nat. Struct. Biol.* **2003**, *10*, 468–474. [CrossRef] [PubMed]
132. Berczi, A.; Moller, I.M. Control of the activity of plant plasma membrane MgATPase by the viscosity of the aqueous phase. *Physiol. Plant.* **1993**, *89*, 409–415. [CrossRef]
133. Demchenko, A.P.; Ruskyn, O.I.; Saburova, E.A. Kinetics of the lactate dehydrogenase reaction in high-viscosity media. *Biochim. Biophys. Acta* **1989**, *998*, 196–203. [CrossRef]
134. Pocker, Y.; Janjic, N. Origin of viscosity effects in carbonic anhydrase catalysis. Kinetic studies with bulky buffers at limiting concentrations. *Biochemistry* **1988**, *27*, 4114–4120. [CrossRef] [PubMed]
135. Ando, T.; Asai, H. The effects of solvent viscosity on the kinetic parameters of myosin and heavy meromyosin ATPase. *J. Bioenerg. Biomembr.* **1977**, *9*, 283–288. [CrossRef] [PubMed]
136. Malyan, A.N. The effect of medium viscosity on kinetics of ATP hydrolysis by the chloroplast coupling factor CF1. *Photosynth. Res.* **2016**, *128*, 163–168. [CrossRef] [PubMed]
137. Ng, K.; Rosenberg, A. The coupling of catalytically relevant conformational fluctuations in subtilisin BPN' to solution viscosity revealed by hydrogen isotope exchange and inhibitor binding. *Biophys. Chem.* **1991**, *41*, 289–299. [CrossRef]
138. Kyushiki, H.; Ikai, A. The effect of solvent viscosity on the rate-determining step of fatty acid synthetase. *Proteins Struct. Funct. Genet.* **1990**, *8*, 287–293. [CrossRef] [PubMed]
139. Goguadze, N.G.; Hammerstad-Pedersen, J.M.; Khoshtariya, D.E.; Ulstrup, J. Conformational dynamics and solvent viscosity effects in carboxypeptidase-A-catalyzed benzoylglycylphenyllactate hydrolysis. *Eur. J. Biochem.* **1991**, *200*, 423–429. [CrossRef] [PubMed]
140. Sashi, P.; Bhuyan, A.K. Viscosity dependence of some protein and enzyme reaction rates: Seventy-five years after Kramers. *Biochemistry* **2015**, *54*, 4453–4461. [CrossRef] [PubMed]
141. Bakhtina, M.; Lee, S.; Wang, Y.; Dunlap, C.; Lamarche, B.; Tsai, M.D. Use of viscogens, dNTPαS, and rhodium(III) as probes in stopped-flow experiments to obtain new evidence for the mechanism of catalysis by DNA polymerase β. *Biochemistry* **2005**, *44*, 5177–5187. [CrossRef] [PubMed]
142. Jones, H.B.L.; Wells, S.A.; Prentice, E.J.; Kwok, A.; Liang, L.L.; Arcus, V.L.; Pudney, C.R. A complete thermodynamic analysis of enzyme turnover links the free energy landscape to enzyme catalysis. *FEBS J.* **2017**, *284*, 2829–2842. [CrossRef]
143. Okada, A. Fractional power dependence of mean lifetime of electron transfer reaction on viscosity of solvent. *J. Chem. Phys.* **1999**, *111*, 2665–2677. [CrossRef]
144. Ivković-Jensen, M.M.; Kostić, N.M. Effects of viscosity and temperature on the kinetics of the electron-transfer reaction between the triplet state of zinc cytochrome c and cupriplastocyanin. *Biochemistry* **1997**, *36*, 8135–8144. [CrossRef]
145. Feng, C.; Kedia, R.V.; Hazzard, J.T.; Hurley, J.K.; Tollin, G.; Enemark, J.H. Effect of solution viscosity on intramolecular electron transfer in sulfite oxidase. *Biochemistry* **2002**, *41*, 5816–5821. [CrossRef]
146. Curtis, J.E.; Dirama, T.E.; Carri, G.A.; Tobias, D.J. Inertial suppression of protein dynamics in a binary glycerol-trehalose glass. *J. Phys. Chem. B* **2006**, *110*, 22953–22956. [CrossRef]
147. Walser, R.; van Gunsteren, W.F. Viscosity dependence of protein dynamics. *Proteins Struct. Funct. Genet.* **2001**, *42*, 414–421. [CrossRef]
148. Perkins, J.; Edwards, E.; Kleiv, R.; Weinberg, N. Molecular dynamics study of reaction kinetics in viscous media. *Mol. Phys.* **2011**, *109*, 1901–1909. [CrossRef]
149. Damjanovich, S.; Bot, J.; Somogyi, B.; Sümegi, J. Effect of glycerol on some kinetic parameters of phosphorylase b. *BBA Enzymol.* **1972**, *284*, 345–348. [CrossRef]
150. Liu, Y.; Meng, Z.; Shi, R.; Zhan, L.; Hu, W.; Xiang, H.; Xie, Q. Effects of temperature and additives on the thermal stability of glucoamylase from Aspergillus niger. *J. Microbiol. Biotechnol.* **2015**, *25*, 33–43. [CrossRef] [PubMed]
151. Paz-Alfaro, K.J.; Ruiz-Granados, Y.G.; Uribe-Carvajal, S.; Sampedro, J.G. Trehalose-mediated thermal stabilization of glucose oxidase from Aspergillus niger. *J. Biotechnol.* **2009**, *141*, 130–136. [CrossRef] [PubMed]
152. Kumar, S.; Satyanarayana, T. Purification and kinetics of a raw starch-hydrolyzing, thermostable, and neutral glucoamylase of the thermophilic mold Thermomucor indicae-seudaticae. *Biotechnol. Prog.* **2003**, *19*, 936–944. [CrossRef] [PubMed]

153. Faber-Barata, J.; Sola-Penna, M. Opposing effects of two osmolytes—trehalose and glycerol—on thermal inactivation of rabbit muscle 6-phosphofructo-1-kinase. *Mol. Cell. Biochem.* **2005**, *269*, 203–207. [CrossRef] [PubMed]
154. Lerbret, A.; Affouard, F.; Hédoux, A.; Krenzlin, S.; Siepmann, J.; Bellissent-Funel, M.-C.C.; Descamps, M. How strongly does trehalose interact with lysozyme in the solid state? Insights from molecular dynamics simulation and inelastic neutron scattering. *J. Phys. Chem. B* **2012**, *116*, 11103–11116. [CrossRef] [PubMed]
155. Zelent, B.; Bialas, C.; Gryczynski, I.; Chen, P.; Chib, R.; Lewerissa, K.; Corradini, M.G.; Ludescher, R.D.; Vanderkooi, J.M.; Matschinsky, F.M. Tryptophan Fluorescence Yields and Lifetimes as a Probe of Conformational Changes in Human Glucokinase. *J. Fluoresc.* **2017**, *27*, 1621–1631. [CrossRef]
156. Priev, A.; Almagor, A.; Yedgar, S.; Gavish, B. Glycerol decreases the volume and compressibility of protein interior. *Biochemistry* **1996**, *35*, 2061–2066. [CrossRef]
157. Almagor, A.; Priev, A.; Barshtein, G.; Gavish, B.; Yedgar, S. Reduction of protein volume and compressibility by macromolecular cosolvents: Dependence on the cosolvent molecular weight. *Biochim. Biophys. Acta—Protein Struct. Mol. Enzymol.* **1998**, *1382*, 151–156. [CrossRef]
158. Somogyi, B.; Karasz, F.E.; Trón, L.; Couchma, P.R. The effect of viscosity on the apparent decomposition rate on enzyme-ligand complexes. *J. Theor. Biol.* **1978**, *74*, 209–216. [CrossRef]
159. Yedgar, S.; Tetreau, C.; Gavish, B.; Lavalette, D. Viscosity dependence of O_2 escape from respiratory proteins as a function of cosolvent molecular weight. *Biophys. J.* **1995**, *68*, 665–670. [CrossRef]
160. Lavalette, D.; Tetreau, C. Viscosity-dependent energy barriers and equilibrium conformational fluctuations in oxygen recombination with hemerythrin. *Eur. J. Biochem.* **1988**, *177*, 97–108. [CrossRef] [PubMed]
161. Findsen, E.W.; Friedman, J.M.; Ondrias, M.R. Effect of solvent viscosity on the heme-pocket dynamics of photolyzed (carbonmonoxy) hemoglobin. *Biochemistry* **1988**, *27*, 8719–8724. [CrossRef] [PubMed]
162. Kumar, R.; Jain, R.; Kumar, R. Viscosity-dependent structural fluctuation of the M80-containing Ω-loop of horse ferrocytochrome c. *Chem. Phys.* **2013**, *418*, 57–64. [CrossRef]
163. Kleinert, T.; Doster, W.; Leyser, H.; Petry, W.; Schwarz, V.; Settles, M. Solvent composition and viscosity effects on the kinetics of CO binding to horse myoglobin. *Biochemistry* **1998**, *37*, 717–733. [CrossRef]
164. Rivera-Moran, M.A. The Viscosity of Trehalose Solutions and Its Effect on Enzyme Kinetics of Pyruvate Kinase from Geobacillus Stearothermophilus. Master's Thesis, Licenciatura en Biofísica, Instituto de Física, Universidad Autónoma de San Luis Potosí, San Luis Potosí, México, 2014.
165. Hottiger, T.; De Virgilio, C.; Hall, M.N.; Boller, T.; Wiemken, A. The role of trehalose synthesis for the acquisition of thermotolerance in yeast. II. Physiological concentrations of trehalose increase the thermal stability of proteins in vitro. *Eur. J. Biochem.* **1994**, *219*, 187–193. [CrossRef]
166. Uversky, V.N.; Dunker, A.K. *Intrinsically Disordered Protein Analysis*; Methods in Molecular Biology; Uversky, V.N., Dunker, A.K., Eds.; Humana Press: Totowa, NJ, USA, 2012; Volume 895, ISBN 978-1-61779-926-6.
167. Takagi, A.; Kamijo, M.; Ikeda, S. Darier disease. *J. Dermatol.* **2016**, *43*, 275–279. [CrossRef]
168. Pace, C.N.; Hebert, E.J.; Shaw, K.L.; Schell, D.; Both, V.; Krajcikova, D.; Sevcik, J.; Wilson, K.S.; Dauter, Z.; Hartley, R.W.; et al. Conformational stability and thermodynamics of folding of ribonucleases Sa, Sa2 and Sa3. *J. Mol. Biol.* **1998**, *279*, 271–286. [CrossRef]
169. Pradeep, L.; Udgaonkar, J.B. Diffusional barrier in the unfolding of a small protein. *J. Mol. Biol.* **2007**, *366*, 1016–1028. [CrossRef]
170. Santoro, M.M.; Bolen, D.W. Unfolding free energy changes determined by the linear extrapolation method. 1. Unfolding of phenylmethanesulfonyl alpha-chymotrypsin using different denaturants. *Biochemistry* **1988**, *27*, 8063–8068. [CrossRef]
171. Ghosh, K.; De Graff, A.M.R.; Sawle, L.; Dill, K.A. Role of proteome physical chemistry in cell behavior. *J. Phys. Chem. B* **2016**, *120*, 9549–9563. [CrossRef] [PubMed]
172. Ajito, S.; Iwase, H.; Takata, S.; Hirai, M. Sugar-mediated stabilization of protein against chemical or thermal denaturation. *J. Phys. Chem. B* **2018**, *122*, 8685–8697. [CrossRef] [PubMed]
173. Sampedro, J.G.; Cortés, P.; Muñoz-Clares, R.A.; Fernández, A.; Uribe, S. Thermal inactivation of the plasma membrane H+-ATPase from Kluyveromyces lactis. Protection by trehalose. *Biochim. Biophys. Acta Protein Struct. Mol. Enzymol.* **2001**, *1544*, 64–73. [CrossRef]
174. Sekiguchi, S.; Hashida, Y.; Yasukawa, K.; Inouyey, K. Stabilization of bovine intestine alkaline phosphatase by sugars. *Biosci. Biotechnol. Biochem.* **2012**, *76*, 95–100. [CrossRef] [PubMed]

175. Yang, J.L.; Mu, H.; Lü, Z.R.; Yin, S.J.; Si, Y.X.; Zhou, S.M.; Zhang, F.; Hu, W.J.; Meng, F.G.; Zhou, H.M.; et al. Trehalose has a protective effect on human brain-type creatine kinase during thermal denaturation. *Appl. Biochem. Biotechnol.* **2011**, *165*, 476–484. [CrossRef]
176. Barreca, D.; Bellocco, E.; Galli, G.; Laganà, G.; Leuzzi, U.; Magazù, S.; Migliardo, F.; Galtieri, A.; Telling, M.T.F. Stabilization effects of kosmotrope systems on ornithine carbamoyltransferase. *Int. J. Biol. Macromol.* **2009**, *45*, 120–128. [CrossRef]
177. Neves, V.A.; Da Silva, M.A. Polyphenol oxidase from yacon roots (Smallanthus sonchifolius). *J. Agric. Food Chem.* **2007**, *55*, 2424–2430. [CrossRef]
178. Baptista, R.P.; Cabral, J.M.S.; Melo, E.P. Trehalose delays the reversible but not the irreversible thermal denaturation of cutinase. *Biotechnol. Bioeng.* **2000**, *70*, 699–703. [CrossRef]
179. Meyer-Fernandes, J.R.; Arrese, E.L.; Wells, M.A. Allosteric effectors and trehalose protect larval Manduca sexta fat body glycogen phosphorylase B against thermal denaturation. *Insect Biochem. Mol. Biol.* **2000**, *30*, 473–478. [CrossRef]
180. Liyaghatdar, Z.; Emamzadeh, R.; Rasa, S.M.M.; Nazari, M. Trehalose radial networks protect Renilla luciferase helical layers against thermal inactivation. *Int. J. Biol. Macromol.* **2017**, *105*, 66–73. [CrossRef]
181. Sola-Penna, M.; Meyer-Fernandes, J.R. Protective role of trehalose in thermal denaturation of yeast pyrophosphatase. *Zeitschrift für Naturforsch. Sect. C J. Biosci.* **1994**, *49*, 327–330. [CrossRef] [PubMed]
182. Barreca, D.; Laganà, G.; Magazù, S.; Migliardo, F.; Gattuso, G.; Bellocco, E. FTIR, ESI-MS, VT-NMR and SANS study of trehalose thermal stabilization of lysozyme. *Int. J. Biol. Macromol.* **2014**, *63*, 225–232. [CrossRef] [PubMed]
183. Yadav, J.K.; Prakash, V. Thermal stability of α-amylase in aqueous cosolvent systems. *J. Biosci.* **2009**, *34*, 377–387. [CrossRef] [PubMed]
184. Zhang, Y.; Xin, Y.; Yang, H.; Zhang, L.; Xia, X.; Tong, Y.; Chen, Y.; Wang, W. Thermal inactivation of xanthine oxidase from Arthrobacter M3: Mechanism and the corresponding thermostabilization strategy. *Bioprocess Biosyst. Eng.* **2014**, *37*, 719–725. [CrossRef]
185. Gheibi, N.; Saboury, A.A.; Haghbeen, K.; Moosavi-Movahedi, A.A. The effect of some osmolytes on the activity and stability of mushroom tyrosinase. *J. Biosci.* **2006**, *31*, 355–362. [CrossRef]
186. Guerrero-Mendiola, C.; Oria-Hernández, J.; Ramírez-Silva, L. Kinetics of the thermal inactivation and aggregate formation of rabbit muscle pyruvate kinase in the presence of trehalose. *Arch. Biochem. Biophys.* **2009**, *490*, 129–136. [CrossRef]
187. Felix, C.F.; Moreira, C.C.; Oliveira, M.S.; Sola-Penna, M.; Meyer-Fernandes, J.R.; Scofano, H.M.; Ferreira-Pereira, A. Protection against thermal denaturation by trehalose on the plasma membrane H+-ATPase from yeast. Synergetic effect between trehalose and phospholipid environment. *Eur. J. Biochem.* **1999**, *266*, 660–664. [CrossRef]
188. Ruiz-Granados, Y.; De La Cruz-Torres, V.; Sampedro, J. The oligomeric state of the plasma membrane H+-ATPase from Kluyveromyces lactis. *Molecules* **2019**, *24*, 958. [CrossRef]
189. Zoldák, G.; Zubrik, A.; Musatov, A.; Stupák, M.; Sedlák, E. Irreversible thermal denaturation of glucose oxidase from Aspergillus niger is the transition to the denatured state with residual structure. *J. Biol. Chem.* **2004**, *279*, 47601–47609. [CrossRef]
190. Tang, X.; Pikal, M.J. Measurement of the kinetics of protein unfolding in viscous systems and implications for protein stability in freeze-drying. *Pharm. Res.* **2005**, *22*, 1176–1185. [CrossRef]
191. Baptista, R.P.; Pedersen, S.; Cabrita, G.J.M.; Otzen, D.E.; Cabral, J.M.S.; Melo, E.P. Thermodynamics and mechanism of cutinase stabilization by trehalose. *Biopolymers* **2008**, *89*, 538–547. [CrossRef] [PubMed]
192. Rhee, Y.M.; Pande, V.S. Solvent viscosity dependence of the protein folding dynamics. *J. Phys. Chem. B* **2008**, *112*, 6221–6227. [CrossRef]
193. Klimov, D.K.; Thirumalai, D. Viscosity dependence of the folding rates of proteins. *Phys. Rev. Lett.* **1997**, *79*, 317–320. [CrossRef]
194. Qiu, L.; Hagen, S.J. A limiting speed for protein folding at low solvent viscosity. *J. Am. Chem. Soc.* **2004**, *126*, 3398–3399. [CrossRef] [PubMed]
195. Hagen, S.J. Solvent viscosity and friction in protein folding dynamics. *Curr. Protein Pept. Sci.* **2010**, *11*, 385–395. [CrossRef]
196. de Sancho, D.; Sirur, A.; Best, R.B. Molecular origins of internal friction effects on protein-folding rates. *Nat. Commun.* **2014**, *5*, 4307. [CrossRef]

197. Sekhar, A.; Vallurupalli, P.; Kay, L.E. Folding of the four-helix bundle FF domain from a compact on-pathway intermediate state is governed predominantly by water motion. *Proc. Natl. Acad. Sci. USA* **2012**, *109*, 19268–19273. [CrossRef]
198. Paz-Alfaro, K.J. Thermal Inactivation of Glucose Oxidase from Aspergillus niger. Stabilization by Trehalose. Master's Thesis, Licenciatura en Nutrición, Universidad Autónoma del Estado de Hidalgo, Pachuca, Hidalgo, México, 2009.
199. Pace, C.N. Measuring and increasing protein stability. *Trends Biotechnol.* **1990**, *8*, 93–98. [CrossRef]
200. Niranjani, G.; Murugan, R. Theory on the mechanism of DNA renaturation: Stochastic nucleation and zipping. *PLoS ONE* **2016**, *11*, e0153172. [CrossRef]
201. Sikorav, J.L.; Orland, H.; Braslau, A. Mechanism of thermal renaturation and hybridization of nucleic acids: Kramers' process and universality in watson-crick base pairing. *J. Phys. Chem. B* **2009**, *113*, 3715–3725. [CrossRef]
202. Lannan, F.M.; Mamajanov, I.; Hud, N.V. Human telomere sequence DNA in water-free and high-viscosity solvents: G-quadruplex folding governed by Kramers rate theory. *J. Am. Chem. Soc.* **2012**, *134*, 15324–15330. [CrossRef] [PubMed]
203. Dupuis, N.F.; Holmstrom, E.D.; Nesbitt, D.J. Tests of Kramers' theory at the single-molecule level: Evidence for folding of an isolated RNA tertiary interaction at the viscous speed limit. *J. Phys. Chem. B* **2018**, *122*, 8796–8804. [CrossRef] [PubMed]
204. Hori, N.; Denesyuk, N.A.; Thirumalai, D. Frictional effects on RNA folding: Speed limit and Kramers turnover. *J. Phys. Chem. B* **2018**, *122*, 11279–11288. [CrossRef] [PubMed]

© 2020 by the authors. Licensee MDPI, Basel, Switzerland. This article is an open access article distributed under the terms and conditions of the Creative Commons Attribution (CC BY) license (http://creativecommons.org/licenses/by/4.0/).

Article

Continuous Production of 2-Phenylethyl Acetate in a Solvent-Free System Using a Packed-Bed Reactor with Novozym® 435

Shang-Ming Huang [1], Hsin-Yi Huang [1], Yu-Min Chen [1], Chia-Hung Kuo [2,*] and Chwen-Jen Shieh [1,*]

[1] Biotechnology Center, National Chung Hsing University, Taichung 402, Taiwan; zxzxmj2323@hotmail.com (S.-M.H.); hyhuang8273@dragon.nchu.edu.tw (H.-Y.H.); sunnycg710@gmail.com (Y.-M.C.)
[2] Department of Seafood Science, National Kaohsiung University of Science and Technology, Kaohsiung 811, Taiwan
* Correspondence: kuoch@nkust.edu.tw (C.-H.K.); cjshieh@nchu.edu.tw (C.-J.S.); Tel.: +886+7+361-7141 (ext. 23646) (C.-H.K.); +886+4+2284-0450 (ext. 5121) (C.-J.S.)

Received: 1 June 2020; Accepted: 24 June 2020; Published: 26 June 2020

Abstract: 2-Phenylethyl acetate (2-PEAc), a highly valued natural volatile ester, with a rose-like odor, is widely added in cosmetics, soaps, foods, and drinks to strengthen scent or flavour. Nowadays, 2-PEAc are commonly produced by chemical synthesis or extraction. Alternatively, biocatalysis is a potential method to replace chemical synthesis or extraction for the production of natural flavour. Continuous synthesis of 2-PEAc in a solvent-free system using a packed bed bioreactor through immobilized lipase-catalyzed transesterification of ethyl acetate (EA) with 2-phenethyl alcohol was studied. A Box–Behnken experimental design with three-level-three-factor, including 2-phenethyl alcohol (2-PE) concentration (100–500 mM), flow rate (1–5 mL min^{-1}) and reaction temperature (45–65 °C), was selected to investigate their influence on the molar conversion of 2-PEAc. Then, response surface methodology and ridge max analysis were used to discuss in detail the optimal reaction conditions for the synthesis of 2-PEAc. The results indicated both 2-PE concentration and flow rate are significant factors in the molar conversion of 2-PEAc. Based on the ridge max analysis, the maximum molar conversion was 99.01 ± 0.09% under optimal conditions at a 2-PE concentration of 62.07 mM, a flow rate of 2.75 mL min^{-1}, and a temperature of 54.03 °C, respectively. The continuous packed bed bioreactor showed good stability for 2-PEAc production, enabling operation for at least 72 h without a significant decrease of conversion.

Keywords: Lipase; transesterification; 2-phenylethyl acetate; packed-bed reactor; solvent-free; ethyl acetate

1. Introduction

Floral scent is an important ingredient to enhance flavor, and its main component is esters. They are commonly applied to many products, such as perfume, cosmetics, natural food additives, pharmaceuticals, and even in oral use [1]. Among aromatic compounds, 2-phenylethyl acetate (2-PEAc) is one of the most important chemicals of flower fragrance [2,3]. 2-PEAc, $C_{10}H_{12}O_2$, is a transparent, colorless oily liquid with the fragrance of rose and honey. A rose emission scent is mainly contributed by 2-PEAc, cis-3-hexenyl acetate, geranyl acetate, and citronellyl acetate when a rose flower has opened [4,5]. Nowadays, 2-PEAc is also widely used in the blending of flower and fruit flavors, belonging to a food additive [6]. In the traditional method for preparing 2-PEAc, although it can be extracted from plants or chemically synthesized, the concentration of 2-PEAc aromatic esters extracted by this conventional method is low, the extraction process is complicated, and the cost is high [7]. Essential oils extracted from plants require a large number of raw materials. Moreover, the extraction process is complicated because there are always many factors that affect the yield [8]. Since the extraction method of aromatic compounds is more expensive and the yield is low, the chemical synthesis becomes the main source of 2-PEAc. The chemical method is usually through acetylation of 2-phenethyl alcohol (2-PE) or esterification of 2-PE with acetic acid to obtain 2-PEAc [8,9]. However, chemical synthesis has disadvantages such as non-specific reactions, long reaction times, many byproducts, environmental pollution, and so on [10]. In recent years, the alternative methods for producing natural aromatic esters by microbial fermentation or metabolic engineering were studied [11–14]. Although microbial fermentation had a high transform rate, it was challenging to separate and recover the 2-PEAc from fermentation broth.

Biocatalysis applied in ester synthesis is useful and its synthesized product can be identical to the natural product. Biocatalysis offers several advantages, such as high specificity, selectivity, low energy consumption, and high yield [15–17]. Currently, an esterification or transesterification reaction catalyzed by lipase (triacylglycerol ester hydrolase, EC 3.1.1.3) in the organic solvent has been performed to produce esters [18,19]. Lipase-catalyzed reactions have been applied to the synthesis of emulsifiers [20], wax esters [21], structural lipids [22], and biodiesel [23]. Enzymatic catalysis is performed under moderate reaction conditions (pH, temperature, and atmospheric pressure), and the substrate is more specific for producing high-quality natural products [24]. Regarding industrial development, immobilized biocatalysts are favored because of their convenience for separation, recycling, and reuse. Recently, the immobilization of lipase on hydrophobic supports has shown hyper-activation activity [25,26]. However, when the substrate is very large or hydrophilic, lipase-catalyzed ester synthesis may cause steric hindrance, thereby reducing lipase activity [27].

Solvent, as the name suggests, can dissolve the reactants to form a uniform reaction system; the solvent can adjust the concentration and temperature of the reactants to control the rate and direction of the chemical reaction; the solvent can also be used to extract and separate specific compounds. Therefore, in traditional chemical reactions, the choice of solvent is often an important issue. However, solvents, especially organic solvents, are the main source of environmental pollution. As humanity's environmental awareness is increasingly awakened, green chemistry has gradually formed a new scientific philosophy. A solvent-free system for lipase-catalyzed reaction employing a reactant as the solvent is a simple mixture of reactants. Solvent-free systems present advantages, such as offering greater safety, reduction in solvent extraction costs, increased reactant concentrations, consequently volume productivity, and being friendly to the environment [28]. To date, the solvent-free system has been developed successfully for the synthesis of many esters, such as octyl ferulate [29], octyl hydroxyphenylpropionate [30], caffeic acid phenethyl ester [31], ethyl valerate (green apple flavour) and hexyl acetate (pear flavour) [32]. So far, the use of solvent-free systems for enzymatic synthesis of 2-PEAc has not been reported. Ethyl acetate (EA) has been used as reactant for synthesis of DHA/EPA (Omega-3 fatty acids) ethyl ester [33]. Therefore, we choose EA as a reactant, it also acts as a solvent in the lipase-catalyzed synthesis of 2-PEAc.

A fixed bed reactor is also called a packed bed reactor, and it is filled with a solid catalyst or solid reactants to achieve a multi-phase reaction. The solids are usually in the form of particles, stacked into a bed of a certain height (or thickness), the bed is stationary, and the fluid reacts through the bed. A packed bed reactor has low mechanical wear on the catalyst. The flow of the fluid in the bed is close to the plug flow result in using a smaller amount of catalyst and a smaller reactor volume can obtain a larger production capacity. Since the residence time can be strictly controlled and the temperature distribution can be adjusted appropriately, it is particularly beneficial for achieving high selectivity and conversion. The packed bed reactor using immobilized enzyme has been used for the production of galactooligosaccharides [34], high fructose syrup [35], hexyl laurate [25], and geraniol esters [36]. Four types of bioreactor, namely (1) stirred-tank, (2) packed-bed, (3) membrane, and (4) fluidized-bed, are currently used in the bio-, chemical, and food industry. Small-scale factories usually use batch reactors, but factories with large annual output must use continuous reactors, which is more economical. Commercially, they are best used continuously to minimize labor and overhead costs.

Therefore, this study applies the advantages of the solvent-free system and designs of packed bed reactors with immobilized lipase for the synthesis of 2-PEAc.

In this study, a green strategy for synthesis of 2-PEAc by transesterification of ethyl acetate with 2-phenyl alcohol catalyzed by lipase was studied. The 2-PEAc was synthesized using Novozym® 435 in a packed-bed reactor with a solvent-free system, the reaction parameters affecting the synthesis of 2-PEAc were evaluated, and the response surface methodology (RSM) using a three-level-three-factor Box–Behnken design was conducted to determine the optimal condition of 2-phenethyl alcohol (2-PE) concentration, mixture flow rate and reaction temperature on molar conversion of 2-PEAc.

2. Results and Discussion

2.1. Prime Experiment

The enzyme-catalyzed reaction in a solvent-free system simply mixes the reactants to carry out the enzyme-catalyzed reaction without the organic solvent [37]. It possesses some advantages of organic phase enzyme-catalyzed reactions, such as the reaction thermodynamic balance moving from hydrolysis to synthesis, and the enzyme has a high degree of stability. It also overcomes the shortcomings of the high toxicity of organic solvents, flammability and volatility, emitting pollution to the environment, and the high cost of recovery and recycling. At the same time, the solvent-free synthesis provides a new molecular environment different from the traditional solvent for the reaction. In the solvent-free system, the enzyme directly acts on the substrate, which improves the substrate and product concentration and the reaction selectivity. The purification process is easy with fewer steps. Solvent-free system biosynthesis is more suitable for the development of the food industry because of its obvious advantages such as mild reaction conditions, high substrate concentration, and high reaction rate. Initially, the synthesis efficiency of 2-PEAc was investigated at a flow rate in the range of 1 to 5 mL min^{-1}. Figure 1 shows the lipase-catalyzed synthesis of 2-PEAc from 2-PE and EA by transesterification. The reaction was carried out in a continuous packed bed reactor at 50 °C in a solvent-free system containing 100 mM 2-PE (dissolved in EA) and 1 g immobilized enzyme. Figure 2 shows that at the flow rate of 1 mL min^{-1}, 3 mL min^{-1}, and 5 mL min^{-1}, the conversion rates of 2-PEAc after reaction for 2 min reached 100%, 95%, and 80%, respectively. The molar conversion increased by decreasing the flow rate. However, the conversion rate reached 100% and was almost unchanged when the flow rate was less than 1 mL min^{-1}.

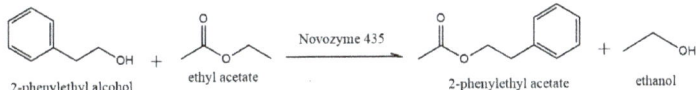

Figure 1. The reaction diagram of 2-phenylethyl acetate (2-PEAc) synthesis catalyzed by Novozym® 435.

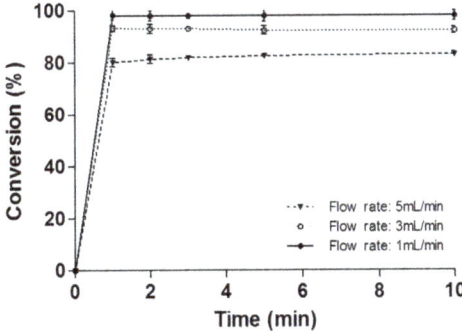

Figure 2. Effect of flow rate on the lipase-catalyzed synthesis of 2-PEAc in a packed-bed reactor at a substrate concentration of 100 mM 2-phenethyl alcohol (2-PE) and temperature of 55 °C.

2.2. Model Fitting

The major objective of this study was to develop and evaluate a statistical approach to better understand the relationship between variables and lipase-catalyzed transesterification reactions in solvent-free systems. According to the prime experimental result, the actual yields of 2-PEAc and the selected variables, including reaction temperature 50–70 °C, 2-PE concentration of 100–500 mM, and flow rate of 1–5 mL min^{-1}, are shown in Table 1. The RSREG program of SAS was employed to fit the second-order polynomial equation (Equation (1)) to the experimental data. Among the various treatments, the highest molar conversion 98.66% was obtained by treatment 10 with a substrate concentration of 100 mM 2-PE, a flow rate of 1 mL min^{-1}, and a reaction temperature of 55 °C. In contrast, the lowest molar conversion 62.05% was obtained by treatment 6 with a substrate concentration of 500 mM 2-PE, a flow rate of 5 mL min^{-1}, and a reaction temperature of 55 °C. The second-order polynomial equation obtained was shown as follows:

$$Y\ (\%) = 84.53616 - 0.035396X_1 + 1.83595X_2 + 0.64601X_3 - 0.015505X_1X_2 + 0.000830699X_1X_3 \\ + 0.032430X_2X_3 - 0.0000213020X_1^2 - 0.54497X_2^2 - 0.00781849X_3^2 \quad (1)$$

where X_1 is the concentration of 2-PE (100–500 mM), X_2 is the flow rate (1–5 mL min^{-1}), and X_3 is the reaction temperature (50–70 °C).

Table 1. Three-level-three-factor Box–Behnken design and the experiment data for response surface analysis.

Treatment No. [a]	Factor			Molar Conversion (%)
	X_1 Concentration of 2-PE (mM)	X_2 Flow Rate (mL min^{-1})	X_3 Observed Molar Conversion (%)	
1	0(300)	1(5)	−1(45)	76.53 ± 1.70
2	1(500)	0(3)	−1(45)	74.58 ± 1.62
3	−1(100)	0(3)	−1(45)	97.30 ± 0.41
4	0(300)	−1(1)	−1(45)	95.44 ± 0.80
5	−1(100)	1(5)	0(55)	94.26 ± 0.95
6	1(500)	1(5)	0(55)	62.05 ± 1.01
7	0(300)	0(3)	0(55)	90.03 ± 1.80
8	0(300)	0(3)	0(55)	89.50 ± 1.99
9	0(300)	0(3)	0(55)	89.23 ± 2.07
10	−1(100)	−1(1)	0(55)	98.66 ± 0.01
11	1(500)	−1(1)	0(55)	91.25 ± 0.33
12	0(300)	1(5)	1(65)	79.10 ± 1.22
13	1(500)	0(3)	1(65)	81.93 ± 1.86
14	−1(100)	0(3)	1(65)	98.00 ± 0.32
15	0(300)	−1(1)	1(65)	95.42 ± 0.16

[a] The treatments were run in a random order.

Analysis of variance (ANOVA) data (Table 2) shows that the second-order polynomial model can represent well the actual relationship between response and significant variables ($R^2 = 0.9971$ and $p < 0.0001$). In contrast, the ANOVA results of responses revealed that there was no significant fit as $p > 0.05$. Therefore, this model was adequate to predict the synthesis yields of 2-PEAc within the range of variables employed. Furthermore, Figure 3 shows that a highly linear relationship between predicted and experimental variables on molar conversion was obtained in this study. The actual 15 experimental results of molar conversion were similar to the prediction of the second-order polynomial equation. The effects of the three variables on molar conversion were analyzed by a joint test (Table 3). The results indicated that all three variables had significant effects on molar conversion ($p < 0.05$). Among them, the concentration of 2-PE (X_1) and flow rate (X_2) were more important variables toward molar conversion ($p < 0.001$) in this study.

Table 2. Analysis of variance (ANOVA) analysis for continuous lipase-catalyzed 2-PEAc synthesis.

Factors	Degree of Freedom	Sum of Squares	Prob > F
Linear	3	1374.688873	<0.0001
Quadratic	3	20.478243	0.0267
Cross product	3	166.589386	0.0002
Total Model	9	1561.756502	<0.0001
Lack of fit			
Pure error	3	4.206608	0.1088
Total error	2	0.335839	
Linear	5	4.542448	
		$R^2 = 0.9971$	

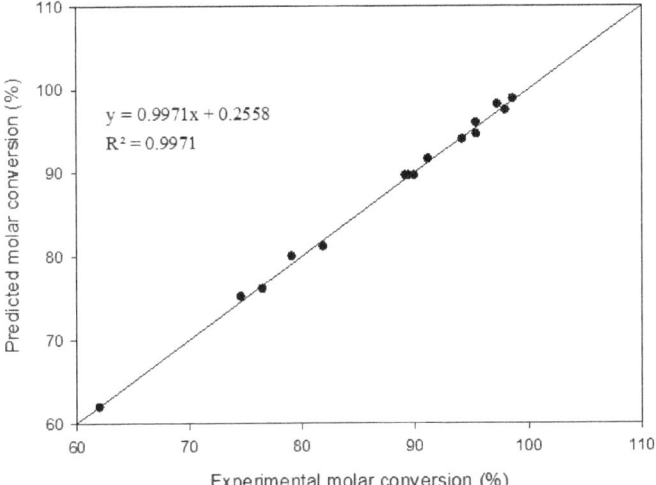

Figure 3. A linear relationship between the predicted and experimental molar conversion of 2-PEAC.

Table 3. Analysis of continuous synthesis variance of 2-PEAc in joint test ANOVA for substrate concentration pertaining to the response (initial rate).

Factors	Degree of Freedom	Sum of Squares	Prob > F
Concentration of 2-PE (X_1)	4	234.012890	<0.0001
Flow rate (X_2)	4	191.319574	<0.0001
Temperature (X_3)	4	7.255284	0.0213

2.3. Mutual Effect of Parameters

As shown in Figure 4A, the response surface plots show the effect of the concentration of 2-PE and flow rate on the molar conversion of 2-PEAc catalyzed by Novozym® 435 at 55 °C. When the concentration of 2-PE decreased from 500 to 100 mM, the molar conversion increased by approximately 10% and 30% at flow rates of 1 mL min^{-1} and 5 mL min^{-1}, respectively. The results indicated that the molar conversion increased as 2-PE concentration decreased and the flow rate increased.

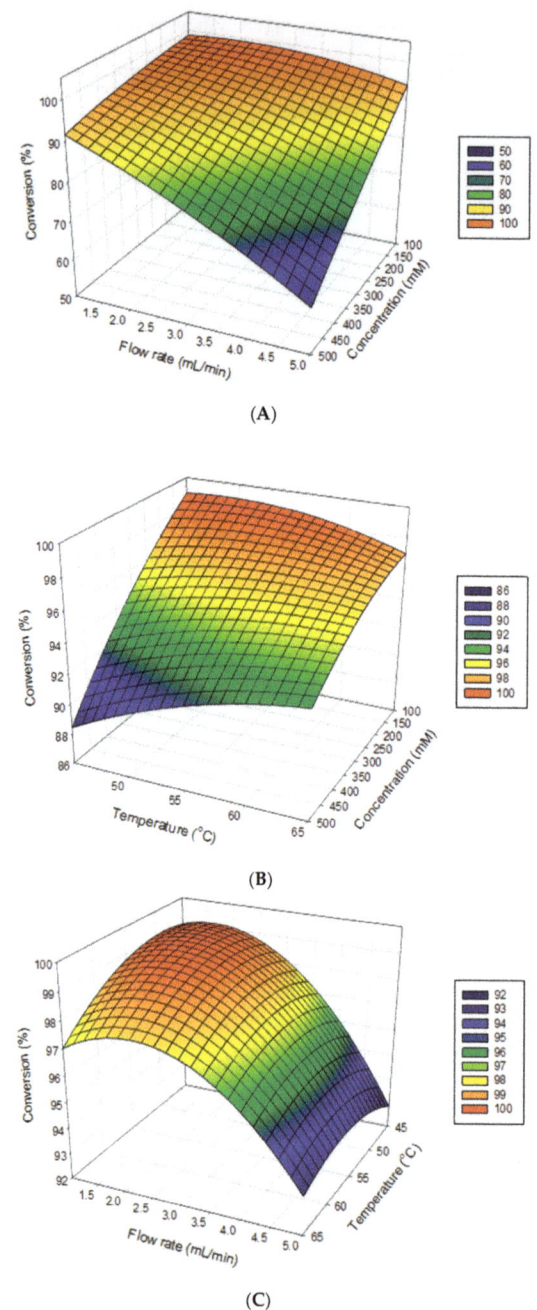

Figure 4. Response surface plots show the mutual effects of reaction (**A**) flow rate and concentration of 2-PE at a temperature of 55 °C; (**B**) temperature and concentration of 2-PE at a flow rate of 1 mL min^{-1}; (**C**) flow rate and temperature at a 2-PE concentration of 100 Mm.

Figure 4B shows the effects of concentration of 2-PE and reaction temperature on the molar conversion of 2-PEAc at 1 mL min^{-1} flow rate, indicating that the higher molar conversion was observed depending on a decreased level of 2-PE concentration and an increased level of reaction temperature. Moreover, Figure 4C illustrates the effects of flow rate and reaction temperature on molar conversion of 2-PEAc at a fixed 2-PE concentration of 100 mM, indicating that the higher molar conversion was obtained by decreasing flow rate when reaction temperature increased from 45 °C to 60 °C. By keeping each constant and examining a series of contour plots generated from the prediction model, Equation (1), the relationship between the reaction factors and response, can be better understood. From the joint test (Table 3), it was found that concentration of PE (X_1) and flow rate (X_2) were the most important variables for 2-PEAc synthesis ($p < 0.0001$, while the temperature (X_3) has less effect on 2-PEAc synthesis ($p = 0.0213$), as compared to the concentration of PE and flow rate.

The similar results observed in contour plots are shown in Figure 5. At any given concentration of 2-PE from 100 to 500 mM, Figure 5a–c indicated that a decrease both in flow rate and 2-PE concentration led to a higher molar conversion of 2-PEAc at different fixed reaction temperatures. The effects of varying concentrations of 2-PE and reaction temperature are shown in Figure 5d–f. At any given flow rate from 1 to 5 mL min^{-1}, a decrease in an adequate reaction temperature was effective for the molar conversion of 2-PEAc. The effects of different flow rates and the reaction temperatures are shown in Figure 5g–i. Reaction temperature has less effect on the molar conversion of 2-PEAc. In addition, Figure 5g shows that a better molar conversion occurred at around 150 mM 2-PE, 3 mL min^{-1}, and 55 °C experimental conditions. It should be noted that in a continuous lipase catalytic system, the reduction in flow rate and substrate molecules would lead to an increase in molar conversion. This phenomenon may be that excess substrate is not easily reacted at a high concentration of 2-PE or flow rate, resulting in a decrease in molar conversion. Furthermore, the higher temperature had little effect on the conversion at 60 °C, indicating that the temperature higher than 60 °C may denature the enzyme and reduce the molar conversion.

2.4. Attaining Optimization

In this study, the optimal synthesis of 2-PEAc was determined by ridge max analysis, which calculated the maximum response when the radius increased from the original design center. As shown in Table 4, the ridge max analysis predicted that the maximum conversion was 100.22% at 62.07 mM 2-PE and 2.75 mL min^{-1} flow rate, respectively. The validity of the prediction model is tested by conducting experiments under the best conditions for prediction. According to the statistics of the ridge maximum results, the maximum molar conversion rate was 99.01 ± 0.09%, indicating that the observed values are consistent with the predicted values from Equation (1). Presently, the continuous-flow bioreactor is suitable for maximizing the reaction efficiency, minimizing the demand for power and volume, and fully mixing the enzyme and substrate to attend a faster reaction rate. Obviously, from the statistical analysis results of RSM and ridge max, the optimal reaction conditions of this study include higher molar conversion by continuous lipase-catalyzed synthesis system.

Table 4. The estimated ridge of maximum response for variable production rate.

Coded Radius	Estimated Response (%)	Observed Response (%)	Uncoded Factor Values		
			X_1 (mM)	X_2 (mL/min)	X_3 (°C)
0.0	89.59 ± 0.55	89.59 ± 0.41	300.00	3.00	55.00
0.2	92.02 ± 0.54	91.23 ± 0.30	269.05	2.75	55.17
0.4	94.11 ± 0.53	92.65 ± 0.22	235.09	2.54	55.25
0.6	95.87 ± 0.51	95.12 ± 0.14	196.17	2.40	55.18
0.7	96.65 ± 0.51	95.80 ± 0.09	174.46	2.38	55.07
0.8	97.39 ± 0.52	97.19 ± 0.35	151.65	2.40	54.90
1.0	98.80 ± 0.59	98.74 ± 0.04	105.61	2.54	54.49
1.2	100.22 ± 0.73	99.01 ± 0.09	62.07	2.75	54.03

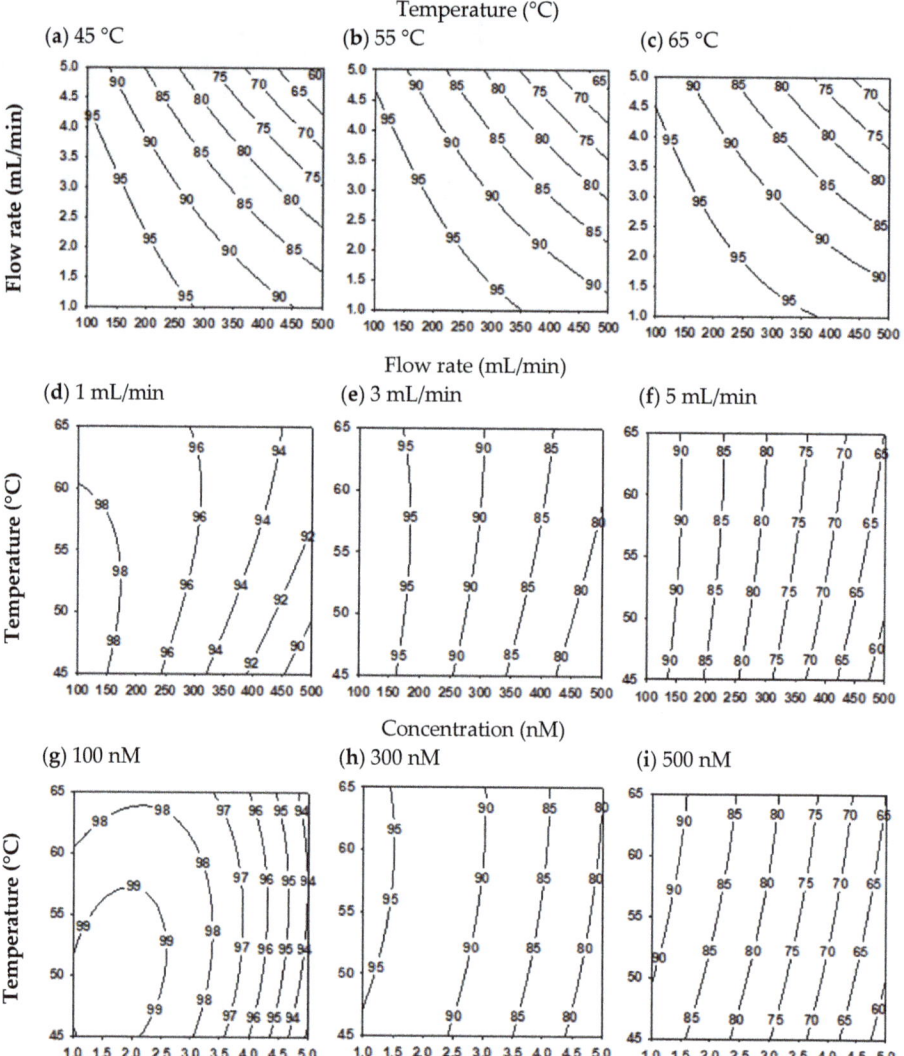

Figure 5. Contour plots show response behavior with varying concentration of 2-PE and flow rate at (**a**) 45 °C, (**b**) 55 °C and (**c**) 65 °C; concentration of 2-PE and temperature at flow rate (**d**) 1 mL min^{-1}, (**e**) 3 mL min^{-1} and (**f**) 5 mL min^{-1}; flow rate and temperature at (**g**) 100 mM, (**h**) 300 mM and (**i**) 500 mM 2-PE; numbers on contours denote molar conversion of 2-PEAc (%) at given reaction conditions.

2.5. Operational Stability

The purpose of this part of the experiment was to determine the stability of this enzyme for long-time operation. Since the reaction produces a by-product of ethanol in the batch reactor system leading to enzyme inhibition, the inhibition would make lipase less active. The enzymatic synthesis of pentyl oleate with immobilized lipase was inhibited by a high concentration of 1-penthanol [38]. This inhibition effect has been found in reactions among butyric and lauric acid with ethanol [39,40]. Therefore, the stability of the enzyme in the batch reactor will not be better than that in the continuous packed bed reactor, because the product in the packed bed reactor is continuously removed from the column, which can avoid the excessive concentration of by-product ethanol causing the enzyme inhibition. The Novozym® 435 in the continuous packed bed is operated for 3 days under the conditions of a 2-PE concentration of 257.44 mM, a flow rate of 1.049 mL min^{-1}, and a reaction temperature of 55.59 °C. The experimental results are shown in Figure 6. After 3 days of operation, the molar conversion is still 94%, and there is no tendency to decline. The experimental results are similar to those reported by Royon et al., who used packed bed reactors to investigate the production and stability of Novozym® 435 catalyzed cotton seed oil to produce biodiesel. After operation for 20 days, the yield remained at 95% [41]. Halim et al. used a packed bed reactor with immobilized lipase to produce biodiesel from waste palm oil with transesterification. After 120 h of operation, the production of fatty acid methyl esters was 79%, and there was no obvious decline [42]. Chang et al. discussed the use of Novozym® 435 as the catalyst for the production of biodiesel in the continuous packed bed reactor. After 7.5 days of operation, there was still a 75% yield with no tendency to decline [43]. Our results show that Novozym® 435 is suitable for the continuous packed bed reactor to catalyze the synthesis of 2-PEAc. Compared to the batch reactor, the stability of long-term operation can be attributed to the by-product ethanol not accumulating in the continuous packed bed reactor.

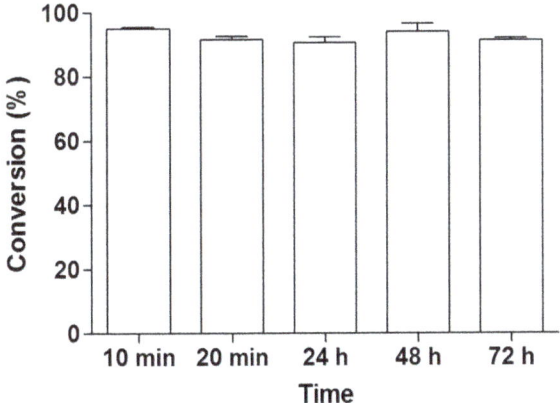

Figure 6. Continuous packed bed reactor with Novozym® 435 to synthesize 2-PEAc. The reaction was carried out for 3 days at an enzyme amount of 10,000 propyl laurate units (PLU), a 2-PE concentration of 257 mM, a flow rate of 1 mL min^{-1} and reaction temperature of 56 °C.

Therefore, the continuous packed bed reactor is very suitable for the industrial mass production of rose flavoured 2-PEAc.

3. Materials and Methods

3.1. Materials

Immobilized lipase Novozym® 435 (10,000 propyl laurate units, PLU, g^{-1}) from *Candida antarctica* B (EC 3.1.1.3) was supported on a macroporous acrylic resin. 2-PE, 2-PEAc, and EA were purchased from Sigma Chemical Co. (St Louis, MO, USA). Molecular sieve 4 Å was obtained from Davison Chemical (Baltimore, MD, USA). All chemicals employed were of analytical reagent grade.

3.2. Continuous Lipase-Catalyzed Synthesis of 2-Phenylethyl Acetate (2-PEAc)

All materials were dehydrated through a 4 Å molecular sieve overnight. Novozym® 435 was used as a biocatalyst for the esterification of 2-PE and EA (Figure 1). Before reaction, different concentrations of 2-PE (100–500 mmol L^{-1}) in EA were thoroughly mixed in a feeding flask. The reaction was implemented in a stainless-steel tube packed-bed reactor. The reactor is 25 cm in length with an inner diameter of 0.46 cm and contains 1 g of Novozym® 435. The reaction mixture was pumped continuously into the reactor under designed conditions. The flow rates (1–5 mL min^{-1}) were controlled by a Hitachi L-7100 Quaternary Gradient Pump (Hitachi, Tokyo, Japan). The diagram of the stainless-steel tube packed-bed reactor is shown in Figure 7. Based on the reactor design in this study, the reactor is a plug flow model. Assuming that there is no boundary layer near the inner wall of the tube, the mean residence time could be evaluated from 4.15 to 0.83 min at a flow rate of 1 to 5 mL min^{-1}.

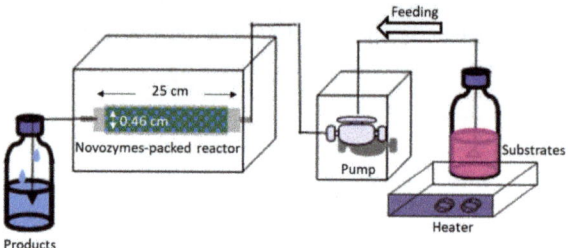

Figure 7. The diagram of the Novozym® 435 packed-bed bioreactor.

3.3. Experimental Design

In this study, the three-level-three-factor Box–Behnken design was employed, and 15 experiments were carried out in a continuous packed-bed reactor and a solvent-free system. The experimental variables were 2-phenethyl alcohol (2-PE) concentration (100–500 mmol L^{-1}), mixture flow rate (1–5 mL min^{-1}) and reaction temperature (45–65 °C), as shown in Table 1.

3.4. Quantitation of 2-PEAc

After the synthesis reaction, 2-PEAc in the reaction mixture was quantitated by injecting a 1 µL aliquot of the mixture into an Agilent 7890A gas chromatography (Agilent Technologies, Santa Clara, CA, USA) equipped with a flame ionization detector (FID) and MTX-65 TG fused silica capillary column (30 m × 0.25 mm i.d., film thickness 0.1 µm; Restek Corp., Bellefonte, PA, USA). The injector and FID temperatures were set to 230 and 250 °C. The oven temperature was maintained at 80 °C for 2 min, raised to 100 °C at a rate of 20 °C min^{-1} for 3 min, and finally raised to 230 °C at a rate of 50 °C min^{-1}, then held for 1.5 min. Nitrogen is a carrier gas with a constant flow rate of 3.5 mL min^{-1}. The molar conversion (%) is defined as (the number of moles of 2-PEAc produced per initial 2-PE mole added) × 100.

3.5. Statistical Analysis

The experimental data (Table 1) were analyzed by response surface regression (RSREG) procedure of SAS (SAS Institute, Cary, NC, USA) to fit the second-order polynomial equation (Equation (2)).

$$y = \beta_{k0} + \sum_{i=1}^{3} \beta_i x_i + \sum_{i=1}^{3} \beta_{ii} x_i^2 + \sum_{i=1}^{2}\sum_{j=i+1}^{3} \beta_{ij} x_i x_j \tag{2}$$

where y is the response (molar conversion %), β_{k0} is a constant, β_{ki}, β_{kii} and β_{kij} are coefficients and x_i and x_j are the non-coded independent variables. The ridge-max option was used to calculate the estimated ridge of the maximum response when the radius increases from the original design center.

4. Conclusions

The continuous Novozyme® 435-catalyzed synthesis of 2-PEAc in the solvent-free system was studied. The synthesis of 2-PEAc was optimized by the Box–Behnken design and RSM method, and the experimental design method was suitable for the optimization of operating conditions. Productivity was significantly affected by 2-PE concentration and flow rate, and the maximum conversion was 99.01 ± 0.09% at 62.07 mM 2-PE and 2.75 mL min^{-1} flow rate, and 54.03 °C, respectively. Finally, this solvent-free esterification system can be applied to the environmentally friendly production of natural flavour compounds, such as rose aromatic esters.

Author Contributions: Conceptualization, C.-J.S. and C.-H.K.; investigation, Y.-M.C.; data curation, S.-M.H. and Y.-M.C.; writing—original draft preparation, S.-M.H., H.-Y.H. and C.-H.K.; writing—review and editing, C.-J.S. and C.-H.K. All authors have read and agreed to the published version of the manuscript.

Funding: The authors are very grateful to the Ministry of Science and Technology of Taiwan, ROC for supporting this research (Grants No. 107-2320-B-005 -012 -MY3).

Conflicts of Interest: There is no conflict of interest regarding the publication of this article.

References

1. McGinty, D.; Vitale, D.; Letizia, C.S.; Api, A.M. Fragrance material review on phenethyl acetate. *Food Chem. Toxicol.* **2012**, *50*, S491–S497. [CrossRef] [PubMed]
2. Sá, A.G.A.; Meneses, A.C.D.; Araújo, P.H.H.D.; Oliveira, D.D. A review on enzymatic synthesis of aromatic esters used as flavor ingredients for food, cosmetics and pharmaceuticals industries. *Trends Food Sci. Technol.* **2017**, *69*, 95–105. [CrossRef]
3. Bayout, I.; Bouzemi, N.; Guo, N.; Mao, X.; Serra, S.; Riva, S.; Secundo, F. Natural flavor ester synthesis catalyzed by lipases. *Flavour Fragr. J.* **2020**, *35*, 209–218. [CrossRef]
4. Shalit, M.; Guterman, I.; Volpin, H.; Bar, E.; Tamari, T.; Menda, N.; Adam, Z.; Zamir, D.; Vainstein, A.; Weiss, D.; et al. Volatile ester formation in roses. Identification of an acetyl-coenzyme A. Geraniol/Citronellol acetyltransferase in developing rose petals. *Plant Physiol.* **2003**, *131*, 1868–1876. [CrossRef]
5. Guterman, I.; Masci, T.; Chen, X.; Negre, F.; Pichersky, E.; Dudareva, N.; Weiss, D.; Vainstein, A. Generation of phenylpropanoid pathway-derived volatiles in transgenic plants: Rose alcohol acetyltransferase produces phenylethyl acetate and benzyl acetate in petunia flowers. *Plant Mol. Biol.* **2006**, *60*, 555–563. [CrossRef]
6. Hirata, H.; Ohnishi, T.; Watanabe, N. Biosynthesis of floral scent 2-phenylethanol in rose flowers. *Biosci. Biotechnol. Biochem.* **2016**, *80*, 1865–1873. [CrossRef]
7. Arora, P.K. *Microbial Technology for the Welfare of Society*; Springer: Singapore, 2019.
8. Martinez-Avila, O.; Sanchez, A.; Font, X.; Barrena, R. Bioprocesses for 2-phenylethanol and 2-phenylethyl acetate production: Current state and perspectives. *Appl. Microbiol. Biotechnol.* **2018**, *102*, 9991–10004. [CrossRef]
9. Adler, P.; Hugen, T.; Wiewiora, M.; Kunz, B. Modeling of an integrated fermentation/membrane extraction process for the production of 2-phenylethanol and 2-phenylethylacetate. *Enzyme Microb. Technol.* **2011**, *48*, 285–292. [CrossRef]
10. Khan, N.R.; Rathod, V.K. Enzyme catalyzed synthesis of cosmetic esters and its intensification: A review. *Process Biochem.* **2014**, *50*, 1793–1806. [CrossRef]
11. Majetic, C.J.; Raguso, R.A.; Ashman, T.L. The impact of biochemistry vs. population membership on floral scent profiles in colour polymorphic *Hesperis matronalis*. *Ann. Bot.* **2008**, *102*, 911–922. [CrossRef]
12. Białecka-Florjańczyk, E.; Krzyczkowska, J.; Stolarzewicz, I.; Kapturowska, A. Synthesis of 2-phenylethyl acetate in the presence of *Yarrowia lipolytica* KKP 379 biomass. *J. Mol. Catal. B Enzym.* **2012**, *74*, 241–245. [CrossRef]
13. Guo, D.; Zhang, L.; Pan, H.; Li, X.; Guo, D.; Zhang, L.; Pan, H.; Li, X. Metabolic engineering of *Escherichia coli* for production of 2-phenylethylacetate from L-phenylalanine. *MicrobiologyOpen* **2017**, *6*, e00486. [CrossRef] [PubMed]
14. Zhang, B.; Xu, D.; Duan, C.; Yan, G. Synergistic effect enhances 2-phenylethyl acetate production in the mixed fermentation of *Hanseniaspora vineae* and *Saccharomyces cerevisiae*. *Process Biochem.* **2020**, *90*, 44–49. [CrossRef]
15. Xavier Malcata, F.; Reyes, H.R.; Garcia, H.S.; Hill, C.G.; Amundson, C.H. Immobilized lipase reactors for modification of fats and oils—A review. *J. Am. Oil Chem. Soc.* **1990**, *67*, 890–910. [CrossRef]
16. Fomuso, L.B.; Akoh, C.C. Lipase-catalyzed acidolysis of olive oil and caprylic acid in a bench-scale packed bed bioreactor. *Food Res. Int.* **2002**, *35*, 15–21. [CrossRef]
17. Facin, B.R.; Melchiors, M.S.; Valério, A.; Oliveira, J.V.; Oliveira, D.D. Driving immobilized lipases as biocatalysts: 10 years state of the art and future prospects. *Ind. Eng. Chem. Res.* **2019**, *58*, 5358–5378. [CrossRef]

18. Nielsen, N.S.; Yang, T.; Xu, X.; Jacobsen, C. Production and oxidative stability of a human milk fat substitute produced from lard by enzyme technology in a pilot packed-bed reactor. *Food Chem.* **2006**, *94*, 53–60. [CrossRef]
19. Mathpati, A.C.; Kalghatgi, S.G.; Mathpati, C.S.; Bhanage, B.M. Immobilized lipase catalyzed synthesis of n-amyl acetate: Parameter optimization, heterogeneous kinetics, continuous flow operation and reactor modeling. *J. Chem. Technol. Biotechnol.* **2018**, *93*, 2906–2916. [CrossRef]
20. Ye, R.; Hayes, D.G.; Burton, R.; Liu, A.; Harte, F.M.; Wang, Y. Solvent-free lipase-catalyzed synthesis of technical-grade sugar esters and evaluation of their physicochemical and bioactive properties. *Catalysts* **2016**, *6*, 78. [CrossRef]
21. Kuo, C.H.; Chen, H.H.; Chen, J.H.; Liu, Y.C.; Shieh, C.J. High yield of wax ester synthesized from cetyl alcohol and octanoic acid by lipozyme RMIM and Novozym 435. *Int. J. Mol. Sci.* **2012**, *13*, 11694–11704. [CrossRef]
22. Chojnacka, A.; Gładkowski, W. Production of structured phosphatidylcholine with high content of myristic acid by lipase-catalyzed acidolysis and interesterification. *Catalysts* **2018**, *8*, 281. [CrossRef]
23. Kim, K.H.; Lee, O.K.; Lee, E.Y. Nano-immobilized biocatalysts for biodiesel production from renewable and sustainable resources. *Catalysts* **2018**, *8*, 68.
24. Ortiz, C.; Ferreira, M.L.; Barbosa, O.; dos Santos, J.C.S.; Rodrigues, R.C.; Berenguer-Murcia, Á.; Briand, L.E.; Fernandez-Lafuente, R. Novozym 435: The "perfect" lipase immobilized biocatalyst? *Catal. Sci. Technol.* **2019**, *9*, 2380–2420. [CrossRef]
25. Ju, H.Y.; Yang, C.K.; Yen, Y.H.; Shieh, C.J. Continuous lipase-catalyzed synthesis of hexyl laurate in a packed-bed reactor: Optimization of the reaction conditions in a solvent-free system. *J. Chem. Technol. Biotechnol.* **2009**, *84*, 29–33. [CrossRef]
26. de Meneses, A.C.; Almeida Sá, A.G.; Lerin, L.A.; Corazza, M.L.; de Araújo, P.H.H.; Sayer, C.; de Oliveira, D. Benzyl butyrate esterification mediated by immobilized lipases: Evaluation of batch and fed-batch reactors to overcome lipase-acid deactivation. *Process Biochem.* **2019**, *78*, 50–57. [CrossRef]
27. Chapman, J.; Ismail, A.; Dinu, C. Industrial Applications of Enzymes: Recent Advances, Techniques, and Outlooks. *Catalysts* **2018**, *8*, 238. [CrossRef]
28. Ghamgui, H.; Karra-Chaabouni, M.; Gargouri, Y. 1-Butyl oleate synthesis by immobilized lipase from *Rhizopus oryzae*: A comparative study between n-hexane and solvent-free system. *Enzyme Microb. Technol.* **2004**, *35*, 355–363. [CrossRef]
29. Huang, S.M.; Wu, P.Y.; Chen, J.H.; Kuo, C.H.; Shieh, C.J. Developing a high-temperature solvent-free system for efficient biocatalysis of octyl ferulate. *Catalysts* **2018**, *8*, 338. [CrossRef]
30. Lee, C.C.; Chen, H.C.; Ju, H.Y.; Chen, J.H.; Kuo, C.H.; Chung, Y.L.; Liu, Y.C.; Shieh, C.J. Green and efficient production of octyl hydroxyphenylpropionate using an ultrasound-assisted packed-bed bioreactor. *J. Ind. Microbiol. Biotechnol.* **2012**, *39*, 655–660. [CrossRef]
31. Chen, H.C.; Kuo, C.H.; Twu, Y.K.; Chen, J.H.; Chang, C.M.J.; Liu, Y.C.; Shieh, C.J. A continuous ultrasound-assisted packed-bed bioreactor for the lipase-catalyzed synthesis of caffeic acid phenethyl ester. *J. Chem. Technol. Biotechnol.* **2011**, *86*, 1289–1294. [CrossRef]
32. Karra-Châabouni, M.; Ghamgui, H.; Bezzine, S.; Rekik, A.; Gargouri, Y. Production of flavour esters by immobilized *Staphylococcus simulans* lipase in a solvent-free system. *Process Biochem.* **2006**, *41*, 1692–1698. [CrossRef]
33. Kuo, C.H.; Huang, C.Y.; Lee, C.L.; Kuo, W.C.; Hsieh, S.L.; Shieh, C.J. Synthesis of DHA/EPA ethyl esters via lipase-catalyzed acidolysis using Novozym® 435: A kinetic study. *Catalysts* **2020**, *10*, 565. [CrossRef]
34. Rodriguez-Colinas, B.; Fernandez-Arrojo, L.; Santos-Moriano, P.; Ballesteros, A.O.; Plou, F.J. Continuous packed bed reactor with immobilized β-galactosidase for production of galactooligosaccharides (GOS). *Catalysts* **2016**, *6*, 189. [CrossRef]
35. Neifar, S.; Cervantes, F.V.; Bouanane-Darenfed, A.; BenHlima, H.; Ballesteros, A.O.; Plou, F.J.; Bejar, S. Immobilization of the glucose isomerase from *Caldicoprobacter algeriensis* on Sepabeads EC-HA and its efficient application in continuous high fructose syrup production using packed bed reactor. *Food Chem.* **2020**, *309*, 125710. [CrossRef] [PubMed]
36. Salvi, H.M.; Kamble, M.P.; Yadav, G.D. Synthesis of geraniol esters in a continuous-flow packed-bed reactor of immobilized lipase: Optimization of process parameters and kinetic modeling. *Appl. Biochem. Biotechnol.* **2018**, *184*, 630–643. [CrossRef]

37. Sun, J.; Yu, B.; Curran, P.; Liu, S.Q. Lipase-catalysed transesterification of coconut oil with fusel alcohols in a solvent-free system. *Food Chem.* **2012**, *134*, 89–94. [CrossRef]
38. Cavallaro, V.; Tonetto, G.; Ferreira, M.L. Optimization of the enzymatic synthesis of pentyl oleate with lipase immobilized onto novel structured support. *Fermentation* **2019**, *5*, 48. [CrossRef]
39. Pires-Cabral, P.; Da Fonseca, M.; Ferreira-Dias, S. Synthesis of ethyl butyrate in organic media catalyzed by *Candida rugosa* lipase immobilized in polyurethane foams: A kinetic study. *Biochem. Eng. J.* **2009**, *43*, 327–332. [CrossRef]
40. Gawas, S.D.; Jadhav, S.V.; Rathod, V.K. Solvent free lipase catalysed synthesis of ethyl laurate: Optimization and kinetic studies. *Appl. Biochem. Biotechnol.* **2016**, *180*, 1428–1445. [CrossRef]
41. Royon, D.; Daz, M.; Ellenrieder, G.; Locatelli, S. Enzymatic production of biodiesel from cotton seed oil using t-butanol as a solvent. *Bioresour. Technol.* **2007**, *98*, 648–653. [CrossRef]
42. Halim, S.F.A.; Kamaruddin, A.H.; Fernando, W. Continuous biosynthesis of biodiesel from waste cooking palm oil in a packed bed reactor: Optimization using response surface methodology (RSM) and mass transfer studies. *Bioresour. Technol.* **2009**, *100*, 710–716. [CrossRef] [PubMed]
43. Chang, C.; Chen, J.H.; Chieh-ming, J.C.; Wu, T.T.; Shieh, C.J. Optimization of lipase-catalyzed biodiesel by isopropanolysis in a continuous packed-bed reactor using response surface methodology. *New Biotechnol.* **2009**, *26*, 187–192. [CrossRef] [PubMed]

© 2020 by the authors. Licensee MDPI, Basel, Switzerland. This article is an open access article distributed under the terms and conditions of the Creative Commons Attribution (CC BY) license (http://creativecommons.org/licenses/by/4.0/).

Article

Characterization of Electrode Performance in Enzymatic Biofuel Cells Using Cyclic Voltammetry and Electrochemical Impedance Spectroscopy

Adama A. Bojang and Ho Shing Wu *

Department of Chemical Engineering and Materials Science, Yuan Ze University, 135 Yuan Tung Road Chung Li, Taoyuan 32003, Taiwan; zazafj1990@gmail.com
* Correspondence: cehswu@saturn.yzu.edu.tw; Tel.: +886-3-4638800-2564

Received: 20 June 2020; Accepted: 9 July 2020; Published: 13 July 2020

Abstract: The main objective of this study was to examine the quantitative performance of the electrochemical redox reaction of glucose by glucosidase and oxygen with laccase in a phosphate buffer solution at pH 7.0. The characterization of electrode performance was performed by using electrochemical analysis such as cyclic voltammetry (CV) and electrochemical impedance spectroscopy (EIS). The use of such electrochemical analysis (CV and EIS) enables a better understanding of the redox process, the charge transfer resistance, and, hence, the potential mass transfer among the electrode materials in phosphorus buffer solution. The experimental results show that the maximum power densities of the bioanode and the biocathode electrodes were 800 µA/cm^2 and 600 µA/cm^2, respectively. Both the bioanode and biocathode show high internal resistance. The occurrence of peak-separation shows an excellent mass-transfer mechanism and better chemical reactivity in the electrode.

Keywords: cyclic voltammetry; electrochemical impedance spectroscopy; carbon nanotubes; redox mediators

1. Introduction

Since the invention of fossil fuels, from which energy is generated by the burning method, they have caused some severe impacts on the environment. The world needs a lot of energy for daily activities to be continued, which results in a high dependency on fossil fuels and, due to the scarcity of this natural resource, the destruction of the environment and other global warning effects [1]. An alternative source of energy is needed because of the demand for energy with an increasing consumption rate, a decrease in the supply of natural fossil fuels, and the challenges of the destruction of the ecosphere and ecology [1,2]. From the fuel cell or electrochemical industry is a high demand for researching a clean and safe energy source that is environmentally attainable and efficient in terms of use and production output. Fuel cells of all categories (biofuel cells) are one of the electrochemical systems that generate energy and store it for future applications, including batteries [3].

The study of the chemical reaction and electrical conductivity is called electrochemistry. It further includes the dynamics of chemical reactions, which are due to the electrical conductivity across a medium and, in turn, the generation of current or energy from the chemical reactions. This study uses two basic electrochemical principles, which are electrochemical impedance spectroscopy (EIS) and Cyclic Voltammetry (CV) [4]. EIS is a useful tool in electrochemistry that is used in the characterization of enzymatic biofuel cells. It analyzes the maximum power production, which is limited by the high resistance of the enzymatic biofuel cells. The combination of carbon nanotubes (CNTs) with enzymes (glucose oxidase and laccase) as the electrode produces the maximum power output. Likewise, CV was used on a carbon paper electrode in an aqueous solution containing phosphate buffer solution with glucose dissolved in it as a supporting electrolyte [5].

EIS is used in electrochemical analysis for the characterization of an electrode, presently, by the charge transfer and resistances of the solution and electrode materials. EIS analysis follows a steady-state principle, as the changes in the current of the electrode in the biofuel cell are transmitted in signals of small magnitude. Moreover, EIS has some merits over other electrochemical techniques because it can measure the impedance of the biofuel cell system without causing distress to the overall operation; that is why it is called a non-intrusive or non-destructive technique. The EIS technique is currently used in many electrochemistry applications, such as for studying corrosion [6] and in biofuel cell systems [7]. Meanwhile, voltammetry or CV is also a useful tool in electrochemistry. It examines the redox-reaction mechanism (oxidation and reduction) of the electrode. It scans the potential (voltage) by sweeping the electrode from a lower potential to a higher potential. By doing this, an equilibrium is a retort, and the potential voltage and current output of the electrode are observed.

The classification of a fuel cell is generally based on the conventional mechanism; as such, biofuel cells belong to a group of traditional fuel cells that involve the use of biological enzyme catalysts that generate current from a biochemical reaction in the form of a redox reaction. Since biological enzymes have many applications in fuel cell systems, this and other characteristics make them famous for use in vast quantities for the production of energy in some portable devices, and also, they are very compatible with the ecosphere and sustainable [8]. Biofuel cells are also integrated in biological-implantable devices since they can work as micro-reactors, used for life support purposes, used in fields of biomedicine, or used in particular environmental operations to ease pollution and global warming's effects since they are environmentally friendly [9]. Biological or enzymatic biofuel cells are as valuable as other energy systems such as batteries in many ways and forms. The use of metal catalysts is somewhat costly compared to that of enzymatic catalysts used in biofuel cells, which are cheaper, provide fast redox reaction conditions, and are renewable. This results in a good economic benefit over traditional fuel cells [10]. The biological enzyme catalyst in the biofuel cells provides energy via a redox-chemical reaction, which allows the operation of these cells in different fields [11]. This study is aimed at characterizing the electrode performance in enzymatic biofuel cells using the CV technique and EIS technique.

2. Results and Discussion

2.1. Effect of Scan Rate

Both the bioanodic and biocathodic current peaks increase with an increasing scan rate, as illustrated in Figures 1 and 2, respectively. The bioanodic current peak tends toward a less positive voltage potential, while the biocathodic current peak tends toward a more positive voltage potential. Generally, the current (i_{pa}) is proportional to the square of the scan rate (Figures 1b and 2b). The current peak separations between the bioanodic and biocathodic current peak potentials at a scan rate of 50 mV/s are higher than those at 59 mV/s, which is one of the primary indications of a quasi-reversible system according to previous work [12].

Generally, a current/peak ratio i_{pa}/i_{pc} of approximate unity reflects immovability of both the bioanode and biocathode catalytic enzyme on the surface area of the electrode. Any oxidation or reduction reactions are actually slow, which was practically demonstrated for both electrodes during the CV. The effect of the scan rates on the oxidation and reduction of GOx and LAc in the presence of glucose and PBS was highlighted. The peak current ratios i_{pa}/i_{pc} increase as the scan rate increases for the black electrode. For both the GOx and LAc electrodes, the peak ratio increases at a scan rate of 20 mV/s and decreases evenly at the scan rate 30–50 mV/s (Figures 1c and 2c). This indicates that a chemical reaction occurred between the GOx and its combinations (carbon nanotubes (CNT), M, and polypyrrole (PPY)), as occurred for LAc, too. These observed phenomena were studied extensively in previous work [12–14]. According to Figures 1 and 2, the cyclic voltammograms show the shifting of the anodic and cathodic peak potentials with an increasing trend, which indicates efficient mass transfer between the electrodes, and the oxidation and reduction of the enzyme were best achieved with the CNT combination since it has good electrical conductivity.

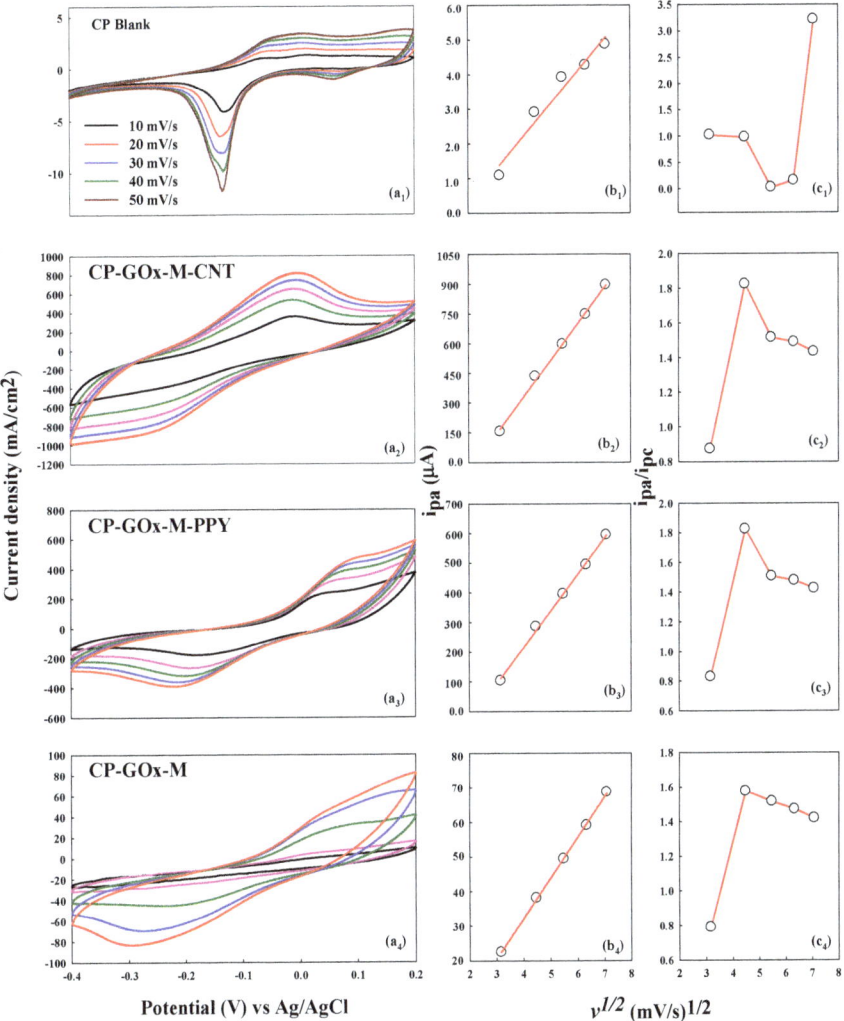

Figure 1. (a_1–a_4) Cyclic voltammogram for the blank electrode and bioanode electrode with different combinations at different scan rates, 10–50 mV/s, (b_1–b_4) i_{pa} in proportion to the square root of the scan rates, and (c_1–c_4) ratio of $i_{pa}:i_{pc}$ to the square root of the scan rates.

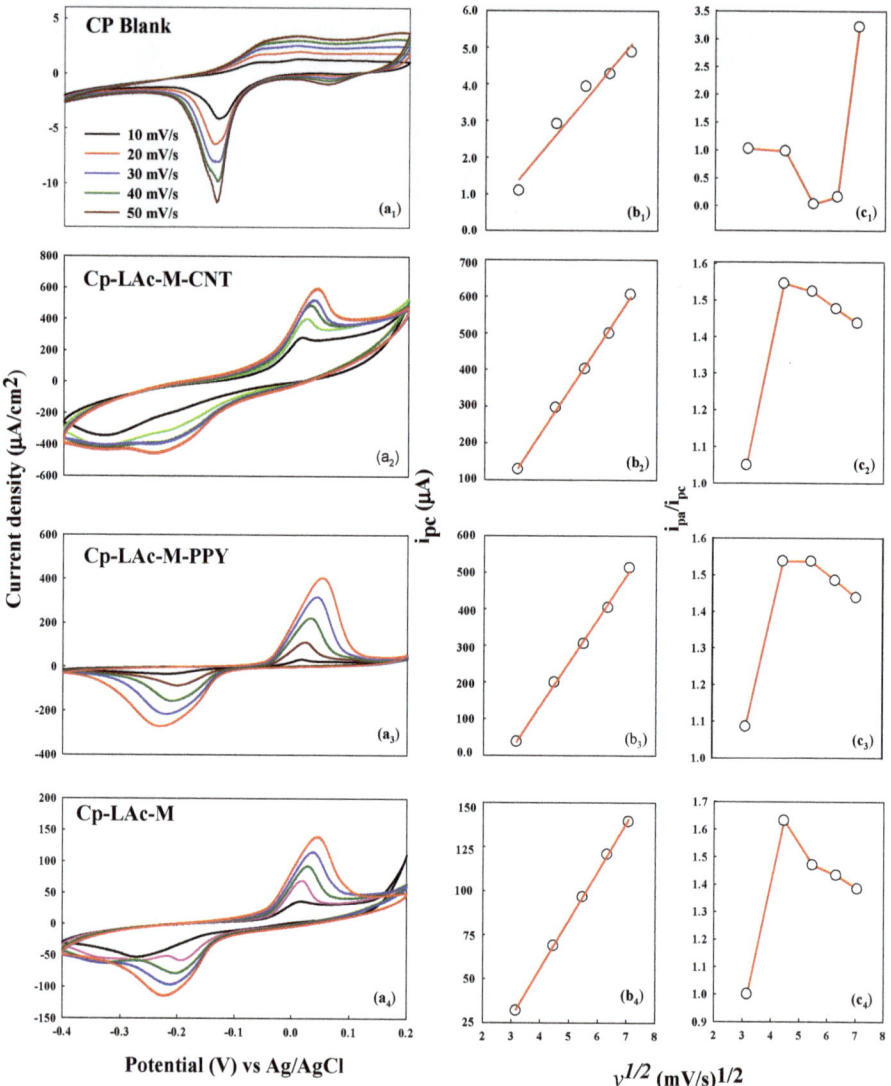

Figure 2. (a_1–a_4) Cyclic voltammogram for the blank electrode and biocathode electrode with different combinations at different scan rates 10–50 mV/s, (b_1–b_4) i_{pa} in proportionality to the square root of the scan rates, and (c_1–c_4) ratio of i_{pa}:i_{pc} to the square root of the scan rates.

The determination of the oxidation current peak from the scan rate with its square root is detailed in Figures 1b and 2b. The intercept is reported as starting from the origin of the voltage potential in the CV analysis. The oxidation current peak is represented in a linear plot. The current is proportional to the square root of the scan rate, as described in Figures 1 and 2. The roles of ferricyanide on the bioanode and 2,2-azino-bis (3-ethylbenzothiazoline-6-sulphonic acid (ABTS) on the biocathode can be limited by mass transport. As established from the slope of the oxidation current in proportion to $v^{1/2}$ and the postulation of the reversibility of the ferricyanide redox reaction, the diffusion coefficient (D) of the bioanode and the biocathode (electrodes) were deduced using the transmuted Randles–Sevcik equation. For the blank electrode in Figure 1b$_1$, $D = \left(\frac{slope}{2.69\times10^5 \, AC}\right)^2$, where slope = 0.1891 cm^2/V s^{-1}, C = 0.1 mol/dm^3, and A = 1 cm^2; therefore, $D = \left(\frac{0.1891}{2.69\times10^5 \times 1\times0.1}\right)^2 = 7.02 \times 10^{-6}$ cm^2/s. The other combinations of bioanode and biocathode were calculated and are listed in Table 1.

Figure 1 shows that an increase in the scan rate increases the mass transfer for the GOx bioanode with different combinations of mediators (GOx-M-CNT, GOx-M-PPY, and GOx-M). The shapes of the peaks increase at high currents and reduce uniformly at low currents, which indicates excellent oxidation and reduction, respectively, and the best electrode that showed better mass transfer and reversibility was GOx-M-CNT, followed by GOx-M-PPY.

From the results, it can be concluded that GOx-M-CNT and LAc-M-CNT are the best electrodes, whose diffusion coefficients are better in the CNT combination for both GOx (9.2 × 10^{-6} cm^2/s) and LAc (9.0 × 10^{-6} cm^2/s), as shown in Table 1 [12]. Moreover, the blank carbon paper electrode shows little material diffusion or low conductivity because there were no essential supporting materials such CNT and mediators to aid this effect and, hence, a low diffusion coefficient (7.02 × 10^{-6} cm^2/s) resulted, as shown in Table 1.

It was determined, after this calculation, that the peak current is proportional to the diffusion coefficient ($i_p \propto D$) in the Randles–Sevcik equation. If the peak current is high, the diffusion rate is large. Therefore, in the additional analysis, the best-modified electrode is the one that has a better current peak coefficient. Not only do the concentration of the electrolyte, diffusion, or electro-active species affect the current peak, but also, the scan rate plays a more significant role in the determination of i_p. In this redox reaction, the current is defined as the charge over time or, generally, the electrons passing per time. For a fast scan rate, which is directly proportional to the electrons caught per time, a higher voltage scan rate will lead to a higher $i_{p,}$, and the total electrons passing remains intact [15]. The electron transfer in this reaction is calculated using the Randles–Sevcik equation. All the calculations give results under 1. Consequently, to simplify the calculations, the electrons transferred are used and equal to 1 (n = 1) in this study. The number of electrons reassigned in the redox occurrence is typically 1 [16].

The case of the biocathode (Figure 2) exposed to different scan rates with different combinations (LAc-M-CNT, LAc-M-PPY, and LAc-M) was studied. It shows an increase in the peak current, but what is unique in both combinations is the arrangement of the endpoint of the peaks (from −0.4 to 0.2 potential vs. Ag/AgCl) at the same point for both cases; this could be due to the limited diffusion of the electrode, since LAc is not a natural reducer of oxygen. Nevertheless, the best electrode was LAc-M-CNT (9.0 × 10^{-6} cm^2/s), followed by LAc-M-PPY (8.9 × 10^{-6} cm^2/s) (Table 1), since these two electrodes give an excellent mass transfer. This phenomenon was observed by Barriere [17].

Table 1. Calculation of bioanode and biocathode voltammograms.

Electrode	Epa (V)	Epc (V)	i_{pa} (A)	i_{pc} (A)	Slope cm/v s^{-1}	D (cm^2/s)
Blank	0.005	−0.147	7.17 × 10^{-6}	−3.141 × 10^{-6}	0.1891	7.02 × 10^{-6}
GOx-M-CNT	0.082	−0.256	5.798 × 10^{-4}	−3.174 ×10^{-4}	0.2479	9.2 × 10^{-6}
GOx-M-PPY	0.073	0.255	5.740 × 10^{-4}	−3.145 × 10^{-4}	0.2464	9.3 × 10^{-6}
GOx-M	0.064	−0.194	5.644 × 10^{-4}	−3.579 × 10^{-4}	0.2342	8.8 × 10^{-6}
LAc-M-CNT	0.044	−0.233	2.171 × 10^{-4}	−1.519 ×10^{-4}	0.2395	9.0 × 10^{-6}
LAc-M-PPY	0.049	0.234	2.434 × 10^{-4}	−1.793 × 10^{-4}	0.2384	8.9 × 10^{-6}
LAc-M	0.054	−0.203	2.563 × 10^{-4}	−1.889 × 10^{-4}	0.2314	8.7 × 10^{-6}

Note: these results were calculated at different scan rates (10–50 mV/s), and the slope of the plot of i_{pa} vs. $v^{1/2}$ (scan rate) in Figures 1b and 2b corresponds to the diffusion coefficient.

As illustrated in Figures 1 and 2, respectively, the pattern of the current dimension in both the forward and backward scans (−0.4 to 0.2 V vs. Ag/AgCl) of both the bioanode and biocathode shows a good separation pattern. This effect is due to the presence of carbon materials or substances, which have good capacitance, as mentioned by researchers [18]. For the bioanode, the current density for Cp-GOx-M-CNT was significantly high compared to that for other combinations. Likewise, the same goes for Cp-LAc-M-CNT from the biocathode perspective. Nevertheless, the i_{pa}/i_{pc} ratio of

Cp-GOx-M-CNT is close to unity; Cp-GOX-M-CNT is characterized as possessing a quasi-reversible character as indicated by the appearance of the peak-to-peak voltage potential, $\Delta E_p > 59/n$ mV. From the analytical point of view, the current density for different scan rates (10–50 mV) shows a quasi-reversible process for Cp-GOx-M-CNT, Cp-GOx-M-PPY, Cp-LAc-M-CNT, Cp-LAc-M-PPY, and Cp-LAc-M, as shown in Figures 1 and 2. The elevation of the ΔEp resulted from an increase in the scan rate from 10 to 50 mV/s, respectively. With the assumption that the number of electrons transferred (n) is 1, the bioanodic and biocathodic current peaks relative to the scan rate are a function of the linear plot (i_p with $v^{1/2}$). Kinetically, the electron transfer is not sufficiently fast to produce sufficient concentrations of the reacting species and products as determined to be essential by the Nernst equation [19]. An increase in the scan rate from 30 mV/s leads to an increase in the current density or capacitance that is faster or that is more than the faradaic current. If the capacitance current decreases the faradaic current, then the identification of the highest current peak is minimal. An additional increase in the scan rate at 50 mV/s causes the faradaic-current to overlap with the capacitance current. Generally, Cp-GOx-M-CNT for the bioanode electrode and Cp-LAc-M-CNT for the biocathode electrode did possess a quasi-reversible mechanism, which shows the even influence of the electron transfer and mass transfer rates.

2.2. Kinetic and Electrical Characteristics of Bioanode and Biocathode Electrodes

The estimation of the average various electron transfer ratio constant ($k°$) is performed by using the Nicholson method, using Equation (1).

$$\psi = K^0 \left(\frac{\pi D n v F}{RT} \right)^{-1/2} \tag{1}$$

where ψ stands for the kinetic factor gain according to Nicholson [18,20,21]. Since a tangible measurement is not performed, the transfer value is taken to be 0.5. Moreover, for simplicity, the determination of the coefficient of ψ is established and deduced from a pattern relation between ψ and ΔE_p, which was done by using Equation (2) with the representation of the curve range of ΔE_p.

$$\psi = \frac{a + b(\Delta E_p)}{1 + c(\Delta E_p)} \tag{2}$$

These assumption-based constant factors are estimated from a non-linear regression-fitting, as $a = -0.54050$, $b = 0.00140$, and $c = -0.01700$. Using graphing, plotting ψ against $v^{1/2}$ will lead to the deduction of the kinetic factor K^0, which is determined from the slope of the plot. A greater value of K^0 means less time for the attainment of equilibrium related to a lower $k°$ value. As illustrated in Figure 3, Cp-GOx-M-CNT has a larger K^0 value, which means it can achieve equilibrium faster than all the other combinations. It can also be seen that the diffusion coefficient also plays a vital role in the electrode kinetics.

2.3. Effect of Immobilization on the Hydrophobic Electrode

To demonstrate the enzymatic performance of the electrode, primarily, a blank carbon paper electrode was analyzed and then surveyed with enzyme immobilized electrodes. The consequences are equated and conveyed in the subsequent CV curve. If the peak current of oxidation or reduction does not appear, it indicates that there is no redox reaction process on the electrode.

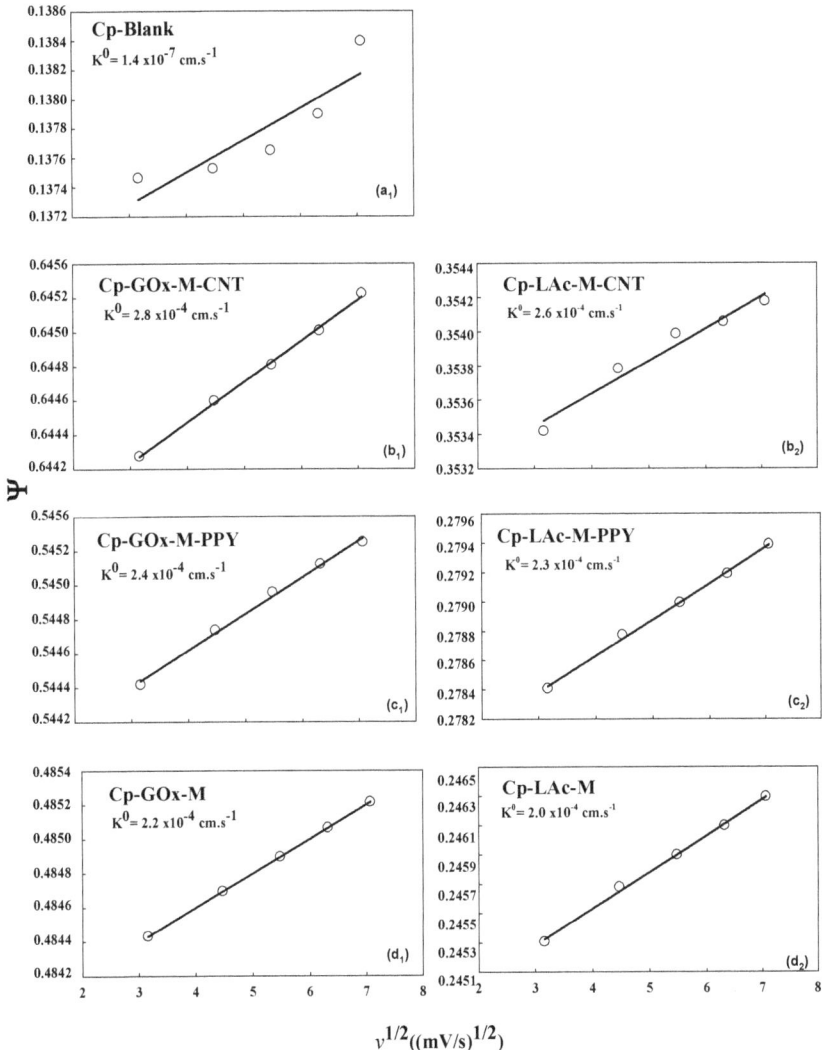

Figure 3. Plot of ψ vs. $v^{1/2}$ for (a_1) carbon paper, (b_1–d_1) bioanode electrode, and (b_2–d_2) biocathode electrode at a scan rate of 10–50 mv/s.

2.3.1. Bioanode

Due to the behavior of the hydrophobic anode electrode's performance with the covalent-bonding immobilization method, the CV scan was carried out in five random prepared samples (Figure 4). It was found that the GOx modified carbon nanotube (CNT) and M (red) combination has a high oxidation and reduction peak compared to the others. On the other hand, GOx modified with a mediator and polypyrrole (blue) gives a higher oxidation peak than GOx with M (pink), GOx (green), and blank (black).

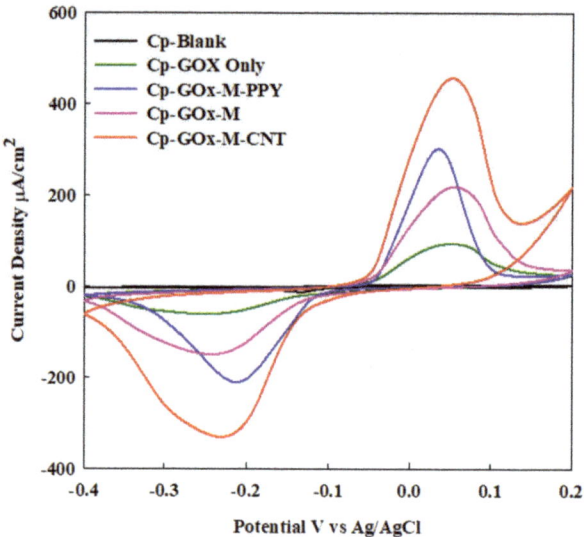

Figure 4. Voltammograms of hydrophobic bioanode electrodes under 0.1 M PBS, pH 7, 37 °C, and N_2 saturation.

2.3.2. Biocathode

The equivalent amendment of the bioanode electrode was also equipped for the biocathode in the presence of oxygen (Figure 5). The highest value of the oxidation peak current was attributed to LAc modified CNT and M (red). The presence of mediators (ABTS) produced a very remarkable result here, as shown by LAc modified M and PPY (blue), and LAc only yielded a good result, which is surprising. A blank electrode gave the smallest cathodic peak with the carbon nanotube (black).

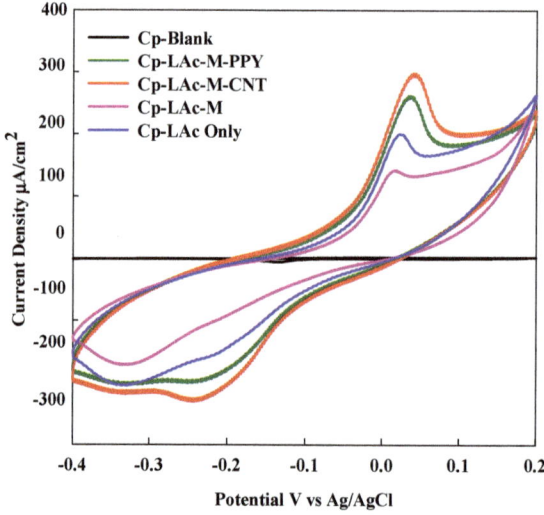

Figure 5. Voltammograms of the hydrophobic biocathode electrodes under 0.1 M PBS, pH 7, 37 °C, and O_2 saturation.

2.3.3. Effect of Immobilization on the Hydrophilic Electrode

The covalent-bonding method was proposed to be employed for hydrophilic provision to produce a superior enzyme-support collaboration compared to the hydrophobic electrode. Owing to the hydrophilic carbon paper having no PTFE (polytetrafluoroethylene) coating treatment on the surface, this support is calculatingly engaged for better performance of the enzymatic biofuel cell. The comparable modification for a hydrophobic electrode was designed and examined with half-cell analysis by exhaustive CV.

Bioanode

Voltammograms of the hydrophilic bioanode electrodes are given in Figure 6. The electrode modified with CP-GOx-M-CNT (red) has a meaningfully broader potential range than the others. It likewise produced the best peak in current density. On the other hand, the second biggest i_p was from an electrode modified with CP-GOx-M-PPy (blue).

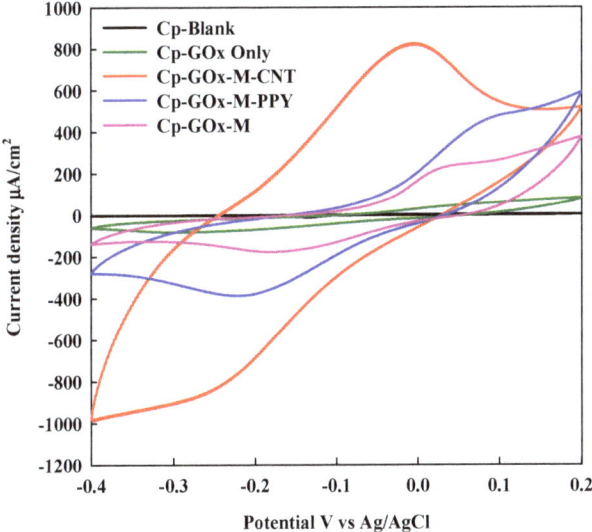

Figure 6. Voltammograms of hydrophilic bioanode electrodes under 0.1 M PBS, pH 7, 37 °C, and N_2 saturation.

Although the CP-GOx-M remained in the third place, this performance indicates that the GOx-M-CNT modification can improve the performance of the electrode better than GOx-M-PPy using the covalent bonding immobilization method. The electrode with the presence of GOx performed poorly. This finding demonstrated that the enzyme generating current with the help of a substrate such as a CNT can produce high current density. The smallest peak was observed for the blank electrode. Moreover, the benefits of CNTs have been reported in previous work [22–24].

Biocathode

According to the voltammograms of the hydrophilic biocathodes (Figure 7), the highest reduction peak current was attained by the electrode coated with LAc-M-CNT (red). The performance of the LAc-M-PPY electrode was excellent, too (green). On the other hand, depending on the irreversibility of the electrode, the most durable electrode was LAc-M. This finding demonstrates that the M and CNT combination not only successfully increases the reversibility of the electrode but increases the diffusion rate of the electron transfer. Most importantly, according to the above analysis, the peaks rise evenly, which shows efficient mass transfer among the modified electrodes.

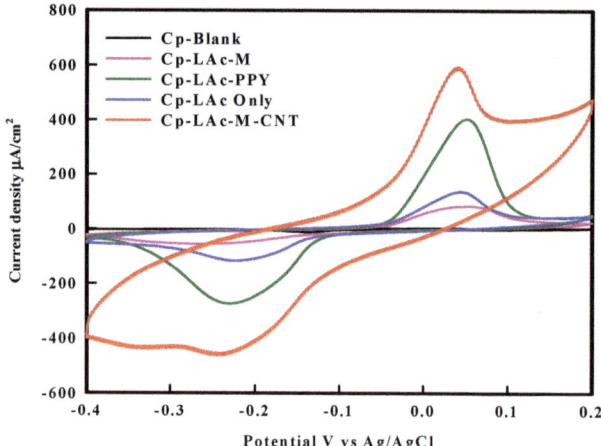

Figure 7. Voltammograms of the hydrophilic biocathode electrode under 0.1 M PBS, pH 7, 37 °C, and O_2 saturation.

2.4. Characterization of the Electrode Using Electrochemical Impedance Spectroscopy

2.4.1. Bioanode Impedance

Enzymatic biofuel cells have an excellent anodic internal resistance, which plays a significant role in their polarity of resistance. The mechanism of oxidation occurring at the bioanodic electrode surface causes the anodic charge-transfer resistance to decrease significantly. This effect shows the catalytic process and contrivance of the bioanode in terms of electron transfer, as illustrated in previous studies [25,26]. In this work, GOx was immobilized with different conductive materials (CNT, carbon nanoball (CNB), and PPY), and each was analyzed using EIS. Figure 8 shows that GOx-M-CNT has higher charge resistance, which indicates better oxidation and stability than GOx-M-CNB. The best charge resistance is observed for GOx-M-CNT (1.05×10^{-9} Ω), as shown in Table 2.

The circuit characterization of the enzymatic electrode surface shows two distinct layers, which involve the inner and the outer layers, respectively (Figure 8). This type of circuit diagram generally represents a permeable bioelectrode surface, which indicates several holes on the surface of the electrode. The conducting layers are pointed at these holes. Additionally, the immobilized layers are arranged around each other. It is noticeable that there is no semicircle distance curve, which means there is an efficient mass transfer, and for the bioanode impedance, the Warburg impedance data are absent. Warburg (**W**) is not present in the equivalent circuit. The outer layer is considered as the measurement electrode layer. R_1 signifies the electron-transfer resistance of the electrode surface; other factor constants such as R_2 and C_2 are located on the inner-layer surface of the electrode, and their effects do not change. The capacitance of the outer surface of the electrode is taken to be C_1 (Table 2). The desirability is due to the positively charged surface and the negatively charged electrode surface, which permit a redox chemical reaction. Negatively charged redox molecules are attracted to the positively charged electrode surface. Meanwhile, electrostatic movement, diffusional effects, and redox concentration affect the surface of the electrode, as these conditions increase the electron movements near the electrode surface, and the electron-transfer resistance is reduced. Table 2 shows that GOx-M-CNT demonstrated a lower charge transfer resistance. This finding is due to the increase in surface area. The charge transfer resistance decreases in the order of GOx-M-CNT > GOx-M-CNB >

GOx-M (Table 2). This phenomenon of layer-by-layer covered electrode surfaces is also observed in studies that have employed this Nyquist plot type [7,27–29].

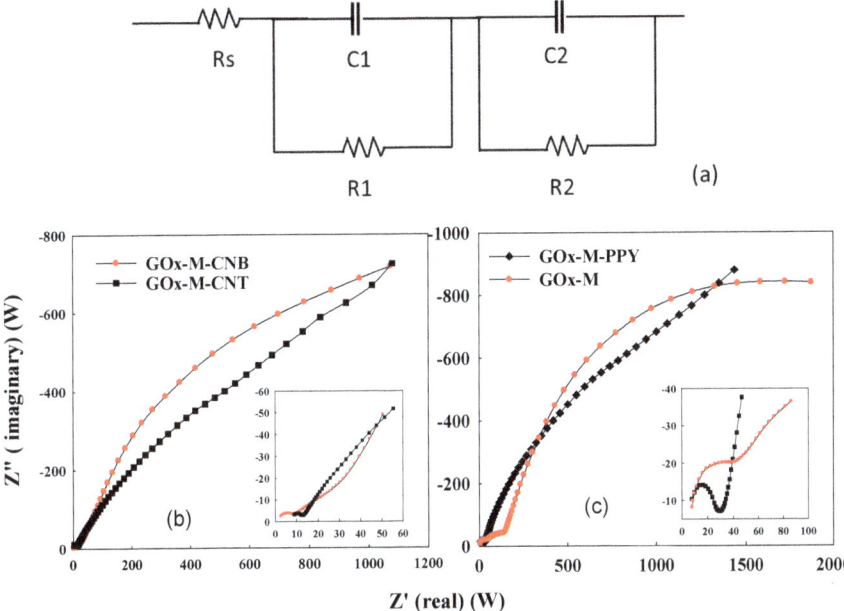

Figure 8. (a) Representation of Randle circuit and Nyquist plot for (b) GOx-M-CNB and GOx-M-CNT, and (c) GOx-M-PPY and GOx-M bioanode impedance of enzymatic biofuel cell (EBFC).

Table 2. Impedance data for the bioanode.

Electrode	R_s (Ω)	R_1 (Ω)	R_2 (Ω)	C_1 (Ω)	C_2 (Ω)
GOx-M-CNT	1.7	0.002	1.05×10^{-9}	1.84×10^{-7}	2.94×10^{-6}
GOx-M-CNB	1.7	0.001	1.05×10^{-6}	1.99×10^{-6}	2.88×10^{-5}
GOx-M-PPY	1.9	0.010	1.06×10^{-7}	2.85×10^{-3}	2.76×10^{-4}
GOx-M	1.10	0.10	1.00×10^{-4}	2.77×10^{-4}	2.66×10^{-4}

However, the kinetic rate of the enzymatic reaction according to the characterization of the redox-oxidation formation rate is representative of the enzymatic biofuel cell (EBFC). The bio-anodic charge-transfer resistance measurements show that this effect is well established in the literature [26,30]. The electrochemical deposition of the bioanode electrode increases the impact on the surface area of the bioanode electrode and hence reduces the charge transfer resistance. The effect of the enzymes on the surface of the electrode catalyzes the oxidation of glucose and also decreases the charge transfer resistance.

Effects of Bioanode Mediators

Some enzymes can perform well with some inorganic mediators. In this case, the mediator selected for the bioanode was $K_3Fe(CN)_6$, as it serves as the electron-transfer agent or intermediate between the biological enzyme (GOx) and the bioanode electrode. Interestingly, few enzymes and microorganisms (bacteria) help in terms of the electron-transfer mechanism in bioanode systems. The analysis of mediators in biofuel cells can be done using EIS, through which interpretations can be shown from low to high frequencies [31]. It can be observed that the Bode phase angle plot in the frequency region is attributed to the charge-transfer resistance (R_{ES}) of the $K_3Fe(CN)_6$. Usually, the concentration of mediators in the bio-anolyte is deficient, and thus, they confer low charge-transfer resistance, but the mediators offer excellent redox-electron transfer (Figure 9b). Generally, R_A denotes the charge-transfer resistance for the bioanode, while R_{other} is the charge-transfer resistance of some metal-salts that are dissolved in the electrolyte (PBS).

The developed biochemical mediators help the bio-anolyte by decreasing the charge-transfer impedance for the enzymatic redox-oxidation and, kinetically, improve the rate of electron transfer between the bio-enzyme and the bioanode electrode. It is observed that the magnitude of the frequencies reduced in the lower region. This finding is due to the modification of the electrode with both a CNT and CNB. This effect can be extrapolated as the charge-transfer resistance impedance of the redox-oxidation substance (substrate) (R_A) (Figure 9a). CNT has a better charge transfer resistance than CNB. With this evidence, one can say that the presence of mediators, either organic or inorganic in a redox-chemical reaction, aids the reaction in terms of speed. Some experts have reported that the reaction with mediators is 10–20 times faster than with most electrochemical redox-oxidation, which involves some rate-limiting-step on the bioanode [25,26,31–33].

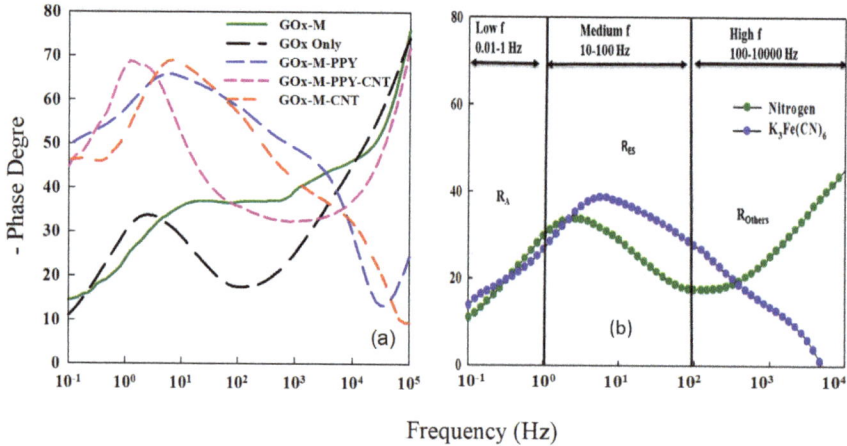

Figure 9. Bode phase angle plot of (**a**) GOx enzyme with different combinations of mediators, (**b**) the bioanode for potassium ferrocyanide and nitrogen.

Effect of Assembling Agents

Nafion solution was used to allow the ion exchange between the bioanode and biocathode solution. Therefore, to verify the effect of the Nafion, we examined the fuel cell with Nafion and without Nafion solution as an assembling agent, and the results show that the EBFC with Nafion solution shows a high-power output than that without Nafion solution, this reflected on the EIS responds on the EBFC, as shown in Figure 10. The Bode plot shows that the impedance of the biocathode and bioanode with Nafion solution is higher than that without the Nafion solution. The impedance of the bioanode and biocathode with and without Nafion solution decreases over time owing to the surge in the porosity of the surface layers. This effect was also studied in previous work [26,31].

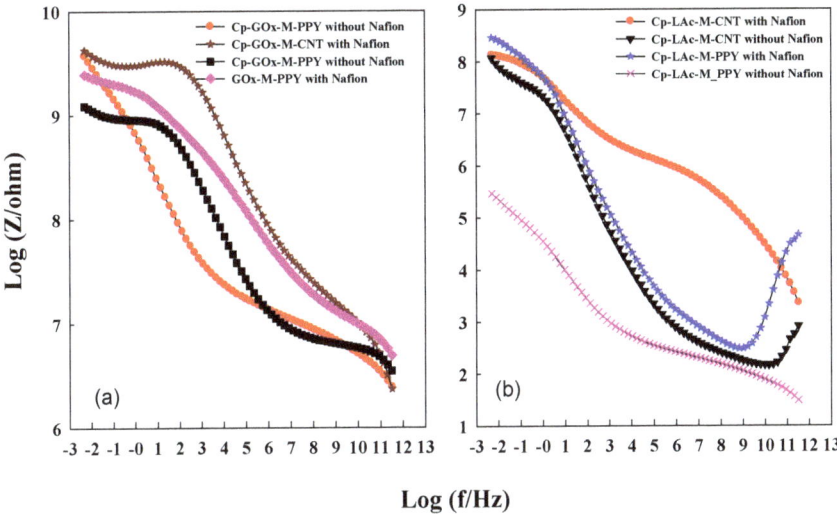

Figure 10. Bode plot of (**a**) bioanode impedance and (**b**) biocathode impedance for EBFC with and without Nafion solution.

2.4.2. Characterization of the Biocathode

The oxygen-reduction at the biocathode by the LAc enzyme is a less known focal limiting issue in the enactment of EBFC, even though there has been much enhancement in the development of a well-organized catalyst [34,35]. In this biocathode system, both the charge transfer resistance and the mass oxygen transfer are the primary limiting determinants for the enactment of EBFC. For the bioanode, the maximum current is reached because of the maximum biochemical conversion rate [36,37]. However, it is different in the biocathode because the main limiting factors are the poor mass transfer and bad solubility of oxygen. Figure 11a shows the Bode plot for the ABTS cathode mediator, which contains a maximum of two factors or constants, which are related to the metallic salt present in the electrolyte at higher frequencies and ABTS reduction at medium frequencies. Meanwhile, as shown in the Bode plot, the oxygen biocathode contains three factor that are related to the oxygen reduction at lower frequencies. According to Ramasamy et al. [26], there is a substantial time constant in the medium frequency region, which is linked to the metal salts in the electrolytes and interrelated with the charge-transfer impedance of the mediator, in this case, ABTS. This can also be observed for the biocathode, with different combinations of the LAc enzyme with the various mediators, and it is observed that the LAc-M-CNT combination shows a more suitable impedance than the others.

Effect of Biocathode Material Type

The type of biocathode substance is of considerable significance regarding steadiness. Martin et al. mentioned the importance of electrochemical viability for improving the EBFC, predominantly in the case of air-based biocathode EBFCs [38]. As shown in Figure 12, the carbon paper electrode was modified with ABTS mediator. CP with CNT shows a higher charge resistance than all the other combinations. Therefore, it can be concluded that the LAc-M-CNT biocathode has the best catalytic performance concerning oxygen-reduction chemical redox reactions.

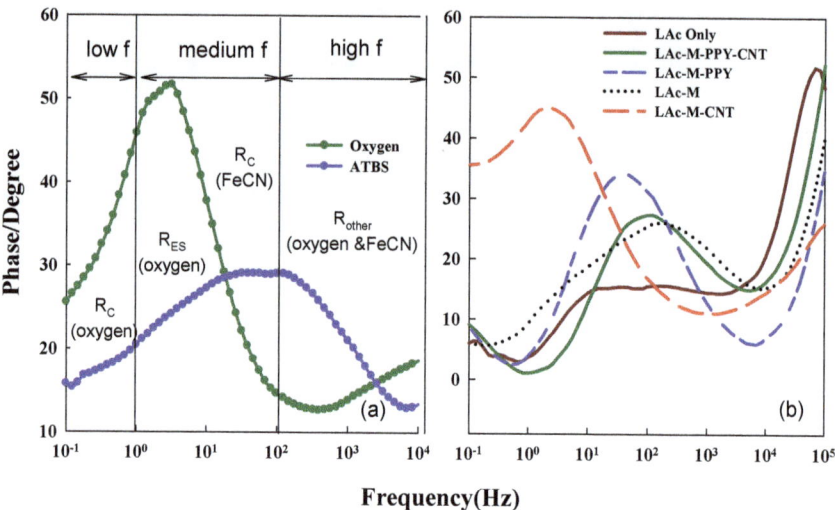

Figure 11. Bode plot of (**a**) biocathode with 2,2-azino-bis (3-ethylbenzothiazoline-6-sulphonic acid (ABTS) and oxygen-based EBFC; (**b**) Bode phase angle plot of LAc with different electrodes.

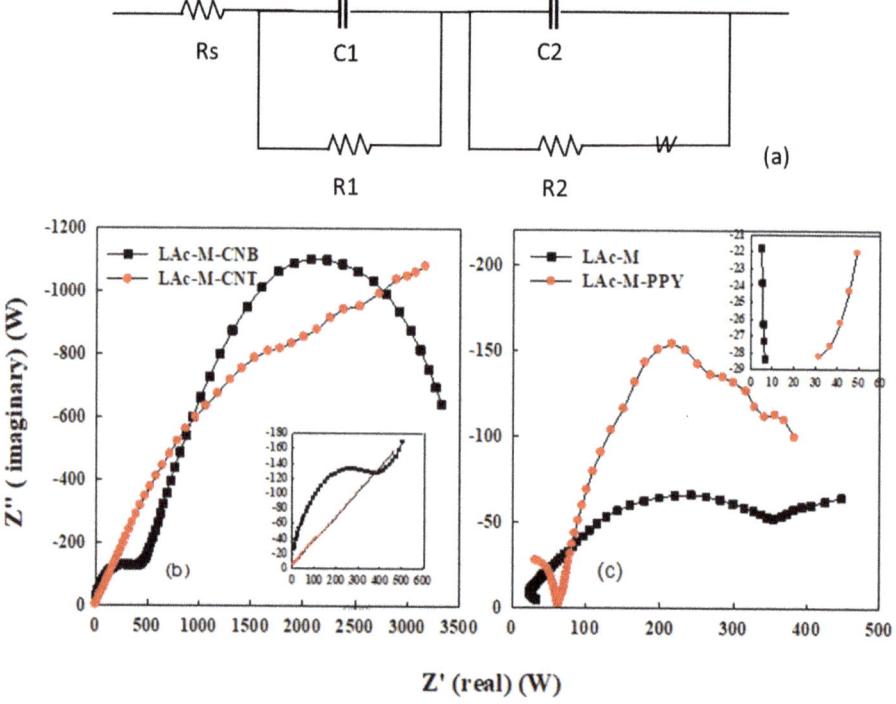

Figure 12. (**a**) Representation of Randle circuit and Nyquist plot for (**b**) LAc-M-CNB and LAc-M-CNT and (**c**) LAc-M-PPY and LAc-M biocathode impedance for EBFC.

As shown in Figure 12, there is one invisible semicircle and a linear domain, and this linear domain represents the mass-transfer resistance (W), which is formed by redox migration toward the electrode surface. Similar to the bioanode, the equivalent circuit of the biocathode shows an outer and inner layer. The presence of Warburg impedance indicates that there is a low mass transfer from the biocathode because it is still a challenge for the O_2 to be reduced by LAc enzyme, as explained by scholars [8,11]. In this experiment, a condition occurred in which the redox experimentation on the electrode surface strengthened the interface. Among biological enzymes, LAc is the best at reducing kinetically slow oxygen. As shown in Table 3, LAc-M-CNT shows an excellent charge transfer resistance (1.45×10^{-6} Ω). The resistance of the solution also plays a vital role in the mass transfer process, which depends on the pH and other assembling agents.

Table 3. Impedance data for the biocathode.

Electrode	R_s (Ω)	R_1 (Ω)	R_2 (Ω)	C_1 (Ω)	C_2 (Ω)	W (Ω)
LAc-M-CNT	1.7	2.3×10^{-7}	1.45×10^{-6}	1.0×10^{-7}	2.22×10^{-6}	0.5
LAc-M-CNB	1.7	2.4×10^{-8}	1.55×10^{-5}	1.20×10^{-5}	2.45×10^{-4}	0.03
LAc-M-PPY	1.20	2.54×10^{-6}	1.63×10^{-4}	1.36×10^{-6}	2.66×10^{-3}	0.034
LAc-M	1.22	2.60×10^{-9}	1.74×10^{-3}	1.40×10^{-6}	2.88×10^{-2}	0.004

3. Experimentation

3.1. Chemicals

Phosphate buffer solution (PBS), carbon nanotubes (CNTs), polypyrrole (PPY), 2,2-azino-bis (3-ethylbenzothiazoline-6-sulphonic acid) (ABTS), potassium ferrocyanide (III) $K_3Fe(CN_6)$, glucosidase (GOx), and laccase (Lac) were purchased from Sigma-Aldrich (St. Louis, MO, USA). Glucose, N-(3-dimethylaminopropyl)-N′-ethylcarbodiimide, and N-hydroxysuccinimide were purchased from Fisher Chemical (London, UK).

3.2. Electrolyte Solution

In the PBS (pH 7, 0.2 M) preparation, monosodium phosphate (0.477 mol) and disodium sulfate (0.523 mol) were dissolved in distilled water. The pH of the mixture was measured using a pH meter (JENWAY 3510, Barloworld Scientific Ltd., Dunmow, Essex, UK), and adjusted by using sulfuric acid or sodium hydroxide to attain a neutral pH. Furthermore, water was added to increase the volume of the phosphate buffer to 1 L, at pH 7. Generally, PBS was used as the stock solution (1 M). Later, the stock solution was diluted to form other buffer quantities with the required concentrations such as 0.4, 0.5, or 0.1 M. This research purposely used Sorensen's phosphate buffer solution (pKa = 7.2, pH 5.8–8.0, 0.1 M), which consists of NaH_2PO_4 and Na_2SO_4. Citrate buffer solution (CBS, pH 5) was prepared by forming two stock solutions, which included 100 mL of citric acid (0.1 M) and sodium citrate (0.1 M), respectively.

3.3. Preparation of Enzyme Solution and Immobilization Technique

Before enzyme preparation, the carbon nanotube was treated with acid to induce the COOH groups on the sidewalls of the CNT and then to immobilize enzyme by covalent bonding. In the making of the enzyme solution, the most vital aspect was to verify the enzyme concentration by checking the concentration of the enzyme on the package because an enzyme with the same CAS number (i.e., GOx) may be extracted at different levels.

3.3.1. Oxidation of CNT with HNO_3/H_2SO_4

The CNT was obtained from Conyuan Bio-Chemical Technology Co. of Taiwan (denoted as T2040, Taoyuan, Taiwan). It underwent an oxidation reaction to activate the acidic groups on the side walls. The mode of oxidation was impregnation with sulfuric acid and nitric acid in a ratio of 3:1, respectively. The solution underwent reflux at 90 °C for 24 h in a round flask, which was connected to a condenser. Later, the oxidized CNT was washed using deionized water until the pH was 7 in solution. The oxidized CNT was oven-dried at 104 °C under vacuum conditions for 24 h. Finally, the oxidized CNT was analyzed using FTIR and XRD to observe the acidic groups (COOH) on the sidewalls of the carbon nanotube.

3.3.2. Preparation of Bioanode and Biocathode Enzyme Solutions

The preparation of the bioanode or biocathode enzyme solution was performed according to the following steps. Firstly, GOx and Lac were immobilized on the surface of the CNT by direct covalent bonding. Secondly, 30 mM N-ethyl-N'-(3-dimethylaminopropyl) carbodiimide and 90 mM N-hydroxysuccinimide were mixed in PBS (pH 7.0, 50 mM). The solution was pipetted onto the nanocomposite surface of the CNT. After 1 h, the CNT on the carbon paper surface was dried using nitrogen gas. Then, a PBS solution (pH 7.0, 50 mM) containing 4 mg/mL of GOx (or Lac) was dropped on the surface and allowed to react for 4 h [39]; 16 mg of ABTS (or $K_3Fe(CN)_6^{3-}$) was also dissolved in PBS (pH 7.0, 50 mM), and 30 mL of each mediator was dropped on the respective electrode and then dried for 1 h. ABTS and $K_3Fe(CN)_6^{3-}$, as the mediators (M), were used with Lac and GOx, respectively.

3.3.3. Cyclic Voltammetry

The CV apparatus used was a CH 600 electrochemical analyzer (Bio-analytical systems, Virginia, NV, USA), coupled to an Acer computer (Hsinchu, Taiwan). A three electrode cell was used in the voltammetry experiment with modified carbon paper electrodes (1 cm × 1 cm) as working electrodes, a platinum wire as an auxiliary electrode, and a Ag/AgCl electrode (3M KCl) as a reference electrode (Bio-analytical systems, USA). The working electrode was modified with the different combinations of bioanode enzyme, biocathode enzyme, potassium ferrocyanide (mediator), and CNT for each set of experiments. The experiments were carried out at 37 °C.

Electrochemical analysis using CV was utilized to determine the diffusion coefficients of different combinations of mediators with the electrode enzyme. In order to produce good results, some steps were followed. For example, each scan of CV analysis from an initial potential to the final potential was considered as a segment. Therefore, two segments made one complete cyclic curve. EIS experiments were run using a CH600 electrochemical analyzer and were analyzed using the Zview software from Scribner Associate Inc (Version 2016, Scribner Associates, Budapest, Hungary).

3.3.4. Randles–Sevcik Equation

In CV, the Randles–Sevcik equation is used to describe the effect of the scan rate on the peak current i_p. The significant parameters of cyclic voltammograms are the enormities of the anodic peak current (i_{pa}) and cathodic peak current (i_{pc}), the anodic peak potential (E_{pa}), and the cathodic peak potential (E_{pc}). The peak current for a reversible system is described by the Randles–Sevcik equation for the forward sweep of the first cycle:

$$i_p = (2.69 \times 10^5) n^{3/2} A D^{1/2} C v^{1/2} \tag{3}$$

Electrochemical irreversibility is caused by the slow electron exchange of the redox species with the working electrode. In this case, the equation will be:

$$i_p = (2.99 \times 10^5) \alpha^{1/2} A D^{1/2} C v^{1/2} \tag{4}$$

where i_p is the peak current (A), n is the electron transfer, A is the electrode area (cm^2), D is the diffusion coefficient (cm^2/s), C is the concentration (moles/cm^3), v is the scan rate (V/s), and α is the transfer coefficient.

In a reversible redox reaction, increases in i_p with the scan rate are related to the concentration of a solution. The concept of concentration is essential in the study of the electrochemistry of electrodes. For fast chemical redox reactions, i_{pa} and i_{pc} should be indistinguishable [40]. According to the Randles–Sevcik equation, the current peak i_p is related to the square root of the scan rate $v^{1/2}$. The electro-active substances are also directly related to the square root of the scan rate. The electro-active materials are essential in the determination of the concentration and diffusion coefficient D. A plot of the current peak (i_p) with $v^{1/2}$ gives a linear regression, which shows the presence of a redox reaction mechanism. Hence, the Randles–Sevcik equation validates the diffusional mechanisms, which are a free adsorption process. Thus, the chemical reaction should have fast kinetics, and the concentrations of species or substances and the area of an electrode should be provided. If the redox chemical reaction is reversible, then multiple evaluations could be applicable over a broad potential range. The ratio of the forward to backward reactions of the redox reaction is interpreted as current peaks i_{pa}/i_{pc}. The separation of the peak is denoted as ΔE_p, whose values are provided by the sweeping potentials of the electrode. Mostly reflecting a reversible chemical or electrochemical scheme, the value of i_{pa}/i_{pc} is known to be unity. However, the peak separation must be in concordance with Equation (5), while the potential difference between peaks and half-wave potentials must agree with Equation (6). The redox process is not reversible if these circumstances are not in accord with Equations (5) and (6). Hence, the electron-transfer mechanism is more complicated in some experiments [41]

$$\Delta E_p = E_{pa} - E_{pc} = \frac{59}{n} \ (mV) \tag{5}$$

$$\Delta E_p = E_p - E_{2/p} = \frac{59}{n} \ (mV) \tag{6}$$

4. Conclusions

This study has established the use of GOx/LAc modified electrodes as bioanodes and biocathodes for biofuel cells. The CNT-based electrodes gave high catalytic currents for the O_2 reduction process in redox mediators with astonishing operative constancy. High sensitivity and stability, together with straightforward preparation, make GOx/LAc electrodes a promising candidate for constructing simple electrochemical biofuel cells or sensors for oxygen. The cyclic voltammograms show that hydrophilic carbon paper electrodes produce excellent electrode performance compared to a hydrophobic carbon paper electrode. The best hydrophilic electrode was Cp-GOx-M-CNT, with a current density of 800 µA/cm^2, for the hydrophilic bioanode and Cp-LAc-M-CNT, with a current density of 600 µA/cm^2, for the hydrophilic biocathode. Moreover, the best-performing hydrophobic electrode was Cp-GOx-M-CNT, with a current density of 500 µA/cm^2, for the hydrophobic bioanode and Cp-LAc-M-CNT, with a current density of 300 µA/cm^2, for the hydrophobic biocathode. The experimental approaches of CV and EIS, which were conducted in this study, deliver prospects for qualitative and quantitative representation, even under physiologically significant conditions. EIS analysis shows that the internal resistance of the solution sometimes affects the interaction of the enzymes and the mediators in terms of electron transfer.

Finally, the well-designed GOx/LAc-based electrodes are very suitable for their possible applications, while this construction of enzyme electrodes will provide ideas for the improvement of a novel group of biofuel cells and will be valuable for the expansion of bio-reactors, bio-sensors, and micro-reactors.

Author Contributions: Writing—original draft preparation, A.A.B.; writing—review and editing, H.S.W.; supervision, H.S.W. All authors have read and agreed to the published version of the manuscript.

Funding: We thank the Ministry of Science and Technology of Taiwan for financially supporting this study under grant number MOST 107-2218-E-155-001.

Conflicts of Interest: The authors do not have any conflicts of interest to declare.

Abbreviations

ABTS	2,2-azino-bis (3-ethylbenzothiazoline-6-sulphonic acid)
C1	internal capacitance
C2	internal capacitance
CNB	carbon nanoball
CNT	carbon nanotube
CV	cyclic voltammetry
Cp	carbon paper
EBFC	enzymatic biofuel cell
FTIR	Fourier transform infrared spectroscopy
GOx	glucose oxidase
LAc	laccase
M	mediator
PBS	phosphate buffer solution
PPY	polypyrrole
R1	internal resistance
R2	internal resistance
R_A	charge-transfer resistance for the bioanode
R_{other}	charge-transfer resistance of metal-salts
Rs	resistance of the solution
XRD	X-ray diffractometer
W	Warburg impedance

References

1. Mahlambi, M.M.; Ngila, C.J.; Mamba, B.B. Recent developments in environmental photocatalytic degradation of organic pollutants: The case of titanium dioxide nanoparticles;a review. *J. Nanomater.* **2015**, *2015*, 790173. [CrossRef]
2. Calabrese Barton, S.; Gallaway, J.; Atanassov, P. Enzymatic Biofuel Cells for Implantable and Microscale Devices. *Chem. Rev.* **2004**, *104*, 4867–4886. [CrossRef] [PubMed]
3. Liu, C.; Alwarappan, S.; Chen, Z.; Kong, X.; Li, C.-Z. Membraneless enzymatic biofuel cells based on graphene nanosheets. *Biosens. Bioelectron.* **2010**, *25*, 1829–1833. [CrossRef] [PubMed]
4. Munauwarah, R.; Bojang, A.A.; Wu, H.S. Characterization of enzyme immobilized carbon electrode using covalent-entrapment with polypyrrole. *J. Chin. Inst. Eng.* **2018**, *41*, 710–719. [CrossRef]
5. Dai, D.J.; Chan, D.S.; Wu, H.S. Modified carbon nanoball on electrode surface using plasma in enzyme-based biofuel cells. *Energy Procedia* **2012**, *14*, 1804–1810. [CrossRef]
6. Tozar, A.; Karahan, İ.H. Structural and corrosion protection properties of electrochemically deposited nano-sized Zn–Ni alloy coatings. *Appl. Surf. Sci.* **2014**, *318*, 15–23. [CrossRef]
7. Wagner, N.; Schnurnberger, W.; Müller, B.; Lang, M. Electrochemical impedance spectra of solid-oxide fuel cells and polymer membrane fuel cells. *Electrochim. Acta* **1998**, *43*, 3785–3793. [CrossRef]
8. Xu, S.; Minteer, S.D. Enzymatic Biofuel Cell for Oxidation of Glucose to CO_2. *ACS Catal.* **2012**, *2*, 91–94. [CrossRef]
9. Zebda, A.; Gondran, C.; Le Goff, A.; Holzinger, M.; Cinquin, P.; Cosnier, S. Mediatorless high-power glucose biofuel cells based on compressed carbon nanotube-enzyme electrodes. *Nat. Commun.* **2011**, *2*, 370. [CrossRef]
10. Kim, J.; Jia, H.; Wang, P. Challenges in biocatalysis for enzyme-based biofuel cells. *Biotechnol. Adv.* **2006**, *24*, 296–308. [CrossRef]
11. Rasmussen, M.; Abdellaoui, S.; Minteer, S.D. Enzymatic biofuel cells: 30 years of critical advancements. *Biosens. Bioelectron.* **2016**, *76*, 91–102. [CrossRef] [PubMed]
12. Kasa, T.; Solomon, T. Cyclic voltammetric and electrochemical simulation studies on the electro-oxidation of catechol in the presence of 4, 4-bipyridine. *Am. J. Phys. Chem.* **2016**, *5*, 45–55. [CrossRef]

13. Klis, M.; Maicka, E.; Michota, A.; Bukowska, J.; Sek, S.; Rogalski, J.; Bilewicz, R. Electroreduction of laccase covalently bound to organothiol monolayers on gold electrodes. *Electrochim. Acta* **2007**, *52*, 5591–5598. [CrossRef]
14. Kjeang, E.; Djilali, N.; Sinton, D. Microfluidic fuel cells: A review. *J. Power Sources* **2009**, *186*, 353–369. [CrossRef]
15. Zanello, P. *Inorganic Electrochemistry: Theory, Practice and Applications*; Royal Society of Chemistry: London, UK, 2003.
16. Skoog, D.A.; Holler, F.J.; Crouch, S.R. *Instrumental Analysis*; Brooks/Cole, Cengage Learning Belmont: Belmont, CA, USA, 2007; Volume 47.
17. Barrière, F.; Kavanagh, P.; Leech, D. A laccase–glucose oxidase biofuel cell prototype operating in a physiological buffer. *Electrochim. Acta* **2006**, *51*, 5187–5192. [CrossRef]
18. Lavagnini, I.; Antiochia, R.; Magno, F. An Extended Method for the Practical Evaluation of the Standard Rate Constant from Cyclic Voltammetric Data. *Electroanalysis* **2004**, *16*, 505–506. [CrossRef]
19. Ajeel, M.A.; Taeib Aroua, M.K.; Ashri Wan Daud, W.M. Reactivity of carbon black diamond electrode during the electro-oxidation of Remazol Brilliant Blue R. *RSC Adv.* **2016**, *6*, 3690–3699. [CrossRef]
20. Laurent, N.; Haddoub, R.; Flitsch, S.L. Enzyme catalysis on solid surfaces. *Trends Biotechnol.* **2008**, *26*, 328–337. [CrossRef]
21. Abreu, C.M.; Izquierdo, M.; Keddam, M.; Nóvoa, X.R.; Takenouti, H. Electrochemical behaviour of zinc-rich epoxy paints in 3% NaCl solution. *Electrochim. Acta* **1996**, *41*, 2405–2415. [CrossRef]
22. Gao, Y.; Kyratzis, I. Covalent Immobilization of Proteins on Carbon Nanotubes Using the Cross-Linker 1-Ethyl-3-(3-dimethylaminopropyl)carbodiimide—A Critical Assessment. *Bioconjugate Chem.* **2008**, *19*, 1945–1950. [CrossRef]
23. Wang, S.G.; Zhang, Q.; Wang, R.; Yoon, S.F.; Ahn, J.; Yang, D.J.; Tian, J.Z.; Li, J.Q.; Zhou, Q. Multi-walled carbon nanotubes for the immobilization of enzyme in glucose biosensors. *Electrochem. Commun.* **2003**, *5*, 800–803. [CrossRef]
24. Mubarak, N.M.; Wong, J.R.; Tan, K.W.; Sahu, J.N.; Abdullah, E.C.; Jayakumar, N.S.; Ganesan, P. Immobilization of cellulase enzyme on functionalized multiwall carbon nanotubes. *J. Mol. Catal. B Enzym.* **2014**, *107*, 124–131. [CrossRef]
25. Kashyap, D.; Dwivedi, P.K.; Pandey, J.K.; Kim, Y.H.; Kim, G.M.; Sharma, A.; Goel, S. Application of electrochemical impedance spectroscopy in bio-fuel cell characterization: A review. *Int. J. Hydrog. Energy* **2014**, *39*, 20159–20170. [CrossRef]
26. Ramasamy, R.P.; Ren, Z.; Mench, M.M.; Regan, J.M. Impact of initial biofilm growth on the anode impedance of microbial fuel cells. *Biotechnol. Bioeng.* **2008**, *101*, 101–108. [CrossRef] [PubMed]
27. Lee, J.A.; Hwang, S.; Kwak, J.; Park, S.I.; Lee, S.S.; Lee, K.-C. An electrochemical impedance biosensor with aptamer-modified pyrolyzed carbon electrode for label-free protein detection. *Sens. Actuators B Chem.* **2008**, *129*, 372–379. [CrossRef]
28. Sharma, V.; Tanwar, V.K.; Mishra, S.K.; Biradar, A.M. Electrochemical impedance immunosensor for the detection of cardiac biomarker Myogobin (Mb) in aqueous solution. *Thin Solid Film.* **2010**, *519*, 1167–1170. [CrossRef]
29. Uygun, Z.O.; Sezgintürk, M.K. A novel, ultra sensible biosensor built by layer-by-layer covalent attachment of a receptor for diagnosis of tumor growth. *Anal. Chim. Acta* **2011**, *706*, 343–348. [CrossRef]
30. Kondaveeti, S.; Lee, S.-H.; Park, H.-D.; Min, B. Bacterial communities in a bioelectrochemical denitrification system: The effects of supplemental electron acceptors. *Water Res.* **2014**, *51*, 25–36. [CrossRef]
31. Ramaraja, P.; Ramasamy, N.S. Electrochemical Impedance Spectroscopy for Microbial Fuel Cell Characterization. *J. Microb. Biochem. Technol.* **2013**. [CrossRef]
32. Reuillard, B.; Le Goff, A.; Agnes, C.; Holzinger, M.; Zebda, A.; Gondran, C.; Elouarzaki, K.; Cosnier, S. High power enzymatic biofuel cell based on naphthoquinone-mediated oxidation of glucose by glucose oxidase in a carbon nanotube 3D matrix. *Phys. Chem. Chem. Phys.* **2013**, *15*, 4892–4896. [CrossRef]
33. Kang, Z.; Jiao, K.; Yu, C.; Dong, J.; Peng, R.; Hu, Z.; Jiao, S. Direct electrochemistry and bioelectrocatalysis of glucose oxidase in CS/CNC film and its application in glucose biosensing and biofuel cells. *RSC Adv.* **2017**, *7*, 4572–4579. [CrossRef]

34. Yunpu, W.; Leilei, D.A.I.; Liangliang, F.A.N.; Shaoqi, S.; Yuhuan, L.I.U.; Roger, R. Review of microwave-assisted lignin conversion for renewable fuels and chemicals. *J. Anal. Appl. Pyrolysis* **2016**, *119*, 104–113. [CrossRef]
35. Hamelers, H.V.M.; Ter Heijne, A.; Sleutels, T.H.J.A.; Jeremiasse, A.W.; Strik, D.P.B.T.B.; Buisman, C.J.N. New applications and performance of bioelectrochemical systems. *Appl. Microbiol. Biotechnol.* **2010**, *85*, 1673–1685. [CrossRef] [PubMed]
36. Freguia, S.; Tsujimura, S.; Kano, K. Electron transfer pathways in microbial oxygen biocathodes. *Electrochim. Acta* **2010**, *55*, 813–818. [CrossRef]
37. Ter Heijne, A.; Strik, D.P.B.T.B.; Hamelers, H.V.M.; Buisman, C.J.N. Cathode Potential and Mass Transfer Determine Performance of Oxygen Reducing Biocathodes in Microbial Fuel Cells. *Environ. Sci. Technol.* **2010**, *44*, 7151–7156. [CrossRef] [PubMed]
38. Martin, E.; Tartakovsky, B.; Savadogo, O. Cathode materials evaluation in microbial fuel cells: A comparison of carbon, Mn_2O_3, Fe_2O_3 and platinum materials. *Electrochim. Acta* **2011**, *58*, 58–66. [CrossRef]
39. Amatore, C.; Da Mota, N.; Lemmer, C.; Pebay, C.; Sella, C.; Thouin, L. Theory and Experiments of Transport at Channel Microband Electrodes under Laminar Flows. 2. Electrochemical Regimes at Double Microband Assemblies under Steady State. *Anal. Chem.* **2008**, *80*, 9483–9490. [CrossRef]
40. Dutta, S.; Wu, K.C.W. Enzymatic breakdown of biomass: Enzyme active sites, immobilization, and biofuel production. *Green Chem.* **2014**, *16*, 4615–4626. [CrossRef]
41. Hauch, A.; Georg, A. Diffusion in the electrolyte and charge-transfer reaction at the platinum electrode in dye-sensitized solar cells. *Electrochim. Acta* **2001**, *46*, 3457–3466. [CrossRef]

© 2020 by the authors. Licensee MDPI, Basel, Switzerland. This article is an open access article distributed under the terms and conditions of the Creative Commons Attribution (CC BY) license (http://creativecommons.org/licenses/by/4.0/).

Article

Regioselective Hydroxylation of Naringin Dihydrochalcone to Produce Neoeriocitrin Dihydrochalcone by CYP102A1 (BM3) Mutants

Thi Huong Ha Nguyen [1,†], Su-Min Woo [1,†], Ngoc Anh Nguyen [1], Gun-Su Cha [2], Soo-Jin Yeom [1,3], Hyung-Sik Kang [1,3,*] and Chul-Ho Yun [1,3,*]

[1] School of Biological Sciences and Biotechnology, Graduate School, Chonnam National University, Yongbong-ro 77, Gwangju 61186, Korea; huongha0207@gmail.com (T.H.H.N.); dntnals0@gmail.com (S.-M.W.); ngocanh61093@gmail.com (N.A.N.); soojin258@chonnam.ac.kr (S.-J.Y.)
[2] Namhae Garlic Research Institute, 2465-8 Namhaedaero, Gyeongsangnamdo 52430, Korea; gscha450@gmail.com
[3] School of Biological Sciences and Technology, Chonnam National University, Yongbong-ro 77, Gwangju 61186, Korea
* Correspondence: kanghs@jnu.ac.kr (H.-S.K.); chyun@jnu.ac.kr (C.-H.Y.); Tel.: +82-62-530-2195 (H.-S.K.); +82-62-530-2194 (C.-H.Y.)
† These authors equally contributed.

Received: 23 June 2020; Accepted: 21 July 2020; Published: 23 July 2020

Abstract: Naringin dihydrochalcone (DC) is originally derived from the flavonoid naringin, which occurs naturally in citrus fruits, especially in grapefruit. It is used as an artificial sweetener with a strong antioxidant activity with potential applications in food and pharmaceutical fields. At present, enzymatic and chemical methods to make products of naringin DC by hydroxylation reactions have not been developed. Here, an enzymatic strategy for the efficient synthesis of potentially valuable products from naringin DC, a glycoside of phloretin, was developed using *Bacillus megaterium* CYP102A1 monooxygenase. The major product was identified to be neoeriocitrin DC by NMR and LC-MS analyses. Sixty-seven mutants of CYP102A1 were tested for hydroxylation of naringin DC to produce neoeriocitrin DC. Six mutants with high activity were selected to determine the kinetic parameters and total turnover numbers (TTNs). The k_{cat} value of the most active mutant was 11 min^{-1} and its TTN was 315. The productivity of neoeriocitrin DC production increased up to 1.1 mM h^{-1}, which corresponds to 0.65 g L^{-1} h^{-1}. In this study, we achieved a regioselective hydroxylation of naringin DC to produce neoeriocitrin DC.

Keywords: CYP102A1; naringin dihydrochalcone; neoeriocitrin dihydrochalcone; regioselective hydroxylation

1. Introduction

Dihydrochalcone (DC) is a bicyclic flavonoid family with two aromatic rings and a saturated C3 bridge [1]. DC compounds are mainly found in citrus fruits, grapefruits, and apples, and they play an important role in resisting biotic or abiotic stresses in plant [2,3]. To date, more than 200 DC compounds have been identified from over 30 plant families [4]. As DC compounds show strong antioxidant activities, a large number of studies have researched the potential benefits of DC compounds to human health. They were demonstrated to be effective in preventing different physiopathological processes [3], notably diabetes [5] and bone resorption [6]. In recent years, scientists have more often been attracted by in vitro and in vivo biological activities of DC compounds.

Naringin DC (3,5-dihydroxy-4[3-(4-hydroxyphenyl)propanoyl]phenyl 2-O-(6-deoxy-α-L-mannopyranosyl)-β-L-glucopyranoside) (Figure 1) is known as a widely used artificial sweetener [7,8].

Naringin DC is produced when naringin is treated with a strong base, such as potassium hydroxide, and then catalytically hydrogenated. Naringin is a flavanone-7-O-glycoside between the flavanone naringenin and the disaccharide neohesperidose. Naringin DC has a sweet value approximately 300 times higher than that of sucrose [9]. Naringin DC has high antioxidant activity, which performs better free-radical scavenging than its corresponding flavanone naringin [10]. Besides, naringin DC is a glycoside of phloretin that shows an inhibitory effect on active transport of glucose into cells by SGLT1 and SGLT2 [8]. Naringin DC was suggested as a promising therapeutic agent for Alzheimer's disease treatment against multiple effects that reduce Aβ levels, suppress neuroinflammation, and enhance neurogenesis [8]. The antioxidant and noncalorie sweetener abilities of naringin DC can make it be a potential compound for applications in food, beverages, and pharmaceuticals [11].

Cytochrome P450 (CYP or P450) is known as one of the largest enzyme families found in all organisms. P450s catalyze the oxidation of various endogenous and xenobiotic compounds [12]. Due to their diversity of substrates, P450s are attractive as biocatalysts for producing chemicals, including bioactive compounds and pharmaceuticals [13,14]. CYP102A1 (P450 BM3) from *Bacillus megaterium* is a self-sufficient monooxygenase enzyme, which is naturally fused to its redox partner, a mammalian-like diflavin reductase. Engineered CYP102A1 mutants have been extensively obtained through rational design and directed evolution to catalyze the oxidation of several non-natural substrates, environmental chemicals, and pharmaceuticals [15–19]. It was also suggested that the engineered CYP102A1 can be developed as a potential biocatalyst for biotechnology applications [20,21].

In this study, we have tried to find an enzymatic strategy for the production of products from naringin DC. A large set of CYP102A1 mutants were used for the efficient synthesis of potentially valuable products from naringin DC. To the best of our knowledge, the enzymatic hydroxylation of naringin DC has not previously been reported. This work is the first report on enzymatic synthesis of neoeriocitrin DC, a major product of naringin DC (Figure 1).

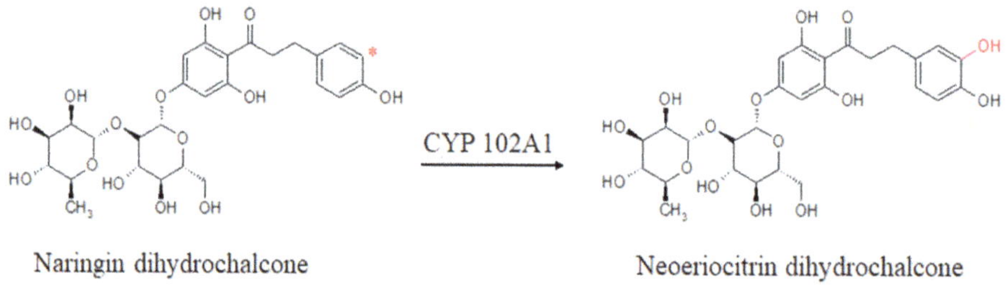

Figure 1. Chemical structures of naringin DC and neoeriocitrin DC. The conversion of the substrate, naringin DC, to its corresponding product, neoeriocitrin DC, is catalyzed by CYP102A1 in the presence of NADPH. An enzymatic reaction site on naringin DC is marked by a star.

2. Results and Discussion

2.1. Hydroxylation of Naringin DC by CYP102A1 Mutants

First, to determine the ability of CYP102A1 to hydroxylate naringin DC, the catalytic activity of the wild type (WT) and its 60 mutants [12,14,19,22–25] toward naringin DC were tested at 200 µM substrate for 30 min at 37 °C (Figure 2). The 60 mutants used for first screening were selected based on our previous works showing their improved catalytic activities on a number of non-natural substrates, such as natural products and pharmaceuticals (each mutant bears amino acid substitutions relative to WT CYP102A1, as summarized in Supplementary Table S1). To obtain more highly active mutants, the randomized DNA library obtained from the M16V2 library (see Materials and Methods) was screened using a colorimetric (blue) colony-based method and HPLC analysis (Figure 3). Finally, seven mutants were selected from approximately 500 blue colonies (from M524 to M850 in Supplementary Table S1).

The selection was based on the mutants' expression levels and catalytic activity of naringin DC hydroxylation (Figure 2). The mutants M601 (7.1 min^{-1}), M620 (8.2 min^{-1}), M788 (5.3 min^{-1}), and M850 (8.0 min^{-1}) showed higher catalytic activity than M16V2 (3.4 min^{-1}).

In the HPLC chromatogram, one minor and one major product were observed (Figure 3). Among all tested mutants, 26 mutants showed apparent but very low activity toward naringin DC (<0.5 min^{-1}). CYP102A1 WT did not show any apparent activities. Meanwhile, seven mutants (G1, M179, M601, M620, M221, M788, and M850) showed high catalytic activity for naringin DC (>5 min^{-1}) (Figure 2). Mutants G1 (10.2 min^{-1}) and M221 (9.8 min^{-1}) showed approximately three-fold higher catalytic activity toward naringin DC than that of M16V2.

Six mutants were selected for further experiments to determine the kinetic parameters and total turnover numbers (TTNs). M16V2 was selected as it was used a template to make the DNA library. M179 showed a medium activity and the other four mutants (G1, M221, M620, and M850) had high activities.

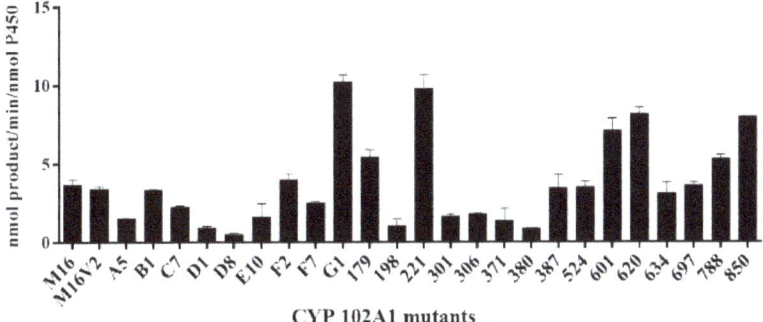

Figure 2. Catalytic activity of naringin DC hydroxylation by CYP102A1 mutants. The reactions contained 200 µM naringin DC as a substrate in 100 mM potassium phosphate buffer (pH 7.4) and 0.20 µM CYP102A1. NADPH-generating systems were added to initiate the reaction, and the reaction mixtures were incubated for 30 min at 37 °C.

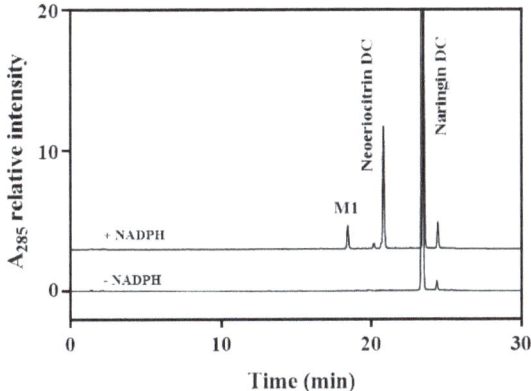

Figure 3. HPLC chromatogram of naringin DC and its products formed by CYP102A1 mutant M221. The peaks of reaction mixtures of HPLC chromatograms were identified by comparing their retention times with those of neoeriocitrin DC (t_R = 20.8 min) and naringin DC (t_R = 23.4 min). The retention time of M1 was 18.4 min.

2.2. Optimal Expression of CYP102A1 M221 Mutant

To find the best *Escherichia coli* strain for protein expression of the M221 mutant, the plasmid M221 (in pCW vector) was transformed to a set of competent E. coli cells (DH5α-F'IQ, BL21, SHuffle T7, Rosetta, MG1655, and JM109). The P450 expression levels at different induction periods (12 to 28 h) and incubation temperatures (20 and 25 °C) were analyzed. The MG1655 strain had a higher capability of producing recombinant M221 protein than other strains. Thus, the MG1655 strain was selected for the next experiments. The protein expression level of the M221 enzyme was determined by the evaluation of CO difference spectrum, a typical Fe^{2+} CO versus Fe^{2+} spectrum of the heme group [26], and obtained after 12 to 28 h culture at 20 and 25 °C. OD_{600} and the protein expression level are proportional from 12 to 20 h (Figure 4A). After that, the cell growth rate reached the stationary phase, but the P450 expression level showed difference rates between 20 and 25 °C. At 20 °C, the P450 expression achieved stability up to 28 h. Meanwhile, the P450 expression increased up to 24 h and reached the highest level of approximately 0.5 nmol of M221/mL (24 h point time) at 25 °C. At 28 h culture at 25 °C, the P450 level decreased. The MG1655 strain showed the capability of producing the highest expression level of M221 among tested strains (Figure 4B).

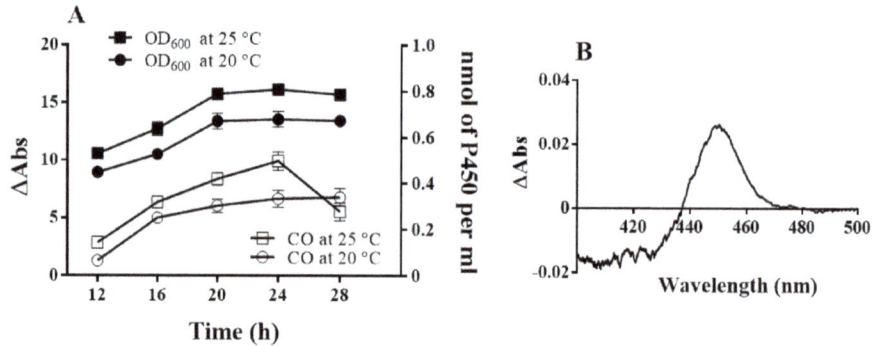

Figure 4. Optimal expression of CYP102A1 M221 mutant in the *E. coli* strain MG1655. (**A**) The cultures were grown at 20 or 25 °C up to 28 h. OD_{600} and P450 concentration were measured at indicated time. (**B**) CO-difference spectrum of M221 at 24 h culture time at 25 °C.

2.3. Characterizing a Major Product of Naringin DC by CYP102A1 Mutants

Products and the substrate were characterized by results of the HPLC (Figure 3), LC-MS (Figures 5 and 6), and NMR spectroscopy (Figure 7). Naringin DC's minor and major products made by CYP102A1 mutants were M1 and neoeriocitrin DC, respectively. The formation of a monohydroxylated product as a major product was confirmed by LC-MS (Figure 5). The minor product (M1) has *m/z* 596, which indicates two protons were deleted from the monohydroxylated product (*m/z* 598). However, the M1 product formation rate is too low (1.5 min^{-1}) compared to that of the major product (8.4 min^{-1}) (Figure 3). Here, we mainly focus on the production of the major catechol product.

The major product was prepared by preparative HPLC (Figure 6), and its chemical structure was identified by NMR (Figure 7). The chemical shifts and splitting patterns of the major product's ^{1}H and ^{13}C NMR spectra are shown (see also Supplementary Figure S1–S3 for NMR spectra).

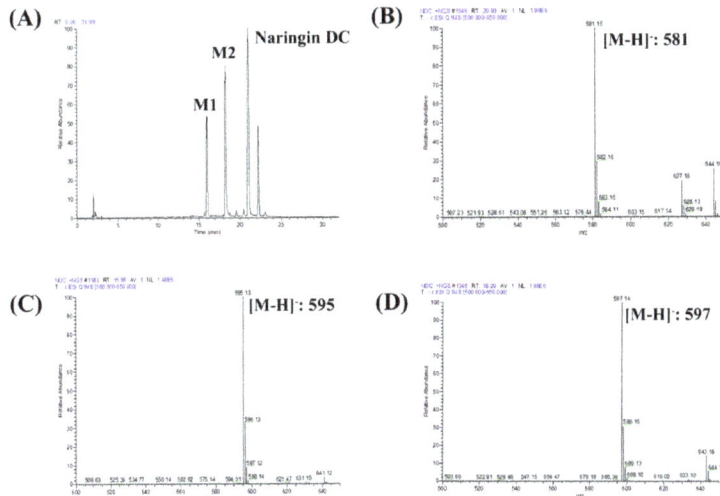

Figure 5. LC-MS analysis of naringin DC and its products produced by CYP102A1 M221 mutant. (**A**) LC-MS chromatogram of naringin DC and its products; (**B**) Naringin DC shows *m/z* 582; (**C**) The minor product (M1) shows *m/z* 596; (**D**) The major product (M2) of *m/z* 598 was found to be monohydroxylated.

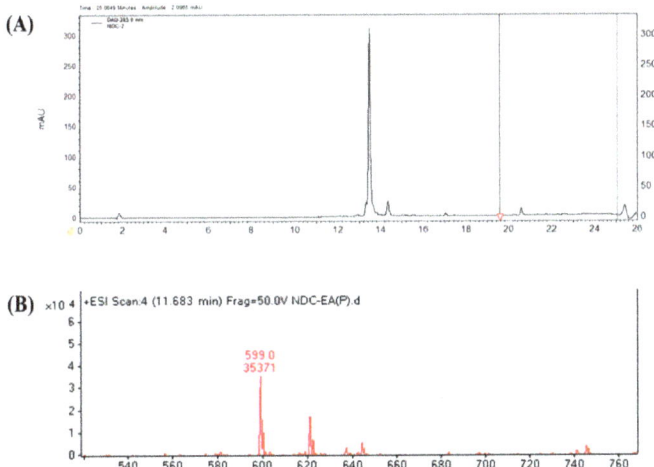

Figure 6. HPLC and LC-MS analyses of naringin DC's major product produced by CYP102A1 M221 mutant. (**A**) preparative-HPLC chromatogram of the major product of naringin DC (C18 column, 10 × 150 mm, gradient from 10% to 100% methanol, 3 mL/min). (**B**) LC-MS analyses of the major product DC [M+H]$^+$ 599 (*m/z* 598).

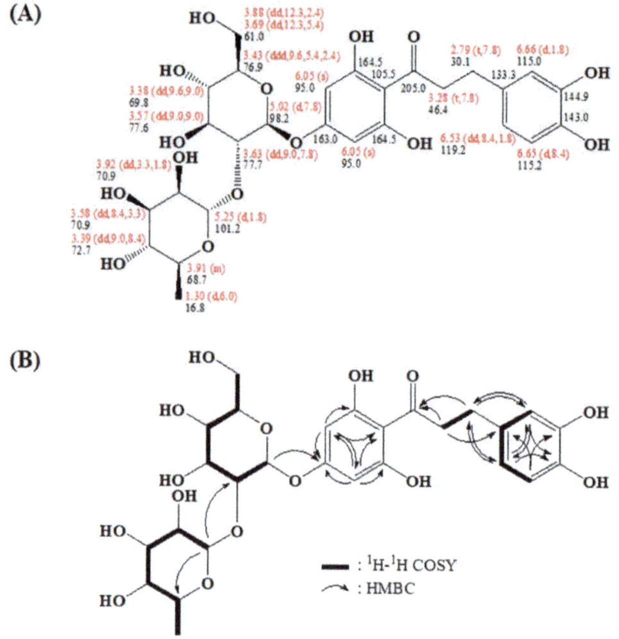

Figure 7. (**A**) Chemical structure of the major product (neoeriocitrin DC) from naringin DC. Red: ^{1}H chemical shift values (multiplicity and coupling constants [Hz] in parenthesis); Black: ^{13}C chemical shift values which were assigned from 2D HMBC NMR spectra. (**B**) 2D NMR spectra of the major product, neoeriocitrin DC. (See also Supplementary Figure S1–S3 for NMR spectra.)

2.4. Kinetic Parameters and TTNs of Naringin DC Hydroxylation by CYP102A1 Mutants

Six mutants (M16V2, G1, M179, M221, M620, and M850) that showed high rates of naringin DC product formation among tested mutants were selected and used to measure the kinetic parameters of naringin DC hydroxylation (Table 1). WT CYP102A1 did not exhibit appreciable activity by which to determine reliable kinetic parameters. M16V2 was used as a control because it was used as a template for the DNA library. The k_{cat} values of mutants G1 (11 min^{-1}) and M221 (10.8 min^{-1}) increased compared to M16V2 by 41% and 38%, respectively. The K_m values of G1, M179, and M221 decreased to half, and mutants M620 and M850 showed 1.5–2.8-fold increases in K_m values when compared to that of M16V2. The catalytic efficiencies (k_{cat}/K_m) of neoeriocitrin DC formation by mutants G1, M179, and M221 were 0.151, 0.070, and 0.137 (min^{-1}μM^{-1}), which were more efficient than that of M16V2 by 3.1-, 1.4-, and 2.8-fold, respectively. M620 and M850 showed decreased catalytic efficiencies due to increased K_m values.

Table 1. Kinetic parameters of naringin DC hydroxylation by CYP102A1. The chimeric mutant M16V2 and selected mutants (G1, M179, M221, M620, and M850).

Enzymes	k_{cat} (min^{-1})	K_m (μM)	k_{cat}/K_m (min^{-1}μM^{-1})
M16V2	7.8 ± 0.5	160 ± 25	0.049 ± 0.008
G1	11.0 ± 0.3	73 ± 5	0.151 ± 0.011
M179	5.3 ± 0.2	76 ± 9	0.070 ± 0.008
M221	10.8 ± 0.2	79 ± 6	0.137 ± 0.011
M620	5.7 ± 0.7	441 ± 101	0.013 ± 0.003
M850	4.5 ± 0.4	243 ± 43	0.019 ± 0.004

Four mutants (M16V2, G1, M221, and M850) were selected and used to measure the TTNs of naringin DC hydroxylation. When the assays were carried out at the reaction times of 20 min, 30 min, 1 h, 2 h, and 4 h, overall product formation was in the range of 105 to 315 TTNs (Figure 8). All mutants showed increased neoeriocitrin DC formation rate, which was 1.1–1.8-fold higher than M16V2 during indicated reaction time. In addition, the results showed that neoeriocitrin DC is stable at least up to 2 h and then the product seems to degrade. For a 1 h reaction with M221 or G1, the productivity was 1.1 mM h^{-1}, which corresponds to 0.65 g L^{-1} h^{-1}.

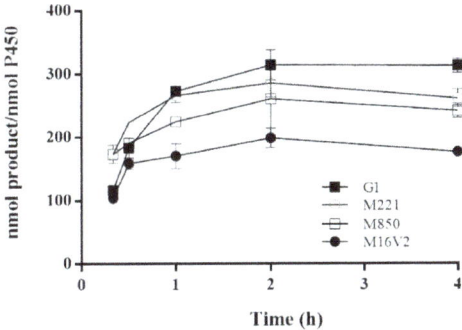

Figure 8. TTNs of naringin DC hydroxylation by CYP102A1 mutants. The reactions contained 500 µM naringin DC substrate and 0.40 µM of M16V2, G1, M221, or M850 in a 100-mM potassium phosphate buffer (pH 7.4). NADPH-generating systems were added to initiate the reaction, and the reaction mixtures were incubated for 20 min, 30 min, 1 h, 2 h, and 4 h at 37 °C.

2.5. Spectral Titration of Naringin DC toward CYP102A1 Mutants

The differences of binding affinity between M16V2 and selected mutants toward the substrate naringin DC were analyzed by spectral binding titration (Figure 9). Binding of naringin DC to CYP102A1 G1, M179, and M221 produced a typical Type II spectral shift, with an increase at 420 nm and a decrease at 390 nm, indicating an increase in the low-spin fraction of the enzyme. The spectrally determined dissociation constants (K_d) of naringin DC to G1, M179, and M221 were 1.6, 2.2, and 1.4 µM, respectively (Figure 9). This result indicates that naringin DC can bind to the active sites of the mutants with a high affinity (K_d of 1–2 µM). However, M16V2 did not show an apparent spectral change.

It was suggested that the catechol moieties of polyphenol compounds are important for their biological antiadipogenesis, antiobesity, and anticancer functions [27]. The biological activities of resveratrol and its catechol product, piceatannol, were reported to have antioxidation, antiobesity, anti-inflammatory and anticancer abilities [28–30]. Piceatannol was shown to be more potent than resveratrol in inhibitory effects on adipogenesis, obesity, and carcinogenesis [31]. Polydatin, a glycoside of resveratrol, was reported to have many biomedical properties related to antioxidation, antiplatelet aggregation, cardioprotective activity, and anti-inflammatory and immune-regulating functions [32]. Astringin, a catechol product of polydatin, was found to have a more potential antioxidative activity than polydatin [33] and a potential cancer chemopreventive activity [34]. Moreover, 7,3'4'-trihydroxyisoflavone (7,3'4'-THIF) (a catechol product of daidzein), but not daidzein itself, inhibited UVB-induced skin tumor in hairless mice. Thus, 7,3'4'-THIF is considered a new candidate chemoprotective agent [35]. Recently, we found that 3-OH phloretin, a catechol product of phloretin, shows an inhibitory effect on adipocyte differentiation [36]. All of the catechol products mentioned above can be efficiently produced using bacterial P450s [19,22,36,37].

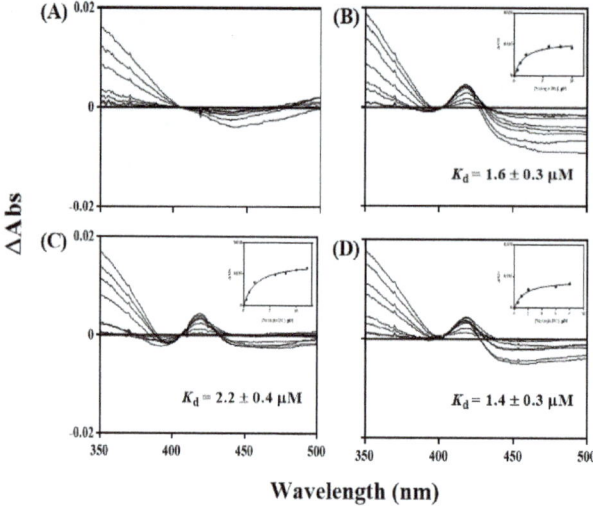

Figure 9. Binding titration of naringin DC to CYP102A1. The assay contained 0–20 μM naringin DC substrate in 100 mM potassium phosphate buffer (7.4) and 1.0 μM of M16V2 (**A**), G1 (**B**), M179 (**C**), and M221 (**D**). The inset of each panel (**B**–**D**) shows a plot of induced Soret absorbance change ($\Delta A_{390-420}$) versus the relevant concentration of naringin DC. Spectrally determined dissociation constants (K_d) were also shown.

It is known that glycosylation is an essential mechanism for diverse biological functions and the structure of natural flavonoids in plants [38,39]. Glycosylation of flavonoids can modify color and taste properties [39,40]. Furthermore, it leads to their strong solubility and stability in water [41–43], which helps to improve physiological and pharmacological properties that increase compound bioavailability [43,44]. Naringin DC is a glycoside of phloretin that shows several beneficial antioxidant, anticancer [45,46], and antiobesity [47] effects. Phloretin is widely used as a cosmeceutical ingredient for UV protection [48]. However, the biological functions of a catechol product of naringin DC, neoeriocitrin DC, have not been reported until now. Therefore, the production of neoeriocitrin DC from naringin DC by CYP102A1 reported here should be a good strategy to obtain it. Furthermore, toxicological research is needed if neoeriocitrin DC is applied as a sweetener.

In this study, we found that neoeriocitrin DC, can be produced by an enzymatic biotransformation using CYP102A1. It is now possible to study neoeriocitrin DC's biological functions, such as its antiobesity, anti-inflammatory, and anticancer abilities. Further investigation is necessary to improve the production of neoeriocitrin DC to meet the minimum space-time yield and a minimum final product concentration for industrial application [49]. Although we tried whole-cell biocatalysis experiments with *E. coli* expressing CYP102A1 genes for improved production of neoeriocitrin DC, no products of naringin DC were obtained (results not shown). Surface display [50] and export of CYP102A1 to the periplasmic space [51] of *E. coli* might be good whole cell systems for the industrial application. This result indicates that the carbohydrate moiety of naringin DC, may inhibit its transport into the *E. coli* cells because phloretin, the aglycone of naringin DC, can enter into the cells and be hydroxylated to 3-OH product [36].

3. Materials and Methods

3.1. Materials

Glucose-6-phosphate, glucose-6-phosphate dehydrogenase from baker's yeast, naringin DC, and β-nicotinamide adenine dinucleotide phosphate (NADP$^+$) were purchased from Sigma-Aldrich (St. Louis, MO, USA). Other chemicals and solvents with the highest grade were obtained from commercial suppliers

3.2. Optimal Expression of CYP102A1 M221 Mutant

The plasmid of M221 (in pCW vector) was transformed to a set of competent *E. coli* cells (DH5α-F'IQ, BL21, SHuffle T7, Rosetta, MG1655, and JM109) and spread on Luria-Bertani agar plate with ampicillin (100 µg/mL). The single colony was grown in 5 mL of Luria-Bertani medium supplemented with ampicillin (100 µg/mL) with shaking at 170 rpm overnight while maintaining 37 °C. The aliquots of cell culture (1% *v/v*) were inoculated in 50 mL of Terrific Broth medium supplemented with ampicillin (100 µg/mL). The cells were grown at 37 °C with shaking at 170 rpm to an OD$_{600}$ of approximately 0.6–0.8. Then, isopropyl-β-D-thiogalactopyranoside (0.5 mM) and δ-aminolevulinic acid (1.0 mM) were added for enzyme expression. After the cultures were allowed to grow at 20 or 25 °C with 150 rpm, OD$_{600}$ and CO spectra were measured at culture times of 12, 16, 20, 24, and 28 h. CYP102A1 (P450) concentrations of whole cells were determined from the CO-difference spectra using an extinction molecular coefficient, $\varepsilon = 91$ mM/cm [26].

3.3. CYP102A1 Mutants Used to Screen Highly Active Naringin DC Hydroxylases

An extensive set of CYP102A1 mutants was generated in previous work [12,14,19,22–25], and the WT BM3 and 60 mutants were used for screening highly active mutants towards naringin DC.

To make more active mutants having naringin DC hydroxylase activity, a random mutagenesis was performed to make a DNA library of the M16V2 heme domain. The chimeric protein M16V2 was originally made by exchanging the reductase domain of M16 with that of CYP102A1 natural variant V2, as described [21,24]. The error-prone PCR was performed on the CYP102A1 heme domain (1st–430th amino acid residues) of the M16V2 to make a DNA library. Oligonucleotide primers were used to introduce the BamHI/SacI restriction sites: BamHI forward, 5'-ataGGATCCatgacaattaaagaaatg cctc-3' and SacI reverse, 5'-ataGAGCTCgtagtttgtatgatcttcaaagtcaaag tg-3'. DNA libraries of random mutants were constructed using a reaction mixture (50 µL) of 10 pmol of each primer, 0.2 mM dNTP (0.05 mM each of dATP, dGTP, dCTP, and dTTP), Taq DNA polymerase (5 units/µL), MgCl$_2$ (2.5 or 5 mM), and MnCl$_2$ (0.1 or 0.15 mM) in 10 mM Tris-HCl containing 50 mM KCl (pH 8.4, 25 °C).

The PCR reaction was started at 95 °C for 5 min and run through 26 thermocycles of 95 °C for 60 s, 58 °C for 60 s, and 72 °C for 90 s. After completing the reaction, the reaction medium was held at 72 °C for 5 min and subsequently soaked at 4 °C. The amplified PCR library fragments were purified and cloned into the pCWBM3M16V2/BamHI/SacI vector using the restriction sites of BamHI and SacI. The mutation rate (2.9 mutations per 1290 bp) was validated by sequencing 12 randomly selected clones before screening of expression level and activity. The size of the screened mutant library was approximately 1.0×10^6.

Seven mutants selected based on a blue colony-based colorimetric method and hydroxylation activity toward naringin DC were expressed in the *E. coli* strain DH5αF'-IQ. The CYP102A1 is expressed in cytosol and partially purified as supernatant lysate after removal of cell debris and membrane fractions [24]. The lysate was used to determine the CYP102A1 (P450) concentrations from the CO-difference spectra [26]. For the M16V2 and its mutants, a typical culture yielded 300 to 700 nM of P450.

3.4. Hydroxylation of Naringin Dihydrochalcone by CYP102A1 Mutants

The reaction mixtures contained 200 µM naringin DC substrate in 100 mM potassium phosphate buffer (pH 7.4) and 0.20 µM CYP102A1. An aliquot of a NADPH-generating system (10 mM glucose-6-phosphate, 0.5 mM NADP$^+$, and 1.0 UI yeast glucose-6-dehydrogenase/mL) was added to the initial reaction. The reaction mixture was incubated for 30 min at 37 °C and stopped by 600 mL ice-cold ethyl acetate.

The naringin DC and its products were analyzed by HPLC using a Gemini C18 column (4.6 × 150 mm, 5 µm, 110 Å; Phenomenex, Torrance, CA, USA) with the mobile phase A (water containing 0.5% methanol and 0.1% formic acid) and the mobile phase B (acetonitrile) [52]. The flow rate of the elution column was 1.0 mL/min by a gradient pump (LC-20AD; Shimadzu, Kyoto, Japan) with the following gradient: 0–3 min controlled at 9% mobile phase B, 3–20 min gradually increased reaching to 30% mobile phase B, 20–21 min decreased to 9% mobile phase B, and 21–30 min controlled at 9% mobile phase B and detected by UV at 285 nm.

The kinetic parameters of CYP102A1 mutants were determined by reaction, including 10–500 µM of naringin DC in 100 mM potassium phosphate buffer (pH 7.4) and 0.20 µM enzymes. The NADPH-generating systems were added to the initial reaction and the reaction mixtures were incubated for 30 min at 37 °C. A stock of substrate solution was prepared in methanol and diluted in the enzymatic reactions to the final organic solvent concentration of <1% (v/v). The kinetic parameter results were analyzed using GraphPad Prism software (Graph, San Diego, CA, USA).

The TTNs of CYP102A1 mutants were determined by reaction contained in 500 µM naringin DC in 100 mM potassium phosphate buffer (pH 7.4) and 0.40 µM enzymes. The NADPH-generating systems were added to the initial reaction and the reaction mixtures were incubated for 20 min, 30 min, 1 h, 2 h, and 4 h at 37 °C.

3.5. LC-MS Analysis

To identify the minor and major products of naringin DC produced by CYP102A1 mutants, a liquid chromatography–mass spectrometry (LC–MS) analysis was performed, and the LC profile and fragmentation patterns of the authentic compounds (naringin DC and neoeriocitrin DC) were compared on a Thermo Scientific AccelaTM and TSQ QuantumTM Access MAX system with the heated electrospray ionization interface with HESI II probe (Thermo Fisher Scientific, Waltham, MA, USA) (Figure 5). The oxidation reaction of naringin DC by CYP102A1 was performed as described above. The separation was performed on a ZorBax SB-C18 (4.6 × 250 mm, 5 µm, 80 Å; Agilent Technologies, Santa Clara, CA, USA); the gradient mobile phase was 0.5% (v/v) methanol and 0.1% formic acid (v/v) in water (A) in acetonitrile (B), delivered at a flow rate of 1.0 mL/min. The initial composition of mobile phase B was 9%; after 3 min the mobile phase B composition increased to 30% over 17 min, decreased to 9% over 2 min, and finally re-equilibrated to the initial conditions over 13 min. Thus, the total run time was 35 min. The temperatures of the column and autosampler were kept at 40 and 4 °C, respectively, and the injection volumes were 5 µL for all samples tested here. The electrospray ionization procedure was performed in the negative ion mode. The spray voltage was 3500 V and vaporizer temperature was 300 °C. Capillary temperature was 200 °C. Nitrogen sheath gas and auxiliary gas pressures were 40 psi and 12 psi, respectively. All data were acquired with full scan mass spectrometry (full scan) or single ion monitoring (SIM) in the negative ion detection mode using XcaliburTM 3.0 software.

To confirm the purity of the naringin DC product produced by the CYP102A1 mutants and purified by preparative HPLC, an LC–MS analysis of products was performed to compare LC profiles and fragmentation patterns with those of the authentic compounds (naringin DC and neoeriocitrin DC) (Figure 6). The mutant M221 was included with 200 µM naringin DC for 50 min at 37 °C with the NADPH-generating system and performed using Applied Biosystems' QTRAP-3200 mass spectrometer (Waltham, MA, USA) having LC-MS solution software.

3.6. NMR Spectroscopy

After the major product of naringin DC was produced by the M221, separated by preparative HPLC (C18 column, 10 × 150 mm, gradient from 10% to 100% methanol, 3 mL/min), and collected in an ice bucket, the solvent was then removed by freezer-dryer. NMR investigations were performed at ambient temperature on a JNM-ECA600 600MHz FT-NMR spectrometer (JEOL Ltd., Tokyo, Japan). CD$_3$OD was used as solvent, and chemical shifts for proton NMR spectra were measured in parts per million (ppm) relative to tetramethylsilane.

3.7. Spectral Binding Titration

Spectral determinations of K_d values for the binding of substrates to the P450s were performed as described [53]. The binding affinity of naringin DC to four CYP102A1 mutants was determined (at 23 °C) by titrating 1.0 µM enzyme in 100 mM potassium phosphate buffer (pH 7.4). The absorption difference between 350 and 500 nm was plotted against the substrate concentration (0–20 µM). The K_d values were estimated using GraphPad Prism software (GraphPad Software, San Diego, CA, USA).

4. Conclusions

An enzymatic strategy for the efficient synthesis of a potentially valuable product from naringin DC, a glycoside of phloretin, was developed using *Bacillus megaterium* CYP102A1 monooxygenase. At present, no enzymatic or chemical methods to make products of naringin DC by hydroxylation reactions have not been reported. In this study, a set of CYP102A1 mutants was used to catalyze the hydroxylation of naringin DC. We found that the major product is neoeriocitrin DC by NMR and LC-MS analyses. Sixty seven mutants of CYP102A1 were tested for hydroxylation of naringin DC to produce neoeriocitrin DC. Six mutants with high activity were selected to determine the kinetic parameters and total turnover numbers (TTNs). The k_{cat} value of the most active mutant was 11 min^{-1} and its TTN was 315. The productivity of neoeriocitrin DC production increased up to 1.1 mM h^{-1}, which corresponds to 0.65 g L^{-1} h^{-1}. We achieved an efficient regioselective hydroxylation of naringin DC to produce neoeriocitrin DC, a catechol product.

Supplementary Materials: The following are available online at http://www.mdpi.com/2073-4344/10/8/823/s1, Figure S1. ^{1}H NMR spectra of the major product, neoeriocitrin DC; Figure S2. 2D HMBC NMR spectra of the major product, neoeriocitrin DC; Figure S3. 2D COSY NMR spectra of the major product, neoeriocitrin DC; Table S1: The amino acid sequence of M16V2 and CYP102A1 mutants.

Author Contributions: Conceptualization, S.-J.Y., H.-S.K., and C.-H.Y.; investigation, T.H.H.N., S.-M.W., N.A.N., and G.-S.C.; writing—original draft preparation, T.H.H.N. and C.-H.Y.; supervision, S.-J.Y., H.-S.K., and C.-H.Y.; funding acquisition, H.-S.K. and C.-H.Y. All authors have read and agreed to the published version of the manuscript.

Funding: This research was funded by the Next-Generation BioGreen 21 program (SSAC, grant no.: PJ01333101); Rural Development Administration and the Basic Research Lab Program (NRF-2018R1A4A1023882); National Research Foundation of Korea, Republic of Korea.

Conflicts of Interest: The authors declare no conflict of interest.

References

1. Szliszka, E.; Czuba, Z.P.; Mazur, B.; Paradysz, A.; Krol, W. Chalcones and Dihydrochalcones Augment TRAIL-Mediated Apoptosis in Prostate Cancer Cells. *Molecules* **2010**, *15*, 5336–5353. [CrossRef] [PubMed]
2. Xiao, Z.; Zhang, Y.; Chen, X.; Wang, Y.; Chen, W.; Xu, Q.; Li, P.; Ma, F. Extraction, identification, and antioxidant and anticancer tests of seven dihydrochalcones from Malus "Red Splendor" fruit. *Food Chem.* **2017**, *231*, 324–331. [CrossRef] [PubMed]
3. Gaucher, M.; Dugé de Bernonville, T.; Lohou, D.; Guyot, S.; Guillemette, T.; Brisset, M.-N.; Dat, J.F. Histolocalization and physico-chemical characterization of dihydrochalcones: Insight into the role of apple major flavonoids. *Phytochemistry* **2013**, *90*, 78–89. [CrossRef] [PubMed]

4. Rozmer, Z.; Perjési, P. Naturally occurring chalcones and their biological activities. *Phytochem. Rev.* **2016**, *15*, 87–120. [CrossRef]
5. Ehrenkranz, J.R.L.; Lewis, N.G.; Kahn, C.R.; Roth, J. Phlorizin: A review. *Diabetes/Metab. Res. Rev.* **2005**, *21*, 31–38. [CrossRef]
6. Dugé de Bernonville, T.; Guyot, S.; Paulin, J.-P.; Gaucher, M.; Loufrani, L.; Henrion, D.; Derbré, S.; Guilet, D.; Richomme, P.; Dat, J.F.; et al. Dihydrochalcones: Implication in resistance to oxidative stress and bioactivities against advanced glycation end-products and vasoconstriction. *Phytochemistry* **2010**, *71*, 443–452. [CrossRef]
7. Tang, N.; Yan, W. Solubilities of Naringin Dihydrochalcone in Pure Solvents and Mixed Solvents at Different Temperatures. *J. Chem. Eng. Data* **2016**, *61*, 4085–4089. [CrossRef]
8. Yang, W.; Zhou, K.; Zhou, Y.; An, Y.; Hu, T.; Lu, J.; Huang, S.; Pei, G. Naringin Dihydrochalcone Ameliorates Cognitive Deficits and Neuropathology in APP/PS1 Transgenic Mice. *Front. Aging Neurosci.* **2018**, *10*, 169. [CrossRef]
9. Río, J.A.D.; Fuster, M.D.; Sabater, F.; Porras, I.; García-Lidón, A.; Ortuño, A. Selection of citrus varieties highly productive for the neohesperidin dihydrochalcone precursor. *Food Chem.* **1997**, *59*, 433–437. [CrossRef]
10. Nakamura, Y.; Watanabe, S.; Miyake, N.; Kohno, H.; Osawa, T. Dihydrochalcones: Evaluation as Novel Radical Scavenging Antioxidants. *J. Agric. Food Chem.* **2003**, *51*, 3309–3312. [CrossRef]
11. Liu, B.; Zhu, X.; Zeng, J.; Zhao, J. Preparation and physicochemical characterization of the supramolecular inclusion complex of naringin dihydrochalcone and hydroxypropyl-β-cyclodextrin. *Food Res. Int.* **2013**, *54*, 691–696. [CrossRef]
12. Park, S.-H.; Kim, D.-H.; Kim, D.; Kim, D.-H.; Jung, H.-C.; Pan, J.-G.; Ahn, T.; Kim, D.; Yun, C.-H. Engineering Bacterial Cytochrome P450 (P450) BM3 into a Prototype with Human P450 Enzyme Activity Using Indigo Formation. *Drug Metab. Dispos.* **2010**, *38*, 732–739. [CrossRef] [PubMed]
13. Guengerich, F.P. Cytochrome P450 enzymes in the generation of commercial products. *Nat. Rev. Drug Discov.* **2002**, *1*, 359–366. [CrossRef] [PubMed]
14. Kim, D.-H.; Kim, K.-H.; Kim, D.-H.; Liu, K.-H.; Jung, H.-C.; Pan, J.-G.; Yun, C.-H. Generation of human metabolites of 7-ethoxycoumarin by bacterial cytochrome P450 BM3. *Drug Metab. Dispos.* **2008**, *36*, 2166–2170. [CrossRef]
15. Narhi, L.O.; Fulco, A.J. Characterization of a catalytically self-sufficient 119,000-dalton cytochrome P-450 monooxygenase induced by barbiturates in Bacillus megaterium. *J. Biol. Chem.* **1986**, *261*, 7160–7169.
16. Girvan, H.M.; Waltham, T.N.; Neeli, R.; Collins, H.F.; McLean, K.J.; Scrutton, N.S.; Leys, D.; Munro, A.W. Flavocytochrome P450 BM3 and the origin of CYP102 fusion species. *Biochem. Soc. Trans.* **2006**, *34*, 1173–1177. [CrossRef]
17. Urlacher, V.B.; Girhard, M. Cytochrome P450 monooxygenases: An update on perspectives for synthetic application. *Trends Biotechnol.* **2012**, *30*, 26–36. [CrossRef]
18. Whitehouse, C.J.C.; Bell, S.G.; Wong, L.-L. P450(BM3) (CYP102A1): Connecting the dots. *Chem. Soc. Rev.* **2012**, *41*, 1218–1260. [CrossRef]
19. Le, T.-K.; Jang, H.-H.; Nguyen, H.; Doan, T.; Lee, G.-Y.; Park, K.; Ahn, T.; Joung, Y.; Kang, H.-S.; Yun, C.-H. Highly regioselective hydroxylation of polydatin, a resveratrol glucoside, for one-step synthesis of astringin, a piceatannol glucoside, by P450 BM3. *Enzym. Microb. Technol.* **2016**, *97*, 34–42. [CrossRef]
20. Bernhardt, R. Cytochromes P450 as versatile biocatalysts. *J. Biotechnol.* **2006**, *124*, 128–145. [CrossRef]
21. Kang, J.-Y.; Kim, S.-Y.; Kim, D.; Kim, D.; Sun-Mi, S.; Park, S.-H.; Kim, K.-H.; Jung, H.; Pan, J.-G.; Joung, Y.; et al. Characterization of diverse natural variants of CYP102A1 found within a species of Bacillus megaterium. *AMB Express* **2011**, *1*, 1. [CrossRef] [PubMed]
22. Kim, D.-H.; Ahn, T.; Jung, H.-C.; Pan, J.-G.; Yun, C.-H. Generation of the human metabolite piceatannol from the anticancer-preventive agent resveratrol by bacterial cytochrome P450 BM3. *Drug Metab. Dispos.* **2009**, *37*, 932–936. [CrossRef] [PubMed]
23. Kim, K.-H.; Kang, J.-Y.; Kim, D.; Park, S.-H.; Park, S.; Kim, D.; Park, K.; Lee, Y.J.; Jung, H.; Pan, J.-G.; et al. Generation of Human Chiral Metabolites of Simvastatin and Lovastatin by Bacterial CYP102A1 Mutants. *Drug Metab. Dispos. Biol. Fate Chem.* **2010**, *39*, 140–150. [CrossRef]
24. Kang, J.-Y.; Ryu, S.H.; Park, S.-H.; Cha, G.S.; Kim, D.-H.; Kim, K.-H.; Hong, A.W.; Ahn, T.; Pan, J.-G.; Joung, Y.H.; et al. Chimeric cytochromes P450 engineered by domain swapping and random mutagenesis for producing human metabolites of drugs. *Biotechnol. Bioeng.* **2014**, *111*, 1313–1322. [CrossRef] [PubMed]

25. Jang, H.-H.; Ryu, S.-H.; Le, T.-K.; Doan, T.T.M.; Nguyen, T.H.H.; Park, K.D.; Yim, D.-E.; Kim, D.-H.; Kang, C.-K.; Ahn, T.; et al. Regioselective C-H hydroxylation of omeprazole sulfide by Bacillus megaterium CYP102A1 to produce a human metabolite. *Biotechnol. Lett.* **2017**, *39*, 105–112. [CrossRef] [PubMed]
26. Omura, T.; Sato, R. The carbon monoxide-binding pigment of liver microsomes. I. Evidence for its hemoprotein nature. *J. Biol. Chem.* **1964**, *239*, 2370–2378.
27. Tungmunnithum, D.; Thongboonyou, A.; Pholboon, A.; Yangsabai, A. Flavonoids and Other Phenolic Compounds from Medicinal Plants for Pharmaceutical and Medical Aspects: An Overview. *Medicines* **2018**, *5*, 93. [CrossRef]
28. Ovesná, Z.; Kozics, K.; Bader, Y.; Saiko, P.; Handler, N.; Erker, T.; Szekeres, T. Antioxidant activity of resveratrol, piceatannol and 3,3′,4,4′,5,5′-hexahydroxy-trans-stilbene in three leukemia cell lines. *Oncol. Rep.* **2006**, *16*, 617–624. [CrossRef]
29. Lai, T.N.H.; Herent, M.-F.; Quetin-Leclercq, J.; Nguyen, T.B.T.; Rogez, H.; Larondelle, Y.; André, C.M. Piceatannol, a potent bioactive stilbene, as major phenolic component in Rhodomyrtus tomentosa. *Food Chem.* **2013**, *138*, 1421–1430. [CrossRef]
30. Arai, D.; Kataoka, R.; Otsuka, S.; Kawamura, M.; Maruki-Uchida, H.; Sai, M.; Ito, T.; Nakao, Y. Piceatannol is superior to resveratrol in promoting neural stem cell differentiation into astrocytes. *Food Funct.* **2016**, *7*, 4432–4441. [CrossRef]
31. Kershaw, J.; Kim, K.-H. The Therapeutic Potential of Piceatannol, a Natural Stilbene, in Metabolic Diseases: A Review. *J. Med. Food* **2017**, *20*, 427–438. [CrossRef] [PubMed]
32. Du, Q.-H.; Peng, C.; Zhang, H. Polydatin: A review of pharmacology and pharmacokinetics. *Pharm. Biol.* **2013**, *51*, 1347–1354. [CrossRef] [PubMed]
33. Mérillon, J.-M.; Fauconneau, B.; Teguo, P.W.; Barrier, L.; Vercauteren, J.; Huguet, F. Antioxidant Activity of the Stilbene Astringin, Newly Extracted from Vitis vinifera Cell Cultures. *Clin. Chem.* **1997**, *43*, 1092–1093. [CrossRef]
34. Waffo-Teguo, P.; Lee, D.; Cuendet, M.; Mérillon, J.; Pezzuto, J.M.; Kinghorn, A.D. Two new stilbene dimer glucosides from grape (Vitis vinifera) cell cultures. *J. Nat. Prod.* **2001**, *64*, 136–138. [CrossRef]
35. Lee, D.E.; Lee, K.W.; Byun, S.; Jung, S.K.; Song, N.; Lim, S.H.; Heo, Y.-S.; Kim, J.E.; Kang, N.J.; Kim, B.Y.; et al. 7,3′,4′-Trihydroxyisoflavone, a Metabolite of the Soy Isoflavone Daidzein, Suppresses Ultraviolet B-induced Skin Cancer by Targeting Cot and MKK4. *J. Biol. Chem.* **2011**, *286*, 14246–14256. [CrossRef]
36. Nguyen, N.A.; Jang, J.; Le, T.-K.; Nguyen, T.H.H.; Woo, S.-M.; Yoo, S.-K.; Lee, Y.J.; Park, K.D.; Yeom, S.-J.; Kim, G.-J.; et al. Biocatalytic Production of a Potent Inhibitor of Adipocyte Differentiation from Phloretin Using Engineered CYP102A1. *J. Agric. Food Chem.* **2020**, *68*, 6683–6691. [CrossRef]
37. Pandey, B.P.; Roh, C.; Choi, K.-Y.; Lee, N.; Kim, E.J.; Ko, S.; Kim, T.; Yun, H.; Kim, B.-G. Regioselective hydroxylation of daidzein using P450 (CYP105D7) from Streptomyces avermitilis MA4680. *Biotechnol. Bioeng.* **2010**, *105*, 697–704. [PubMed]
38. Thibodeaux, C.J.; Melançon, C.E.; Liu, H. Natural-product sugar biosynthesis and enzymatic glycodiversification. *Angew. Chem. Int. Ed. Engl.* **2008**, *47*, 9814–9859. [CrossRef]
39. Gutmann, A.; Bungaruang, L.; Weber, H.; Leypold, M.; Breinbauer, R.; Nidetzky, B. Towards the synthesis of glycosylated dihydrochalcone natural products using glycosyltransferase-catalysed cascade reactions. *Green Chem.* **2014**, *16*, 4417–4425. [CrossRef]
40. Bowles, D.; Isayenkova, J.; Lim, E.-K.; Poppenberger, B. Glycosyltransferases: Managers of small molecules. *Curr. Opin. Plant Biol.* **2005**, *8*, 254–263. [CrossRef]
41. Wang, X. Structure, mechanism and engineering of plant natural product glycosyltransferases. *FEBS Lett.* **2009**, *583*, 3303–3309. [CrossRef] [PubMed]
42. Plaza, M.; Pozzo, T.; Liu, J.; Gulshan Ara, K.Z.; Turner, C.; Nordberg Karlsson, E. Substituent effects on in vitro antioxidizing properties, stability, and solubility in flavonoids. *J. Agric. Food Chem.* **2014**, *62*, 3321–3333. [CrossRef] [PubMed]
43. Slámová, K.; Kapešová, J.; Valentová, K. "Sweet Flavonoids": Glycosidase-Catalyzed Modifications. *Int. J. Mol. Sci.* **2018**, *19*, 2126. [CrossRef] [PubMed]
44. Xu, L.; Qi, T.; Xu, L.; Lu, L.; Xiao, M. Recent progress in the enzymatic glycosylation of phenolic compounds. *J. Carbohydr. Chem.* **2016**, *35*, 1–23. [CrossRef]

45. Devi, M.A.; Das, N.P. In vitro effects of natural plant polyphenols on the proliferation of normal and abnormal human lymphocytes and their secretions of interleukin-2. *Cancer Lett.* **1993**, *69*, 191–196. [CrossRef]
46. Nelson, J.A.; Falk, R.E. The efficacy of phloridzin and phloretin on tumor cell growth. *Anticancer. Res.* **1993**, *13*, 2287–2292.
47. Alsanea, S.; Gao, M.; Liu, D. Phloretin Prevents High-Fat Diet-Induced Obesity and Improves Metabolic Homeostasis. *AAPS J.* **2017**, *19*, 797–805. [CrossRef]
48. Wu, Y.; Zheng, X.; Xu, X.-G.; Li, Y.-H.; Wang, B.; Gao, X.-H.; Chen, H.-D.; Yatskayer, M.; Oresajo, C. Protective effects of a topical antioxidant complex containing vitamins C and E and ferulic acid against ultraviolet irradiation-induced photodamage in Chinese women. *J. Drugs Dermatol.* **2013**, *12*, 464–468.
49. Julsing, M.K.; Cornelissen, S.; Bühler, B.; Schmid, A. Heme-iron oxygenases: Powerful industrial biocatalysts? *Curr. Opin. Chem. Biol.* **2008**, *12*, 177–186. [CrossRef]
50. Yim, S.-K.; Kim, D.-H.; Jung, H.-C.; Pan, J.-G.; Kang, H.-S.; Ahn, T.; Yun, C.-H. Surface display of heme- and diflavin-containing cytochrome P450 BM3 in Escherichia coli: A whole cell biocatalyst for oxidation. *J. Microbiol. Biotechnol.* **2010**, *20*, 712–717. [CrossRef]
51. Kaderbhai, M.A.; Ugochukwu, C.C.; Kelly, S.L.; Lamb, D.C. Export of Cytochrome P450 105D1 to the Periplasmic Space of Escherichia coli. *Appl. Environ. Microbiol.* **2001**, *67*, 2136–2138. [CrossRef] [PubMed]
52. Remsberg, C.M.; Yáñez, J.A.; Vega-Villa, K.R.; Miranda, N.D.; Andrews, P.K.; Davies, N.M. HPLC-UV Analysis of Phloretin in Biological Fluids and Application to Pre-Clinical Pharmacokinetic Studies. *J. Chromatogr. Sep. Tech.* **2010**, *1*, 101. [CrossRef]
53. Hosea, N.A.; Miller, G.P.; Guengerich, F.P. Elucidation of Distinct Ligand Binding Sites for Cytochrome P450 3A4. *Biochemistry* **2000**, *39*, 5929–5939. [CrossRef] [PubMed]

© 2020 by the authors. Licensee MDPI, Basel, Switzerland. This article is an open access article distributed under the terms and conditions of the Creative Commons Attribution (CC BY) license (http://creativecommons.org/licenses/by/4.0/).

MDPI
St. Alban-Anlage 66
4052 Basel
Switzerland
Tel. +41 61 683 77 34
Fax +41 61 302 89 18
www.mdpi.com

Catalysts Editorial Office
E-mail: catalysts@mdpi.com
www.mdpi.com/journal/catalysts

www.ingramcontent.com/pod-product-compliance
Lightning Source LLC
LaVergne TN
LVHW070150100526
838202LV00015B/1924